ELEMENTARY MATRIX ALGEBRA

ELEMENTARY MATRIX ALGEBRA

Third Edition

FRANZ E. HOHN

DOVER PUBLICATIONS, INC.
Mineola, New York

Bibliographical Note

This Dover edition, first published in 2002, is an unabridged republication of the Third Edition of the work originally published in 1973 by The Macmillan Company, New York.

Library of Congress Cataloging-in-Publication Data

Hohn, Franz Edward, 1915–1977.
    Elementary matrix algebra / Franz E. Hohn.—3rd ed.
      p. cm.
    Originally published: 3rd ed. New York : Macmillan, [1973].
    Includes bibliographical references and index.
    ISBN 0-486-42534-7 (pbk.)
     1. Matrices. I. Title.

QA188 .H63 2002
512.9'434—dc21

2002031299

Manufactured in the United States of America
Dover Publications, Inc., 31 East 2nd Street, Mineola, N.Y. 11501

PREFACE TO THE THIRD EDITION

The third edition of *Elementary Matrix Algebra* is designed for the same audience and has the same scope as earlier editions. As before, full and clear exposition, complemented by many illustrative examples, is a basic characteristic of this book. The purpose is to provide a text that can teach by itself. The detailed exposition and the many examples, which the student can study on his own, make it possible to progress very rapidly through the following pages.

There are, however, two major changes. The first of these is that substantial geometrical material has been introduced early in the text. This helps to give a good intuitive basis for the following treatment of vector spaces. The second and more important change is that the chapter on determinants has been deferred until it is essential for following work. This serves to place maximum emphasis on the basic concepts and methods of linear algebra.

Other changes include simplification of the notation wherever possible, a more coherent arrangement of the material, and expansion of the lists of exercises, many of which refer to applications.

Chapter 1, "Introduction to Matrix Algebra," presents the most basic laws of matrix algebra. Chapter 2, "Linear Equations," introduces the sweepout process for obtaining the complete solution of any given system of linear equations, homogeneous or nonhomogeneous. This permits extensive use of concrete examples and exercises to facilitate the learning of following abstract concepts. In Chapter 3, "Vector Geometry in $\mathscr{E}^3$," and Chapter 4, "Vector Geometry in n-Dimensional Space," matrix algebra is used to present useful geometric ideas, techniques, and terminology as well as to lay a strong intuitive foundation for Chapter 5, "Vector Spaces," in which the approach is fully general. Since the reader can solve linear equations and invert matrices, computational exercises can be used to facilitate the learning of the abstract ideas. Chapter 6, "The Rank of a Matrix," reviews and unifies much that precedes and provides a complete treatment of the structure of the solution space of a system of linear equations. Chapter 7, "Determinants," presents the most commonly used properties of determinants. It also uses this tool to provide alternative solutions to algebraic and geometric problems treated earlier, thus providing valuable review. Chapter 8, "Linear Transformations," treats both linear operators and linear transformations of coordinates, using geometrical ideas extensively to give the material intuitive content. Unitary transformations and projections receive special emphasis. Chapter 9, "The Characteristic Value Problem," presents this topic in extensive detail and makes use of virtually all the preceding material. In Chapter 10, "Quadratic, Bilinear, and Hermitian Forms," the main emphasis is placed on real quadratic forms. Definiteness receives special attention because of its importance in applications.

There is considerably more material here than can be offered in a one-semester course unless the students have already been introduced to linear algebra in their calculus courses. The purpose is to provide a degree of choice so differing interests can be met, to make the book useful for students with varying backgrounds, and to provide a fairly comprehensive reference volume. The approach is frankly computational since the book is intended for those who use matrix algebra as a tool. However, the treatment is fully rigorous and the sequencing of topics is mathematically sound. The intent is to guarantee that the study of this book will make a following, mature study of linear algebra a more rewarding experience than it might otherwise have been.

A book like this cannot be prepared without substantial assistance from others. Thanks for good help go to Mrs. Carolyn Bloemker, who did the typing, to Professor Donald R. Sherbert, who read the proof critically, to the editors of The Macmillan Company, in particular Mr. Leo Malek, who have produced a most attractive volume, to the very capable printers and compositors, and to my wife, Mrs. Marian Hohn, who has been patient and helpful throughout the long processes of revision and production. To all of them, I am deeply grateful.

                                                                                    FRANZ E. HOHN

Urbana, Illinois

PREFACE TO THE SECOND EDITION

This second edition of *Elementary Matrix Algebra* retains the point of view and the scope of the first edition. Some of the topics have been reordered to make the exposition simpler. For example, the partitioning of matrices is now treated in Chapter One so that it can be used to greater advantage. The theorem on the determinant of the product of two matrices is proved without use of the Laplace expansion in Chapter Two. Thus the material on the Laplace expansion may be omitted entirely if that is desired since it is not essential to later developments. Similar changes occur in following chapters. A number of topics which are useful in applications (projections, for example) have been introduced in various places. This has been done in such a way that these topics need not be made part of the classroom work. An exceptionally detailed index, referencing also symbols and those exercises which contain important contributions to the theory, has been provided. Finally, a number of errors have been corrected and a great many new exercises have been included. It is hoped that the net effect of all these changes is to make the book easier to read and more useful as a text and as a reference. Exactly the same type of course can be taught from it as before.

The author wishes to thank his colleagues, Professors Richard L. Bishop and Hiram Paley, for their careful reading of the revised manuscript and of the proof, respectively. Many corrections and improvements are the direct result of their efforts. Professor Gene Golub provided a number of interesting and important exercises. Thanks are also due the many users of the first edition who have sent the author helpful suggestions, exercises, corrections, and letters of encouragement. Finally, the author is most grateful to the staff of The Macmillan Company and to the printers for their exceptionally cordial and able assistance.

FRANZ E. HOHN

Philo, Illinois

PREFACE TO THE FIRST EDITION

This text has been developed over a period of years for a course in Linear Transformations and Matrices given at the University of Illinois. The students have been juniors, seniors, and graduates whose interests have included such diverse subjects as aeronautical engineering, agricultural economics, chemistry, econometrics, education, electrical engineering, high speed computation, mechanical engineering, metallurgy, physics, psychology, sociology, statistics, and pure mathematics.

The book makes no pretense of being in any sense "complete." On the other hand, to meet as well as possible the needs of so varied a group, I have searched the literature of the various applications to find what aspects of matrix algebra and determinant theory are most commonly used. The book presents this most essential material as simply as possible and in a logical order with the objective of preparing the reader to study intelligently the applications of matrices in his special field. The topics are separated so far as possible into distinct, self-contained chapters in order to make the book more useful as a reference volume. With the same purpose in view, the principal results are listed as numbered theorems, identified as to chapter and section, and printed in italics. Formulas are similarly numbered, but the numbers are always enclosed in parentheses so as to distinguish formulas from theorems. Again for reference purposes, I have added appendices on the Σ and Π notations and on the algebra of complex numbers, for many readers will no doubt be in need of review of these matters.

The exercises often present formal aspects of certain applications, but no knowledge of the latter is necessary for working any problem. To keep down the size of the volume, detailed treatment of applications was omitted. The exercises range from purely formal computation and extremely simple proofs to a few fairly difficult problems designed to challenge the reader. I hope that every reader will find it possible to work most of the simpler exercises and at least to study the rest, for many useful results are contained in them, and, also, there is no way to learn the techniques of computation and of proof except through practice. The exercises marked with an asterisk (*) are of particular importance for immediate or later use and should not be overlooked.

In order to make the learning and the teaching of matrix algebra as easy as possible, I have tried always to proceed by means of ample explanations from the familiar and the concrete to the abstract. Abstract algebraic concepts themselves are not the prime concern of the volume. However, I have not hesitated to make an important mathematical point where the need and the motivation for it are clear, for it has been my purpose that in addition to learning useful methods of manipulating matrices, the reader should progress significantly in mathematical maturity as a result of careful study of this book. In fact, since the definitions of fields, groups, and vector spaces as well as of other abstract concepts appear and are used here, I believe that a course of this kind is not only far more practical but is also better preparation for later work in abstract algebra than is the traditional course in the theory of equations. I also believe that some appreciation of these abstract ideas will help the student of applications to read and work in his own field with greater insight and understanding.

The chief claim to originality here is in the attempt to reduce this material to the junior-senior level. Although a few proofs and many exercises are believed to be new, my debt to the standard authors—M. Bôcher, L. E. Dickson, W. L. Ferrar, C. C. McDuffee, F. J. Murnaghan, G. Birkhoff and S. MacLane, O. Schreier and E. Sperner, among others—is a very great one, and I acknowledge it with respect and gratitude.

I am particularly indebted to Professor A. B. Coble for permission to adapt to my needs his unpublished notes on determinants. The credit is his for all merit in the organization of Chapter Two. I am also indebted to Professor William G. Madow who helped to encourage and guide my efforts in their early stages. The value of the critical assistance of Professors Paul Bateman, Albert Wilansky, and Wilson Zaring cannot be overemphasized. Without their severe but kindly criticisms, this book would have been much less acceptable. However, I alone am responsible for any errors of fact or judgment which still persist. For additional critical aid, and for many of the problems, I owe thanks to a host of students and colleagues who have had contact with this effort. Finally, I owe thanks to Mrs. Betty Kaplan and to Mrs. Rachel Dyal for their faithful and competent typing of the manuscript, to Wilson Zaring and Russell Welker for assistance with reading proof, and to the staff of The Macmillan Company for their patient and helpful efforts during the production of this book.

                                                          FRANZ E. HOHN

Urbana, Illinois

CONTENTS

CHAPTER

1 Introduction to Matrix Algebra

1.1 Matrices	1
1.2 Equality of Matrices	2
1.3 Addition of Matrices	3
1.4 Commutative and Associative Laws of Addition	3
1.5 Subtraction of Matrices	4
1.6 Scalar Multiples of Matrices	6
1.7 The Multiplication of Matrices	7
1.8 The Properties of Matrix Multiplication	9
1.9 Exercises	13
1.10 Linear Equations in Matrix Notation	18
1.11 The Transpose of a Matrix	20
1.12 Symmetric, Skew-Symmetric, and Hermitian Matrices	22
1.13 Scalar Matrices	24
1.14 The Identity Matrix	26
1.15 The Inverse of a Matrix	26
1.16 The Product of a Row Matrix into a Column Matrix	29
1.17 Polynomial Functions of Matrices	30
1.18 Exercises	32
1.19 Partitioned Matrices	41
1.20 Exercises	46

2 Linear Equations

2.1 Linear Equations	51
2.2 Three Examples	52

2.3	Exercises	57
2.4	Equivalent Systems of Equations	60
2.5	The Echelon Form for Systems of Equations	63
2.6	Synthetic Elimination	66
2.7	Systems of Homogeneous Linear Equations	70
2.8	Exercises	73
2.9	Computation of the Inverse of a Matrix	75
2.10	Matrix Inversion by Partitioning	78
2.11	Exercises	82
2.12	Number Fields	83
2.13	Exercises	85
2.14	The General Concept of a Field	85
2.15	Exercises	88

3 Vector Geometry in $\mathscr{E}^3$

3.1	Geometric Representation of Vectors in $\mathscr{E}^3$	90
3.2	Operations on Vectors	91
3.3	Isomorphism	94
3.4	Length, Direction, and Sense	94
3.5	Orthogonality of Two Vectors	97
3.6	Exercises	100
3.7	The Vector Equation of a Line	102
3.8	The Vector Equation of a Plane	105
3.9	Exercises	109
3.10	Linear Combinations of Vectors in $\mathscr{E}^3$	111
3.11	Linear Dependence of Vectors; Bases	114
3.12	Exercises	118

4 Vector Geometry in n-Dimensional Space

4.1	The Real n-Space $\mathscr{R}^n$	120
4.2	Vectors in $\mathscr{R}^n$	121
4.3	Lines and Planes in $\mathscr{R}^n$	122
4.4	Linear Dependence and Independence in $\mathscr{R}^n$	125
4.5	Vector Spaces in $\mathscr{R}^n$	126
4.6	Exercises	127
4.7	Length and the Cauchy-Schwarz Inequality	129
4.8	Angles and Orthogonality in $\mathscr{E}^n$	132
4.9	Half-Lines and Directed Distances	134
4.10	Unitary n-Space	135

4.11 Exercises   138
4.12 Linear Inequalities   141
4.13 Exercises   147

5   Vector Spaces

5.1   The General Definition of a Vector Space   149
5.2   Linear Combinations and Linear Dependence   153
5.3   Exercises   159
5.4   Basic Theorems on Linear Dependence   164
5.5   Dimension and Basis   167
5.6   Computation of the Dimension of a Vector Space 172
5.7   Exercises   174
5.8   Orthonormal Bases   177
5.9   Exercises   180
5.10 Intersection and Sum of Two Vector Spaces   183
5.11 Exercises   186
5.12 Isomorphic Vector Spaces   187
5.13 Exercises   190

6   The Rank of a Matrix

6.1   The Rank of a Matrix   191
6.2   Basic Theorems About the Rank of a Matrix   195
6.3   Matrix Representation of Elementary Transformations 198
6.4   Exercises   204
6.5   Homogeneous Systems of Linear Equations   210
6.6   Nonhomogeneous Systems of Linear Equations   217
6.7   Exercises   221
6.8   Another Look at Nonhomogeneous Systems   229
6.9   The Variables One Can Solve for   231
6.10 Basic Solutions   234
6.11 Exercises   237

7   Determinants

7.1   The Definition of a Determinant   239
7.2   Some Basic Theorems   243
7.3   The Cofactor in det A of an Element of A 247
7.4   Cofactors and the Computation of Determinants   250
7.5   Exercises   253
7.6   The Determinant of the Product of Two Matrices   264

7.7	A Formula for A^{-1}	270
7.8	Determinants and the Rank of a Matrix	272
7.9	Solution of Systems of Equations by Using Determinants	274
7.10	A Geometrical Application of Determinants	278
7.11	Exercises	280
7.12	More About the Rank of a Matrix	288
7.13	Definitions	292
7.14	The Laplace Expansion	295
7.15	The Determinant of a Product of Two Square Matrices	299
7.16	The Adjoint Matrix	300
7.17	The Row-and-Column Expansion	302
7.18	The Diagonal Expansion of the Determinant of a Matrix	303
7.19	Exercises	305

8 Linear Transformations

8.1	Mappings	310
8.2	Linear Mappings	313
8.3	Some Properties of Linear Operators on Vector Spaces	317
8.4	Exercises	320
8.5	Linear Transformations of Coordinates	323
8.6	Transformation of a Linear Operator	331
8.7	Exercises	334
8.8	The Algebra of Linear Operators	340
8.9	Groups of Operators	342
8.10	Exercises	345
8.11	Unitary and Orthogonal Matrices	347
8.12	Exercises	348
8.13	Unitary Transformations	351
8.14	Orthogonal Transformations	353
8.15	The Eulerian Angles	354
8.16	The Triangularization of a Real Matrix	356
8.17	Exercises	360
8.18	Orthogonal Vector Spaces	361
8.19	Exercises	366
8.20	Projections	367
8.21	Orthogonal Projections in $\mathscr{U}^n$	371
8.22	Exercises	373

9 The Characteristic Value Problem

9.1	Definition of the Characteristic Value Problem	375
9.2	Four Examples	377

9.3 Two Basic Theorems   381
9.4 Exercises   383
9.5 The Characteristic Polynomial and Its Roots   387
9.6 Similar Matrices   392
9.7 Exercises   393
9.8 The Characteristic Roots of a Hermitian Matrix   395
9.9 The Diagonal Form of a Hermitian Matrix   397
9.10 The Diagonalization of a Hermitian Matrix   400
9.11 Examples   401
9.12 Triangularization of an Arbitrary Matrix   403
9.13 Normal Matrices   404
9.14 Exercises   406
9.15 Characteristic Roots of a Polynomial Function of a Matrix   409
9.16 The Cayley-Hamilton Theorem   411
9.17 The Minimum Polynomial of a Matrix   413
9.18 Powers of Matrices   418
9.19 Exercises   419

10 Quadratic, Bilinear, and Hermitian Forms

10.1 Quadratic Forms   422
10.2 Diagonalization of Quadratic Forms   424
10.3 A Geometrical Application   425
10.4 Definite Forms and Matrices   428
10.5 Exercises   433
10.6 Lagrange's Reduction   438
10.7 Kronecker's Reduction   443
10.8 Sylvester's Law of Inertia for Real Quadratic Forms   445
10.9 A Necessary and Sufficient Condition for Positive
      Definiteness   449
10.10 An Important Example   452
10.11 Exercises   454
10.12 Pairs of Quadratic Forms   455
10.13 Values of Quadratic Forms   457
10.14 Exercises   462
10.15 Bilinear Forms   463
10.16 The Equivalence of Bilinear Forms   465
10.17 Cogredient and Contragredient Transformations   467
10.18 Exercises   469
10.19 Hermitian Forms   471
10.20 Definite Hermitian Forms   474
10.21 Exercises   474

APPENDIX

I   The Notations Σ and Π 477

II   The Algebra of Complex Numbers   491

BIBLIOGRAPHY   501

INDEX   513

ELEMENTARY MATRIX
ALGEBRA

CHAPTER
1

Introduction to Matrix Algebra

1.1 Matrices

There are many situations in both pure and applied mathematics in which we have to deal with rectangular arrays of numbers or functions. An array of this kind may be represented by the symbol

$$(1.1.1) \qquad A = \begin{bmatrix} a_{11} & a_{12} & \cdots & a_{1n} \\ a_{21} & a_{22} & \cdots & a_{2n} \\ \vdots & & & \\ a_{m1} & a_{m2} & \cdots & a_{mn} \end{bmatrix}.$$

The numbers or functions a_{ij} of this array are called its **elements** or **entries** and in this book are assumed to have real or complex values. Such an array, subject to rules of operation to be defined below, is called a **matrix**. We shall denote matrices with pairs of square brackets, but pairs of double bars, $\parallel \quad \parallel$, and pairs of parentheses, ( ), are also used for this purpose. The subscripts i and j of the element a_{ij} of a matrix A identify respectively the **row** and the **column** of A in which a_{ij} is located. When there is no need to distinguish between rows and columns, we call them simply **lines** of the matrix.

A matrix A with m rows and n columns is called a **matrix of order (m, n)** or **an $m \times n$ ("m by n") matrix**. When $m = n$ so that the matrix is square, it is called a **matrix of order n** or an **n-square matrix**. When A is of order n, the elements $a_{11}, a_{22}, \ldots, a_{nn}$ are said to constitute the **main** or **principal diagonal** of A and the elements $a_{n1}, a_{n-1,2}, \ldots, a_{1n}$ constitute its **secondary diagonal**. When A is $m \times 1$, it is called a **column matrix** or **vector**.

It is often convenient to abbreviate the symbol (1.1.1) to the form $[a_{ij}]_{(m,n)}$, which means "the matrix of order (m, n) whose elements are the a_{ij}'s." When the order of the matrix need not be specified or is clear from the context, this is abbreviated further to the form $[a_{ij}]$. Another convenient procedure which we shall follow is to denote matrices by capital letters such as A, B, X, Y, etc., whenever it is not necessary to indicate explicitly the elements or the orders of the matrices in question.

A simple illustration of the matrix concept is the following: The coefficients of x and y in the system of linear equations

(1.1.2)
$$2x + 6y = -1$$
$$4x - y = 3,$$

provide the matrix of order 2:

$$\begin{bmatrix} 2 & 6 \\ 4 & -1 \end{bmatrix},$$

which is called the **coefficient matrix** of the system. The 2×3 matrix

$$\begin{bmatrix} 2 & 6 & -1 \\ 4 & -1 & 3 \end{bmatrix},$$

containing the coefficients of x and y and the constant terms as well, is called the **augmented matrix** of the system. The coefficient and augmented matrices of systems of equations are useful in investigating their solutions, as we shall see later.

1.2 Equality of Matrices

Two matrices $[a_{ij}]_{(m,n)}$ and $[b_{ij}]_{(m,n)}$ are defined to be **equal** if and only if $a_{ij} = b_{ij}$ for each pair of subscripts i and j. In words, *two matrices are equal if and only if they have the same order and have equal corresponding elements throughout*.

From this definition and from the properties of equality in ordinary algebra, there follow four properties of the equality of matrices:

(a) If A and B are any two matrices, either $A = B$ or $A \neq B$ (the **determinative** property).
(b) If A is any matrix, $A = A$ (the **reflexive** property).
(c) If $A = B$, then $B = A$ (the **symmetric** property).
(d) If $A = B$ and $B = C$, then $A = C$ (the **transitive** property).

Many mathematical relationships other than equality of matrices possess these same four properties. (The similarity of triangles is a simple example. Can you think of others?) Any relation between pairs of mathematical objects which possesses these properties is called an **equivalence relation**. Several types of equivalence relations will be defined and used in this book.

Because equality means that matrices are in fact identical, *a matrix may be substituted for any equal matrix in the following operations.*

1.3 Addition of Matrices

If $A = [a_{ij}]_{(m, n)}$ and $B = [b_{ij}]_{(m, n)}$, we define the **sum** $A + B$ to be the matrix $[(a_{ij} + b_{ij})]_{(m,n)}$. That is, the sum of two matrices of the same order is found by adding corresponding elements throughout. For example,

$$\begin{bmatrix} 1 - t & 2 \\ 3 & 1 + t \end{bmatrix} + \begin{bmatrix} 1 + t & -2 \\ -3 & 1 - t \end{bmatrix} = \begin{bmatrix} 2 & 0 \\ 0 & 2 \end{bmatrix}.$$

Two matrices of the same order are said to be **conformable for addition**.
Since the sum of any two $m \times n$ matrices is again an $m \times n$ matrix, we say that the set of all $m \times n$ matrices is **closed with respect to addition**.

1.4 Commutative and Associative Laws of Addition

Throughout this book, the real and the complex numbers and functions thereof will be called **scalars** to distinguish them from the arrays which are called matrices. In scalar algebra, the fact that $a + b = b + a$ for any two scalars a and b is known as the **commutative law of addition**. The fact that $a + (b + c) = (a + b) + c$ for any three scalars a, b, and c is known as the **associative law of addition**. It is not hard to see that these laws extend to matrix addition also.

Let A, B, C be arbitrary matrices of the same order. Then, using the definition of the sum of two matrices and the commutative law of addition of scalars, we have in the abbreviated notation

$$A + B = [a_{ij} + b_{ij}] = [b_{ij} + a_{ij}] = B + A.$$

Similarly, applying the associative law for the addition of scalars, we have

$$A + (B + C) = [a_{ij} + (b_{ij} + c_{ij})] = [(a_{ij} + b_{ij}) + c_{ij}] = (A + B) + C.$$

We have thus proved

Theorem 1.4.1: *The addition of matrices is both commutative and associative; that is, if A, B, and C are conformable for addition,*

(1.4.1) $$A + B = B + A,$$

(1.4.2) $$A + (B + C) = (A + B) + C.$$

The reader who finds the above notation a little too condensed should write out the details in full for matrices of order, say, (2, 3).

These two laws, applied repeatedly if necessary, enable us to arrange the terms of a sum in any order we wish, and to group them in any fashion we wish. In particular, they justify the absence of parentheses in an expression like $A + B + C$, which (for given A, B, C of the same order) has a uniquely defined meaning.

Another important property of matrix addition is given in

Theorem 1.4.2: $A + C = B + C$ *if and only if* $A = B$.

The fact that $A + C = B + C$ implies $A = B$ is called the **cancellation law for addition**.

Indeed, $A + C = B + C$ if and only if $a_{ij} + c_{ij} = b_{ij} + c_{ij}$ in every case. But $a_{ij} + c_{ij} = b_{ij} + c_{ij}$ if and only if $a_{ij} = b_{ij}$, by the cancellation law of addition in the complex domain. This implies $A + C = B + C$ if and only if $A = B$.

1.5 Subtraction of Matrices

A matrix all of whose elements are zero is called a **zero matrix** and is denoted by 0, or by 0_n, or by $0_{m \times n}$ when the order needs emphasis. The basic property of the matrix $0_{m \times n}$ is that, for all $m \times n$ matrices A,

(1.5.1) $$A + 0 = A;$$

that is, the zero matrix is an **identity element for addition**.

The **negative** of an $m \times n$ matrix $A = [a_{ij}]$ is defined to be $-A = [-a_{ij}]$.

That is, the negative of A is formed by changing the sign of every element of A. The reason for this definition is, of course, to guarantee that

(1.5.2)
$$A + (-A) = 0.$$

Thus $-A$ is an **inverse** of A **with respect to addition**.

Instead of $A + (-A) = 0$, we agree to write $A - A = 0$. In general, we define

(1.5.3)
$$A - B = A + (-B).$$

This implies that the difference $A - B$ may be found by subtracting corresponding elements.

For example,

$$\begin{bmatrix} 2 & 1 & -3 \\ -4 & 0 & 1 \end{bmatrix} - \begin{bmatrix} 1 & -1 & 2 \\ 0 & 1 & 0 \end{bmatrix} = \begin{bmatrix} 1 & 2 & -5 \\ -4 & -1 & 1 \end{bmatrix}.$$

An important consequence of the preceding definitions is that, if X, A, and B are all of the same order, then a solution of the equation

(1.5.4)
$$X + A = B$$

is

$$X = B - A.$$

In fact, replacement of X by $B - A$ in $X + A$ yields

$$\begin{aligned}
(B - A) + A &= (B + (-A)) + A \\
&= B + ((-A) + A) \\
&= B + 0 \\
&= B,
\end{aligned}$$

which shows that $B - A$ is indeed a solution of (1.5.4). [Note that (1.5.2), (1.4.2), (1.4.1), and (1.5.1) were all employed in the proof.] Moreover, it is the *only* solution, for, if Y is any solution, then

$$Y + A = B,$$

so that, by substitution,

$$(Y + A) + (-A) = B + (-A),$$
$$Y + (A + (-A)) = B - A,$$
$$Y + 0 = B - A,$$
$$Y = B - A.$$

(Each use of a basic law should be identified here by the reader.) In summary, $B - A$ is the *unique* solution of the equation $X + A = B$.

In particular, this proves that the equations

$$X + A = A, \quad X + A = 0$$

have respectively the unique solutions 0 and $-A$; that is, for a given order (m, n), *the additive identity element is unique,* and *each matrix A has a unique inverse with respect to addition,* namely the matrix $-A$ defined above.

In summary, with respect to addition, the set of all $m \times n$ matrices has the same properties as the set of complex numbers. This is because matrix addition is effected by mn independent scalar additions—one in each of the mn positions.

1.6   Scalar Multiples of Matrices

If $A = [a_{ij}]$ and if α is a scalar, we define $\alpha A = A\alpha = [\alpha a_{ij}]$. In words, *to multiply a matrix A by a scalar α, multiply every element of A by α.* This definition is, of course, suggested by the fact that, if we add n A's, we obtain a matrix whose elements are those of A each multiplied by n. For example,

$$\begin{bmatrix} a & b \\ c & d \end{bmatrix} + \begin{bmatrix} a & b \\ c & d \end{bmatrix} = \begin{bmatrix} 2a & 2b \\ 2c & 2d \end{bmatrix} = 2\begin{bmatrix} a & b \\ c & d \end{bmatrix}.$$

The operation of multiplying a matrix by a scalar has these basic properties:

$$1 \cdot A = A,$$
$$(\alpha + \beta)A = \alpha A + \beta A,$$

(1.6.1)

$$\alpha(A + B) = \alpha A + \alpha B,$$
$$\alpha(\beta A) = (\alpha\beta)A.$$

All four are readily proved by appealing to the definition. In the case of the second, we have $(\alpha + \beta)A = [(\alpha + \beta)a_{ij}] = [\alpha a_{ij} + \beta a_{ij}] = [\alpha a_{ij}] + [\beta a_{ij}] = \alpha A + \beta A$. We leave it as an exercise to the reader to prove the other laws in a similar fashion.

1.7   The Multiplication of Matrices

Frequently in the mathematical treatment of a problem, the work can be simplified by the introduction of new variables. Translations of axes, effected by equations of the form

$$x = x' + h,$$
$$y = y' + k,$$

and rotations of axes, effected by

$$x = x' \cos \theta - y' \sin \theta,$$
$$y = x' \sin \theta + y' \cos \theta,$$

are the most familiar examples. A rotation of axes is a special case of a change of variables of the type

(1.7.1)
$$x = a_{11}x' + a_{12}y',$$
$$y = a_{21}x' + a_{22}y',$$

in which the a's are constants. Substitutions of the latter kind are called **linear homogeneous transformations** of the variables and are of great usefulness. The properties of these transformations suggest the law that should be adopted for the multiplication of matrices, which we now proceed to illustrate.

Let us consider, for example, the effect on the system of linear functions

(1.7.2)
$$2x + 3y,$$
$$3x - 4y,$$
$$-5x + 6y,$$

resulting from an application of the linear transformation (1.7.1). Substitution from (1.7.1) into (1.7.2) yields the new system of linear functions

(1.7.3)
$$(2a_{11} + 3a_{21})x' + (2a_{12} + 3a_{22})y',$$
$$(3a_{11} - 4a_{21})x' + (3a_{12} - 4a_{22})y',$$
$$(-5a_{11} + 6a_{21})x' + (-5a_{12} + 6a_{22})y'.$$

From the three systems of linear expressions, (1.7.1), (1.7.2), and (1.7.3), we obtain three coefficient matrices. Since the third matrix is in a sense the "product" of the first two, we shall relate them by the following *matrix equation*:

$$
\begin{bmatrix} 2 & 3 \\ 3 & -4 \\ -5 & 6 \end{bmatrix} \cdot \begin{bmatrix} a_{11} & a_{12} \\ a_{21} & a_{22} \end{bmatrix} = \begin{bmatrix} (2a_{11} + 3a_{21}) & (2a_{12} + 3a_{22}) \\ (3a_{11} - 4a_{21}) & (3a_{12} - 4a_{22}) \\ (-5a_{11} + 6a_{21}) & (-5a_{12} + 6a_{22}) \end{bmatrix}.
$$

The question now is, "What rule for 'multiplying' matrices does this equation imply?" The element $(2a_{11} + 3a_{21})$ in the *first row and first column* of the matrix on the right may be obtained by multiplying the elements of the *first row* of the extreme left matrix respectively by the corresponding elements of the *first column* of the second matrix on the left and then adding the results: (first × first) + (second × second). If we multiply the elements of the *second row* of the extreme left matrix respectively by the corresponding elements of the *first column* of the second matrix and add, we obtain the entry $(3a_{11} - 4a_{21})$ in the *second row and first column* on the right. A similar procedure is followed for every other entry on the right. (The reader should check them all.)

This example suggests the following general definition. Let A be an $m \times p$ matrix and let B be a $p \times n$ matrix. The **product** AB is then defined to be the $m \times n$ matrix whose element in the ith row and jth column is found by multiplying corresponding elements of the ith row of A and of the jth column of B, and then adding the results. Symbolically, we may write

$$
\begin{bmatrix} a_{11} & a_{12} & \cdots & a_{1p} \\ a_{21} & a_{22} & \cdots & a_{2p} \\ \vdots & & & \\ a_{m1} & a_{m2} & \cdots & a_{mp} \end{bmatrix} \cdot \begin{bmatrix} b_{11} & b_{12} & \cdots & b_{1n} \\ b_{21} & b_{22} & \cdots & b_{2n} \\ \vdots & & & \\ b_{p1} & b_{p2} & \cdots & b_{pn} \end{bmatrix} = \begin{bmatrix} c_{11} & c_{12} & \cdots & c_{1n} \\ c_{21} & c_{22} & \cdots & c_{2n} \\ \vdots & & & \\ c_{m1} & c_{m2} & \cdots & c_{mn} \end{bmatrix},
$$

where

$$
c_{ij} = a_{i1} b_{1j} + a_{i2} b_{2j} + \cdots + a_{ip} b_{pj} = \sum_{k=1}^{p} a_{ik} b_{kj}.
$$

(The arrows have been used for emphasis and are not customarily part of the notation.)

Two things should be noted particularly. First, the product AB has the same number of rows as the matrix A and the same number of columns as the matrix B. Second, the number of columns in A and the number of rows in B must be the same since otherwise there will not always be corresponding elements to multiply together. When the number of columns of a matrix A is the same as the number of rows of a matrix B, A **is said to be conformable to B for multiplication.**

These matters are illustrated further in the examples which follow:

(a) $\begin{bmatrix} 1 & -1 & 2 \\ 3 & 0 & 1 \end{bmatrix}_{(2,3)} \cdot \begin{bmatrix} 1 & 2 & 0 \\ 0 & -1 & 1 \\ 1 & 2 & -1 \end{bmatrix}_{(3,3)}$

$= \begin{bmatrix} (1\cdot1+(-1)\cdot0+2\cdot1) & (1\cdot2+(-1)(-1)+2\cdot2) & (1\cdot0+(-1)\cdot1+2\cdot(-1)) \\ (3\cdot1+0\cdot0+1\cdot1) & (3\cdot2+0\cdot(-1)+1\cdot2) & (3\cdot0+0\cdot1+1\cdot(-1)) \end{bmatrix}$

$= \begin{bmatrix} 3 & 7 & -3 \\ 4 & 8 & -1 \end{bmatrix}_{(2,3)}.$

(b) $[x_1, x_2]_{(1,2)} \cdot \begin{bmatrix} 1 & 2 & 4 \\ -1 & 3 & -2 \end{bmatrix}_{(2,3)}$

$= [(x_1 - x_2), \quad (2x_1 + 3x_2), \quad (4x_1 - 2x_2)]_{(1,3)}.$

(c) $[x_1, x_2, x_3]_{(1,3)} \cdot \begin{bmatrix} y_1 \\ y_2 \\ y_3 \end{bmatrix}_{(3,1)} = [(x_1 y_1 + x_2 y_2 + x_3 y_3)]_{(1,1)}.$

1.8 The Properties of Matrix Multiplication

In the product AB we say that B is **premultiplied** by A and that A is **postmultiplied** by B. This terminology is essential since ordinarily $AB \neq BA$. In fact, if A has order (m, p) and B has order (p, n) with $m \neq n$, the product AB is defined but the product BA is not. Thus the fact that A is conformable to B for multiplication *does not imply* that B is conformable to A for multiplication. Even if $m = n$, we need not have $AB = BA$. That is, *matrix multiplication is not in general commutative.* We give some numerical examples in which the reader should verify every detail:

(a) $\begin{bmatrix} 0 & 1 & 2 & 3 \\ 3 & 2 & 1 & 0 \end{bmatrix} \cdot \begin{bmatrix} 0 & 3 \\ 1 & 2 \\ 2 & 1 \\ 3 & 0 \end{bmatrix} = \begin{bmatrix} 14 & 4 \\ 4 & 14 \end{bmatrix},$

but

$$\begin{bmatrix} 0 & 3 \\ 1 & 2 \\ 2 & 1 \\ 3 & 0 \end{bmatrix} \cdot \begin{bmatrix} 0 & 1 & 2 & 3 \\ 3 & 2 & 1 & 0 \end{bmatrix} = \begin{bmatrix} 9 & 6 & 3 & 0 \\ 6 & 5 & 4 & 3 \\ 3 & 4 & 5 & 6 \\ 0 & 3 & 6 & 9 \end{bmatrix}.$$

(b)

$$\begin{bmatrix} 2 & -1 \\ -1 & 2 \end{bmatrix} \cdot \begin{bmatrix} 1 & 4 \\ -1 & 1 \end{bmatrix} = \begin{bmatrix} 3 & 7 \\ -3 & -2 \end{bmatrix},$$

but

$$\begin{bmatrix} 1 & 4 \\ -1 & 1 \end{bmatrix} \cdot \begin{bmatrix} 2 & -1 \\ -1 & 2 \end{bmatrix} = \begin{bmatrix} -2 & 7 \\ -3 & 3 \end{bmatrix}.$$

This last example shows that multiplication is not commutative even in the case of square matrices.

The fact that "multiplication is not, in general, commutative" does not mean that we *never* have $AB = BA$. There are, in fact, important special cases when this equality holds. Examples will appear later in this chapter.

The familiar rule of scalar algebra that if a product is zero, then one of the factors must be zero, also fails to hold for matrix multiplication. An example is the product

$$\begin{bmatrix} 1 & 2 & 0 \\ 1 & 1 & 0 \\ -1 & 4 & 0 \end{bmatrix} \cdot \begin{bmatrix} 0 & 0 & 0 \\ 0 & 0 & 0 \\ 1 & 4 & 9 \end{bmatrix} = \begin{bmatrix} 0 & 0 & 0 \\ 0 & 0 & 0 \\ 0 & 0 & 0 \end{bmatrix}.$$

Here neither factor is a zero matrix, although the product is.

When a product $AB = 0$ but neither A nor B is 0, then the factors A and B are called **divisors of zero**. Thus, in the algebra of matrices, there exist divisors of zero, whereas in the algebra of complex numbers there do not.

We illustrate a final contrast with the laws of scalar algebra by means of the following example. Let

$$A = \begin{bmatrix} 1 & 2 & 0 \\ 1 & 1 & 0 \\ -1 & 4 & 0 \end{bmatrix}. \quad B = \begin{bmatrix} 1 & 2 & 3 \\ 1 & 1 & -1 \\ 2 & 2 & 2 \end{bmatrix}, \quad C = \begin{bmatrix} 1 & 2 & 3 \\ 1 & 1 & -1 \\ 1 & 1 & 1 \end{bmatrix}.$$

Then

$$AB = \begin{bmatrix} 3 & 4 & 1 \\ 2 & 3 & 2 \\ 3 & 2 & -7 \end{bmatrix} = AC.$$

Thus we can have $AB = AC$ without having $B = C$. In other words, we cannot ordinarily cancel A from $AB = AC$ even if $A \neq 0$. However, there is an important special case when the cancellation is possible, as we shall see later.

In summary, then, three fundamental properties of multiplication in scalar algebra do not carry over to matrix algebra:

(a) The commutative law $AB = BA$ does not hold true generally.
(b) From $AB = 0$, we cannot conclude that at least one of A and B must be zero; that is, there exist divisors of zero.
(c) From $AB = AC$ or $BA = CA$ we cannot in general conclude that $B = C$, even if $A \neq 0$; that is, the cancellation law does not hold in general in multiplication.

These rather staggering losses might make one wonder whether matrix multiplication is not a nearly useless operation. This is, of course, not the case, for, as we shall prove, the most vital properties—the associative and the distributive laws—still remain. However, it should be clear at this point why we have been, and must continue to be, so careful to prove the validity of the matrix operations which we employ.

Theorem 1.8.1: *The multiplication of matrices is associative.*

Let

$$A = [a_{ij}]_{(m,n)}, \qquad B = [b_{jk}]_{(n,p)}, \qquad C = [c_{kr}]_{(p,q)}.$$

Then the theorem says that

$$(AB)C = A(BC).$$

Applying the definition of multiplication, we see first that

$$AB = \left[\sum_{j=1}^{n} a_{ij} b_{jk} \right]_{(m,p)}.$$

Here i ranges from 1 to m and denotes the row of the element in parentheses, whereas k ranges from 1 to p and denotes its column.

We apply the definition now to AB and C. The new summation will be on the column subscript k of AB, which is the row subscript of C, so

$$(AB)C = \left[\sum_{k=1}^{p} \left(\sum_{j=1}^{n} a_{ij} b_{jk} \right) c_{kr} \right]_{(m,q)}.$$

Multiplying the factor c_{kr} into each sum in parentheses, we obtain

$$(AB)C = \left[\sum_{k=1}^{p} \left(\sum_{j=1}^{n} a_{ij} b_{jk} c_{kr} \right) \right]_{(m,q)},$$

in which the row subscript i ranges from 1 to m while the column subscript r ranges from 1 to q.

In the same way we find

$$A(BC) = \left[\sum_{j=1}^{n} \left(\sum_{k=1}^{p} a_{ij} b_{jk} c_{kr} \right) \right]_{(m,q)}.$$

Since the order of summation is arbitrary in a finite sum, we have

$$\sum_{k=1}^{p} \left(\sum_{j=1}^{n} a_{ij} b_{jk} c_{kr} \right) = \sum_{j=1}^{n} \left(\sum_{k=1}^{p} a_{ij} b_{jk} c_{kr} \right)$$

for each pair of values of i and r, so that $(AB)C = A(BC)$.

If the uses made of the $\sum$ sign in this proof are unfamiliar to the reader, he may refer to an explanation of these matters in Appendix I. It would also help to write out the proof in full for 2×2 matrices.

Theorem 1.8.2: *Matrix multiplication is distributive with respect to addition.*

To make this explicit, let

$$A = [a_{ik}]_{(m,n)}, \qquad B = [b_{kj}]_{(n,p)}, \qquad C = [c_{kj}]_{(n,p)}.$$

Here A is conformable to B and also to C for multiplication, and B is conformable to C for addition. Then the theorem says that

$$A(B + C) = AB + AC.$$

Indeed

$$A(B + C) = [a_{ik}]_{(m,n)}[(b_{kj} + c_{kj})]_{(n,p)}$$

$$= \left[\sum_{k=1}^{n} a_{ik}(b_{kj} + c_{kj}) \right]_{(m,p)}$$

$$= \left[\sum_{k=1}^{n} a_{ik} b_{kj} + \sum_{k=1}^{n} a_{ik} c_{kj} \right]_{(m,p)}$$

$$= \left[\sum_{k=1}^{n} a_{ik} b_{kj} \right]_{(m,p)} + \left[\sum_{k=1}^{n} a_{ik} c_{kj} \right]_{(m,p)}$$

$$= AB + AC.$$

The theorem also says that, assuming conformability,

$$(D + E)F = DF + EF.$$

This second distributive law is distinct from the first, since matrix multiplication is not in general commutative. It is proved in the same manner as the first, however, and details are left to the reader.

The proofs of the last two theorems involve a detailed examination of the elements of the matrices involved. They are thus essentially scalar in nature. As the theory develops, we shall increasingly employ proofs involving only manipulations with matrices. Such proofs are typically more compact than are scalar-type proofs of the same results and, hence, are to be preferred. The reader's progress in learning *matrix* algebra will be accelerated if in the exercises he avoids the use of scalar-type proof whenever this is possible. For example, to prove that, for conformable matrices,

$$(A + B)(C + D) = AC + BC + AD + BD,$$

we do not again resort to a scalar type of proof. We simply note that, by the first distributive law above, $(A + B)(C + D) = (A + B)C + (A + B)D$, so that, by the second distributive law,

$$(A + B)(C + D) = AC + BC + AD + BD.$$

1.9 Exercises

Throughout this book, exercises marked with an asterisk (*) develop an important part of the theory and should not be overlooked.

In many problems, conditions of conformability for addition or multiplication must be satisfied for the problem to have meaning. These conditions are usually rather obvious, so that we shall frequently omit statement of them. The reader is then expected to make the necessary assumptions in working the exercises.

In these exercises, the words "prove" and "show" are to be taken as synonymous.

1. In each case, find all solutions (x, y) of the given equation:

(a) $\begin{bmatrix} x + y & 2 \\ 1 & x - y \end{bmatrix} = \begin{bmatrix} 3 & 2 \\ 1 & 5 \end{bmatrix}$,

(b) $\begin{bmatrix} x + y & 2 \\ 1 & 2x - 2y \end{bmatrix} = \begin{bmatrix} 2 & x + y \\ x - y & 3 \end{bmatrix}$

2. Given that

$$A_1 = \begin{bmatrix} 1 & 0 \\ 0 & 1 \end{bmatrix}, \quad A_2 = \begin{bmatrix} 0 & 1 \\ 1 & 0 \end{bmatrix}, \quad A_3 = \begin{bmatrix} 1 & 0 \\ 0 & -1 \end{bmatrix}, \quad A_4 = \begin{bmatrix} 0 & 1 \\ -1 & 0 \end{bmatrix},$$

solve for x_1, x_2, x_3, and x_4 :

$$x_1 A_1 + x_2 A_2 + x_3 A_3 + x_4 A_4 = \begin{bmatrix} 2 & -1 \\ -1 & 2 \end{bmatrix}.$$

3. Prove that $(A + B) - C = A + (B - C)$ and name each property used in the proof. Why is $(A - B) + C \neq A - (B + C)$ in general?

*4. Prove in detail the second distributive law:

$$(D + E)F = DF + EF.$$

*5. Prove that if $\alpha A = 0$ where α is a scalar, then either $\alpha = 0$ or $A = 0$. Prove also that $A = B$ if and only if $\alpha A = \alpha B$ for all scalars α.

6. Given that $\alpha A = \beta A$ and $A \neq 0$, prove that $\alpha = \beta$ (α and β scalars).

*7. Prove:

(a) $\alpha A \cdot \beta B = \alpha \beta \cdot AB$ (α, β scalars).
(b) $(-1)A = -A$.
(c) $(-A)(-B) = AB$.
(d) $A(\alpha B) = (\alpha A)B = \alpha(AB)$.

8. Perform the matrix multiplications:

(a) $\begin{bmatrix} 1 & -1 & 1 \\ 2 & 0 & 1 \\ 3 & -1 & 2 \end{bmatrix} \begin{bmatrix} 1 & 2 \\ -1 & 1 \\ 1 & 3 \end{bmatrix}$,

(b) $\begin{bmatrix} 1 & 2 & 3 & 4 \end{bmatrix} \begin{bmatrix} 1 \\ 2 \\ 3 \\ 4 \end{bmatrix}$,

(c) $\begin{bmatrix} 1 & 0 & 0 \\ 0 & 1 & 0 \\ 0 & 0 & 1 \end{bmatrix} \begin{bmatrix} a_1 & a_2 & a_3 \\ b_1 & b_2 & b_3 \\ c_1 & c_2 & c_3 \end{bmatrix}$,

(d) $\begin{bmatrix} \alpha_1 & 0 & 0 \\ 0 & \alpha_2 & 0 \\ 0 & 0 & \alpha_3 \end{bmatrix} \begin{bmatrix} a_1 & a_2 \\ b_1 & b_2 \\ c_1 & c_2 \end{bmatrix}$,

(e) $\begin{bmatrix} a_1 & a_2 \\ b_1 & b_2 \\ c_1 & c_2 \end{bmatrix} \begin{bmatrix} \alpha_1 & 0 \\ 0 & \alpha_2 \end{bmatrix}$,

(f) $\begin{bmatrix} 1 \\ 2 \\ 3 \\ 4 \end{bmatrix} \begin{bmatrix} 1 & 2 & 3 & 4 \end{bmatrix}$,

(g) $\begin{bmatrix} 2 & 1 & -1 \end{bmatrix} \begin{bmatrix} 4 & -1 & 2 \\ -1 & 0 & 1 \\ 2 & 1 & 0 \end{bmatrix} \begin{bmatrix} 2 \\ 1 \\ -1 \end{bmatrix}$,

(h) $\begin{bmatrix} 1 & 0 \\ i & 1 \end{bmatrix} \begin{bmatrix} 1 & i \\ -i & 0 \end{bmatrix} \begin{bmatrix} 1 & -i \\ 0 & 1 \end{bmatrix}$, where $i^2 = -1$.

9. (a) Using

$$A = \begin{bmatrix} 1 & -1 & 1 \\ 2 & 0 & 1 \end{bmatrix}, \quad B = \begin{bmatrix} 1 & -1 & 0 \\ 0 & 1 & -1 \\ 1 & 1 & 1 \end{bmatrix}, \quad C = \begin{bmatrix} 1 & 0 \\ 0 & 1 \\ 1 & 1 \end{bmatrix},$$

test the rule $(AB)C = A(BC)$.

(b) Under what conditions is a matrix product $ABCD$ defined? According to the associative law, what are the various ways of computing it?

(c) If you have to compute a product $AB \cdots MX$ where X is a vector, how will you apply the associative law so as to reduce the computation to a minimum? Illustrate by computing the product

$$\begin{bmatrix} 2 & 1 & 1 \\ -1 & 4 & 2 \\ 3 & 0 & -1 \end{bmatrix} \begin{bmatrix} -1 & -1 & 2 \\ 0 & 3 & 0 \\ 1 & -1 & 2 \end{bmatrix} \begin{bmatrix} 2 & -1 \\ 4 & 0 \\ 1 & 5 \end{bmatrix} \begin{bmatrix} 1 \\ 1 \end{bmatrix}.$$

10. (a) Explain why in matrix algebra

$$(A + B)^2 \neq A^2 + 2AB + B^2$$

and

$$(A + B)(A - B) \neq A^2 - B^2,$$

except in special cases. Under what circumstances would equality hold?

(b) Expand $(A + B)^3$.

11. Prove that, if A has identical rows and AB is defined, AB has identical rows also.

*12. Given that A is a square matrix, define $A^{p+1} = A^p \cdot A$ for $p \geq 1$. By induction on q, prove that, for each positive integer p,

$$A^p A^q = A^{p+q} \quad \text{and} \quad (A^p)^q = A^{pq}$$

for all positive integers q. (Note that neither the definition nor the proof is of the scalar type.)

13. If $A = \begin{bmatrix} 0 & i \\ i & 0 \end{bmatrix}$, compute A^2, A^3, and A^4. (Here $i^2 = -1$). Give a general rule for A^n. (You may treat the cases n even and n odd separately.)

14. Compute A^2, B^2, and B^4, where

$$A = \begin{bmatrix} a & 1 - a \\ 1 + a & -a \end{bmatrix}, \quad B = \begin{bmatrix} b & -\left(\frac{1}{b} + b\right) \\ b & -b \end{bmatrix}.$$

Define what is meant by an nth root of a square matrix and comment on what these examples imply about the number of nth roots of a matrix.

15. Evaluate:

$$\begin{bmatrix} 0 & 1 \\ 0 & 0 \end{bmatrix}^2, \quad \begin{bmatrix} 0 & 1 & 0 \\ 0 & 0 & 1 \\ 0 & 0 & 0 \end{bmatrix}^3, \quad \begin{bmatrix} 0 & 1 & 0 & 0 \\ 0 & 0 & 1 & 0 \\ 0 & 0 & 0 & 1 \\ 0 & 0 & 0 & 0 \end{bmatrix}^4.$$

Then state and prove a general rule illustrated by these three examples.

16. Let $AB = C$ where A and B are of order n. If, in the ith row of A, $a_{ik} = 1$, where i and k are fixed, but all other elements of the ith row are zero, what can be said about the ith row of C? What is the analogous fact for columns?

17. Let A and B be of order n and let

$$C_1 = \alpha_1 A + \beta_1 B,$$
$$C_2 = \alpha_2 A + \beta_2 B,$$

where α_1, α_2, β_1, and β_2 are scalars such that $\alpha_1 \beta_2 \neq \alpha_2 \beta_1$. Show that $C_1 C_2 = C_2 C_1$ if and only if $AB = BA$.

18. Let

$$A = \begin{bmatrix} 3 & 4 & 2 \\ -2 & -1 & -1 \\ -1 & -3 & -1 \end{bmatrix}, \quad B = \begin{bmatrix} -1 & -1 & -1 \\ 2 & 2 & 2 \\ 1 & 1 & 1 \end{bmatrix}.$$

Compare the products AB and BA.

19. For what real values of x is

$$\begin{bmatrix} x & 4 & 1 \end{bmatrix} \begin{bmatrix} 2 & 1 & 0 \\ 1 & 0 & 2 \\ 0 & 2 & 4 \end{bmatrix} \begin{bmatrix} x \\ 4 \\ 1 \end{bmatrix} > 0?$$

For what value of x is this product a minimum and what is the minimum value?

20. If $AB = BA$, the matrices A and B are said to be **commutative** or to **commute**. Show that for all values of a, b, c, d, the matrices

$$A = \begin{bmatrix} a & -b \\ b & a \end{bmatrix} \quad \text{and} \quad B = \begin{bmatrix} c & -d \\ d & c \end{bmatrix}$$

commute.

21. What must be true about a, b, c, and d if the matrices

$$\begin{bmatrix} a & b \\ c & d \end{bmatrix} \quad \text{and} \quad \begin{bmatrix} 1 & 1 \\ -1 & 1 \end{bmatrix}$$

are to commute?

22. If $AB = -BA$, the matrices A and B are said to be **anticommutative** or to **anticommute**. Show that each of the matrices

$$\sigma_x = \begin{bmatrix} 0 & 1 \\ 1 & 0 \end{bmatrix}, \quad \sigma_y = \begin{bmatrix} 0 & -i \\ i & 0 \end{bmatrix}, \quad \sigma_z = \begin{bmatrix} 1 & 0 \\ 0 & -1 \end{bmatrix} \quad (i^2 = -1),$$

anticommutes with the others. These are the **Pauli spin matrices**, which are used in the study of electron spin in quantum mechanics.

23. The matrix $AB - BA$ (A and B of order n) is called the **commutator** of A and B. Using Exercise 22, show that the commutators of σ_x and σ_y, σ_y and σ_z, and σ_z and σ_x are respectively $2i\sigma_z$, $2i\sigma_x$, and $2i\sigma_y$.

24. If $A \circ B = AB - BA$, prove that

(a) $A \circ (B \circ C) = (A \circ B) \circ C$ if and only if $B \circ (A \circ C) = 0$,
(b) $A \circ (B \circ C) + B \circ (C \circ A) + C \circ (A \circ B) = 0$.

25. A matrix A such that $A^p = 0$ for some positive integer p is called **nilpotent**. Show that every 2×2 nilpotent matrix A such that $A^2 = 0$ may be written in the form

$$\begin{bmatrix} \lambda\mu & \mu^2 \\ -\lambda^2 & -\lambda\mu \end{bmatrix},$$

where λ and μ are scalars, and that every such matrix is nilpotent. If A is real, must λ and μ also be real?

26. If

$$A_i = \begin{bmatrix} \cos\theta_i & -\sin\theta_i \\ \sin\theta_i & \cos\theta_i \end{bmatrix}, \quad i = 1, 2,$$

show that A_1 and A_2 commute. What is the connection with transformations used in plane analytic geometry?

27. Given that

$$\begin{bmatrix} 1 & 0 & 0 \\ 0 & 2 & 0 \\ 0 & 0 & -3 \end{bmatrix} \cdot A \cdot \begin{bmatrix} 1 & 0 & 0 \\ 0 & 0 & 1 \\ 0 & 1 & 0 \end{bmatrix} = \begin{bmatrix} 1 & 2 & 3 \\ 4 & 5 & 4 \\ 3 & 2 & 1 \end{bmatrix},$$

find the matrix A.

*28. The sum of the main diagonal elements a_{ii}, $i = 1, 2, \ldots, n$, of a square matrix A of order n is called the **trace** of A:

$$\text{tr } A = a_{11} + a_{22} + \cdots + a_{nn}.$$

(a) If A and B are of order n, show that

$$\text{tr}(A + B) = \text{tr } A + \text{tr } B.$$

(b) If C is of order (m, n) and G is of order (n, m), show that

$$\text{tr } CG = \text{tr } GC.$$

29. Given that A, B, C all have order n, use Exercise 28(b) to show that

$$\text{tr } ABC = \text{tr } BCA = \text{tr } CAB; \qquad \text{tr } ACB = \text{tr } BAC = \text{tr } CBA.$$

What is the generalization of this observation? Show that ordinarily tr $ABC \neq$ tr ACB.

*30. A square matrix of the form

$$D_n = \begin{bmatrix} d_{11} & 0 & \cdots & 0 \\ 0 & d_{22} & \cdots & 0 \\ \vdots & & & \\ 0 & 0 & \cdots & d_{nn} \end{bmatrix},$$

that is, one in which $d_{ij} = 0$ if $i \neq j$, is called a **diagonal matrix** of order n. (Note that this does *not* say $d_{ii} \neq 0$.) Let A be any matrix of order (p, q) and evaluate the products $D_p A$ and $A D_q$. Describe the results in words. What happens in the special cases $d_{11} = d_{22} = \cdots = \alpha$ and $d_{11} = d_{22} = \cdots = 1$?

*31. (a) Show that any two diagonal matrices of the same order commute.
   (b) Give a formula for D^p where D is diagonal and p is a positive integer.
   (c) Show that if D is diagonal with *non-negative* elements, if p is a fixed positive integer, and if A is a fixed matrix, then $AD^p = D^p A$ if and only if $AD = DA$.

32. A matrix A such that $A^2 = A$ is called **idempotent**. Determine all diagonal matrices of order n which are idempotent. How many are there?

*33. Show by induction that, if A is square and $AB = \lambda B$, where λ is a scalar, then $A^p B = \lambda^p B$ for every positive integer p.

*34. Given that $AB = BA$, show that, for all positive integers r and s, $A^r B^s = B^s A^r$.

35. Prove by induction that if B, C are of order n and if $A = B + C$, $C^2 = 0$, and $BC = CB$, then for every positive integer k, $A^{k+1} = B^k(B + (k + 1)C)$.

1.10   Linear Equations in Matrix Notation

In a great variety of applications of mathematics, there appear systems of linear equations of the general form

(1.10.1)
$$\begin{aligned}
a_{11}x_1 + a_{12}x_2 + \cdots + a_{1n}x_n &= b_1 \\
a_{21}x_1 + a_{22}x_2 + \cdots + a_{2n}x_n &= b_2 \\
&\vdots \\
a_{m1}x_1 + a_{m2}x_2 + \cdots + a_{mn}x_n &= b_m,
\end{aligned}$$

where the number m of equations is not necessarily equal to the number n of unknowns. In view of the definition of matrix multiplication, such a system of equations may be written as the single matrix equation

$$(1.10.2) \qquad \begin{bmatrix} a_{11} & a_{12} & \cdots & a_{1n} \\ a_{21} & a_{22} & \cdots & a_{2n} \\ \vdots & & & \\ a_{m1} & a_{m2} & \cdots & a_{mn} \end{bmatrix} \cdot \begin{bmatrix} x_1 \\ x_2 \\ \vdots \\ x_n \end{bmatrix} = \begin{bmatrix} b_1 \\ b_2 \\ \vdots \\ b_m \end{bmatrix}.$$

In fact, if we compute the matrix product on the left, this equation becomes

$$(1.10.3) \qquad \begin{bmatrix} (a_{11}x_1 + a_{12}x_2 + \cdots + a_{1n}x_n) \\ (a_{21}x_1 + a_{22}x_2 + \cdots + a_{2n}x_n) \\ \vdots \\ (a_{m1}x_1 + a_{m2}x_2 + \cdots + a_{mn}x_n) \end{bmatrix} = \begin{bmatrix} b_1 \\ b_2 \\ \vdots \\ b_m \end{bmatrix}.$$

Since these two matrices are equal if and only if all their corresponding elements are equal, this single equation is equivalent to the system of (1.10.1). If we now put

$$A = \begin{bmatrix} a_{11} & a_{12} & \cdots & a_{1n} \\ a_{21} & a_{22} & \cdots & a_{2n} \\ \vdots & & & \\ a_{m1} & a_{m2} & \cdots & a_{mn} \end{bmatrix}, \qquad X = \begin{bmatrix} x_1 \\ x_2 \\ \vdots \\ x_n \end{bmatrix}, \qquad B = \begin{bmatrix} b_1 \\ b_2 \\ \vdots \\ b_m \end{bmatrix}$$

the bulky equation (1.10.2) may be written in the highly compact form

$$(1.10.4) \qquad\qquad AX = B.$$

When a system of equations is written in the form (1.10.2) or (1.10.4), in which one matrix equation replaces the entire system of scalar equations, it is said to be represented in **matrix notation**.

Single column matrices, such as X and B in the above discussion, are called **vectors**. The relation between this concept of a vector and the usual one will be developed later. For the present, *a vector is simply a column matrix.* The elements of such a matrix are commonly called its **components**. The vector X above is called an ***n*-vector** since it has n components. By the same token, B is an m-vector. Frequently, to save space, an n-vector is written in the form $\{a_1, a_2, \ldots, a_n\}$, the curly braces being used to identify it as a column matrix. In some books **row matrices**, that is, matrices consisting of a single row, are also called vectors.

Using this terminology, we see that, whenever we know a set of values of $x_1, x_2, \ldots, x_n$ which simultaneously satisfy the scalar equations (1.10.1), we also know the components of a vector which satisfies the matrix equation (1.10.4), and conversely. Such a vector is called a **solution** of (1.10.4). The problems of solving the system of scalar equations (1.10.1) and of solving the matrix equation (1.10.4) are thus seen to be equivalent. In Chapter 2, we discuss in detail the computation of the solutions of a system (1.10.1).

At times it is convenient to represent the system (1.10.1) in the vector form

$$(1.10.5) \qquad x_1 A_1 + x_2 A_2 + \cdots + x_n A_n = B,$$

where A_j denotes the jth column of the coefficient matrix A. For example, we can write the system

$$2x_1 - 3x_2 + 4x_3 = 1$$

$$3x_1 - 4x_2 + 6x_3 = 5$$

in matrix notation as

$$\begin{bmatrix} 2 & -3 & 4 \\ 3 & -4 & 6 \end{bmatrix} \begin{bmatrix} x_1 \\ x_2 \\ x_3 \end{bmatrix} = \begin{bmatrix} 1 \\ 5 \end{bmatrix}$$

or as

$$x_1 \begin{bmatrix} 2 \\ 3 \end{bmatrix} + x_2 \begin{bmatrix} -3 \\ -4 \end{bmatrix} + x_3 \begin{bmatrix} 4 \\ 6 \end{bmatrix} = \begin{bmatrix} 1 \\ 5 \end{bmatrix}.$$

Substitution reveals that every vector of the form

$$\begin{bmatrix} x_1 \\ x_2 \\ x_3 \end{bmatrix} = t \begin{bmatrix} -2 \\ 0 \\ 1 \end{bmatrix} + \begin{bmatrix} 15 \\ 7 \\ -2 \end{bmatrix}$$

is a solution of this system. In the next chapter, we see how such solutions may be computed.

1.11   The Transpose of a Matrix

The matrix A^{T} of order (n, m) obtained by interchanging rows and columns in a matrix A of order (m, n) is called the **transpose** of A. For example, the transpose of

$$\begin{bmatrix} 2 & 0 & -1 \\ 1 & 1 & 4 \end{bmatrix} \qquad \text{is} \qquad \begin{bmatrix} 2 & 1 \\ 0 & 1 \\ -1 & 4 \end{bmatrix}.$$

Theorem 1.11.1: *If A^{T} and B^{T} are the transposes of A and B, and if α is a scalar, then:*

(1.11.1)   (a) $(A^{\mathsf{T}})^{\mathsf{T}} = A$.
          (b) $(A + B)^{\mathsf{T}} = A^{\mathsf{T}} + B^{\mathsf{T}}$.
          (c) $(\alpha A)^{\mathsf{T}} = \alpha A^{\mathsf{T}}$.
          (d) $(AB)^{\mathsf{T}} = B^{\mathsf{T}} A^{\mathsf{T}}$.

The first three of these rules are easy to think through. Detailed proofs are left to the reader. Only (d) will be proved here. Let $A = [a_{ik}]_{(m,n)}$, $B = [b_{kj}]_{(n,p)}$. Then $AB = [c_{ij}]_{(m,p)}$, where $c_{ij} = \sum_{k=1}^{n} a_{ik} b_{kj}$. Here i, ranging from 1 to m, identifies the row, and j, ranging from 1 to p, identifies the column of the element c_{ij}.

Now B^{T} is of order (p, n) and A^{T} is of order (n, m) so that B^{T} is conformable to A^{T} for multiplication. To compute the element γ_{ji} in the jth row and the ith column of $B^{\mathsf{T}} A^{\mathsf{T}}$, we must multiply the jth row of B^{T} into the ith column of A^{T}. Observing that the second subscript of an element in B or A identifies the row and the first subscript identifies the column in which it appears in B^{T} or A^{T}, we see that

$$B^{\mathsf{T}} A^{\mathsf{T}} = [\gamma_{ji}]_{(p,m)} = \left[\sum_{k=1}^{n} b_{kj} a_{ik} \right]_{(p,m)} = \left[\sum_{k=1}^{n} a_{ik} b_{kj} \right]_{(p,m)}.$$

Here j ranges from 1 to p and identifies the row of γ_{ji}, whereas i ranges from 1 to m and identifies the column. Thus $\gamma_{ji} = c_{ij}$ but with the meanings of i and j for rows and columns just opposite in the two cases, so that $B^{\mathsf{T}} A^{\mathsf{T}} = (AB)^{\mathsf{T}}$. The reader would do well to construct a numerical example.

The preceding proof illustrates the fact that the expression for the typical element of a matrix may at times assume a complicated form, but that an essential aspect of the expression is a *pair of subscripts* which may be used to identify the row and the column of the element. Such subscripts are called **free subscripts**. Thus, in the expression,

$$c_{ij} = \sum_{k=1}^{n} a_{ik} b_{kj},$$

the free subscripts are i and j. Moreover, i is used to designate the row and j the column of the element. The index k is an **index of summation**. It has nothing to do with the row–column position of the element c_{ij}. On the other hand, in the expression

$$\gamma_{ji} = \sum_{k=1}^{n} b_{kj} a_{ik}$$

used in the foregoing, the free subscripts j and i designate respectively the row and the column of the element γ_{ji}. This illustrates the fact that, whereas the index i is frequently used as a row index and j as a column index, this is not necessary nor is it always convenient. Any convenient index symbol may be used for either purpose. Indeed, in the sum represented by c_{ij} in the foregoing, the index k is used both ways—as a column index in a_{ik} and as a row index in b_{kj}. It is this freedom to use indices in both ways that enabled us to prove the foregoing theorem in a simple fashion.

As another example illustrating these matters, let

$$a_{ij} = \sum_{k=1}^{n} r_{ik} s_{jk},$$

where i denotes the row and j the column of the element a_{ij}. If i ranges from 1 to m and j from 1 to p, then we can interpret r_{ik} as the entry in the ith row and kth column of a matrix $R_{m \times n}$. In s_{jk}, k must play the role of a *row* index if we want to represent the matrix $A = [a_{ij}]_{m \times p}$ as a product. Assuming that normally the first subscript is the row subscript, it follows that the element s_{jk} is the element in the jth column and kth row of a matrix $(S_{p \times n})^{\mathsf{T}}$. That is,

$$A = RS^{\mathsf{T}}.$$

Consider finally the sum

$$c_{ij} = \sum_{k_2=1}^{n} \sum_{k_1=1}^{n} a_{ik_1} a_{k_1 k_2} a_{k_2 j}, \qquad i, j = 1, 2, \ldots, n,$$

which we can rewrite as

$$c_{ij} = \sum_{k_2=1}^{n} \left(\sum_{k_1=1}^{n} a_{ik_1} a_{k_1 k_2} \right) a_{k_2 j}.$$

The sum in parentheses is the element in row i and column k_2 of A^2. Hence the second sum represents the element in row i and column j of $A^2 \cdot A$. That is, we have

$$[c_{ij}]_{n \times n} = A^3.$$

1.12 Symmetric, Skew-Symmetric, and Hermitian Matrices

A **symmetric matrix** is a square matrix A such that $A = A^{\mathsf{T}}$. A **skew-symmetric matrix** is a square matrix A such that $A = -A^{\mathsf{T}}$. These definitions may also be stated in terms of the individual elements: A is symmetric if and only if $a_{ij} = a_{ji}$ for all pairs of subscripts; it is skew-symmetric if and only if $a_{ij} = -a_{ji}$ for all pairs of subscripts. The reader should demonstrate the equivalence of the alternative definitions in each case.

The following are examples of symmetric and skew-symmetric matrices respectively:

$$\text{(a)} \quad \begin{bmatrix} 0 & 1 & 2 \\ 1 & 2 & 3 \\ 2 & 3 & 4 \end{bmatrix}; \quad \text{(b)} \quad \begin{bmatrix} 0 & 1 & 2 \\ -1 & 0 & 3 \\ -2 & -3 & 0 \end{bmatrix}.$$

Example (b) illustrates the fact that the main diagonal elements of a skew-symmetric matrix must all be zero. Why is this true?

A matrix is called a **real matrix** if and only if all its elements are real. In the applications, real symmetric matrices occur most frequently. However, matrices of complex elements are also of great importance. When the elements of such a matrix A are replaced by their complex conjugates, the resulting matrix is called the **conjugate** of A and is denoted by $\overline{A}$. Evidentally a matrix A is real if and only if $A = \overline{A}$. Transposing $\overline{A}$, we obtain the **transposed conjugate** or **tranjugate** $(\overline{A})^{\mathsf{T}}$ of A. This will be denoted by the symbol A^*. (A^* is sometimes called the *adjoint* of A, but *not in this book*.) For example, if

$$A = \begin{bmatrix} 1-i & 2 \\ i & 1+i \end{bmatrix},$$

then

$$\overline{A} = \begin{bmatrix} 1+i & 2 \\ -i & 1-i \end{bmatrix},$$

and

$$A^* = \begin{bmatrix} 1+i & -i \\ 2 & 1-i \end{bmatrix}.$$

When $A = A^*$, that is, when $a_{ij} = \overline{a}_{ji}$ for all pairs of subscripts, A is called a **Hermitian matrix** (after the French mathematician, Charles Hermite, 1822–1901). When matrices of complex elements appear in the applications, for example, in the theory of atomic physics, they are often Hermitian. The matrices

$$\begin{bmatrix} 0 & i \\ -i & 0 \end{bmatrix} \quad \text{and} \quad \begin{bmatrix} 4 & 1-i \\ 1+i & 2 \end{bmatrix}$$

are simple examples of Hermitian matrices, as is readily verified. Why must the diagonal elements of a Hermitian matrix all be real numbers?

If the elements of A are real, $A^\mathsf{T} = A^*$, so that the property of being real and symmetric is a special case of the property of being Hermitian. The following pages will contain a great many results about Hermitian matrices. The reader interested only in the real case may interpret the word "Hermitian" as "real symmetric," and all will be well.

1.13 Scalar Matrices

A square matrix of the form

$$
\begin{bmatrix}
\alpha & 0 & 0 & \cdots & 0 \\
0 & \alpha & 0 & \cdots & 0 \\
\vdots & & & & \\
0 & 0 & 0 & \cdots & \alpha
\end{bmatrix}_n
$$

in which each element of the main diagonal equals the scalar α and all other elements are zero, is called a **scalar matrix** of order n. In matrix multiplication, a scalar matrix behaves like a scalar, as the following equations show:

$$
\begin{bmatrix}
\alpha & 0 & \cdots & 0 \\
0 & \alpha & \cdots & 0 \\
\vdots & & & \\
0 & 0 & \cdots & \alpha
\end{bmatrix}_m
\cdot
\begin{bmatrix}
a_{11} & a_{12} & \cdots & a_{1n} \\
a_{21} & a_{22} & \cdots & a_{2n} \\
\vdots & & & \\
a_{m1} & a_{m2} & \cdots & a_{mn}
\end{bmatrix}
$$

$$
=
\begin{bmatrix}
a_{11} & a_{12} & \cdots & a_{1n} \\
a_{21} & a_{22} & \cdots & a_{2n} \\
\vdots & & & \\
a_{m1} & a_{m2} & \cdots & a_{mn}
\end{bmatrix}
\cdot
\begin{bmatrix}
\alpha & 0 & \cdots & 0 \\
0 & \alpha & \cdots & 0 \\
\vdots & & & \\
0 & 0 & \cdots & \alpha
\end{bmatrix}_n
= \alpha
\begin{bmatrix}
a_{11} & a_{12} & \cdots & a_{1n} \\
a_{21} & a_{22} & \cdots & a_{2n} \\
\vdots & & & \\
a_{m1} & a_{m2} & \cdots & a_{mn}
\end{bmatrix}.
$$

The scalar matrices are even more fundamentally like scalars, however, for if α and β are any two scalars, and if

(1.13.1)
$$
\alpha + \beta = \gamma,
$$
$$
\alpha\beta = \delta,
$$

then

(1.13.2)

$$\begin{bmatrix} \alpha & 0 & \cdots & 0 \\ 0 & \alpha & \cdots & 0 \\ \vdots & & & \\ 0 & 0 & \cdots & \alpha \end{bmatrix}_n + \begin{bmatrix} \beta & 0 & \cdots & 0 \\ 0 & \beta & \cdots & 0 \\ \vdots & & & \\ 0 & 0 & \cdots & \beta \end{bmatrix}_n = \begin{bmatrix} \gamma & 0 & \cdots & 0 \\ 0 & \gamma & \cdots & 0 \\ \vdots & & & \\ 0 & 0 & \cdots & \gamma \end{bmatrix}_n$$

$$\begin{bmatrix} \alpha & 0 & \cdots & 0 \\ 0 & \alpha & \cdots & 0 \\ \vdots & & & \\ 0 & 0 & \cdots & \alpha \end{bmatrix}_n \cdot \begin{bmatrix} \beta & 0 & \cdots & 0 \\ 0 & \beta & \cdots & 0 \\ \vdots & & & \\ 0 & 0 & \cdots & \beta \end{bmatrix}_n = \begin{bmatrix} \delta & 0 & \cdots & 0 \\ 0 & \delta & \cdots & 0 \\ \vdots & & & \\ 0 & 0 & \cdots & \delta \end{bmatrix}_n .$$

These equations show that corresponding to the arithmetic of the real numbers, for example, there is a strictly analogous arithmetic of scalar matrices of a fixed order n in which the scalar matrix with main diagonal elements α corresponds to the real number α. In the same way, corresponding to the arithmetic of complex numbers, there is an analogous arithmetic of scalar matrices the main diagonal elements of which are complex numbers. Many other examples of such a correspondence between a set of scalars and a corresponding set of scalar matrices could be constructed.

There are three essential points to be noted in this situation. First, we deal with a collection $\mathscr{S}$ of scalars such that, if α and β belong to $\mathscr{S}$, and if $\alpha + \beta = \gamma$, $\alpha\beta = \delta$, then γ and δ also belong to $\mathscr{S}$. (In the two examples cited in the foregoing, the sum and the product of two real numbers are again real numbers, and the sum and the product of two complex numbers are again complex numbers.) Second, we deal with a collection $\mathscr{M}$ of scalar matrices of a fixed order n such that to each scalar α in $\mathscr{S}$ there corresponds the unique scalar matrix of $\mathscr{M}$ the main diagonal elements of which are all equal to α. Conversely, to each scalar matrix in $\mathscr{M}$ with diagonal elements α there corresponds the unique scalar α of $\mathscr{S}$. That is, the scalars in $\mathscr{S}$ are in *one-to-one correspondence* with the scalar matrices in $\mathscr{M}$. The third point to be noted is that, if α and β are in the set $\mathscr{S}$ of scalars, and if equations (1.13.1) hold, then so do equations (1.13.2). That is, to the sum and the product of two scalars in $\mathscr{S}$ there correspond respectively the sum and the product of the corresponding scalar matrices of $\mathscr{M}$. This collection of ideas we identify by saying that the set $\mathscr{S}$ of scalars and the set $\mathscr{M}$ of corresponding matrices are **isomorphic,** or that there is an **isomorphism** between $\mathscr{S}$ and $\mathscr{M}$. [Isomorphic means *of the same* (iso-) *form* (morphos).]

The notion of isomorphism developed here is a particular application of a general concept of isomorphism which is one of the powerful tools of modern abstract algebra.

1.14   The Identity Matrix

The scalar matrix of order n corresponding to the scalar 1 will be denoted by the symbol I_n, or simply by I if the order need not be emphasized:

$$I_n = \begin{bmatrix} 1 & 0 & 0 & \cdots & 0 \\ 0 & 1 & 0 & \cdots & 0 \\ \vdots & & & & \\ 0 & 0 & 0 & \cdots & 1 \end{bmatrix}_n .$$

It is called the **identity matrix** or the **unit matrix** of order n because it plays in matrix algebra the role corresponding to that played by the integer 1 in scalar algebra, as the isomorphism just explained would lead us to expect. However, the role extends beyond the domain of scalar matrices, for if A is any $m \times n$ matrix, we have

$$I_m A = A I_n = A,$$

as is readily verified.

From the isomorphism explained in the foregoing, it is clear that, since 1 is the only scalar satisfying the equations $xa = ax = a$ for every scalar a, I_n is likewise the only scalar matrix of order n satisfying the matrix equations $XA = AX = A$ for every scalar matrix A of order n. It therefore follows that I_n is the only matrix of order n satisfying the equations $XA = AX = A$ for every matrix A, scalar or not, of order n. This latter conclusion may also be proved directly. In fact, let B be any matrix of order n such that $AB = BA = A$ for *every* matrix A of order n. Letting $A = I_n$ in particular, we have $I_n B = B I_n = I_n$ or $B = I_n$; that is, the identity matrix is uniquely defined by the property that it behaves like the scalar 1 in matrix multiplication.

In conclusion, we note that every scalar matrix is related to the identity matrix as follows:

$$\begin{bmatrix} \alpha & 0 & \cdots & 0 \\ 0 & \alpha & \cdots & 0 \\ \vdots & & & \\ 0 & 0 & \cdots & \alpha \end{bmatrix}_n = \alpha I_n .$$

This is a convenient way of reducing bulky notation.

1.15   The Inverse of a Matrix

For any scalar $\alpha \neq 0$, there exists another scalar α^{-1}, the reciprocal or inverse of α, such that

$$\alpha \alpha^{-1} = \alpha^{-1} \alpha = 1.$$

The analogous equation for scalar matrices, indicated by the isomorphism just pointed out, is

$$(\alpha^{-1}I_n)(\alpha I_n) = (\alpha I_n)(\alpha^{-1}I_n) = I_n.$$

Hence we say that the matrix $\alpha^{-1}I_n$ is the *inverse* of the matrix αI_n, and we write

$$\alpha^{-1}I_n = (\alpha I_n)^{-1}, \quad (\alpha \neq 0).$$

For example,

$$2 \cdot \tfrac{1}{2} = \tfrac{1}{2} \cdot 2 = 1$$

and

$$\begin{bmatrix} 2 & 0 \\ 0 & 2 \end{bmatrix} \begin{bmatrix} \tfrac{1}{2} & 0 \\ 0 & \tfrac{1}{2} \end{bmatrix} = \begin{bmatrix} \tfrac{1}{2} & 0 \\ 0 & \tfrac{1}{2} \end{bmatrix} \begin{bmatrix} 2 & 0 \\ 0 & 2 \end{bmatrix} = \begin{bmatrix} 1 & 0 \\ 0 & 1 \end{bmatrix}.$$

From the isomorphism between scalars and scalar matrices, it follows that a nonzero scalar matrix has only one inverse which is a scalar matrix.

These observations lead us at once to a more general problem of which the foregoing is just a special case. This problem is to determine, when a matrix A of order n is given, any and all matrices B of the same order as A which have the property $AB = BA = I_n$. If such a matrix B exists, we call it an **inverse** of A.

Suppose now that we have both $AB = BA = I$ and $AC = CA = I$. Then, premultiplying both sides of $AC = I$ by B, we find $B(AC) = BI = B$. Similarly, from $BA = I$, we obtain $(BA)C = IC = C$. But $B(AC) = (BA)C$ by the associative law. Hence $B = C$. Thus we have

Theorem 1.15.1: *A square matrix A has at most one inverse; that is, the inverse is unique if it exists.*

In view of this result, it is proper to refer to *the* inverse of a matrix A and to denote it by the convenient symbol A^{-1}, when the inverse exists.

In some cases, the inverse is readily found by inspection. For instance, we have, in the case of a diagonal matrix D,

$$D = \begin{bmatrix} d_1 & 0 & \cdots & 0 \\ 0 & d_2 & \cdots & 0 \\ \vdots & & & \\ 0 & 0 & \cdots & d_n \end{bmatrix} \quad \text{and} \quad D^{-1} = \begin{bmatrix} d_1^{-1} & 0 & \cdots & 0 \\ 0 & d_2^{-1} & \cdots & 0 \\ \vdots & & & \\ 0 & 0 & \cdots & d_n^{-1} \end{bmatrix}$$

provided that all the d_j are different from zero, for it is easily verified that, in this case, $DD^{-1} = D^{-1}D = I_n$.

That an inverse does not always exist is shown by the following example. The product

$$\begin{bmatrix} a & b \\ c & d \end{bmatrix} \cdot \begin{bmatrix} 0 & 1 \\ 0 & 0 \end{bmatrix} = \begin{bmatrix} 0 & a \\ 0 & c \end{bmatrix}$$

shows that there is no way to choose a, b, c, and d so as to make the right member equal to I_2. Hence the matrix $\begin{bmatrix} 0 & 1 \\ 0 & 0 \end{bmatrix}$ has no inverse.

That a nonsquare matrix does not have an inverse in the aforementioned sense follows from the fact that the products AB and BA cannot be equal in this case, whereas $AB = BA$ is required by the definition. Separate "left" and "right" inverses can be defined, however (see Exercise 53, Section 1.18). When A^{-1} exists, the matrix A is said to be **invertible** or **nonsingular**. When A^{-1} does not exist, A is said to be **singular**.

Ordinarily, one cannot determine simply by casual inspection whether or not a given square matrix has an inverse. However, computational procedures to be developed in the next chapter will answer this question and will also provide the inverse if it exists.

There are other important results about inverses that one can prove without knowing how to compute the inverse. For example, we have

Theorem 1.15.2: *If the $n \times n$ matrices A, B,..., M, N, all have inverses, then their product $AB \cdots MN$ has the inverse $N^{-1}M^{-1} \cdots B^{-1}A^{-1}$; that is, the inverse of a product is the product of the inverses in the reverse order.*

In fact, by the associative law

$$\begin{aligned} (AB \cdots MN)(N^{-1}M^{-1} \cdots B^{-1}A^{-1}) &= (AB \cdots M)(NN^{-1})(M^{-1} \cdots B^{-1}A^{-1}) \\ &= (AB \cdots M)(M^{-1} \cdots B^{-1}A^{-1}) \\ &= \cdots \\ &= AA^{-1} \\ &= I. \end{aligned}$$

The product $(N^{-1}M^{-1} \cdots B^{-1}A^{-1})(AB \cdots MN)$ reduces to I in a similar fashion, so the theorem is proved.

For example, the inverse of the product

$$AB = \begin{bmatrix} 2 & 0 \\ 0 & 3 \end{bmatrix} \cdot \begin{bmatrix} 1 & -1 \\ 0 & 1 \end{bmatrix}$$

is

$$B^{-1}A^{-1} = \begin{bmatrix} 1 & 1 \\ 0 & 1 \end{bmatrix} \cdot \begin{bmatrix} \frac{1}{2} & 0 \\ 0 & \frac{1}{3} \end{bmatrix} = \begin{bmatrix} \frac{1}{2} & \frac{1}{3} \\ 0 & \frac{1}{3} \end{bmatrix},$$

as is readily checked.

Finally, we note that the inverse is useful in solving certain types of matrix equations. Thus, if $A_{n \times n} X_{n \times p} = B_{n \times p}$ and, if A^{-1} exists, then $X = A^{-1}B$.

Again, if A^{-1} and B^{-1} exist, then

$$AWB = C$$

if and only if

$$W = A^{-1}CB^{-1},$$

so that the second equation gives the unique solution for W of the first. Note the importance of distinguishing between pre- and postmultiplication here.

1.16   The Product of a Row Matrix into a Column Matrix

The product of a row matrix into a column matrix is a matrix of order 1:

$$[x_1, x_2, \ldots, x_n]_{(1,n)} \cdot [y_1, y_2, \ldots, y_n]^{\mathsf{T}}_{(n,1)}$$
$$= [(x_1 y_1 + x_2 y_2 + \cdots + x_n y_n)]_{(1,1)}.$$

Let the elements of the row matrix and the column matrix, and hence the element of their product, belong to a set of scalars $\mathscr{S}$ such as was described in Section 1.13. Now the set of all 1×1 matrices whose elements belong to $\mathscr{S}$ is isomorphic to $\mathscr{S}$. In fact, if α and β are any two scalars of $\mathscr{S}$, we have, for the corresponding 1×1 matrices, $[\alpha] + [\beta] = [\alpha + \beta]$ and $[\alpha][\beta] = [\alpha\beta]$, so that the matrices behave exactly like scalars with respect to addition and multiplication. Furthermore, the 1×1 matrices have no important properties not possessed by scalars. Hence *we redefine the product of a row matrix into a column matrix to be the scalar corresponding to the 1×1 product matrix, thus:*

$$[x_1, x_2, \ldots, x_n] \cdot [y_1, y_2, \ldots, y_n]^{\mathsf{T}} = (x_1 y_1 + x_2 y_2 + \cdots + x_n y_n).$$

This kind of product arises most commonly in the form $X^{\mathsf{T}} Y$, where X and Y are both column matrices, that is, vectors. It is called the **scalar**

product of the two vectors. Whereas one may treat every 1×1 matrix product as a scalar, one can replace a scalar by a product of matrices only if the rule of conformability is satisfied.

The scalar product is intimately involved in matrix multiplication since

$$A_{m \times n} B_{n \times p} = [A_i B^{(j)}]_{m \times p};$$

that is, the ij-entry of the product is the scalar product of the ith row of A and the jth column of B.

There is another, closely related, function of two vectors X and Y. Reflecting the fact that X and Y may have complex elements, the **inner product** of X and Y, *in that order*, is defined to be X^*Y. The inner product is commonly denoted by the symbol $\langle X, Y \rangle$ so that

$$\langle X, Y \rangle = X^*Y = [\bar{x}_1, \bar{x}_2, \ldots, \bar{x}_n] \cdot [y_1, y_2, \ldots, y_n]^\mathsf{T}$$
$$= \bar{x}_1 y_1 + \bar{x}_2 y_2 + \cdots + \bar{x}_n y_n.$$

In the particular case when X and Y are restricted to the real domain, this reduces to

$$\langle X, Y \rangle = X^\mathsf{T} Y = x_1 y_1 + x_2 y_2 + \cdots + x_n y_n;$$

that is, in the real case, the inner product and the scalar product are the same.

There is a more general concept of inner product, which includes the present definition as a special case and which is developed in more advanced treatments of linear algebra.

1.17 Polynomial Functions of Matrices

Consider a polynomial

$$f(x) = a_p x^p + a_{p-1} x^{p-1} + \cdots + a_1 x + a_0$$

of degree p in an indeterminate x, with either real or complex coefficients. With any such polynomial, we can associate a **polynomial function** of an $n \times n$ matrix A:

$$f(A) = a_p A^p + a_{p-1} A^{p-1} + \cdots + a_1 A + a_0 I_n$$

simply by replacing the constant term a_0 by the scalar matrix $a_0 I_n$ and by replacing x by A throughout. For example, if

$$f(x) = x^2 - 3x - 2,$$

then, for any matrix A of order n,

$$f(A) = A^2 - 3A - 2I_n.$$

Now suppose $f(x)$ and $g(x)$ have only real coefficients. Then if

$$f(x) + g(x) = h(x) \qquad \text{and} \qquad f(x) \cdot g(x) = q(x),$$

$h(x)$ and $q(x)$ are again polynomials with real coefficients. Moreover, since scalars commute with matrices, since the distributive and associative laws hold, and since the exponent laws hold for positive integral powers of A, we may conclude also that

$$f(A) + g(A) = h(A) \qquad \text{and} \qquad f(A) \cdot g(A) = q(A).$$

Hence the set of all real polynomials in an indeterminate x is isomorphic to the set of all real polynomials in an indeterminate matrix A of order n. A similar statement holds true for polynomials with complex coefficients.

This isomorphism implies that the matrix polynomial $f(A)$ is factorable in the same manner as the scalar polynomial $f(x)$. For example, if

$$f(x) = x^3 - 3x^2 + 2x = x(x - 1)(x - 2),$$

then

$$f(A) = A^3 - 3A^2 + 2A = A(A - I_n)(A - 2I_n).$$

Now suppose that the solutions of

$$f(x) = a_p x^p + a_{p-1} x^{p-1} + \cdots + a_1 x + a_0 = 0$$

are $\alpha_1, \alpha_2, \ldots, \alpha_p$, so that

$$f(x) = a_p(x - \alpha_1)(x - \alpha_2) \cdots (x - \alpha_p).$$

Hence, by the isomorphism just explained,

$$(1.17.1) \qquad f(A) = a_p(A - \alpha_1 I)(A - \alpha_2 I) \cdots (A - \alpha_p I),$$

so that the matrix equation $f(A) = 0$ has for solutions the *scalar* matrices $\alpha_1 I, \ldots, \alpha_p I$. Moreover, it has no other *scalar* matrices as solutions. However, since the product of matrices (1.17.1) may be zero even though no factor is zero, the equation $f(A) = 0$ may also have nonscalar matrices as solutions. For example, the equation

$$A^2 + I_2 = 0$$

has for solutions the scalar matrices

$$\begin{bmatrix} i & 0 \\ 0 & i \end{bmatrix}, \qquad \begin{bmatrix} -i & 0 \\ 0 & -i \end{bmatrix},$$

corresponding to the solutions i and $-i$ of the equation $x^2 + 1 = 0$. It also has (among others) the nonscalar solution

$$\begin{bmatrix} 1 & 2 \\ -1 & -1 \end{bmatrix},$$

as the reader should verify.

The fact that in scalar algebra a polynomial equation of degree p in a single unknown x has exactly p solutions therefore does not hold true in matrix algebra, where the number of solutions is always at least p. The notion of isomorphism has thus led us to discover another important difference between matrix algebra and scalar algebra.

1.18   Exercises

1. Illustrate the definition of the equality of two matrices by writing the system of equations $x_j = a_j, j = 1, 2, \ldots, 12$, as a single matrix equation. Can you do this in more than one way?

2. Rewrite this system of equations as a single matrix equation of the form (1.10.4):

$$x_1 - 3x_2 + 4 = 2x_1 + 3x_2 - 4x_3 - 7$$
$$2x_1 + 5x_2 - x_3 = x_2 + 8$$
$$-2x_1 + 9 = x_2 + x_3.$$

Then rewrite it in the form (1.10.5).

3. (a) Interpret the matrix equation

$$\begin{bmatrix} 1 & -4 \\ 3 & 2 \end{bmatrix} \begin{bmatrix} 5 \\ -2 \end{bmatrix} = \begin{bmatrix} 13 \\ 11 \end{bmatrix}$$

as a fact about a system of linear equations in two unknowns.

(b) What is the geometrical significance of the matrix equation

$$\begin{bmatrix} x \\ y \end{bmatrix} = \begin{bmatrix} x_1 \\ y_1 \end{bmatrix} + t \begin{bmatrix} x_2 - x_1 \\ y_2 - y_1 \end{bmatrix}?$$

What is the corresponding equation in three dimensions?

4. Verify that every 5-vector X of the form

$$X = t_1 \begin{bmatrix} -1 \\ -1 \\ 1 \\ 0 \\ 0 \end{bmatrix} + t_2 \begin{bmatrix} 0 \\ -2 \\ 0 \\ 4 \\ 1 \end{bmatrix},$$

where t_1 and t_2 are arbitrary parameters, is a solution of the equation

$$\begin{bmatrix} 1 & -1 & 0 & -1 & 2 \\ 2 & 1 & 3 & 0 & 2 \\ -3 & 2 & -1 & 1 & 0 \end{bmatrix} X = 0.$$

5. (a) Verify that $B^\mathsf{T} A^\mathsf{T} = (AB)^\mathsf{T}$ when

$$A = \begin{bmatrix} 1 & -1 & 2 \\ 2 & 1 & 0 \end{bmatrix}, \qquad B = \begin{bmatrix} 1 & 2 \\ 2 & 0 \\ -1 & 1 \end{bmatrix}.$$

(b) Prove that $(A^\mathsf{T} B^\mathsf{T})^\mathsf{T} = BA$.

*6. Generalize statements (b) and (d) of Theorem 1.11.1 and prove the generalized forms by induction.

*7. Show that ordinarily $AA^\mathsf{T} \neq A^\mathsf{T} A$, but that AA^T and $A^\mathsf{T} A$ are both symmetric. Also show that if A has real elements, the principal diagonal elements of AA^T and $A^\mathsf{T} A$ are non-negative.

8. What may be said about the main diagonal elements of AA^T when A is a real skew-symmetric matrix? What about those of A^2?

9. Show that if A is symmetric or skew-symmetric, $AA^\mathsf{T} = A^\mathsf{T} A$ and A^2 is symmetric.

10. Show that every square matrix A can be be represented uniquely in the form

$$A = A^{(S)} + A^{(SS)}$$

where $A^{(S)}$ is symmetric and $A^{(SS)}$ is skew-symmetric. (*Hint:* If $A = A^{(S)} + A^{(SS)}$, then $A^\mathsf{T} = A^{(S)} - A^{(SS)}$, etc.)

11. Show by means of an example that even though A and B are both symmetric and are of the same order, AB need not necessarily be symmetric.

*12. Prove that if A and B of order n are both symmetric or both skew-symmetric *and commute*, then AB is symmetric.

13. Given that $A = [a_{ij}]_{n \times n}$ and $B = [b_{ij}]_{n \times n}$, represent each of the following as a matrix product:

$$\left[\sum_{k=1}^{n} a_{ik} b_{jk} \right], \qquad \left[\sum_{k=1}^{n} a_{ki} b_{jk} \right], \qquad \left[\sum_{k=1}^{n} b_{ik} a_{kj} \right], \qquad \left[\sum_{k=1}^{n} b_{kj} a_{ik} \right].$$

Assume in each case that i designates the row and j the column of the element in the product.

14. Express as matrix products if i denotes the row and j the column of the entry; $i = 1, 2, \ldots, m; j = 1, 2, \ldots, n$:

(a)
$$\left[\sum_{r=1}^{q} \left(\sum_{k=1}^{p} a_{ik} b_{kr} \right) c_{rj} \right].$$

(b)
$$\left[\sum_{r=1}^{q} c_{jr} \left(\sum_{k=1}^{p} a_{ki} b_{kr} \right) \right].$$

15. Given that $t_k = \operatorname{tr} A^k$, where A is of order n, prove that, for every positive integer k,

$$\left[\frac{\partial t_k}{\partial a_{ij}} \right]^{\mathbf{T}} = k A^{k-1}.$$

*16. Denoting by E_j an n-vector with 1 in the jth row, all other components being 0, interpret each of the products

$$E_j^{\mathbf{T}} A, \qquad A E_j, \qquad E_i^{\mathbf{T}} A E_j, \qquad E_j^{\mathbf{T}} E_k, \qquad E_j E_k^{\mathbf{T}},$$

where A is an arbitrary matrix of order n. The vectors E_j are called the **elementary n-vectors.**

17. Let $W_n = \sum_{j=1}^{n} E_j$. Then how are $A W_n$ and $W_n^{\mathbf{T}} A$ related to A? Evaluate $W_n^{\mathbf{T}} X$, where X is an arbitrary n-vector.

18. Suppose that $AB = C$, and let X denote a matrix with a single column whose elements are the sums of the elements in the corresponding rows of B. Similarly, let Y denote a column matrix whose elements are the row sums of C. Then show that $AX = Y$. (This fact may be used to check matrix multiplication and is useful in machine computation. *Hint:* Use Exercise 17.)

19. Evaluate each of the following products, then express the results in words. The E's are the n-vectors defined in Exercise 16.

$$(E_{i_1} + E_{i_2} + \cdots + E_{i_k})^{\mathbf{T}} A_{n \times m}, \qquad B_{p \times n}(E_{j_1} + E_{j_2} + \cdots + E_{j_r}),$$
$$(E_{i_1} + E_{i_2} + \cdots + E_{i_k})^{\mathbf{T}} C(E_{j_1} + E_{j_2} + \cdots + E_{j_r}).$$

20. Given that $X = [x_1, x_2, \ldots, x_n]^{\mathbf{T}}$, show that

$$X^{\mathbf{T}} E_j = x_j,$$

and hence that

$$X = \sum_{j=1}^{n} (X^{\mathbf{T}} E_j) E_j.$$

Why can one not write $(X^{\mathbf{T}} E_j) E_j = X^{\mathbf{T}} (E_j E_j)$?

21. Denoting by E_{ij} an $n \times n$ matrix with a 1 in the ith row and jth column and 0's elsewhere, evaluate the products

$$E_{ij} E_{jk}, \qquad E_{ij} E_{km}, \qquad A E_{ij}, \qquad E_{ij} A, \qquad E_{ij} A E_{jk}, \qquad E_{ij} A E_{km}.$$

Show that

$$E_{ij} = E_i E_j^\mathsf{T}, \qquad E_{jj}^2 = E_{jj},$$

and

$$(E_{ij} E_{jk})^\mathsf{T} = E_{ki} = E_{kj} E_{ji}.$$

22. Show that, for arbitrary $n \times n$ matrices A and B,

$$A = \sum_{i,j=1}^{n} a_{ij} E_{ij}, \qquad B = \sum_{j,k=1}^{n} b_{jk} E_{jk},$$

and, hence, with the aid of the preceding exercise, that $(AB)^\mathsf{T} = B^\mathsf{T} A^\mathsf{T}$.

*23. (a) Show that $A = I_n$ if and only if $AX = X$ for all n-vectors X.

(b) Show that if A and B are both $m \times n$ matrices, then $A = B$ if and only if $AX = BX$ for all n-vectors X.

(c) Show that if A is an $m \times n$ matrix, then $A = 0$ if and only if $AX = 0$ for all n-vectors X.

24. Show that an n-square matrix is a scalar matrix if and only if it commutes with every n-square matrix B.

*25. Show that, if α is a scalar, $(\alpha A)^* = \bar{\alpha} A^*$. Show also that $A = \bar{\bar{A}}$, $(\bar{A})^\mathsf{T} = (\overline{A^\mathsf{T}})$, $(A^*)^* = A$, $(A + B)^* = A^* + B^*$, $\overline{AB} = \bar{A}\bar{B}$, and $(AB)^* = B^* A^*$.

*26. Show that, for every matrix A, A^*A and AA^* are Hermitian matrices. What may be said about the diagonal entries of A^*A and AA^*? Show also that, if H is Hermitian, so is B^*HB for every conformable matrix B.

27. If A is square and if $A = -A^*$, that is, if $a_{ij} = -\bar{a}_{ji}$ for all i and j, A is called skew-Hermitian. Show that every square matrix A can be written in the form $A = A^{(H)} + A^{(SH)}$, where $A^{(H)}$ is Hermitian and $A^{(SH)}$ is skew-Hermitian. Show also that the diagonal elements of a skew-Hermitian matrix are pure imaginaries.

28. Prove that H is Hermitian if and only if $H = A + iB$, where A and B are real, A is symmetric, and B is skew-symmetric. What is the corresponding result for skew-Hermitian matrices?

29. Show that if H is Hermitian, then H^*H is real if and only if $AB = -BA$, that is, if and only if A and B anticommute, where A and B are as defined in Exercise 28.

30. Show that $\langle Y, X \rangle = \overline{\langle X, Y \rangle}$.

31. Show that, if α and β are real numbers, then the set of all matrices

$$\begin{bmatrix} \alpha & -\beta \\ \beta & \alpha \end{bmatrix}$$

is isomorphic to the set of all complex numbers $\alpha + i\beta$. Hence show that every such matrix has an inverse except when $\alpha = \beta = 0$. What is the inverse? In this connection review also Exercise 20 in Section 1.9.

32. Every complex number z can be written in the polar form $z = re^{i\theta}$, where r and θ are real. What matrices therefore correspond to z and $\bar{z}$ in the isomorphism of Exercise 31?

33. By recalling DeMoivre's theorem, show how to write n real nth roots of the real matrix

$$\begin{bmatrix} a & -b \\ b & a \end{bmatrix} \quad (a^2 + b^2 \neq 0).$$

34. A matrix A such that $A^2 = I$ is called involutory. Show that A is involutory if and only if

$$(A - I)(A + I) = 0.$$

Does it follow that $A = I$ or $-I$? Why? Defend your answer with an example.

35. Determine all involutory matrices of the form

$$\begin{bmatrix} 1 & 2a & b & c \\ 0 & -1 & d & e \\ 0 & 0 & 1 & 2f \\ 0 & 0 & 0 & -1 \end{bmatrix}.$$

36. Show by an example that one can have $AB = B$ even though A is not an identity matrix. The matrix A is called a **left identity** for B. Similarly, if $BC = B$, then C is a **right identity** for B. Give an example of a right identity that is not I_n.

37. Show that if A is a left identity (or right identity) for *all* n-square matrices, then $A = I_n$.

38. Find, by inspection, inverses for the following matrices:

(a) $\begin{bmatrix} 1 & 1 & 0 \\ 0 & 1 & 1 \\ 0 & 0 & 1 \end{bmatrix}$,

(b)
$$D = \begin{bmatrix} 0 & 0 & \cdots & 0 & d_n \\ 0 & 0 & \cdots & d_{n-1} & 0 \\ \vdots & & & & \\ 0 & d_2 & \cdots & 0 & 0 \\ d_1 & 0 & \cdots & 0 & 0 \end{bmatrix}$$
(The elements $d_1, d_2, \ldots, d_n$ are assumed to be different from zero.)

(c) $\begin{bmatrix} 1 & 0 \\ a & 1 \end{bmatrix}$,

(d) $\begin{bmatrix} 1 & 0 & 0 \\ 0 & 1 & 0 \\ a & b & 1 \end{bmatrix}$,

(e) $\begin{bmatrix} 1 & 0 & \cdots & 0 & 0 \\ 0 & 1 & \cdots & 0 & 0 \\ \vdots & & & & \\ a_1 & a_2 & \cdots & a_{n-1} & 1 \end{bmatrix}$,

(f) $\begin{bmatrix} 1 & 0 & 0 & 0 \\ 0 & 1 & 0 & 1 \\ 0 & 0 & 1 & 0 \\ 0 & 0 & 0 & 1 \end{bmatrix}$,

(g) $\begin{bmatrix} 0 & 0 & 1 & 0 \\ 0 & 1 & 0 & 0 \\ 1 & 0 & 0 & 0 \\ 0 & 0 & 0 & 1 \end{bmatrix}$,

(h) $\begin{bmatrix} 1 & 1 & 0 & 0 \\ 0 & 1 & 1 & 0 \\ 0 & 0 & 1 & 1 \\ 0 & 0 & 0 & 1 \end{bmatrix}$.

39. Find an inverse for A, where

$$A = \begin{bmatrix} 1 & 1 & 1 \\ 0 & 1 & 1 \\ 0 & 0 & 1 \end{bmatrix},$$

by solving this equation for the b's:

$$AB = I_3.$$

Here $B = [b_{ij}]_{3 \times 3}$.

40. Evaluate the product

$$\begin{bmatrix} a_{11} & a_{12} \\ a_{21} & a_{22} \end{bmatrix} \cdot \begin{bmatrix} a_{22} & -a_{12} \\ -a_{21} & a_{11} \end{bmatrix},$$

and thereby determine an inverse for the first factor of the product. Under what conditions is there no inverse?

41. Given that

$$\begin{bmatrix} 2 & 0 & 1 \\ 0 & 2 & 0 \\ 0 & 0 & 2 \end{bmatrix} W \begin{bmatrix} 1 & 1 \\ 1 & 2 \end{bmatrix} = \begin{bmatrix} 1 & 0 \\ 0 & 1 \\ 0 & 0 \end{bmatrix},$$

solve for W.

42. Find the inverse of this product without first computing the product:

$$\begin{bmatrix} 0 & 0 & 1 \\ 0 & -2 & 0 \\ 1 & 0 & 0 \end{bmatrix} \begin{bmatrix} 0 & 0 & 1 \\ 1 & 0 & 0 \\ 0 & 1 & 0 \end{bmatrix} \begin{bmatrix} 0 & 1 & 0 \\ 0 & 0 & 1 \\ 1 & 1 & 1 \end{bmatrix}.$$

43. Find by inspection the inverse of

$$\begin{bmatrix} 1 & 0 & 0 & 0 & \cdots & 0 & 0 \\ -1 & 1 & 0 & 0 & \cdots & 0 & 0 \\ 0 & -1 & 1 & 0 & \cdots & 0 & 0 \\ \vdots & & & & & & \\ 0 & 0 & 0 & 0 & \cdots & -1 & 1 \end{bmatrix}_n.$$

44. Show that the inverse of the skew-symmetric matrix of order $2n$.

$$\begin{bmatrix} 0 & 1 & 1 & 1 & \cdots & 1 \\ -1 & 0 & 1 & 1 & \cdots & 1 \\ -1 & -1 & 0 & 1 & \cdots & 1 \\ \vdots & & & & & \\ -1 & -1 & -1 & -1 & \cdots & 0 \end{bmatrix} \text{ is } \begin{bmatrix} 0 & -1 & 1 & -1 & \cdots & -1 \\ 1 & 0 & -1 & 1 & \cdots & 1 \\ -1 & 1 & 0 & -1 & \cdots & -1 \\ \vdots & & & & & \\ 1 & -1 & 1 & -1 & \cdots & 0 \end{bmatrix}.$$

(*American Mathematical Monthly*, Vol. 58, 1951, p. 494.)

*45. Prove that, if A is nonsingular, then from $AB = AC$ we can conclude that $B = C$. (B and C need not be square, of course.)

*46. Show that, if A^{-1} exists, then $(A^{\mathsf{T}})^{-1} = (A^{-1})^{\mathsf{T}}$, $(\bar{A})^{-1} = \overline{(A^{-1})}$, and $(A^*)^{-1} = (A^{-1})^*$.

*47. We define $A^{-n} = (A^{-1})^n$ when A is nonsingular and n is a positive integer. We define $A^0 = I$ for an arbitrary square matrix A, singular or not. Show that the laws of exponents $A^m A^n = A^{m+n}$ and $(A^m)^n = A^{mn}$ now apply for *all* integral values of m and n when A is nonsingular.

48. If

$$A = \begin{bmatrix} \cosh x & \sinh x \\ \sinh x & \cosh x \end{bmatrix},$$

show that, for all integral values of n,

$$A^n = \begin{bmatrix} \cosh nx & \sinh nx \\ \sinh nx & \cosh nx \end{bmatrix}.$$

*49. Show that the inverse of a nonsingular symmetric matrix is also symmetric.

50. Prove that, if $AB = BA$ and $S^2 = B$, then also, if A^{-1} exists,

$$(A^{-1}SA)^2 = B.$$

51. Given that D is diagonal and nonsingular and that

$$D = (I + A)^{-1} \cdot A,$$

prove that A is diagonal also.

52. Prove that, if $S_m = I + A + A^2 + \cdots + A^m$, and if $I - A$ is nonsingular, then $S_m = (I - A^{m+1})(I - A)^{-1}$. See if you can define $\lim_{m \to \infty} A^{m+1}$ and then show that, when this limit is the zero matrix, $\lim_{m \to \infty} S_m = (I - A)^{-1}$.

53. For an arbitrary $m \times n$ matrix A, any $n \times m$ matrix B such that $AB = I_m$ is called a **right inverse** of A, and any $n \times m$ matrix C such that $CA = I_n$ is called a **left inverse** of A. Show that, if a square matrix A has a left (or a right) inverse B, then A^{-1} exists and is equal to B.

54. If $AB = 0$, where A and B are of order n but neither is the zero matrix, A and B are called **divisors of zero**. If $A^p = 0$ for some positive integer p, then A is called **nilpotent**. Show that all divisors of zero and all nilpotent matrices are singular.

*55. Given that $AB = BA$ and that B^{-1} exists, prove that $AB^r = B^rA$ for all integer values of r. If A^{-1} also exists, prove that $A^rB^s = B^sA^r$ for all integers r and s.

*56. Given that

$$P = [E_{j_1}, E_{j_2}, \ldots, E_{j_n}],$$

where the E_j's are a permutation of the n elementary n-vectors (see Exercise 16), prove that

$$P^{\mathsf{T}} = P^{-1}.$$

57. Prove that, if A and B are symmetric and commute, then if A^{-1} and B^{-1} exist, $A^{-1}B$, AB^{-1}, and $A^{-1}B^{-1}$ are symmetric.

58. Show that arbitrary polynomials $f(A)$ and $g(B)$ in fixed matrices A and B of order n commute if and only if A and B commute. (It suffices to show that $A^rB^s = B^sA^r$ for all positive integers r and s if and only if $AB = BA$.)

59. Show that any two polynomial functions of a square matrix A commute.

60. Explain why the notation A/B is ambiguous when A and B are matrices. Given that $f(A)$ and $g(A)$ are polynomial functions of A and that $[g(A)]^{-1}$ exists, prove that

$$f(A) \cdot [g(A)]^{-1} = [g(A)]^{-1} \cdot f(A).$$

Is the notation $f(A)/g(A)$ ambiguous? Why?

61. Find all scalar matrices which satisfy the matrix equations:

(a) $A^2 - 5A + 7I_2 = 0$.

(b) $2A^3 + 3A^2 - 4A - 6I_2 = 0$.

(c) $A^3 - I_n = 0$.

(d) $A^3 - A = 0$.

62. (a) Determine all diagonal matrices of order n which satisfy a polynomial matrix equation $f(A) = 0$ of degree p.

(b) Determine all solutions of the form $\begin{bmatrix} 0 & a \\ b & 0 \end{bmatrix}$ of the equation $A^3 + A = 0$.

63. Show that the nonscalar matrix

$$A = \begin{bmatrix} 3 & 1 \\ -1 & 2 \end{bmatrix}$$

is a solution of the equation $A^2 - 5A + 7I_2 = 0$.

64. Find all second-order matrices which satisfy the equation $A^2 - 2\alpha A + \beta I_2 = 0$. Specialize to the case $\alpha = 0$, $\beta = -1$, and interpret the results.

65. Prove that if $B = P^{-1}AP$, then $B^k = P^{-1}A^kP$. Hence show that, if

$$f(A) = a_p A^p + a_{p-1} A^{p-1} + \cdots + a_1 A + a_0 I = 0,$$

and if $B = P^{-1}AP$, then also $f(B) = 0$.

66. (a) Use the isomorphism established in Exercise 31 to find two additional matrix roots of the equation given in Exercise 63.

(b) Find all square roots of the form

$$\begin{bmatrix} 0 & a \\ b & 0 \end{bmatrix} \quad (a, b \text{ real})$$

of the matrix

$$\begin{bmatrix} -\lambda^2 & 0 \\ 0 & -\lambda^2 \end{bmatrix},$$

where λ is real. The latter matrix corresponds to a negative number, but you will be finding square roots all elements of which are real. Use the isomorphism concept to explain the two simplest cases.

67. Show that

$$\sigma_x^2 = \sigma_y^2 = \sigma_z^2 = I_2^2 = I_2,$$

where σ_x, σ_y, σ_z are the Pauli spin matrices defined in Section 1.9, Exercise 22. What rule about roots in scalar algebra fails to carry over into matrix algebra?

68. A square matrix A such that $a_{ij} = 0$ whenever $i \geq j$ is called an **upper matrix**. For example,

$$\begin{bmatrix} 0 & -1 & 3 \\ 0 & 0 & 2 \\ 0 & 0 & 0 \end{bmatrix}$$

is an upper matrix. Prove that every upper matrix U is nilpotent.

69. A square matrix B such that B^T is upper is called a **lower matrix**. Prove that every lower matrix is nilpotent.

70. Show that the sum and the product of any two $n \times n$ upper matrices whose elements are complex numbers or functions thereof are again upper matrices, that is, that the set of all such matrices is closed under the operations of addition and multiplication.

71. Given that

$$S_n = \begin{bmatrix} 0 & 1 & 0 & \cdots & 0 & 0 \\ 0 & 0 & 1 & \cdots & 0 & 0 \\ \vdots & & & & & \\ 0 & 0 & 0 & \cdots & 0 & 1 \\ 0 & 0 & 0 & \cdots & 0 & 0 \end{bmatrix}_n,$$

give a rule for evaluating $S_n A$, where A is any $n \times n$ matrix. Use this result to develop a rule for evaluating S_n^p, where p is any positive integer.

72. Given that $A = [a_{ij}]_{m \times n}$, evaluate $S_m A, AS_n, S_m^T A, AS_n^T, S_m^T AS_n, S_m AS_n^T, S_m^p AS_n^q$.

73. Given that $f(x) = a_0 + a_1 x + a_2 x^2 + \cdots + a_r x^r + \cdots$, show that

$$f(S_n) = \begin{bmatrix} a_0 & a_1 & a_2 & \cdots & a_{n-2} & a_{n-1} \\ 0 & a_0 & a_1 & \cdots & a_{n-3} & a_{n-2} \\ 0 & 0 & a_0 & \cdots & a_{n-4} & a_{n-3} \\ \vdots & & & & & \\ 0 & 0 & 0 & \cdots & a_0 & a_1 \\ 0 & 0 & 0 & \cdots & 0 & a_0 \end{bmatrix},$$

where S_n is as defined in Exercise 71.

74. Let U denote an upper matrix of order n. Prove that $(I_n + U)^p = I_n$ if and only if $U = 0$. (*Hint:* Expand the left member, then consider successively the **super-diagonals,** that is, the diagonals above the main diagonal, of U.)

75. Prove that, if U is nilpotent and if $U^n = 0$ but $U^{n-1} \neq 0$, then $I + U$ has an inverse,

$$(I + U)^{-1} = I - U + U^2 - U^3 + \cdots + (-1)^{n-1} U^{n-1}.$$

76. Prove that, if U is nilpotent and the polynomial $f(x)$ has no constant term, then $f(U)$ is nilpotent.

77. Prove that, if U is nilpotent and $(I + U)^p = I$, then

$$U(I + f(U)) = 0,$$

where $f(U)$ is a polynomial with constant term zero.

78. Use Exercises 75, 76, 77 to show that, if U is nilpotent and $(I + U)^p = I$, then $U = 0$.

1.19 Partitioned Matrices

At many points in the following chapters we shall find it desirable to subdivide or partition matrices into rectangular blocks of elements. We shall now investigate the consequences of treating these blocks as matrices, that is, of treating the original matrix as a matrix whose elements are matrices. The blocks are called **submatrices** of the original matrix.

Before proceeding formally, we give a few illustrations. If there were any advantages to it, we could write, for example,

$$\begin{bmatrix} 1 & 0 \\ -1 & -3 \\ 2 & 1 \end{bmatrix},$$

as $[A_1,\ A_2]$, where

$$A_1 = \begin{bmatrix} 1 \\ -1 \\ 2 \end{bmatrix} \quad \text{and} \quad A_2 = \begin{bmatrix} 0 \\ -3 \\ 1 \end{bmatrix},$$

or as

$$\begin{bmatrix} B_1 \\ B_2 \\ B_3 \end{bmatrix},$$

where

$$B_1 = [1,\ 0], \quad B_2 = [-1,\ -3], \quad B_3 = [2,\ 1],$$

or as

$$\begin{bmatrix} C_1 \\ C_2 \end{bmatrix},$$

where

$$C_1 = \begin{bmatrix} 1 & 0 \\ -1 & -3 \end{bmatrix} \quad \text{and} \quad C_2 = [2,\ 1].$$

The manner in which a matrix is to be partitioned is often indicated by dashed lines, and the resulting parts of the given matrix are then treated as submatrices of the original matrix. Thus

$$\begin{bmatrix} 1 & 2 & -1 \\ 3 & 0 & 2 \\ 1 & 1 & 1 \end{bmatrix} = \begin{bmatrix} 1 & 2 & -1 \\ 3 & 0 & 2 \\ \hline 1 & 1 & 1 \end{bmatrix} = \begin{bmatrix} A & B \\ C & D \end{bmatrix},$$

where A, B, C, and D denote the submatrices.

As in the preceding example, we shall say that two matrices, partitioned or not, are **equal** if and only if their nonpartitioned forms are equal.

Of course, if we wish to treat these blocks of elements as matrices, we must do so subject to all the laws of computation with matrices. For example, if we wish to write the equation

$$\begin{bmatrix} A_1 & B_1 \\ C_1 & D_1 \end{bmatrix} + \begin{bmatrix} A_2 & B_2 \\ C_2 & D_2 \end{bmatrix} = \begin{bmatrix} (A_1 + A_2) & (B_1 + B_2) \\ (C_1 + C_2) & (D_1 + D_2) \end{bmatrix},$$

we must first make sure that A_1 and A_2 have the same order, and similarly for B_1 and B_2, C_1 and C_2, D_1 and D_2.

In general, let A and B be matrices of the same order. We shall then say that A and B are **identically partitioned** if the resulting matrices of *matrices* contain the same number of rows and the same number of columns and if, in addition, corresponding blocks have the same order. Thus the matrices

$$\begin{bmatrix} 1 & -2 & \vdots & 3 \\ 4 & -1 & \vdots & 6 \\ \hline 1 & 1 & \vdots & 1 \end{bmatrix} \quad \text{and} \quad \begin{bmatrix} a & b & \vdots & c \\ d & e & \vdots & f \\ \hline g & h & \vdots & k \end{bmatrix}$$

are identically partitioned. It is then a simple matter to see that identically partitioned matrices are equal if and only if corresponding submatrices are equal throughout and that they may be added by adding corresponding submatrices throughout.

To show how partitioning into submatrices is used in the multiplication of matrices, we consider first several examples.

(1) Let

$$A = \begin{bmatrix} 1 & 0 & \vdots & 2 \\ 0 & 1 & \vdots & -2 \end{bmatrix} \quad \text{and} \quad B = \begin{bmatrix} 1 & 0 \\ 0 & 1 \\ \hline 3 & -1 \end{bmatrix}.$$

Then we observe that if we treat the submatrices as if they were elements, we have

$$\begin{bmatrix} \begin{bmatrix} 1 & 0 \\ 0 & 1 \end{bmatrix} & \begin{bmatrix} 2 \\ -2 \end{bmatrix} \end{bmatrix} \cdot \begin{bmatrix} \begin{bmatrix} 1 & 0 \\ 0 & 1 \end{bmatrix} \\ [3 \quad -1] \end{bmatrix}$$

$$= \begin{bmatrix} \begin{bmatrix} 1 & 0 \\ 0 & 1 \end{bmatrix} \cdot \begin{bmatrix} 1 & 0 \\ 0 & 1 \end{bmatrix} + \begin{bmatrix} 2 \\ -2 \end{bmatrix} \cdot [3 \quad -1] \end{bmatrix}$$

$$= \begin{bmatrix} \begin{bmatrix} 1 & 0 \\ 0 & 1 \end{bmatrix} + \begin{bmatrix} 6 & -2 \\ -6 & 2 \end{bmatrix} \end{bmatrix} = \begin{bmatrix} 7 & -2 \\ -6 & 3 \end{bmatrix}$$

$$= AB.$$

(2) Let

$$A = \begin{bmatrix} 2 & 1 & \vdots & 0 & 0 \\ 1 & -1 & \vdots & 0 & 0 \\ \hdashline 0 & 0 & \vdots & 0 & 1 \\ 0 & 0 & \vdots & 2 & 0 \end{bmatrix} \quad \text{and} \quad B = \begin{bmatrix} 0 & 0 & \vdots & 1 & 0 \\ 0 & 0 & \vdots & 0 & 1 \\ \hdashline 0 & 2 & \vdots & 0 & 0 \\ -2 & 0 & \vdots & 0 & 0 \end{bmatrix}.$$

Then, again treating the submatrices as though they were elements, we note that because of the zero matrices appearing we have as the product

$$\begin{bmatrix} \begin{bmatrix} 0 & 0 \\ 0 & 0 \end{bmatrix} & \vdots & \begin{bmatrix} 2 & 1 \\ 1 & -1 \end{bmatrix} \cdot \begin{bmatrix} 1 & 0 \\ 0 & 1 \end{bmatrix} \\ \hdashline \begin{bmatrix} 0 & 1 \\ 2 & 0 \end{bmatrix} \cdot \begin{bmatrix} 0 & 2 \\ -2 & 0 \end{bmatrix} & \vdots & \begin{bmatrix} 0 & 0 \\ 0 & 0 \end{bmatrix} \end{bmatrix} = \begin{bmatrix} 0 & 0 & 2 & 1 \\ 0 & 0 & 1 & -1 \\ -2 & 0 & 0 & 0 \\ 0 & 4 & 0 & 0 \end{bmatrix} = AB.$$

Note that when the multiplication is complete, the internal brackets may be dropped.

(3) Let

$$A = \begin{bmatrix} a_{11} & a_{12} & a_{13} \\ a_{21} & a_{22} & a_{23} \\ a_{31} & a_{32} & a_{33} \end{bmatrix} \quad \text{and} \quad B = \begin{bmatrix} b_{11} & b_{12} \\ b_{21} & b_{22} \\ b_{31} & b_{32} \end{bmatrix}.$$

Let us partition A and B, designating the submatrices with double subscripts:

$$A = \begin{bmatrix} a_{11} & a_{12} & \vdots & a_{13} \\ a_{21} & a_{22} & \vdots & a_{23} \\ \hdashline a_{31} & a_{32} & \vdots & a_{33} \end{bmatrix} = \begin{bmatrix} A_{11} & A_{12} \\ A_{21} & A_{22} \end{bmatrix}, \quad B = \begin{bmatrix} b_{11} & b_{12} \\ b_{21} & b_{22} \\ \hdashline b_{31} & b_{32} \end{bmatrix} = \begin{bmatrix} B_{11} \\ B_{21} \end{bmatrix}.$$

Then

$$\begin{bmatrix} A_{11} & A_{12} \\ A_{21} & A_{22} \end{bmatrix} \cdot \begin{bmatrix} B_{11} \\ B_{21} \end{bmatrix}$$

$$= \begin{bmatrix} (A_{11}B_{11} + A_{12}B_{21}) \\ (A_{21}B_{11} + A_{22}B_{21}) \end{bmatrix}$$

$$= \begin{bmatrix} \begin{bmatrix} (a_{11}b_{11} + a_{12}b_{21}) & (a_{11}b_{12} + a_{12}b_{22}) \\ (a_{21}b_{11} + a_{22}b_{21}) & (a_{21}b_{12} + a_{22}b_{22}) \end{bmatrix} + \begin{bmatrix} a_{13}b_{31} & a_{13}b_{32} \\ a_{23}b_{31} & a_{23}b_{32} \end{bmatrix} \\ \begin{bmatrix} (a_{31}b_{11} + a_{32}b_{21}) & (a_{31}b_{12} + a_{32}b_{22}) \end{bmatrix} + \begin{bmatrix} a_{33}b_{31} & a_{33}b_{32} \end{bmatrix} \end{bmatrix}$$

$$= \begin{bmatrix} \begin{bmatrix} a_{11}b_{11} + a_{12}b_{21} + a_{13}b_{31} & a_{11}b_{12} + a_{12}b_{22} + a_{13}b_{32} \\ a_{21}b_{11} + a_{22}b_{21} + a_{23}b_{31} & a_{21}b_{12} + a_{22}b_{22} + a_{23}b_{32} \end{bmatrix} \\ \begin{bmatrix} a_{31}b_{11} + a_{32}b_{21} + a_{33}b_{31} & a_{31}b_{12} + a_{32}b_{22} + a_{33}b_{32} \end{bmatrix} \end{bmatrix}$$

$= AB$ after the internal brackets are dropped.

In each of the above examples, we partitioned the matrices A and B and multiplied them, treating the submatrices as elements. When the multiplication was complete, we dropped the internal brackets, thus in effect undoing the partitioning. The result was in each case the product AB.

The results stated in the last paragraph hold true in general, provided that the partitioning is so carried out that the submatrices to be multiplied are conformable. Let A be of order (m, n) and let B be of order (n, p) so that A is conformable to B for multiplication. Let A_{ij} and B_{ij} be used to denote submatrices. Then we write

$$A = \begin{bmatrix} a_{11} & a_{12} & \cdots & a_{1n} \\ a_{21} & a_{22} & \cdots & a_{2n} \\ \vdots & & & \\ a_{m1} & a_{m2} & \cdots & a_{mn} \end{bmatrix} = \begin{bmatrix} A_{11} & A_{12} & \cdots & A_{1v} \\ A_{21} & A_{22} & \cdots & A_{2v} \\ \vdots & & & \\ A_{\mu 1} & A_{\mu 2} & \cdots & A_{\mu v} \end{bmatrix} \begin{matrix} \}m_1 \text{ rows} \\ \}m_2 \text{ rows} \\ \vdots \\ \}m_\mu \text{ rows} \end{matrix},$$

where $n_1 + n_2 + \cdots + n_v = n$, and $m_1 + m_2 + \cdots + m_\mu = m$.

The matrix B is then partitioned thus:

$$B = \begin{bmatrix} b_{11} & b_{12} & \cdots & b_{1p} \\ b_{21} & b_{22} & \cdots & b_{2p} \\ \vdots & & & \\ b_{n1} & b_{n2} & \cdots & b_{np} \end{bmatrix} \begin{matrix} n_1 \text{ rows } \{ \\ n_2 \text{ rows } \{ \\ \vdots \\ n_v \text{ rows } \{ \end{matrix} \begin{bmatrix} B_{11} & B_{12} & \cdots & B_{1\rho} \\ B_{21} & B_{22} & \cdots & B_{2\rho} \\ \vdots & & & \\ B_{v1} & B_{v2} & \cdots & B_{v\rho} \end{bmatrix},$$

where $p_1 + p_2 + \cdots + p_\rho = p$.

As indicated, the column partitioning of A must be similar to the row partitioning of B so that the matrices A_{ij} will be conformable to the matrices B_{jk} for $j = 1, 2, \ldots, v$. When this is the case, we say that A and B are **conformably partitioned**. The row partitioning of A and the column partitioning of B are arbitrary, being governed only by considerations of convenience for the purpose at hand.

We could now show that

$$AB = [C_{ik}]_{(\mu,\rho)},$$

where

$$C_{ik} = \sum_{j=1}^{v} A_{ij} B_{jk};$$

that is, we could show that, if we multiply the partitioned A and B according to the rule outlined above, the end result would be precisely AB. The proof will be left to the reader to think through. Example 3 gives the key: Show that $[C_{ik}]_{(\mu,\rho)}$ gives each element of AB, in its proper place, by building it up as a sum of groups of products.

1.20   Exercises

1. Compute the product using the indicated partitioning:

$$
\left[
\begin{array}{ccccc}
1 & 2 & 0 & 0 & 0 \\
-2 & 1 & 0 & 0 & 0 \\
\hline
0 & 0 & 1 & 0 & 0 \\
0 & 0 & 0 & 1 & 0 \\
0 & 0 & 0 & 0 & 1
\end{array}
\right]
\left[
\begin{array}{ccccc}
0 & 0 & 0 & 2 & 1 \\
0 & 0 & 0 & 1 & -2 \\
\hline
0 & 0 & 1 & 0 & 0 \\
0 & 1 & 0 & 0 & 0 \\
1 & 0 & 0 & 0 & 0
\end{array}
\right].
$$

2. Compute A^2, using the indicated partitioning, where

$$
A =
\left[
\begin{array}{ccc|cc|c}
1 & 0 & 0 & 0 & 0 & 1 \\
0 & 1 & 0 & 0 & 0 & 1 \\
0 & 0 & 1 & 0 & 0 & 1 \\
\hline
0 & 0 & 0 & 1 & 0 & 0 \\
0 & 0 & 0 & 0 & 1 & 0 \\
\hline
1 & 1 & 1 & 0 & 0 & 1
\end{array}
\right].
$$

3. In the preceding exercise, we have a symmetric matrix, symmetrically partitioned, which is probably the most important case to arise in practice. Write out a generalized scheme for the symmetric partitioning of two symmetric matrices A and B of order n and write a formula for the product AB. Show that, before the internal brackets are removed, the product AB is partitioned in the same way as were A and B.

*4. Let A be of order (m, n) and let B be of order (n, p). Then we can partition A into rows and B into columns thus:

$$A = \begin{bmatrix} A^{(1)} \\ A^{(2)} \\ \vdots \\ A^{(m)} \end{bmatrix}, \quad B = [B_1, B_2, \ldots, B_p],$$

or we can partition A into columns and B into rows:

$$A = [A_1, A_2, \ldots, A_n], \quad B = \begin{bmatrix} B^{(1)} \\ B^{(2)} \\ \vdots \\ B^{(n)} \end{bmatrix}$$

Form the product of the partitioned matrices in each case and observe how these products are related to the product AB.

5. Referring to the preceding exercise, form these products and observe how they are related to the product AB:

(a) A partitioned into columns times B not partitioned;
(b) A not partitioned times B partitioned into rows.

6. Let A_1 and B_1 be square and of the same order. Let the same be true for A_2 and $B_2, \ldots, A_n$ and B_n. Furthermore, let

$$A = \begin{bmatrix} A_1 & 0 & \cdots & 0 \\ 0 & A_2 & \cdots & 0 \\ \vdots & & & \\ 0 & 0 & \cdots & A_n \end{bmatrix}, \quad B = \begin{bmatrix} B_1 & 0 & \cdots & 0 \\ 0 & B_2 & \cdots & 0 \\ \vdots & & & \\ 0 & 0 & \cdots & B_n \end{bmatrix}.$$

Determine the sum and the product of A and B. The matrices A and B here are called **decomposable matrices** or **quasidiagonal matrices.** This problem then shows that the sum and the product of two decomposable matrices of the same kind are decomposable matrices of the same kind. If we abbreviate $A = D[A_1, A_2, \ldots, A_n]$, what can be said about $f(A)$ where $f(A)$ is any polynomial function of A? (A matrix like A is also called a **pseudodiagonal matrix.**)

7. If X_1 is a k_1-vector and X_2 is a k_2-vector, indicate what orders the A_{ij} must have for the product

$$[X_1^{\mathsf{T}}, X_2^{\mathsf{T}}] \cdot \begin{bmatrix} A_{11} & A_{12} \\ A_{21} & A_{22} \end{bmatrix} \cdot \begin{bmatrix} X_1 \\ X_2 \end{bmatrix}$$

to have meaning, and compute the product.

8. Show that, if $A = [A_{ij}]$, where the A_{ij} are *submatrices*, then $[A_{ij}]^{\mathsf{T}} = [A_{ji}^{\mathsf{T}}]$.

9. Given that $A_1^*, A_2^*, \ldots, A_k^*$ are all n-rowed matrices of complex numbers, simplify the expression

$$[A_1^*, A_2^*, \ldots, A_k^*]^*.$$

10. Notice that

$$\begin{bmatrix} 0 & 1 \\ 0 & 0 \end{bmatrix} \cdot \begin{bmatrix} 0 & 0 \\ 1 & 0 \end{bmatrix} = \begin{bmatrix} 1 & 0 \\ 0 & 0 \end{bmatrix},$$

$$\begin{bmatrix} 0 & 1 & 0 \\ 0 & 0 & 1 \\ 0 & 0 & 0 \end{bmatrix} \cdot \begin{bmatrix} 0 & 0 & 0 \\ 1 & 0 & 0 \\ 0 & 1 & 0 \end{bmatrix} = \begin{bmatrix} 1 & 0 & 0 \\ 0 & 1 & 0 \\ 0 & 0 & 0 \end{bmatrix},$$

$$\begin{bmatrix} 0 & 0 & 1 \\ 0 & 0 & 0 \\ 0 & 0 & 0 \end{bmatrix} \cdot \begin{bmatrix} 0 & 0 & 0 \\ 0 & 0 & 0 \\ 1 & 0 & 0 \end{bmatrix} = \begin{bmatrix} 1 & 0 & 0 \\ 0 & 0 & 0 \\ 0 & 0 & 0 \end{bmatrix}.$$

Generalize these to matrices of arbitrary orders and use the E_j's to prove the result.

11. Given that $AB = C$, that is, that

$$A[B_1, B_2, \ldots, B_p] = [C_1, C_2, \ldots, C_p],$$

where B and C are partitioned into columns, prove that

$$A \sum_{j=1}^{p} B_j = \sum_{j=1}^{p} C_j.$$

Compare with Exercise 18, Section 1.18.

12. Consider the matrix

$$S_n = \begin{bmatrix} 0 & 1 & 0 & \cdots & 0 \\ 0 & 0 & 1 & \cdots & 0 \\ \vdots & & & & \\ 0 & 0 & 0 & \cdots & 1 \\ 0 & 0 & 0 & \cdots & 0 \end{bmatrix}_n.$$

Give a rule for evaluating S_n^p, where p is a positive integer. (*Hint:* Write

$$S_n^2 = \begin{bmatrix} E_2^{\mathsf{T}} \\ E_3^{\mathsf{T}} \\ \vdots \\ E_n^{\mathsf{T}} \\ 0 \end{bmatrix} \cdot [0, E_1, E_2, \ldots, E_{n-1}],$$

and use Exercise 16, Section 1.18.)

*13. Let D denote a quasidiagonal matrix of order n, and let A denote any matrix of order n which is partitioned similarly to D. What is the nature of the products DA and AD?

14. Use second-order solutions of $A^2 - 5A + 7I_2 = 0$ to build up quasidiagonal fourth-order solutions of the equation $A^2 - 5A + 7I_4 = 0$.

*15. If $A = [A_1, A_2, \ldots, A_m]$, where the A_j are n-vectors, find representations of A^T, A^*, and the ij-entry of AA^T in terms of the A_j's.

16. Given that

$$A = \begin{bmatrix} 1 & 0 & 0 & 0 & 0 \\ 0 & 0 & 1 & 1 & 0 \\ 0 & 1 & 0 & 0 & 1 \\ 0 & 1 & 0 & 0 & 1 \\ 0 & 0 & 1 & 1 & 0 \end{bmatrix},$$

compute A^2, A^3, and A^4. Then give formulas for A^{2k} and A^{2k+1} and prove them by induction.

17. Given that A_1^{-1}, $A_2^{-1}, \ldots, A_n^{-1}$ exist, what is the inverse of the quasidiagonal matrix

$$\begin{bmatrix} A_1 & 0 & \cdots & 0 \\ 0 & A_2 & \cdots & 0 \\ \vdots & \vdots & & \vdots \\ 0 & 0 & \cdots & A_n \end{bmatrix} ?$$

18. Show that the algebra of complex matrices of the form $A + iB$, where A and B are real matrices of order n, is isomorphic to the algebra of those real matrices of order $2n$ which have the form

$$\begin{bmatrix} A & -B \\ B & A \end{bmatrix}.$$

(See Exercise 31, Section 1.18.)

Show that under this isomorphism "tranjugate" corresponds to transpose and, consequently, "Hermitian" corresponds to "symmetric," and "skew-Hermitian" corresponds to "skew-symmetric."

Show that multiplication by the complex scalar $a + ib$ corresponds to multiplication by the matrix of order $2n$,

$$\begin{bmatrix} aI & -bI \\ bI & aI \end{bmatrix},$$

and that these matrices commute with all of those of the form

$$\begin{bmatrix} A & -B \\ B & A \end{bmatrix}.$$

Finally, show that the matrices of this latter form are exactly those which commute with

$$\begin{bmatrix} 0 & -I \\ I & 0 \end{bmatrix}.$$

(Note that this isomorphism can be used to derive theorems about Hermitian matrices from theorems about symmetric matrices.)

19. Let A, B, and X be n-square matrices. Write $A = R_A + iC_A$ where R_A and C_A are real, and similarly for B and X. Prove that $AX = B$ if and only if

$$\begin{bmatrix} R_A & -C_A \\ C_A & R_A \end{bmatrix} \begin{bmatrix} R_X & -C_X \\ C_X & R_X \end{bmatrix} = \begin{bmatrix} R_B & -C_B \\ C_B & R_B \end{bmatrix}.$$

20. Given that A^{-1} exists, determine X so that

$$\left[\begin{array}{c:c} A^{-1} & 0 \\ \hdashline X & A^{-1} \end{array}\right]$$

is the inverse of

$$\left[\begin{array}{c:c} A & O \\ \hdashline B & A \end{array}\right].$$

21. Find by Exercise 20 the inverses of the matrices

(a) $\left[\begin{array}{cc:cc} 1 & 0 & 0 & 0 \\ 1 & 1 & 0 & 0 \\ \hdashline 0 & 0 & 1 & 0 \\ 0 & 0 & 1 & 1 \end{array}\right]$,   (b) $\left[\begin{array}{cc:cc} 1 & 0 & 0 & 0 \\ 1 & 1 & 0 & 0 \\ \hdashline 1 & 1 & 1 & 0 \\ 1 & 1 & 1 & 1 \end{array}\right].$

22. Given that A_{11}^{-1} and a_{nn}^{-1} exist, determine A^{-1}, where

$$A = \left[\begin{array}{c:c} A_{11} & 0 \\ \hdashline A_{21} & a_{nn} \end{array}\right].$$

CHAPTER
2

Linear Equations

2.1 Linear Equations

A basic aspect of linear algebra is the study of systems of linear equations, that is, systems of equations of the form

(2.1.1)
$$a_{11}x_1 + a_{12}x_2 + \cdots + a_{1n}x_n = b_1$$
$$a_{21}x_1 + a_{22}x_2 + \cdots + a_{2n}x_n = b_2$$
$$\vdots$$
$$a_{m1}x_1 + a_{m2}x_2 + \cdots + a_{mn}x_n = b_m,$$

or, in matrix notation,

(2.1.2)
$$AX = B,$$

where the a_{ij}'s and the b_j's represent numerical coefficients (real or complex numbers) and where $x_1, x_2, \ldots, x_n$ represent unknowns. Such systems arise in virtually every area in which mathematics is applied.

Most readers will be familiar with the solution of such systems of equations by elimination or by determinants. The latter method employs specific formulas that are easy to understand but the computation is prohibitively tedious except in very simple cases. The elimination procedures have the advantage that they can often benefit from clever inspection. However, the use of special tricks is not always possible or useful. Except in very simple cases and even when tricks cannot be used, elimination involves much less computation than does the determinant method. Hence we begin our study of linear equations by developing some systematic elimination procedures and investigating their consequences. These methods apply equally well whether or not the number of equations is equal to the number of unknowns.

An elementary understanding of linear equations and of their solutions, which this chapter is designed to develop, will help us to construct a theory of vector spaces, which will in turn enable us to study, with greater insight, the solution set of an arbitrary system of linear equations. In some cases such a system will have no solution; that is, the solution set will be empty. In other cases, there will be a unique solution. In still other cases, the number of solutions will be infinite. We shall see that for systems of equations with numerical coefficients, these three cases represent the only possibilities.

When a system has no solution, it is said to be **inconsistent**. When it has at least one solution, it is said to be **consistent**. Our objective in this chapter is to develop methods of determining whether or not a given system is consistent and of computing any solutions that may exist. Fortunately, the same procedure may be used to accomplish both purposes.

2.2   Three Examples

These examples illustrate both inconsistent and consistent systems of equations. They employ only operations which the reader has used before. That these familiar operations on systems of equations are valid will be proved in Section 2.4.

Example 1: Find all solutions of the system of equations

$$
\begin{aligned}
3x_1 + 2x_2 - \ x_3 &= -2 \\
x_1 - \ x_2 + 2x_3 &= 3 \\
4x_1 + \ x_2 + \ x_3 &= 0.
\end{aligned}
$$

(2.2.1)

A simple fact is too useful to ignore here. Adding the first two equations, we obtain

$$4x_1 + x_2 + x_3 = 1,$$

whereas the third equation says that

$$4x_1 + x_2 + x_3 = 0.$$

These last two equations imply that $0 = 1$, which is false. Since both equations cannot simultaneously be true, the given system cannot hold for any set of values (x_1, x_2, x_3). That is, there is no solution and the system is inconsistent.

Example 2: Find all solutions of

(2.2.2)
$$
\begin{aligned}
3x_1 + 2x_2 - x_3 &= -2 \\
x_1 + x_2 + x_3 &= 0 \\
-x_1 + x_2 + 2x_3 &= 3 \\
3x_1 + 4x_2 + 2x_3 &= 1.
\end{aligned}
$$

This system of equations illustrates the fact that there may well be more equations than unknowns. Also, we illustrate a more systematic elimination procedure for the first time.

It is convenient to begin by exchanging the first two equations so that we have

$$
\begin{aligned}
x_1 + x_2 + x_3 &= 0 \\
3x_1 + 2x_2 - x_3 &= -2 \\
-x_1 + x_2 + 2x_3 &= 3 \\
3x_1 + 4x_2 + 2x_3 &= 1.
\end{aligned}
$$

This yields an initial equation with leading coefficient 1. (If no equation of the system had had a leading coefficient 1, we would have begun by dividing the first equation by its leading coefficient, as in the next example.)

We now use the first equation to eliminate x_1 from the other three equations. Subtracting three times the first equation from the second and the fourth equations, and adding the first equation to the third, we obtain the new system of equations

$$
\begin{aligned}
x_1 + x_2 + x_3 &= 0 \\
- x_2 - 4x_3 &= -2 \\
2x_2 + 3x_3 &= 3 \\
x_2 - x_3 &= 1.
\end{aligned}
$$

Now we multiply the second equation by -1 so as to get a leading coefficient 1. This makes it easy to eliminate x_2 from all equations but the second. (We could instead have interchanged the second and fourth equations.) Then adding -1 times the resulting second equation once to the first, twice to the third, and once to the fourth, we obtain

$$
\begin{aligned}
x_1 \quad\quad - 3x_3 &= -2 \\
x_2 + \quad 4x_3 &= \quad 2 \\
-5x_3 &= -1 \\
-5x_3 &= -1.
\end{aligned}
$$

Multiplying the third equation by $-\frac{1}{5}$ and employing the resulting equation to eliminate x_3 from all the other equations, we obtain the system

$$
\begin{aligned}
x_1 \quad\quad\quad &= -\tfrac{7}{5} \\
x_2 \quad &= \quad \tfrac{6}{5} \\
x_3 &= \quad \tfrac{1}{5} \\
0 &= \quad 0.
\end{aligned}
$$

It is not hard to check, by substitution in the original set of equations, that this is, in fact, a solution, so that the system is consistent. Moreover, the development here makes clear that this is the only possible solution—in short, that it is *unique*. Indeed, we have shown that if (x_1, x_2, x_3) represents a solution of (2.2.2), that solution must be $(-\tfrac{7}{5}, \tfrac{6}{5}, \tfrac{1}{5})$.

Putting it another way, we have reduced the given system of equations to the simpler system

$$
\begin{aligned}
x_1 &= -\tfrac{7}{5} \\
x_2 &= \quad \tfrac{6}{5} \\
x_3 &= \quad \tfrac{1}{5},
\end{aligned}
$$
(2.2.3)

which we call a solution.

The basic idea of the above procedure is to eliminate from its column all appearances but one of a given variable. For this reason, the method is called the **sweepout process**.

Example 3: Find all solutions of the system of equations

(2.2.4)
$$
\begin{aligned}
2x_1 + 3x_2 + \quad x_3 &= 1 \\
3x_1 - \quad x_2 + 2x_3 &= 5.
\end{aligned}
$$

First let us multiply the leading equation by $\frac{1}{2}$:

$$x_1 + \tfrac{3}{2}x_2 + \tfrac{1}{2}x_3 = \tfrac{1}{2}$$
$$3x_1 - x_2 + 2x_3 = 5.$$

Now subtract three times the new first equation from the second:

$$x_1 + \tfrac{3}{2}x_2 + \tfrac{1}{2}x_3 = \tfrac{1}{2}$$
$$-\tfrac{11}{2}x_2 + \tfrac{1}{2}x_3 = \tfrac{7}{2}.$$

Multiplying the second of these equations by $-\frac{2}{11}$, we obtain

$$x_1 + \tfrac{3}{2}x_2 + \tfrac{1}{2}x_3 = \tfrac{1}{2}$$
$$x_2 - \tfrac{1}{11}x_3 = -\tfrac{7}{11}.$$

Adding $-\frac{3}{2}$ times the second equation to the first, we obtain

(2.2.5)
$$x_1 + \tfrac{7}{11}x_3 = \tfrac{16}{11}$$
$$x_2 - \tfrac{1}{11}x_3 = -\tfrac{7}{11}.$$

Finally, we rewrite these equations as follows:

(2.2.6)
$$x_1 = -\tfrac{7}{11}x_3 + \tfrac{16}{11}$$
$$x_2 = \tfrac{1}{11}x_3 - \tfrac{7}{11}.$$

We have shown that if equations (2.2.4) are true, then equations (2.2.6) are true, that is, that every solution of (2.2.4) must also satisfy (2.2.6). On the other hand, if we substitute these last expressions for x_1 and x_2 into the original equations, then, independently of the value of x_3, they reduce respectively to the identities $1 = 1$ and $5 = 5$. (Check this.) Hence we can choose any particular value for x_3 that we please, compute x_1 and x_2 from (2.2.6), and obtain a particular solution of the original pair. This will be one solution of an infinite set of solutions so obtainable, so this system, too, is consistent. For example, if we put $x_3 = 0$ in (2.2.6), we have the solution $(\tfrac{16}{11}, -\tfrac{7}{11}, 0)$ of (2.2.4). If we put $x_3 = 7$, we have the solution $(-3, 0, 7)$. If we put $x_3 = \tfrac{16}{7}$, we have the solution $(0, -\tfrac{3}{7}, \tfrac{16}{7})$, and so on.

The complete set of solutions of this system may be represented in another way. Suppose we put $x_3 = t$, where t represents an arbitrary parameter, that is, an independent variable. Then we have

(2.2.7)
$$x_1 = -\tfrac{7}{11}t + \tfrac{16}{11}$$
$$x_2 = \tfrac{1}{11}t - \tfrac{7}{11}$$
$$x_3 = t.$$

For each value assigned to t, we get a solution of the system. Moreover, every solution of the system may be obtained by appropriate choice of t, since choosing t is equivalent to choosing x_3.

Such a parametric solution is often useful. Other parametric representations of the set of all solutions are possible. Thus, we could put $x_3 = at + b$, where a and b are any convenient real numbers $(a \neq 0)$, and still have x_1, x_2, and x_3 expressed as linear functions of t. For example, if we put

$$x_3 = 11t + 18,$$

we obtain the parametric solution

(2.2.8)
$$
\begin{aligned}
x_1 &= -7t - 10 \\
x_2 &= t + 1 \\
x_3 &= 11t + 18,
\end{aligned}
$$

the correctness of which is easily checked by substitution. If we write the given system of equations and the solution in matrix form:

$$
\begin{bmatrix} 2 & 3 & 1 \\ 3 & -1 & 2 \end{bmatrix} X = \begin{bmatrix} 1 \\ 5 \end{bmatrix},
$$

$$
X = t \begin{bmatrix} -7 \\ 1 \\ 11 \end{bmatrix} + \begin{bmatrix} -10 \\ 1 \\ 18 \end{bmatrix},
$$

the substitution may be accomplished very efficiently.

These examples illustrate the fact that *a given system of equations may have no solution, exactly one solution, or an infinite number of solutions.* Which of these cases occurs depends on certain relations among the coefficients of the system of equations being solved. Just how this depends on the coefficients is made clear in following chapters.

The technique employed in Examples 2 and 3 illustrates the following general procedure. Use the first equation to eliminate the first variable from all the other equations. Then use the second equation to eliminate another variable (usually but not necessarily the second) from all the other equations, and so on. This results in a system in which each equation contains one selected variable appearing in no other equation. If the system is consistent, the solution is then completed by transposing all remaining terms containing variables other than those selected in the elimination process. Sometimes it is necessary to permute equations at some stage in order to proceed with the elimination in the prescribed manner. We shall show in Section 2.5 how to make this process completely systematic and completely general.

2.3 Exercises

1. Solve these systems of equations by the method employed in Example 2;

(a)
$$x_1 - x_2 + x_3 = 1$$
$$2x_1 + 3x_2 - x_3 = 4$$
$$-x_1 - 2x_2 + 5x_3 = 2,$$

(b)
$$2x_1 + 3x_2 + 4x_3 = 8$$
$$x_1 - x_2 + 2x_3 = 9$$
$$-3x_1 + 2x_2 + x_3 = -4,$$

(c)
$$x_1 + x_2 - x_3 + x_4 = 3$$
$$x_1 + x_3 + x_4 = 2$$
$$x_2 + x_3 - x_4 = -1$$
$$2x_1 - x_2 + x_3 = 2.$$

2. Obtain expressions for the complete solutions of

(a)
$$x_1 - 2x_2 + x_3 = 0$$
$$2x_1 + 3x_2 - x_3 = 4,$$

(b)
$$3x - 4y + 2z = 12$$
$$x + y - 5z = -4$$
$$-2x + 3y + 8z = 1,$$

(c)
$$-3x_1 + x_2 + x_3 + 5x_4 = 6$$
$$2x_2 - 3x_3 + 5x_4 = 4$$
$$- x_3 + 5x_4 = 4,$$

(d)
$$2x_1 + 3x_2 - x_3 - x_4 = 3$$
$$x_1 + 2x_2 + x_3 + 4x_4 = 8$$
$$-x_1 - x_2 + 3x_3 + x_4 = 2.$$

3. Show that the system of equations

$$2x_1 - 3x_2 = 1$$

$$x_1 + 2x_2 = 3$$

$$5x_1 - 7x_2 = 0$$

is inconsistent. There are several ways to do this: Solve two of the equations and show that the solution does not satisfy the third or, better, just start the solution process, as in Example 2, and show that it leads to a contradiction.

4. Determine whether or not each system is consistent:

(a)
$$x_1 + 2x_2 + 3x_3 = 4$$
$$2x_1 + 3x_2 + 4x_3 = 5$$
$$3x_1 + 4x_2 + 5x_3 = 7,$$

(b)
$$x_1 - x_2 + x_3 = 1$$
$$2x_1 + x_2 - 2x_3 = 1$$
$$3x_1 - 2x_2 - x_3 = 0$$
$$4x_1 + x_2 + 3x_3 = 2.$$

5. Given that

$$X = x + y - z$$

$$Y = x - y + z$$

$$Z = -x + y + z,$$

solve for x, y, and z as functions of X, Y, and Z and show that $X + Y + Z = x + y + z$.

6. Express in parametric form the complete solution of the system consisting of the single equation $2x - 5y = 7$. Then do the same for the system consisting of the single equation $5x_1 - 2x_2 + 3x_3 = 4$. In the latter case you will need to use *two* parameters; for example, let $x_2 = s$, $x_3 = t$. Then solve for x_1.

7. Obtain the solution of this system of equations by inspection:

$$x_1 - 2x_2 + 3x_3 + x_4 = 3$$
$$x_2 - x_3 + 2x_4 = 2$$
$$3x_3 + x_4 = 4$$
$$x_4 = 1.$$

8. Find the complete solution in parametric form:

(a) $x_1 - x_2 - x_3 = 1$ (b) $2x - y + z = 0$
       $2x_1 - 2x_2 - 3x_3 = 5,$ $x + 2y - 3z = 0.$

9. Obtain the complete solution of the system

$$a + 2b - 3c = 0$$
$$-3a + b + 2c = 0$$
$$2a - 3b + c = 0.$$

10. What must be the values of b_1, b_2, and b_3 if $[1, 1]^T$ is to be a solution of the following system?

$$\begin{bmatrix} 2 & 3 \\ 5 & -4 \\ -1 & 4 \end{bmatrix} \begin{bmatrix} x \\ y \end{bmatrix} = \begin{bmatrix} b_1 \\ b_2 \\ b_3 \end{bmatrix}.$$

11. How must you choose a, b, and c in order that $(1, 1, 1)$ will be a solution of this system?

$$ax_1 + 2x_2 + 3x_3 = 6$$
$$3x_1 + bx_2 - 5x_3 = 0$$
$$5x_1 - 4x_2 + cx_3 = 7.$$

Show that $(0, 1, 1)$ is not a solution of the system, no matter how a, b, and c are chosen.

12. Show that the system of equations

$$a_{11}x_1 + a_{12}x_2 = b_1$$
$$a_{21}x_1 + a_{22}x_2 = b_2$$

has a unique solution if and only if $a_{11}a_{22} - a_{12}a_{21} \neq 0$.

13. For what values of k will each of the following systems fail to have a *unique* solution? Will they have *any* solutions in these cases? Obtain the solutions in all cases where they exist.

(a) $\quad x + ky = 1$
$\quad\quad kx + y = 1,$

(b) $\quad x + 2y - z = 1$
$\quad\quad 2x - 3y + z = 2$
$\quad\quad kx + 9ky - 4z = 0,$

(c) $\quad x_1 + 2x_2 + 3x_3 = 4$
$\quad\quad 2x_1 + 3x_2 + 4x_3 = 5$
$\quad\quad 3x_1 + 4x_2 + 5x_3 = 2k.$

14. Under what conditions on t will these systems be consistent? What are the solutions in these cases?

(a) $\quad 3x_1 + x_2 + x_3 = t$
$\quad\quad x_1 - x_2 + 2x_3 = 1 - t$
$\quad\quad x_1 + 3x_2 - 3x_3 = 1 + t,$

(b) $\quad tx_1 + x_2 = 0$
$\quad\quad x_1 + tx_2 - x_3 = 1$
$\quad\quad - x_2 + tx_3 = 0.$

15. Given that a parametric solution of

$$\begin{bmatrix} a_{11} & a_{12} & a_{13} \\ a_{21} & a_{22} & a_{23} \end{bmatrix} \begin{bmatrix} x_1 \\ x_2 \\ x_3 \end{bmatrix} = \begin{bmatrix} b_1 \\ b_2 \end{bmatrix}$$

is

$$\begin{bmatrix} x_1 \\ x_2 \\ x_3 \end{bmatrix} = t \begin{bmatrix} \alpha_1 \\ \alpha_2 \\ \alpha_3 \end{bmatrix} + \begin{bmatrix} \beta_1 \\ \beta_2 \\ \beta_3 \end{bmatrix},$$

show that a parametric solution of

$$\begin{bmatrix} a_{11} & a_{12} & a_{13} \\ a_{21} & a_{22} & a_{23} \end{bmatrix} \begin{bmatrix} x_1 \\ x_2 \\ x_3 \end{bmatrix} = \begin{bmatrix} 0 \\ 0 \end{bmatrix}$$

is

$$\begin{bmatrix} x_1 \\ x_2 \\ x_3 \end{bmatrix} = t \begin{bmatrix} \alpha_1 \\ \alpha_2 \\ \alpha_3 \end{bmatrix}$$

and that $[\beta_1, \beta_2, \beta_3]^{\mathsf{T}}$ is a solution of the first equation.

16. Illustrate Exercise 15 by comparing the complete solutions of

(a) $\begin{aligned} x_1 + 2x_2 - x_3 &= 2 \\ 2x_1 + x_2 + x_3 &= 4 \end{aligned}$ and   (b) $\begin{aligned} x_1 + 2x_2 - x_3 &= 0 \\ 2x_1 + x_2 + x_3 &= 0. \end{aligned}$

17. The parametric solution (2.2.8) of (2.2.4) has only integer coefficients. Can you discover by what means a and b were chosen in the expression $x_3 = at + b$ so as to cause this to happen? Are other parametric representations with integer coefficients possible? Can you tell how to find them all? Given two consistent equations in three unknowns, with integer coefficients, is a parametric representation of the solution having only integer coefficients always possible? If your answer is no, give an example that proves your claim. If your answer is yes, prove it.

2.4 Equivalent Systems of Equations

The examples of Section 2.2 appealed to previous experience and were designed to illustrate the basic manipulations used in solving a given system of linear equations by elimination. It is time to make the ideas involved precise.

Consider any system of m linear equations in n unknowns:

(2.4.1)
$$\begin{aligned} a_{11}x_1 + a_{12}x_2 + \cdots + a_{1n}x_n - b_1 &= 0 \\ a_{21}x_1 + a_{22}x_2 + \cdots + a_{2n}x_n - b_2 &= 0 \\ &\ \vdots \\ a_{m1}x_1 + a_{m2}x_2 + \cdots + a_{mn}x_n - b_m &= 0, \end{aligned}$$

where the coefficients a_{ij} and the constant terms b_i, $i = 1, 2, \ldots, m$; $j = 1, 2, \ldots, n$, are real or complex numbers.

Any set of numerical values of $x_1, x_2, \ldots, x_n$ which simultaneously satisfy these equations is called a **particular solution** of the system. Any set of expressions [for example, (2.2.6)] which yields all solutions and only solutions of a given system of equations is called a **complete solution** of that system.

Two systems of equations in the variables $x_1, x_2, \ldots, x_n$ are said to be **equivalent** if and only if every particular solution of either one is also a solution of the other. The process of "solving" a system of equations amounts to deducing from the given system an equivalent system of a prescribed form. Thus (2.2.2) and (2.2.3) are equivalent, as are (2.2.4), (2.2.5), and (2.2.6).

In finding the solutions (2.2.3) and (2.2.5), we performed operations that fall into three categories:

(a) The interchange of two equations of a system.
(b) The multiplication of an equation by an arbitrary nonzero constant.
(c) The addition of an arbitrary multiple of one equation of a system to another equation of the system.

For the proof of the validity of the usual procedure for solving systems of equations, it is only necessary to show that any such operation, applied to a given system, always leads to an equivalent system.

For the purpose of providing this proof, it is convenient to abbreviate equations (2.4.1) thus:

(2.4.2)

$$f_1 = 0$$
$$\vdots$$
$$f_i = 0$$
$$\vdots$$
$$f_j = 0$$
$$\vdots$$
$$f_m = 0,$$

where f_i denotes the expression which constitutes the left member of the ith equation; that is,

$$f_i = a_{i1}x_1 + a_{i2}x_2 + \cdots + a_{in}x_n - b_i.$$

In this notation, replacing $f_i = 0$ by

$$c_i f_i = 0 \quad (c_i \neq 0) \qquad \text{(multiplication of the ith equation by a nonzero constant)}$$

or by

$$f_i + c_j f_j = 0 \quad (j \neq i) \qquad \text{(addition of c_j times the jth equation to the ith equation)}$$

is how one accomplishes operations of the types (b) and (c). The result of substituting a set of values $(a_1, a_2, \ldots, a_n)$ into the expression f_k is $f_k(a_1, a_2, \ldots, a_n)$ and $(a_1, a_2, \ldots, a_n)$ is a solution of (2.4.2) if and only if $f_k(a_1, a_2, \ldots, a_n) = 0$ for $k = 1, 2, \ldots, m$.

The effects of the operations (a), (b), and (c) on the system (2.4.2) may be represented, respectively, by the three systems

(2.4.3)

(a)
$$\begin{cases} f_1 = 0 \\ \vdots \\ f_j = 0 \\ \vdots \\ f_i = 0 \\ \vdots \\ f_m = 0, \end{cases}$$
(b)
$$\begin{cases} f_1 = 0 \\ \vdots \\ c_i f_i = 0 \quad (c_i \neq 0) \\ \vdots \\ f_j = 0 \\ \vdots \\ f_m = 0, \end{cases}$$
(c)
$$\begin{cases} f_1 = 0 \\ \vdots \\ f_i + c_j f_j = 0 \quad (j \neq i) \\ \vdots \\ f_j = 0 \\ \vdots \\ f_m = 0. \end{cases}$$

Now, suppose we know any solution $(a_1, a_2, \ldots, a_n)$ of (2.4.2). Then we have $f_k(a_1, a_2, \ldots, a_n) = 0$, $k = 1, 2, \ldots, m$. This implies that every equation of the three systems (a), (b), (c) is also satisfied by $(a_1, a_2, \ldots, a_n)$. Conversely, if $(a_1, a_2, \ldots, a_n)$ satisfies each equation of system (a), then each equation of system (2.4.2) is satisfied. If it satisfies each equation of system (b), then, because $c_i \neq 0, f_i(a_1, a_2, \ldots, a_n) = 0$, so again each equation of (2.4.2) is satisfied. If it satisfies each equation of system (c), then, because $f_j(a_1, a_2, \ldots, a_n) = 0$ and $f_i(a_1, a_2, \ldots, a_n) + c_j f_j(a_1, a_2, \ldots, a_n) = 0$, it follows that $f_i(a_1, a_2, \ldots, a_n) = 0$, so each equation of (2.4.2) is satisfied. This shows that each of the systems (a), (b), and (c) is equivalent to (2.4.2). We have therefore proved

Theorem 2.4.1: *Interchanging two equations of a system, multiplying an equation of a system by a nonzero constant, and adding any constant multiple of one equation of a system to another equation of the same system all lead to an equivalent system of equations.*

Now, by the definition of equivalence of systems of equations, if a system A is equivalent to B, and B to C, then A must be equivalent to C. Hence repeated applications of these three basic operations necessarily also lead to equivalent systems of equations. Either by this observation or by a direct proof, in the manner of the proof of the preceding theorem, one can establish

Corollary 2.4.2: *The system of linear equations (2.4.2) is equivalent to the following system of equations in which the coefficients $c_1, c_2, \ldots, c_m$ are arbitrary except that the particular coefficient c_i must not be zero:*

$$(2.4.4) \qquad \qquad \begin{aligned} f_1 &= 0 \\ f_2 &= 0 \\ &\vdots \\ f_{i-1} &= 0 \end{aligned}$$

$$c_1 f_1 + c_2 f_2 + \cdots + c_{i-1} f_{i-1} + c_i f_i + c_{i+1} f_{i+1} + \cdots + c_m f_m = 0 \quad (c_i \neq 0)$$

$$\begin{aligned} f_{i+1} &= 0 \\ &\vdots \\ f_m &= 0. \end{aligned}$$

That is, we may replace the ith equation by any nonzero multiple thereof plus arbitrary multiples of the remaining equations. Any or all of the coefficients other than c_i may be zero.

If we can choose the coefficients $c_1, c_2, \ldots, c_m$ $(c_i \neq 0)$ so that $\sum_{j=1}^{m} c_j f_j$ is identically zero in $x_1, x_2, \ldots, x_n$, then we will have reduced the given system to an equivalent system with one less equation in it. This is often a useful device.

In the same way, one can prove

Corollary 2.4.3: *If $i_1, i_2, \ldots, i_m$ is any permutation of the integers 1, 2, ..., m, then the system (2.4.2) is equivalent to the system*

$$
\begin{aligned}
f_{i_1} &= 0 \\
f_{i_2} &= 0 \\
&\;\;\vdots \\
f_{i_m} &= 0.
\end{aligned}
$$
(2.4.5)

That is, one may rearrange the equations of a given system in any order one finds convenient.

These two corollaries justify many of the clever tricks one commonly uses in solving particularly simple systems of equations.

2.5 The Echelon Form for Systems of Linear Equations

We now describe formally the systematic elimination process used in Example 3 of Section 2.2. Consider an arbitrary system of equations represented in the form (2.5.1):

$$
\begin{aligned}
a_{11}x_1 + a_{12}x_2 + \cdots + a_{1n}x_n &= b_1 \\
a_{21}x_1 + a_{22}x_2 + \cdots + a_{2n}x_n &= b_2 \\
&\;\;\vdots \\
a_{m1}x_1 + a_{m2}x_2 + \cdots + a_{mn}x_n &= b_m.
\end{aligned}
$$
(2.5.1)

We rearrange the order of the equations, if necessary, so that the variable x_1 appears with a nonzero coefficient in the first position of the first equation. Then we divide the first equation by the coefficient of x_1 and eliminate x_1 from equations 2 through m by the sweepout process.

Now let x_{i_2} be the first variable (usually x_2) actually appearing in equations 2 through m, and again rearrange the order of the equations, if necessary, so that x_{i_2} appears in the second row. Then we divide the second equation by the coefficient of x_{i_2} and eliminate x_{i_2} from equations 1, 3, 4, ..., m.

Continuing in this fashion, *we eventually stop because we have swept columns corresponding to m of the variables or because we have no more variable terms in the remaining equations.*

We illustrate this observation with an example:

$$x_1 - 2x_2 + x_3 - x_4 = -1$$
$$2x_1 - 4x_2 + 3x_3 - 3x_4 = 4$$
$$-x_1 + 2x_2 + x_3 - x_4 = b_3.$$

First we use x_1 to sweep the first column:

$$x_1 - 2x_2 + x_3 - x_4 = -1$$
$$x_3 - x_4 = 6$$
$$2x_3 - 2x_4 = b_3 - 1.$$

Because of the proportionality of the first two columns of coefficients, this eliminates x_2 also. Hence we must go next to x_3 and use it to sweep the third column:

$$x_1 - 2x_2 = -7$$
$$x_3 - x_4 = 6$$
$$0 = b_3 - 13.$$

This eliminates x_4 also from the third equation.

If now $b_3 \neq 13$, the equations imply a contradiction and are inconsistent. On the other hand, if $b_3 = 13$, the final equation is $0 = 0$, the system is consistent, and we can write the complete solution thus:

$$x_1 = 2x_2 - 7$$
$$x_3 = x_4 + 6.$$

From these last two equations, by assigning values to x_2 and x_4, then computing x_1 and x_3, one can obtain any particular solution whatsoever of the original system.

The example illustrates the fact that in sweeping a column, we may well eliminate more than one variable. It also illustrates the fact that we may eliminate all variables from some of the equations. Because of this, although there were three equations, we were able to employ only two variables for the purpose of sweeping columns. If there are m equations, one can employ at most m variables for this purpose. When this is possible, one can express these m variables in terms of the remaining $n - m$ variables, just as we have done in previous examples.

A system of linear equations will be said to be in **echelon form** (pronounced esh'-uh-lon) if and only if

1. the first nonzero coefficient in each equation is 1, and
2. the number of leading zero coefficients in any equation beyond the first is greater than the number of leading zero coefficients in the preceding equation.

The system of equations is said to be in **reduced echelon form** if and only if, in addition to 1 and 2,

3. the variable appearing in the leading term of any equation appears in no other equation.

The above-described computational process is the process of *deriving a system of equations in reduced echelon form equivalent to a given system.*

The process can be programmed for a computer just as described and is often called, after its inventors, **Gauss-Jordan elimination.**

Let $x_1, x_{i_2}, \ldots, x_{i_r}$ denote the variables used for sweeping columns and let $x_{i_{r+1}}, x_{i_{r+2}}, \ldots, x_{i_n}$ denote the remaining variables. For purposes of further discussion, it is convenient to group all terms containing these latter variables at the right. We thus obtain, from the reduced echelon form, the following system of equations, which, because of the operations performed to obtain it, is equivalent to the original system:

$$
\begin{aligned}
x_1 \qquad &+ \alpha_{1,\,r+1}x_{i_{r+1}} + \alpha_{1,\,r+2}x_{i_{r+2}} + \cdots + \alpha_{1,\,n}x_{i_n} = \beta_1 \\
x_{i_2} \quad &+ \alpha_{2,\,r+1}x_{i_{r+1}} + \alpha_{2,\,r+2}x_{i_{r+2}} + \cdots + \alpha_{2,\,n}x_{i_n} = \beta_2 \\
&\qquad\qquad\qquad\qquad\qquad \vdots \\
x_{i_r} &+ \alpha_{r,\,r+1}x_{i_{r+1}} + \alpha_{r,\,r+2}x_{i_{r+2}} + \cdots + \alpha_{r,\,n}x_{i_n} = \beta_r \\
&\qquad\qquad\qquad\qquad\qquad\qquad\qquad\quad 0 = \beta_{r+1} \\
&\qquad\qquad\qquad\qquad\qquad \vdots \\
&\qquad\qquad\qquad\qquad\qquad\qquad\qquad\quad 0 = \beta_m.
\end{aligned}
$$

Necessarily, $r \leq m$. If we have employed m of the variables to sweep columns, $r = m$ and there are no equations of the form $0 = \beta_j$. The quantities $\beta_{r+1}, \ldots,$ β_m may or may not all be zero in the event that $r < m$.

If at least one of the quantities $\beta_{r+1}, \beta_{r+2}, \ldots, \beta_m$ is *not* zero, then the given equations imply a contradiction and hence are inconsistent. If all the β's just named are zero, then the equations are consistent and we can solve for $x_1, x_{i_2}, \ldots, x_{i_r}$ by transposing all other terms to the right, thereby obtaining what we have called the "complete solution" of the system. *Thus the same procedure gives us a practical test for consistency and also gives the complete solution, when there is one.* The example with which we began illustrates both situations.

In a given case, when something clever can be done, one may find it useful to depart from the strict algorithm outlined here and solve for other variables, rearrange the equations, and so on, always still adhering to the basic pattern of sweeping columns.

2.6 Synthetic Elimination

In solving a system of linear equations, we actually operate only on the coefficients. The variables simply serve to keep the coefficients properly aligned in columns. If we copy the coefficients in columns and keep them carefully aligned, we don't need to copy the variables at all and hence can save a great deal of needless writing. The process is called **synthetic elimination** or the **method of detached coefficients**, because no variables appear in it.

For example, suppose we want to solve, by the sweepout process, the system of equations

$$x_1 + x_2 - x_3 + x_4 = 3$$
$$x_1 \qquad + x_3 + x_4 = 2$$
$$x_2 + x_3 - x_4 = -1$$
$$2x_1 - x_2 + x_3 \qquad = 2.$$

We copy down the coefficients, taking care to locate all zero coefficients in their proper places and then operate on rows of coefficients as though they were the equations they represent, following closely the rules for obtaining the reduced echelon form, but not recopying equations needlessly.

In Table 2.1, R_1, R_2, R_3, and R_4 identify the rows that represent the given equations. The extra column, headed Sum, is used for checking purposes only. The sum entry in any row is the sum of the coefficients and the constant term recorded in that same row. In operating on the rows, we perform the same operations on the sum entries also. Thus when we "subtract row 1 from row 2" to get the fifth row $(R_2' = R_2 - R_1)$, we subtract every entry of row 1 from the corresponding entry of row 2, and also subtract the sum entry of row 1 from that of row 2. The resulting entry in the sum column and row R_2' will agree with the sum of the coefficients and the constant term in row R_2' if no errors have been made. Thus the sum column provides a step-by-step check on accuracy, unless one makes compensating errors, of course.

TABLE 2.1
EXAMPLE OF SYNTHETIC ELIMINATION

Key to Operations	Coefficients				Constant Terms	Sum (for checking)
R_1	1	1	-1	1	3	5
R_2	1	0	1	1	2	5
R_3	0	1	1	-1	-1	0
R_4	2	-1	1	0	2	4
$R_2' = R_2 - R_1$	0	-1	2	0	-1	0
$R_4' = R_4 - 2R_1$	0	-3	3	-2	-4	-6
$R_1'' = R_1 - R_2''$	1	0	1	1	2	5
$*R_2'' = \quad - R_2'$	0	1	-2	0	1	0
$R_3'' = R_3 - R_2''$	0	0	3	-1	-2	0
$R_4'' = R_4' + 3R_2''$	0	0	-3	-2	-1	-6
$R_1''' = R_1'' - R_3'''$	1	0	0	$\frac{4}{3}$	$\frac{8}{3}$	5
$R_2''' = R_2'' + 2R_3'''$	0	1	0	$-\frac{2}{3}$	$-\frac{1}{3}$	0
$*R_3''' = \quad \frac{1}{3}R_3''$	0	0	1	$-\frac{1}{3}$	$-\frac{2}{3}$	0
$R_4''' = R_4'' + 3R_3'''$	0	0	0	-3	-3	-6
$R_1^{(iv)} = R_1''' - \frac{4}{3}R_4^{(iv)}$	1	0	0	0	$\frac{4}{3}$	$\frac{7}{3}$
$R_2^{(iv)} = R_2''' + \frac{2}{3}R_4^{(iv)}$	0	1	0	0	$\frac{1}{3}$	$\frac{4}{3}$
$R_3^{(iv)} = R_3''' + \frac{1}{3}R_4^{(iv)}$	0	0	1	0	$-\frac{1}{3}$	$\frac{2}{3}$
$*R_4^{(iv)} = \quad - \frac{1}{3}R_4'''$	0	0	0	1	1	2

Sweeping out the x_1-column, we need alter only rows 2 and 4, so we do not recopy rows 1 and 3. R_2' and R_4' denote the new second and fourth rows: $R_2' = R_2 - R_1$ and $R_4' = R_4 - 2R_1$.

Now we compute $*R_2'' = -R_2'$ in the next set of rows, then use it to compute R_1'', R_3'', and R_4'' in the fashion indicated in the leftmost column of the table. This amounts to sweeping the x_2-column in the standard way. $*R_2''$ is marked with an asterisk to indicate that it is the first row computed in this set.

Next we compute $*R_3'''$ (marked with an asterisk) and use it to get R_1''', R_2''', and R_4''' by the indicated computations, thus sweeping the x_3-column.

Finally, we compute $*R_4^{(iv)}$ and sweep the x_4-column with the result. This yields the last four rows of the table, from which, by recalling the meanings of the locations of the row entries, we can read off the solution:

$$x_1 = \frac{4}{3}$$
$$x_2 = \frac{1}{3}$$
$$x_3 = -\frac{1}{3}$$
$$x_4 = 1.$$

The successive major stages of the computation are indicated by the primes on the R's. When a row at one such stage is the same as the corresponding row at the previous stage, it need not be recopied. This explains why only two rows appear at the second level. There is, of course, no objection to writing a full set of rows at each level. Except in simple cases, full sets would occur anyhow.

When the number of variables is small and the coefficients are simple, solving by the usual elimination procedure is often easier than using this tabular array. In any other case, the tabular array is economical and helpful. Moreover, it is closely similar to the computer procedure used in solving such problems.

The method of synthetic elimination works equally well when the solution is not unique and it will also detect inconsistent systems of equations. One need only take careful note of the meanings of the entries in the table in order to apply the procedure in these cases. We illustrate with two examples.

Consider first the system of equations represented by

$$\begin{bmatrix} 2 & -1 & 3 \\ 4 & 3 & 2 \\ -1 & 1 & 4 \\ 1 & 5 & 3 \end{bmatrix} \begin{bmatrix} x_1 \\ x_2 \\ x_3 \end{bmatrix} = \begin{bmatrix} 4 \\ 9 \\ 4 \\ 2 \end{bmatrix}.$$

We proceed in Table 2.2, this time dispensing with the sum column for the sake of brevity. Also, we copy down the rows of coefficients in a more convenient order.

TABLE 2.2

AN INCONSISTENT SYSTEM

Key to Operations	Coefficients			Constant Terms
R_1	1	5	3	2
R_2	-1	1	4	4
R_3	2	-1	3	4
R_4	4	3	2	9
$R_2' = R_2 + R_1$	0	6	7	6
$R_3' = R_3 - 2R_1$	0	-11	-3	0
$R_4' = R_4 - 4R_1$	0	-17	-10	1
$R_1'' = R_1 - 5R_2''$	1	0	$-\frac{17}{6}$	-3
$*R_2'' = \frac{1}{6}R_2'$	0	1	$\frac{7}{6}$	1
$R_3'' = R_3' + 11R_2''$	0	0	$\frac{59}{6}$	11
$R_4'' = R_4' + 17R_2''$	0	0	$\frac{59}{6}$	18

The procedure need not be continued to the very end, for the last two lines of the table assert respectively that $\frac{59}{6}x_3 = 11$ and $\frac{59}{6}x_3 = 18$, so the equations are clearly inconsistent.

As a final example, consider the system of three equations in four unknowns represented by

$$\begin{bmatrix} 1 & 1 & -1 & 1 \\ 2 & -1 & 1 & -1 \\ 1 & 4 & -4 & 4 \end{bmatrix} \begin{bmatrix} x_1 \\ x_2 \\ x_3 \\ x_4 \end{bmatrix} = \begin{bmatrix} 2 \\ 1 \\ 5 \end{bmatrix}.$$

TABLE 2.3

A SYSTEM WITH INFINITELY MANY SOLUTIONS

Key to Operations	Coefficients				Constant Terms
R_1	1	1	−1	1	2
R_2	2	−1	1	−1	1
R_3	1	4	−4	4	5
$R'_2 = R_2 - 2R_1$	0	−3	3	−3	−3
$R'_3 = R_3 - R_1$	0	3	−3	3	3
$R''_1 = R_1 - R''_2$	1	0	0	0	1
$*R''_2 = -\frac{1}{3}R'_2$	0	1	−1	1	1
$R''_3 = R'_3 - 3R''_2$	0	0	0	0	0

Here the arithmetic proceeds as shown in Table 2.3. From the last two non-zero rows of Table 2.3 we read the equations

$$x_1 = 1$$
$$x_2 = x_3 - x_4 + 1.$$

That is, x_3 and x_4 may be chosen arbitrarily and their choice determines x_2, but x_1 is inevitably 1.

If we put $x_3 = t$, $x_4 = s$, we may write the solution in parametric form thus:

$$\begin{bmatrix} x_1 \\ x_2 \\ x_3 \\ x_4 \end{bmatrix} = t \begin{bmatrix} 0 \\ 1 \\ 1 \\ 0 \end{bmatrix} + s \begin{bmatrix} 0 \\ -1 \\ 0 \\ 1 \end{bmatrix} + \begin{bmatrix} 1 \\ 1 \\ 0 \\ 0 \end{bmatrix}.$$

The parametric form of the solution is often particularly useful.

2.7 Systems of Homogeneous Linear Equations

A system of linear equations of the form

(2.7.1)

$$a_{11}x_1 + a_{12}x_2 + \cdots + a_{1n}x_n = 0$$
$$a_{21}x_1 + a_{22}x_2 + \cdots + a_{2n}x_n = 0$$
$$\vdots$$
$$a_{m1}x_1 + a_{m2}x_2 + \cdots + a_{mn}x_n = 0,$$

in which the constant terms are all zero, is called a **system of homogeneous linear equations**. Since $(0, 0, \ldots, 0)$ is a solution of every such system, *all systems of homogeneous linear equations are consistent*. The solution $(0, 0, \ldots, 0)$ is called the **trivial solution** because it is inevitably present.

A solution *not* consisting entirely of 0's is called **nontrivial**. The problems of interest are to determine whether or not nontrivial solutions exist and to find them if they do. This is readily accomplished by transforming the system into reduced echelon form. If there are n variables and if the reduced echelon form has $r < n$ nonzero equations, then we can solve for r variables in terms of the remaining $n - r$ variables. By assigning a nonzero value to at least one of these $n - r$ independent variables, we can obtain a particular nontrivial solution. Moreover, since the final system is equivalent to the original system, all solutions are obtainable by assigning values to these $n - r$ variables and every such assignment yields a solution.

If $r = n$, since the constant terms are all 0, the original system is equivalent to

$$x_1 \qquad\qquad = 0$$
$$x_2 \qquad = 0$$
$$\vdots$$
$$x_n = 0,$$

so the trivial solution is the *only* solution.

We illustrate with three examples:

(a)
$$x_1 + x_2 + x_3 = 0$$
$$3x_1 + 2x_2 - 4x_3 = 0$$
$$2x_1 - x_2 + x_3 = 0.$$

The computation

$$
\begin{array}{llrrl}
R_1: & 1 & 1 & 1 & 0 \\
R_2: & 3 & 2 & -4 & 0 \\
R_3: & 2 & -1 & 1 & 0 \\
\hline
R_2': & 0 & -1 & -7 & 0 \\
R_3': & 0 & -3 & -1 & 0 \\
\hline
R_3'': & 0 & 0 & 20 & 0 \\
\hline
\end{array}
$$

shows that $x_3 = 0$, so $x_2 = 0$ and $x_1 = 0$ also; that is, the only solution is the trivial solution.

$$
\text{(b)} \quad
\begin{bmatrix}
1 & -1 & 1 \\
2 & 3 & -1 \\
1 & 4 & -2 \\
1 & 14 & -8
\end{bmatrix}
\begin{bmatrix}
x_1 \\
x_2 \\
x_3
\end{bmatrix} = 0.
$$

In this case, the computation proceeds as follows:

$$
\begin{array}{llrrl}
R_1: & 1 & -1 & 1 & 0 \\
R_2: & 2 & 3 & -1 & 0 \\
R_3: & 1 & 4 & -2 & 0 \\
R_4: & 1 & 14 & -8 & 0 \\
\hline
R_2': & 0 & 5 & -3 & 0 \\
R_3': & 0 & 5 & -3 & 0 \\
R_4': & 0 & 15 & -9 & 0 \\
\hline
R_1'': & 1 & 0 & \frac{2}{5} & 0 \\
R_2'': & 0 & 1 & -\frac{3}{5} & 0 \\
\hline
\end{array}
$$

Since R_3'' and R_4'' consist of all 0's, they have been omitted.
From this we have the complete solution

$$
\begin{aligned}
x_1 &= -\tfrac{2}{5}x_3 \qquad (n = 3, r = 2, n - r = 1). \\
x_2 &= \tfrac{3}{5}x_3
\end{aligned}
$$

Let us put $x_3 = 5t$ so the parametric solution becomes

$$\begin{bmatrix} x_1 \\ x_2 \\ x_3 \end{bmatrix} = t \begin{bmatrix} -2 \\ 3 \\ 5 \end{bmatrix}$$

in vector notation. In this case there are infinitely many nontrivial solutions, each of which is obtainable by assigning an appropriate nonzero value to t.

In certain applications, parameters appear in the coefficients of a system of equations to be studied. One can still use the sweepout procedure, provided care is taken not to divide by zero. For example, consider the system

(2.7.2)
$$x_1 + (2 - \alpha)x_2 + \alpha x_3 = 0$$
$$\alpha x_1 + 2(\alpha - 2)x_2 + 4x_3 = 0.$$

What we actually have here is an infinite set of pairs of equations, one pair for each value of α.

Eliminating x_1 from the second equation, we get this array of coefficients:

$$\begin{bmatrix} 1 & 2 - \alpha & \alpha \\ 0 & \alpha^2 - 4 & 4 - \alpha^2 \end{bmatrix}.$$

If we now assume that $\alpha^2 - 4 \neq 0$, that is, $\alpha \neq \pm 2$, we can divide the second equation by $\alpha^2 - 4$ and hence get the array

$$\begin{bmatrix} 1 & 0 & 2 \\ 0 & 1 & -1 \end{bmatrix}$$

so that the complete solution is

(2.7.3)
$$x_1 = -2t$$
$$x_2 = t$$
$$x_3 = t.$$

Thus, when $\alpha \neq \pm 2$, these systems all have the same complete solution. [Geometrically, if $\alpha \neq \pm 2$, each system (2.7.2) represents a pair of planes on the origin and all these pairs have the same line of intersection (2.7.3).]

It remains to see what happens if $\alpha = \pm 2$. When $\alpha = +2$, the system reduces to a single equation,

$$x_1 + 2x_3 = 0.$$

Here both x_2 and x_3 are arbitrary. If we put $x_2 = s$, $x_3 = t$, we can write the complete solution as

$$x_1 = -2t$$
$$x_2 = \quad s$$
$$x_3 = \quad t.$$

Similarly, when $\alpha = -2$, the system reduces to a single equation

$$x_1 + 4x_2 - 2x_3 = 0,$$

of which the complete solution is given by

$$x_1 = -4s + 2t$$
$$x_2 = \quad s$$
$$x_3 = \quad t.$$

[In these two cases we have one plane rather than two, hence no line of intersection. The complete solutions are simply parametric equations for the planes, each of which lies on the line (2.7.3), however. Note that for values of α very near to ± 2, the planes of a pair very nearly coincide.]

2.8 Exercises

1. Solve each system of equations by deriving the equivalent reduced echelon form, or else show that it is inconsistent.

(a)
$$x_1 - x_2 + x_3 = 4$$
$$2x_1 + x_2 - 3x_3 = -3$$
$$-3x_1 + 2x_2 + x_3 = -6,$$

(b)
$$2x_1 - 3x_2 + x_3 - x_4 = -1$$
$$4x_1 - 6x_2 + 3x_3 + 2x_4 = 3,$$

(c)
$$2x_1 + x_2 = 5$$
$$x_1 - 3x_2 = -1$$
$$3x_1 + 4x_2 = 6,$$

(d)
$$x_1 - 2x_2 + x_3 = 2$$
$$2x_1 - x_2 - x_3 = 7$$
$$4x_1 - 2x_2 + x_3 = 0.$$

2. Solve by the method of synthetic elimination:

(a)
$$\begin{bmatrix} 2 & -1 & 1 \\ 3 & -2 & 1 \\ 5 & 1 & 2 \end{bmatrix} \begin{bmatrix} x_1 \\ x_2 \\ x_3 \end{bmatrix} = \begin{bmatrix} 1 \\ 0 \\ 9 \end{bmatrix},$$

(b)
$$\begin{bmatrix} 1 & -1 & 1 \\ 2 & -3 & 4 \\ 3 & 1 & -5 \end{bmatrix} \begin{bmatrix} x_1 \\ x_2 \\ x_3 \end{bmatrix} = 0,$$

(c)
$$x_1 - x_2 + 2x_3 + x_4 = 1$$
$$2x_1 - x_2 - x_3 + 3x_4 = 2$$
$$-3x_1 + 2x_2 - x_3 - 4x_4 = -3,$$

(d)
$$x_1 + 2x_2 = 3$$
$$2x_2 + 3x_3 = 5$$
$$3x_3 + 4x_4 = 7$$
$$x_1 + 4x_4 = 5.$$

3. For what choice of k will the system

$$2x_1 + x_2 = 5$$
$$x_1 - 3x_2 = -1$$
$$3x_1 + 4x_2 = k$$

be consistent?

4. For what pairs of variables can one solve this system of equations?

$$x_1 + 2x_2 - 3x_3 + 6x_4 = 6$$
$$2x_1 + 4x_2 + 2x_3 - 4x_4 = 4.$$

5. Find the complete solutions of the systems

(a) $x_1 + 2x_2 + 3x_3 + 6x_4 = 0$
    $x_1 - 2x_2 + x_3 - 2x_4 = 0,$

(b) $\quad x_1 - x_2 \qquad - x_4 + 2x_5 = 0$
    $\quad 2x_1 + x_2 + 3x_3 \qquad + 2x_5 = 0$
    $-3x_1 + 2x_2 - x_3 + x_4 \qquad = 0,$

and represent the solution in parametric form.

6. Compare the solution found in Exercise 5(a) with the parametric form of the solution of

$$x_1 + 2x_2 + 3x_3 + 6x_4 = 12$$
$$x_1 - 2x_2 + x_3 - 2x_4 = -2.$$

What general theorem does this suggest?

7. For what values of k will the equation

$$\begin{bmatrix} 1 & 1 & k \\ 1 & k & 1 \\ k & 1 & 1 \end{bmatrix} \begin{bmatrix} x_1 \\ x_2 \\ x_3 \end{bmatrix} = 0$$

have nontrivial solutions? What are these solutions in each case?

8. Show that any one equation of a system (2.4.1) may be replaced by the sum of this equation and any subset of the remaining equations, the results being a system equivalent to (2.4.1). Use this fact to solve for x_1 by inspection, where

$$x_1 + 2x_2 - x_3 = 3$$
$$2x_1 - x_2 + 3x_3 = 2$$
$$3x_1 - x_2 - 2x_3 = 1.$$

9. Under what conditions on m, n, r will the reduced echelon form indicate that a system of linear equations has a unique solution?

10. Under what conditions on the coefficients will a system of equations of this form have a unique solution?

$$a_{11}x_1 + a_{12}x_2 \qquad\qquad\qquad = b_1$$
$$a_{22}x_2 + a_{23}x_3 \qquad\qquad = b_2$$
$$a_{33}x_3 + a_{34}x_4 = b_3$$
$$a_{44}x_4 = b_4.$$

What is this solution? (Many systems of equations appearing in applications have large blocks of coefficients that are zero.)

11. Use a desk calculator to solve this system:

$$x_1 + 0.43210x_2 + 0.61257x_3 = 1$$
$$0.43210x_1 + \qquad x_2 + 0.94761x_3 = 2$$
$$0.61257x_1 + 0.94761x_2 + \qquad x_3 = 3.$$

12. Solve this system of equations:

$$x_1 + (1 - i)x_2 + \qquad ix_3 = 1$$
$$(1 + i)x_1 + \qquad x_2 + (1 + i)x_3 = 1 + i$$
$$-ix_1 + (1 - i)x_2 + \qquad x_3 = 4 + 3i.$$

13. Express the solutions of the system

$$x_1 + 2x_2 + 3x_3 + \alpha x_4 = 0$$
$$x_1 + \alpha x_2 + 3x_3 + 2x_4 = 0$$
$$\alpha x_1 + x_2 + 3x_3 + x_4 = 0$$

as functions of the parameter α.

2.9 Computation of the Inverse of a Matrix

The sweepout procedure which we have used to solve systems of linear equations may also be used to compute the inverse of a matrix. To see how this is done, consider the system of n equations in n variables represented by

$$(2.9.1) \qquad\qquad AX = B.$$

If A^{-1} exists, premultiplication by A^{-1} shows that any X that satisfies (2.9.1) must be given by

(2.9.2) $$X = A^{-1}B.$$

Substitution from (2.9.2) into (2.9.1) shows that we have indeed a solution. That is, if A^{-1} exists, the equation (2.9.1) has the *unique* solution (2.9.2). This implies that when we apply synthetic elimination to the $n \times (n + 1)$ matrix formed by augmenting the matrix A by the column B:

$$[A, B],$$

when A^{-1} exists the end result must be the array

$$[I_n, A^{-1}B].$$

More generally, one can treat the systems of equations

$$AX = B_1, AX = B_2, \ldots, AX = B_p$$

all at the same time by applying sweepout to the array

$$[A, B_1, B_2, \ldots, B_p].$$

This observation applies whether or not A^{-1} exists. When the elimination process is complete, A has been replaced by the coefficient matrix corresponding to the reduced echelon form of these systems. The last p columns of the array tell the story concerning the solutions of the p systems of equations. In the particular case when A^{-1} exists, the final array is, necessarily,

$$[I_n, A^{-1}B_1, A^{-1}B_2, \ldots, A^{-1}B_p].$$

For the computation of the inverse, we make a special choice of the B_j's. We choose them to be the elementary n-vectors, $E_1, E_2, \ldots, E_n$, which are just the columns of I_n:

$$I_n = [E_1, E_2, \ldots, E_n].$$

Recall also that because of the isolated 1's in these vectors, for any matrix $C_{k \times n}$ we have $CE_1 = C_1, CE_2 = C_2, \ldots, CE_n = C_n$. That is, CE_j is the jth column of C for $j = 1, 2, \ldots, n$.

Returning now to the solution of p systems of equations with a common coefficient matrix A of order n, we let $p = n$ and let $B_j = E_j, j = 1, 2, \ldots, n$. Then the initial array is

$$[A, E_1, E_2, \ldots, E_n] = [A, I_n]$$

and, if A^{-1} exists, the final array becomes

$$[I_n, A^{-1}E_1, A^{-1}E_2, \ldots, A^{-1}E_n]$$
$$= [I_n, (A^{-1})_1, (A^{-1})_2, \ldots, (A^{-1})_n] = [I_n, A^{-1}].$$

That is, to invert A we use the sweepout process to reduce $[A, I_n]$ to $[I_n, A^{-1}]$.

We do not need to know that A^{-1} exists before beginning this process. If A^{-1} does not exist, it will not be possible to reduce the left block to I_n. That is, the sweepout process determines whether or not A has an inverse and produces the inverse when it exists.

We illustrate with an example. Find

$$\begin{bmatrix} 3 & 1 & 2 \\ 1 & -4 & 1 \\ 2 & 3 & 0 \end{bmatrix}^{-1}.$$

The arithmetic is given in Table 2.4, where the starred member of each set of rows is computed first.

Note that the technique is first to get a 1 in the 1,1-position, then use it to sweep the rest of the first column. Then we arrange to get a 1 in the 2,2-position and use it to sweep the rest of the second column, etc. Proceeding thus, we eventually obtain an identity matrix on the left, and A^{-1} on the right, or we eventually obtain a row of zeros on the left, in which case an identity matrix cannot be obtained on the left, so that A^{-1} does not exist.

TABLE 2.4

Key to Operations	A to I			I to A^{-1}		
R_1	3	1	2	1	0	0
R_2	1	-4	1	0	1	0
R_3	2	3	0	0	0	1
$*R'_1 = R_2$	1	-4	1	0	1	0
$R'_2 = R_3 - 2R_2$	0	11	-2	0	-2	1
$R'_3 = R_1 - 3R_2$	0	13	-1	1	-3	0
$R''_1 = R'_1 + 4R''_2$	1	0	$\frac{3}{11}$	0	$\frac{3}{11}$	$\frac{4}{11}$
$*R''_2 = \frac{1}{11}R'_2$	0	1	$-\frac{2}{11}$	0	$-\frac{2}{11}$	$\frac{1}{11}$
$R''_3 = R'_3 - 13R''_2$	0	0	$\frac{15}{11}$	1	$-\frac{7}{11}$	$-\frac{13}{11}$
$R'''_1 = R''_1 - \frac{3}{11}R'''_3$	1	0	0	$-\frac{3}{15}$	$\frac{6}{15}$	$\frac{9}{15}$
$R'''_2 = R''_2 + \frac{2}{11}R'''_3$	0	1	0	$\frac{2}{15}$	$-\frac{4}{15}$	$-\frac{1}{15}$
$*R'''_3 = \frac{11}{15}R''_3$	0	0	1	$\frac{11}{15}$	$-\frac{7}{15}$	$-\frac{13}{15}$

In paper-and-pencil computation, the sequence of operations one employs in this process need not be unique. One arranges it so as to exploit any obviously advantageous circumstances. This is illustrated in the example by our using R_2 as R'_1 rather than using $\frac{1}{3}R_1$ as R'_1. On the other hand, when this procedure is programmed for a computer, a completely systematic sequence of steps is used.

To check the work, one verifies that

$$\begin{bmatrix} 3 & 1 & 2 \\ 1 & -4 & 1 \\ 2 & 3 & 0 \end{bmatrix} \begin{bmatrix} -\frac{3}{15} & \frac{6}{15} & \frac{9}{15} \\ \frac{2}{15} & -\frac{4}{15} & -\frac{1}{15} \\ \frac{11}{15} & -\frac{7}{15} & -\frac{13}{15} \end{bmatrix} = \begin{bmatrix} 1 & 0 & 0 \\ 0 & 1 & 0 \\ 0 & 0 & 1 \end{bmatrix}.$$

2.10 Matrix Inversion by Partitioning

In this section, we show how partitioning may be used to compute the inverse of a symmetric matrix in a particularly effective way. The importance of the process lies in the fact that many of the matrices which arise in practice are symmetric. Even when a matrix is not symmetric, finding its inverse can be reduced to the inversion of a symmetric matrix, if that is desired. Thus for all matrices A, $A^{\mathsf{T}}A$ is symmetric and, if A^{-1} exists,

$$(2.10.1) \qquad\qquad A^{-1} = (A^{\mathsf{T}}A)^{-1}A^{\mathsf{T}},$$

so we need only invert a symmetric matrix in order to get A^{-1}.

Let A be the symmetric matrix of order n whose inverse is to be computed. As before, the inverse will be obtained by solving the system of n equations in n unknowns,

$$(2.10.2) \qquad\qquad AX = Y,$$

for X, but the technique will be different. We begin by writing (2.10.2) in partitioned form thus:

$$\begin{matrix} k \\ n-k \end{matrix} \begin{bmatrix} A_{11} & A_{12} \\ \hline A_{21} & A_{22} \end{bmatrix} \cdot \begin{bmatrix} X_1 \\ X_2 \end{bmatrix} \begin{matrix} k \\ n-k \end{matrix} = \begin{bmatrix} Y_1 \\ Y_2 \end{bmatrix} \begin{matrix} k \\ n-k \end{matrix}$$
$$\quad\; k \qquad n-k$$

This is equivalent to replacing (2.10.2) by the equations

$$(2.10.3) \qquad\qquad \begin{aligned} A_{11}X_1 + A_{12}X_2 &= Y_1, \\ A_{21}X_1 + A_{22}X_2 &= Y_2. \end{aligned}$$

In any particular case, we choose k, if possible, so that inverses exist for A_{11} and for other matrices which appear later. Then the first of these equations yields

(2.10.4)
$$X_1 = A_{11}^{-1}Y_1 - A_{11}^{-1}A_{12}X_2.$$

Substituting this into the second equation, we find

(2.10.5)
$$(A_{22} - A_{21}A_{11}^{-1}A_{12})X_2 = Y_2 - A_{21}A_{11}^{-1}Y_1.$$

To keep the notation in hand, let us put

$$B = A_{11}^{-1}A_{12} \quad \text{so that} \quad B^{\mathbf{T}} = A_{21}A_{11}^{-1}$$

since A is symmetric. We write also

$$C = A_{22} - A_{21}A_{11}^{-1}A_{12} = A_{22} - A_{21}B$$

and C is symmetric. Using these abbreviations and now assuming that C^{-1} exists, we rewrite (2.10.4) and (2.10.5), introducing B and C where possible. Then replacing X_2 in equation (2.10.4) by its value from equation (2.10.5), we finally obtain

(2.10.6)
$$X_1 = (A_{11}^{-1} + BC^{-1}B^{\mathbf{T}})Y_1 + (-BC^{-1})Y_2,$$
$$X_2 = (-BC^{-1})^{\mathbf{T}}Y_1 + C^{-1}Y_2.$$

Hence we must have

(2.10.7)
$$A^{-1} = \begin{bmatrix} A_{11}^{-1} + BC^{-1}B^{\mathbf{T}} & -BC^{-1} \\ (-BC^{-1})^{\mathbf{T}} & C^{-1} \end{bmatrix}.$$

Here A^{-1} and its upper left and lower right submatrices are symmetric, a fact which saves much work in the computation of an inverse.

When a symmetric matrix is inverted in this manner, we usually take $k = n - 1$. Then A_{22} reduces to the scalar a_{nn}, C likewise reduces to a scalar, and the computation becomes particularly simple. For example, let

$$A = \begin{bmatrix} 5 & -2 & \vdots & 4 \\ -2 & 1 & \vdots & 1 \\ \cdots & \cdots & & \cdots \\ 4 & 1 & \vdots & 0 \end{bmatrix}.$$

We have

$$A_{11} = \begin{bmatrix} 5 & -2 \\ -2 & 1 \end{bmatrix}, \qquad A_{11}^{-1} = \begin{bmatrix} 1 & 2 \\ 2 & 5 \end{bmatrix}, \qquad A_{12} = \begin{bmatrix} 4 \\ 1 \end{bmatrix},$$

$$B = \begin{bmatrix} 1 & 2 \\ 2 & 5 \end{bmatrix} \cdot \begin{bmatrix} 4 \\ 1 \end{bmatrix} = \begin{bmatrix} 6 \\ 13 \end{bmatrix}, \qquad A_{21} = [4 \quad 1], \qquad A_{22} = 0,$$

$$C = 0 - [4 \quad 1] \cdot \begin{bmatrix} 6 \\ 13 \end{bmatrix} = -37, \qquad C^{-1} = -\frac{1}{37}, \qquad -BC^{-1} = \begin{bmatrix} \dfrac{6}{37} \\ \dfrac{13}{37} \end{bmatrix},$$

$$(BC^{-1})B^{\mathsf{T}} = \begin{bmatrix} \dfrac{-6}{37} \\ \dfrac{-13}{37} \end{bmatrix} \cdot [6 \quad 13] = \begin{bmatrix} \dfrac{-36}{37} & \dfrac{-78}{37} \\ \dfrac{-78}{37} & \dfrac{-169}{37} \end{bmatrix},$$

$$A_{11}^{-1} + BC^{-1}B^{\mathsf{T}} = \begin{bmatrix} \dfrac{1}{37} & \dfrac{-4}{37} \\ \dfrac{-4}{37} & \dfrac{16}{37} \end{bmatrix}.$$

Substituting into (2.10.7) we have, then,

$$A^{-1} = \left[\begin{array}{cc:c} \dfrac{1}{37} & \dfrac{-4}{37} & \dfrac{6}{37} \\ \dfrac{-4}{37} & \dfrac{16}{37} & \dfrac{13}{37} \\ \hdashline \dfrac{6}{37} & \dfrac{13}{37} & \dfrac{-1}{37} \end{array} \right].$$

The computations detailed above may be arranged compactly in the following array from which A^{-1} is easy to write down:

(2.10.8)

A_{11}^{-1}		A_{12}	B	$BC^{-1}B^{\mathsf{T}}$	
1	2	4	6	$\dfrac{-36}{37}$	$\dfrac{-78}{37}$
2	5	1	13	$\dfrac{-78}{37}$	$\dfrac{-169}{37}$
		0	-37		
		A_{22}	C		

In this table the matrix product of the first and second blocks gives the third block of the upper two rows. Then the scalar $A_{12}^{\mathsf{T}} B$, subtracted from the A_{22} entry, gives the value of C. Next BB^{T}, divided by C, gives the entry $BC^{-1}B^{\mathsf{T}}$ of the final block. This last block, added to the first, gives the upper left block of A^{-1}, B divided by $-C$ gives the upper right block, C^{-1} gives the lower right block, and the rest is found by symmetry.

By using the previous example as a starting point and by using the same steps as those employed to construct (2.10.8), we compute the inverse of

$$S = \begin{bmatrix} 5 & -2 & 4 & \vdots & 1 \\ -2 & 1 & 1 & \vdots & -1 \\ 4 & 1 & 0 & \vdots & 0 \\ \hdashline 1 & -1 & 0 & \vdots & 1 \end{bmatrix}$$

by means of the following array:

S_{11}^{-1}			S_{12}	B	$BC^{-1}B^{\mathsf{T}}$		
$\dfrac{1}{37}$	$\dfrac{-4}{37}$	$\dfrac{6}{37}$	1	$\dfrac{5}{37}$	$\dfrac{25}{444}$	$\dfrac{-100}{444}$	$\dfrac{-35}{444}$
$\dfrac{-4}{37}$	$\dfrac{16}{37}$	$\dfrac{13}{37}$	-1	$\dfrac{-20}{37}$	$\dfrac{-100}{444}$	$\dfrac{400}{444}$	$\dfrac{140}{444}$
$\dfrac{6}{37}$	$\dfrac{13}{37}$	$\dfrac{-1}{37}$	0	$\dfrac{-7}{37}$	$\dfrac{-35}{444}$	$\dfrac{140}{444}$	$\dfrac{49}{444}$
			1	$\dfrac{12}{37}$			
			S_{22}	C			

From this array we then have

$$S^{-1} = \begin{bmatrix} \dfrac{1}{12} & \dfrac{-1}{3} & \dfrac{1}{12} & \vdots & \dfrac{-5}{12} \\ \dfrac{-1}{3} & \dfrac{4}{3} & \dfrac{2}{3} & \vdots & \dfrac{5}{3} \\ \dfrac{1}{12} & \dfrac{2}{3} & \dfrac{1}{12} & \vdots & \dfrac{7}{12} \\ \hdashline \dfrac{-5}{12} & \dfrac{5}{3} & \dfrac{7}{12} & \vdots & \dfrac{37}{12} \end{bmatrix}.$$

The reader should check the details. These two examples show how one could invert, by successive steps, a symmetric matrix of arbitrary order. In the examples we have used rational numbers to make it easier for the reader to follow and check the various steps. However, when observational data are involved, one usually has decimal entries, and the computations may be performed by machine. In this connection, an important characteristic of this method is that it permits control of the rounding-off error.

Finally, in the case of observational data, the matrices A_{11} and C are almost always nonsingular, so that the assumptions on which the process rests are well justified.

2.11   Exercises

1. Solve by sweepout:

$$AX = \begin{bmatrix} 0 \\ -1 \\ 2 \end{bmatrix}, \quad AX = \begin{bmatrix} 2 \\ 1 \\ 0 \end{bmatrix}, \quad AX = \begin{bmatrix} -1 \\ 0 \\ 1 \end{bmatrix}$$

where

$$A = \begin{bmatrix} 1 & -2 & 0 \\ 2 & 1 & 2 \\ 0 & 2 & 1 \end{bmatrix}.$$

2. Compute the inverses of these matrices by the sweepout process:

(a) $\begin{bmatrix} 1 & 2 & -1 \\ 2 & 3 & 0 \\ -1 & 0 & 4 \end{bmatrix}$,   (b) $\begin{bmatrix} 1 & 1 & 0 & -1 \\ 1 & 0 & -1 & 1 \\ 0 & -1 & 1 & 1 \\ -1 & 1 & 1 & 0 \end{bmatrix}$,

(c) $\begin{bmatrix} 2 & -1 & 0 \\ 1 & 1 & 2 \\ 3 & -1 & -1 \end{bmatrix}$,   (d) $\begin{bmatrix} 1 & 2 & 2 & 1 \\ 1 & 2 & 1 & 0 \\ 1 & 2 & 0 & 0 \\ 1 & 0 & 0 & 0 \end{bmatrix}.$

3. Show by the sweepout process that these matrices have no inverses:

(a) $\begin{bmatrix} 2 & 1 & 0 \\ -1 & 1 & -3 \\ 3 & 2 & -1 \end{bmatrix}$,   (b) $\begin{bmatrix} 2 & -1 & 0 & 3 \\ 1 & 1 & 2 & 1 \\ 3 & -1 & 4 & 0 \\ 5 & -5 & 0 & 4 \end{bmatrix}$,   (c) $\begin{bmatrix} 0 & 1 & 1 & 1 \\ 1 & 0 & -1 & -1 \\ 1 & 0 & -1 & 1 \\ 0 & 1 & 1 & -1 \end{bmatrix}.$

4. Invert by the method of Section 2.10 and check the result:

$$\begin{bmatrix} 0 & 1 & 2 & 3 \\ 1 & 1 & 2 & 3 \\ 2 & 2 & 2 & 3 \\ 3 & 3 & 3 & 3 \end{bmatrix}.$$

5. Apply (2.10.1) to the matrix

$$\begin{bmatrix} 1 & -1 & 0 \\ -1 & -2 & 1 \\ 1 & -1 & 1 \end{bmatrix}.$$

2.12 Number Fields

All coefficients in systems of equations treated in preceding sections have been assumed to represent real or complex numbers. Real, rational, or complex coefficients have always led respectively to real, rational, or complex solutions or expressions that define the set of all solutions. To examine carefully the relation between the nature of the coefficients and the nature of the solutions we need the concept of a number field.

A nonempty collection or set $\mathscr{F}$ of real or complex numbers that does not consist of the number zero alone will be called a **number field** if and only if the sum, difference, product, and quotient of any two numbers of $\mathscr{F}$ are again numbers of $\mathscr{F}$, division by zero being excepted. According to this definition, each of the following familiar sets of numbers constitutes a field:

1. The set of all rational numbers, that is, the set of all quotients of the form a/b, where a and b are integers but $b \neq 0$.
2. The set of all real numbers.
3. The set of all complex numbers, that is, the set of all numbers of the form $a + bi$, where a and b are real and $i^2 = -1$.

These fields are called respectively the **rational number field**, the **real number field**, and the **complex number field**. They are the most important number fields as far as applications are concerned. Most statistical work employs the real and rational number fields. Much work in physics and engineering employs the complex number field.

The operations of addition, subtraction, multiplication, and division involved in the definition of a field are known as the **four rational operations**.

When the numbers of one field are all members of another field, the first field is called a **subfield** of the second. In the case of the three fields just listed, the rational field is a subfield of the real field, and the real field is a subfield of the field of complex numbers. It is customary to regard any field as a subfield of itself.

Since every number field $\mathscr{F}$ contains $a - a$ for each number a in $\mathscr{F}$, it contains the number 0. Since it contains a/a for some $a \neq 0$, it also contains 1. Since it contains 1, it contains $0 - 1 = -1$, $1 + 1 = 2$, $0 - 2 = -2$, $2 + 1 = 3$, $0 - 3 = -3$, ...; that is, it contains all integers. Since it contains all integers and hence all quotients a/b, where a and b are integers ($b \neq 0$), it contains all rational numbers. Thus *the rational number field is the smallest possible number field and is a subfield of every number field.*

The concept of what a number field is may be clarified by examples of sets of numbers that are *not* fields. The set $\mathscr{Z}$ of all integers is not a number field because the quotient of one integer by another integer (not zero) is not always an integer; that is, it is not always a number of $\mathscr{Z}$. For example, $\frac{3}{5}$ is not an integer. The set of all non-negative real numbers is not a field because not every difference of two such numbers, for example, $\pi - 13$, is a non-negative real number. The set of all complex numbers of the form bi, where b is real, that is, the set of pure imaginary numbers, is not a field because the product of two of these is not a pure imaginary number: $bi \cdot ci = -bc$, which is real.

It is worthwhile to note that there are number fields other than the three familiar ones just mentioned. For example, the set of all numbers of the form $a + b\sqrt{2}$, where a and b are rational numbers, is a number field. In fact, if $\alpha + \beta\sqrt{2}$ and $\gamma + \delta\sqrt{2}$ are any two numbers of this kind, we have

$$(\alpha + \beta\sqrt{2}) \pm (\gamma + \delta\sqrt{2}) = (\alpha \pm \gamma) + (\beta \pm \delta)\sqrt{2}$$
$$(\alpha + \beta\sqrt{2}) \cdot (\gamma + \delta\sqrt{2}) = (\alpha\gamma + 2\beta\delta) + (\beta\gamma + \alpha\delta)\sqrt{2}.$$

Now let $\gamma + \delta\sqrt{2} \neq 0$. Because $\sqrt{2}$ is not rational and $\gamma + \delta\sqrt{2} \neq 0$, we have $\gamma - \delta\sqrt{2} \neq 0$ and $\gamma^2 - 2\delta^2 \neq 0$. Hence

$$\frac{\alpha + \beta\sqrt{2}}{\gamma + \delta\sqrt{2}} = \frac{(\alpha + \beta\sqrt{2})(\gamma - \delta\sqrt{2})}{(\gamma + \delta\sqrt{2})(\gamma - \delta\sqrt{2})} = \frac{\alpha\gamma - 2\beta\delta}{\gamma^2 - 2\delta^2} + \frac{\beta\gamma - \alpha\delta}{\gamma^2 - 2\delta^2}\sqrt{2}$$

Since the set of rational numbers itself forms a number field, the expression on the right is of the form $a + b\sqrt{2}$, with a and b rational numbers. We have thus shown that, when the four rational operations are applied to any two numbers of the form $a + b\sqrt{2}$ with a and b rational, the result is a number of the same kind. Hence the set of all such numbers constitutes a field.

There are, of course, infinitely many other number fields of various types, but in this book our attention is directed almost exclusively to the three fields defined in 1, 2, and 3 above.

Now consider a system of linear equations whose coefficients and constant terms all belong to a particular number field $\mathscr{F}$. Such a system is called a

system of linear equations over $\mathcal{F}$. The sweepout process involves only applying certain of the four rational operations to the coefficients and constant terms at any stage. When a constant multiplier or divisor is used, it is one of the coefficients. Hence every operation leads to a system of equations not only equivalent to the first but also over $\mathcal{F}$. Thus, if there exists a unique solution, it will be over $\mathcal{F}$. When there are infinitely many solutions, the complete solution obtained by sweepout will be over $\mathcal{F}$. One could possibly assign values from other fields to the independent variables in this case, thus obtaining solutions over other number fields than $\mathcal{F}$. This possibility is of no importance in this book.

2.13 Exercises

1. Show that if α ($\neq 0$) belongs to a number field $\mathcal{F}$, so do $n\alpha$ and α^n, where n is any integer.
2. Show that the set of all complex numbers $a + bi$, where a and b are real and where $i^2 = -1$, actually constitutes a number field, as stated in the text.
3. Show that the set of all numbers of the form $a + b\sqrt{p}$, where a and b are arbitrary rational numbers and where p is a prime number, constitutes a number field.
4. Show that if a, b, c belong to a number field $\mathcal{F}$, if $a \neq 0$, and if $ax + b = c$, then x also belongs to $\mathcal{F}$.
5. Show that if $a_1, b_1, c_1, a_2, b_2, c_2$ belong to a number field $\mathcal{F}$, if $a_1 b_2 - a_2 b_1 \neq 0$, and if

$$a_1 x + b_1 y = c_1$$
$$a_2 x + b_2 y = c_2,$$

then x and y also belong to $\mathcal{F}$.

2.14 The General Concept of a Field

There are many sets of objects other than numbers which have most of the abstract properties exhibited by number fields. These sets are called simply "fields," and the computations of Chapters 1 and 2 may be executed in the same basic fashion when the scalars belong to one of these other kinds of fields. The same holds true for the results and techniques of Chapters 5 through 8 except when the real number field is explicitly assumed to be the underlying field of scalars.

   In general a **field** is a set $\mathcal{F}$ of at least two objects with two operations called addition and multiplication such that

A_1:   For all a and b belonging to $\mathcal{F}$, $a + b$ belongs to $\mathcal{F}$.
A_2:   For all a and b belonging to $\mathcal{F}$, $a + b = b + a$.

A_3:   For all a, b, and c belonging to $\mathscr{F}$, $(a + b) + c = a + (b + c)$.

A_4:   There exists an element z belonging to $\mathscr{F}$, the **zero element**, such that for all a belonging to $\mathscr{F}$, $a + z = a$.

A_5:   For each a belonging to $\mathscr{F}$, there exists an element $-a$, the **negative** of a, such that $a + (-a) = z$.

M_1:   For all a, b belonging to $\mathscr{F}$, ab belongs to $\mathscr{F}$.

M_2:   For all a, b, and c belonging to $\mathscr{F}$, $(ab)c = a(bc)$.

M_3:   There exists an element u belonging to $\mathscr{F}$, the **unit element**, such that for all a belonging to $\mathscr{F}$, $au = ua = a$.

M_4:   For each $a \neq z$ belonging to $\mathscr{F}$, there exists an element a^{-1} belonging to $\mathscr{F}$, the **inverse** or **reciprocal** of a, such that $a \cdot a^{-1} = a^{-1} \cdot a = u$.

M_5:   For all a and b belonging to $\mathscr{F}$, $ab = ba$.

D_1:   For all a, b, and c belonging to $\mathscr{F}$, $a(b + c) = ab + ac$.

All number fields are fields in the more general sense defined above. There are many useful fields that are not number fields.

For example, if we define $0 + 0 = 0$, $0 + 1 = 1 + 0 = 1$, $1 + 1 = 0$, and $0 \cdot 0 = 0 \cdot 1 = 1 \cdot 0 = 0$, $1 \cdot 1 = 1$, then the set $\{0, 1\}$ of two elements is a field. It is not hard to verify that all the preceding postulates hold, simply by checking all the possible cases. This finite field is called the **Boolean field**. Although this is a very simple field, it is a very important one in a variety of applications in mathematics, computer science, and electrical engineering.

Another example of a field that is not a number field is the set of all expressions of the form $P(x)/Q(x)$, where $P(x)$ and $Q(x)$ are polynomials in a single variable x with coefficients in a number field $\mathscr{F}$ and where $Q(x) \not\equiv 0$. This field, called the **rational function field** over $\mathscr{F}$, is of great importance in advanced aspects of matrix theory.

Even when $\mathscr{F}$ is not a number field, the elements z and u are customarily denoted by 0 and 1, respectively.

Another algebraic structure that is of major importance in advanced aspects of matrix theory is defined as follows: A **ring** $\mathscr{R}$ is a set with at least two elements and with two operations called addition and multiplication such that for all a, b, and c belonging to $\mathscr{R}$ and for special elements z (the **zero element**) and $-a$ (the **negative** of a),

A_1: $a + b$ belongs to $\mathscr{R}$,

A_2: $a + b = b + a$,

A_3: $(a + b) + c = a + (b + c)$,

A_4: $a + z = a$,

A_5: $a + (-a) = z$,

M_1: ab belongs to $\mathscr{R}$,

M_2: $(ab)c = a(bc)$.

$$D_1: \qquad a(b + c) = ab + ac,$$
$$D_2: \qquad (a + b)c = ac + bc.$$

The following additional laws governing multiplication may or may not hold, depending on the ring in question:

M_3:   There exists a **unit element** u in $\mathcal{R}$ such that for all a in $\mathcal{R}$, $au = ua = a$.

M_4:   For each $a \neq z$ in $\mathcal{R}$, there exists an element a^{-1}, the **inverse** or **reciprocal** of a, such that $aa^{-1} = a^{-1}a = u$.

M_5:   For all a, b in $\mathcal{R}$, $ab = ba$.

If M_3 is satisfied, $\mathcal{R}$ is a **ring with unit**. If M_3 and M_4 are satisfied, $\mathcal{R}$ is a **division ring**. If M_5 is satisfied, $\mathcal{R}$ is a **commutative ring**. If M_3, M_4, and M_5 are satisfied, $\mathcal{R}$ is a field.

Many important rings are noncommutative. For example, the set of all $n \times n$ matrices over a field $\mathcal{F}$ is an example of a noncommutative ring with unit, the unit being I_n. It is not a field because not every $n \times n$ matrix has an inverse.

Some familiar commutative rings are the ring of all integers and the ring of all polynomials in an indeterminate x where the coefficients of the polynomials themselves belong to a ring such as the ring of integers or the field of rational numbers, for example. If $a \neq 1$, the set of all integers of the form na, where n is an arbitrary integer, is an example of a ring without a unit. The ring of even integers is a special case of this.

The concepts of a field and a ring, like those of an isomorphism and of a vector space, reveal the fundamental likeness of a great variety of collections of mathematical objects. Similarly, the concept of an equivalence relation reveals the basic likeness of various different relationships existing among mathematical objects. These examples show how abstract concepts enable us to understand mathematics better by revealing its structure, thereby permitting us to organize mathematical information compactly and systematically. They also suggest ways of extending our present knowledge to new areas, and they often enable us to reduce the proofs of many theorems to the proof of one. For example, if—starting with the abstract definition of a field given above—we can prove a result about fields in general, this result will hold true for the rational number field, the complex field, a field of matrices, and, in fact, for every field that exists. Abstract concepts are thus seen to be powerful tools in mathematical research. Indeed, they have largely made possible the phenomenal expansion of mathematical knowledge in recent decades. Without this knowledge, a major part of our modern scientific development would not have been possible. Thus abstract mathematics, despite its ivory-tower flavor, is "practical" in the fullest sense of the word.

2.15 Exercises

1. Prove that in the Boolean field $\{0, 1\}$, $-1 = 1$. Then solve this system of equations for the x's, the coefficients being presumed to be in the Boolean field:

$$x_1 + x_2 \qquad\qquad = y_1$$
$$x_2 + x_3 \qquad\quad = y_2$$
$$\vdots$$
$$x_{n-1} + x_n = y_{n-1}$$
$$x_n = y_n.$$

This system of equations is important in coding theory.

2. Show that the set of all numbers of the form $\alpha + \beta\sqrt[3]{2} + \gamma\sqrt[3]{4}$, where α, β, γ are rational, constitutes a field.

3. Same as Exercise 2 for $\alpha + \beta\sqrt{2} + \gamma\sqrt{3} + \delta\sqrt{6}$.

4. Show that the set of all quotients $P(x)/Q(x)$, where $P(x)$ and $Q(x)$ are polynomials in a single indeterminate x with coefficients in a field $\mathscr{F}$, and where $Q(x) \neq 0$, constitutes a field. Here we define $P(x)/Q(x) = R(x)/S(x)$ if and only if $P(x)S(x) = Q(x)R(x)$. We also define

$$P(x)/Q(x) + R(x)/S(x) = (P(x)S(x) + Q(x)R(x))/Q(x)S(x)$$

and

$$P(x)/Q(x) \cdot R(x)/S(x) = P(x)R(x)/Q(x)S(x).$$

This is the *rational function field* mentioned in the text.

*5. Prove that in every field $\mathscr{F}$ the elements z and u are unique, that a given element a has precisely one negative, and that a given nonzero element has precisely one reciprocal.

6. Prove that in every ring, and hence in every field, $a + c = b + c$ if and only if $a = b$ (the **cancellation law for addition**).

7. Prove that in every field the equation $ab = 0$ implies that at least one of a and b is zero. Prove also that in every field $ab = ac$ for $a \neq 0$ if and only if $b = c$ (the **cancellation law for multiplication**).

8. Prove that the intersection of two fields with a common nonzero element is also a field.

9. Show that, if p is a fixed integer which is *not* a perfect square, the set of all matrices

$$\begin{bmatrix} a & pb \\ b & a \end{bmatrix},$$

where a and b are rational numbers, constitutes a field. To what number field is it isomorphic?

10. Show that the set of all matrices

$$\begin{bmatrix} a & 2c & 2b \\ b & a & 2c \\ c & b & a \end{bmatrix},$$

where a, b, and c are arbitrary rational numbers, is isomorphic to the number field $\{a + b\sqrt[3]{2} + c\sqrt[3]{4}\}$ of Exercise 2 and hence is also a field.

11. An expression of the form $f(A)/g(A)$, where $f(A)$ and $g(A)$ are polynomials in an indeterminate square matrix A, is called a **rational function of the matrix A**. Define equality, addition, and multiplication as in Exercise 4. Does the set of all such functions of A, with coefficients from the complex field, constitute a field? Why?

12. Consider the set of all $n \times n$ partitioned matrices of the form

$$\left[\begin{array}{c|c} A & 0 \\ \hline B & C \end{array}\right]$$

over a field $\mathscr{F}$ where A is $r \times r$ and r is fixed. Show that this set constitutes a ring with a unit element.

13. A matrix A is **lower triangular** if and only if $a_{ij} = 0$ for $i < j$ and is **upper triangular** if and only if $a_{ij} = 0$ for $i > j$. Prove that the set of lower (upper) triangular matrices of order n over a field $\mathscr{F}$ constitutes a ring with unit over $\mathscr{F}$.

14. Is the set of all symmetric matrices of order n over $\mathscr{F}$ a ring over $\mathscr{F}$?

15. Prove that the set of all matrices over a field $\mathscr{F}$ which commute with a fixed matrix A over $\mathscr{F}$ is a ring.

16. Prove that the set of all matrices of order n of the form

$$\begin{bmatrix} a & b & b & \cdots & b \\ b & a & b & \cdots & b \\ \vdots & & & & \\ b & b & b & \cdots & a \end{bmatrix}$$

over a commutative ring $\mathscr{R}$ also constitutes a commutative ring. Which of these matrices are nonsingular? (You will need to define *nonsingular* here.)

17. Give some examples of finite and infinite sets that are not rings and explain which of the axioms are not satisfied in each case.

18. If, in a ring $\mathscr{R}$, $\lambda m = m$ for all m in $\mathscr{R}$, λ is a **left identity** for $\mathscr{R}$. If $m\rho = m$ for all m in $\mathscr{R}$, ρ is a **right identity** for $\mathscr{R}$. Show that the set of all matrices of the form $\begin{bmatrix} a & b \\ 0 & 0 \end{bmatrix}$, where a and b are arbitrary complex numbers, constitutes a ring in which the element $\begin{bmatrix} 1 & 0 \\ 0 & 0 \end{bmatrix}$ serves as a left identity element but not as a right identity element. Show that this ring does not have a right identity element.

CHAPTER
3

Vector Geometry in $\mathscr{E}^3$

3.1 Geometric Representation of Vectors in $\mathscr{E}^3$

In Chapter 2 we defined an n-vector to be a column matrix with n elements. This usage originates in the fact that real 3-vectors have a ready geometric interpretation in Euclidean space of two and three dimensions ($\mathscr{E}^2$ and $\mathscr{E}^3$) and that this interpretation generalizes readily to spaces of higher dimensions. First we review briefly the three-dimensional case. The two-dimensional case is illustrated in some of the figures.

Each real 3-vector $X = [x_1, x_2, x_3]^{\mathsf{T}}$ determines a unique corresponding point $P:(x_1, x_2, x_3)$ and thus, except when $X = 0$, a unique directed line segment $\overrightarrow{OP}$ from the origin to the point P. The segment $\overrightarrow{OP}$ is also called a **vector**, or a **geometric vector**, and is identified by the vector X which determines it (Figure 3-1). When $X = 0$, $P = O$, so that no proper directed segment is determined. In this case, we regard the origin O as the geometric **zero vector** corresponding to the arithmetic zero vector. Conversely, given any geometric vector $\overrightarrow{OP}$, the coordinates of its terminal point determine the unique, real, arithmetic 3-vector by which $\overrightarrow{OP}$ is named. That is, there is a one-to-one correspondence between real 3-vectors and geometric vectors in $\mathscr{E}^3$. This correspondence enables us to give geometrical meanings to algebraic operations, as is illustrated in the next section.

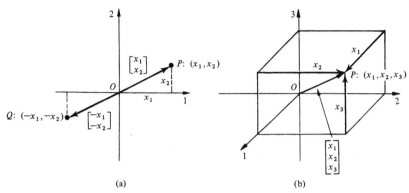

(a)   (b)

FIGURE 3-1. Geometric Representation of Vectors in $\mathscr{E}^2$ and $\mathscr{E}^3$.

3.2 Operations on Vectors

Consider now three directed segments $\overrightarrow{OP}$, $\overrightarrow{OQ}$, and $\overrightarrow{OR}$, corresponding to X, Y, and $X + Y$, respectively. Assume that O, P, and Q are not in the same straight line. Connect R to each of P and Q to complete a quadrilateral $OPRQ$ (Figure 3-2). Since, in Figure 3-2, the triangles OSQ and PTR are congruent, have two sides in one parallel to the corresponding sides in the other, and are similarly placed, the line segments OQ and PR are equal in length and parallel, so that $OPRQ$ is a parallelogram. That is, $\overrightarrow{OR}$ is a diagonal of the parallelogram determined by $\overrightarrow{OP}$ and $\overrightarrow{OQ}$.

Conversely, from Figure 3-2 we see that if $\overrightarrow{OP}$ represents X, if $\overrightarrow{OQ}$

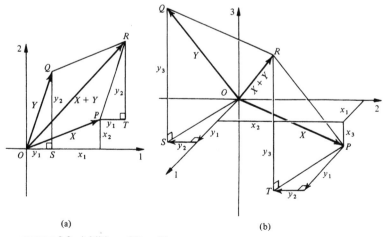

(a)   (b)

FIGURE 3-2. Addition of Two Vectors.

represents Y, and if $\overrightarrow{OR}$ is a diagonal of the parallelogram determined by $\overrightarrow{OP}$ and $\overrightarrow{OQ}$, then the coordinates of R are the sums of corresponding co-ordinates of P and Q so that $\overrightarrow{OR}$ represents $X + Y$. Thus *addition of components of real, arithmetic 3-vectors corresponds to the familiar parallel-ogram law for the addition of geometric vectors.*

We say that $X, Y, Z, \ldots$ are **coplanar vectors** if and only if the directed segments which represent them all lie on the same plane on the origin. A consequence of the parallelogram law of addition is that the vectors $\overrightarrow{OP}$, $\overrightarrow{OQ}$, and $\overrightarrow{OP} + \overrightarrow{OQ}$ are always coplanar, that is, that X, Y, *and their sum $X + Y$ are always coplanar vectors.*

In the special case when O, P, and Q are collinear, that is, when $\overrightarrow{OP}$ and $\overrightarrow{OQ}$ lie on the same straight line, we say that X and Y are **collinear vectors**. In this case the sum $X + Y$ is represented by a directed segment $\overrightarrow{OR}$ which is collinear with the segments $\overrightarrow{OP}$ and $\overrightarrow{OQ}$ (Figure 3-3). The reader should provide full details.

Let c be a real number. Consider directed segments $\overrightarrow{OP}$ and $\overrightarrow{OQ}$, corresponding to the vectors X and cX respectively, neither X nor c being 0. Because O, S, and T are collinear and triangles OSP and OTQ in Figure 3-4 are similar, O, P, and Q must be collinear and $\overrightarrow{OQ} = c\,\overrightarrow{OP}$. Conversely, if $\overrightarrow{OQ} = c\,\overrightarrow{OP}$ for some real number c, similar triangles show that if $\overrightarrow{OP}$ represents X, $\overrightarrow{OQ}$ must represent cX. Thus *multiplication of the vector X by the scalar c corresponds to multiplying the directed segment $\overrightarrow{OP}$ which represents X by the scalar c.*

If X is 0, cX is also 0 for all c, and if c is 0, cX is 0 for all X. In either case, cX is represented by the origin and the preceding conclusion still holds.

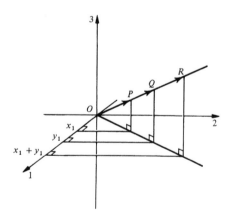

FIGURE 3-3. Addition of Collinear Vectors.

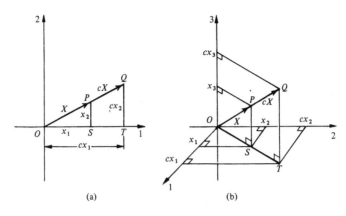

FIGURE 3-4. Multiplication of a Vector by a Scalar.

If $c > 0$, $\overrightarrow{OP}$ and $c\,\overrightarrow{OP}$ have the same sense, while if $c < 0$, they have opposite senses. Figure 3-4 illustrates the case $c > 0$. The reader should draw a figure illustrating the case $c < 0$.

Since arithmetic and geometric addition correspond, as do the multiplication of arithmetic and geometric vectors by scalars, other algebraic laws involving only these two operations have easily obtainable geometric interpretations. For example, the laws $\alpha X + \alpha Y = \alpha(X + Y)$ and $(X + Y) + Z = X + (Y + Z)$ are illustrated in Figure 3-5.

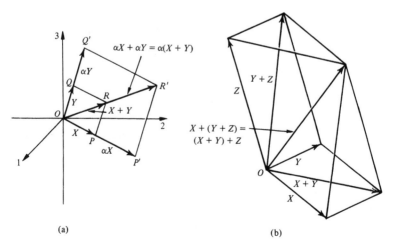

FIGURE 3-5. Geometric Interpretation of Laws of Vector Algebra.

3.3   Isomorphism

Consider again the system of geometric vectors $\overrightarrow{OP}$ in $\mathscr{E}^3$ and the system of real 3-vectors. Each system has an operation of addition and an operation of multiplication by scalars. Moreover, we have seen that there is a one-to-one correspondence between elements of the two systems such that to the sum of two geometric vectors there corresponds the sum of the corresponding arithmetic vectors and to a scalar multiple of a geometric vector there corresponds the same scalar multiple of its arithmetic counterpart. We can, of course, reverse the roles of arithmetic and geometric vectors in the preceding sentence. We describe all this by saying that *the correspondence preserves the operations*: to a sum there corresponds a sum and to a scalar multiple there corresponds a scalar multiple. In consequence, all the abstract laws governing corresponding operations in the two systems are the same. We identify these facts by calling the systems **isomorphic**. The only distinction between the two systems is in the nature of the representation of their elements: arithmetic vectors in the one case, geometric vectors in the other.

Other examples of isomorphic systems will appear later.

3.4   Length, Direction, and Sense

From analytic geometry the reader will recall that an ordered pair of points $P_1 : (x_1, x_2, x_3)$, $P_2 : (y_1, y_2, y_3)$ determines a directed line segment $\overrightarrow{P_1 P_2}$ whose length, the distance between P_1 and P_2, is denoted by $|\overrightarrow{P_1 P_2}|$ (see Figure 3-6), and is given by

(3.4.1) $|\overrightarrow{P_1 P_2}| = \sqrt{(y_1 - x_1)^2 + (y_2 - x_2)^2 + (y_3 - x_3)^2}.$

The **length** $|X|$ **of a vector** X is defined to be the length of the directed segment $\overrightarrow{OP}$ which represents it:

(3.4.2) $|X| = \sqrt{x_1^2 + x_2^2 + x_3^2}.$

In particular, the length of the zero-vector is zero. Since, when X is real,

(3.4.3) $x_1^2 + x_2^2 + x_3^2 = 0$ if and only if $x_1 = x_2 = x_3 = 0,$

the vector X has length 0 if and only if $X = 0$.

When $|X| = 1$, X is called a **unit vector**. Examples are E_1, E_2, E_3, $[1/\sqrt{3}, -1/\sqrt{3}, 1/\sqrt{3}]^\mathsf{T}$, and so on. Since $|X| = 1$ if and only if

(3.4.4) $x_1^2 + x_2^2 + x_3^2 = 1,$

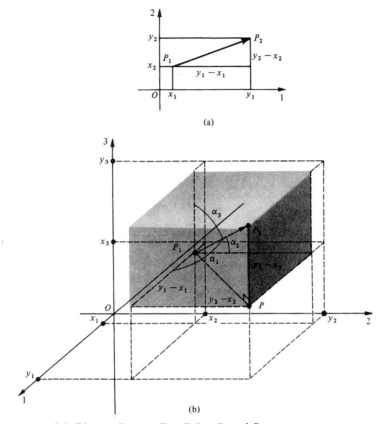

FIGURE 3-6. Distance Between Two Points P_1 and P_2.

the locus of the endpoints P of all unit vectors X is the **unit sphere** with center at the origin.

The **direction angles** α_1, α_2, and α_3 of the directed segment $\overrightarrow{P_1P_2}$ are the smallest undirected angles between $\overrightarrow{P_1P_2}$ and lines on P_1 parallel to the axes and similarly sensed (see Figure 3-6). Let $|\overrightarrow{P_1P_2}| = d$. Then the direction angles are determined by the **direction cosines** of $\overrightarrow{P_1P_2}$:

$$\cos \alpha_1 = \frac{y_1 - x_1}{d},$$

(3.4.5)
$$\cos \alpha_2 = \frac{y_2 - x_2}{d},$$

$$\cos \alpha_3 = \frac{y_3 - x_3}{d},$$

where

$$0 \le \alpha_i \le \pi, \qquad i = 1, 2, 3.$$

Squaring and adding corresponding members of equations (3.4.5), we obtain the identity

(3.4.6) $$\cos^2 \alpha_1 + \cos^2 \alpha_2 + \cos^2 \alpha_3 = 1,$$

so that the three direction angles are not independent. For example, if $\alpha_1 = 45°$ and $\alpha_2 = 60°$, then

$$\left(\frac{1}{\sqrt{2}}\right)^2 + \left(\frac{1}{2}\right)^2 + \cos^2 \alpha_3 = 1.$$

Hence $\cos^2 \alpha_3 = \frac{1}{4}$ so that $\cos \alpha_3 = \pm\frac{1}{2}$ and hence $\alpha_3 = 60°$ or $120°$ (see Figure 3-7).

In the case of a vector X represented by $\overrightarrow{OP}$, formulas (3.4.5) reduce to

(3.4.7)
$$\cos \alpha_1 = \frac{x_1}{|X|},$$
$$\cos \alpha_2 = \frac{x_2}{|X|}, \qquad 0 \le \alpha_i \le \pi,$$
$$\cos \alpha_3 = \frac{x_3}{|X|}.$$

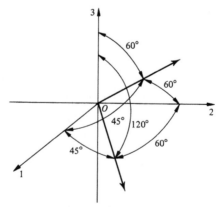

FIGURE 3-7. Related Sets of Direction Angles.

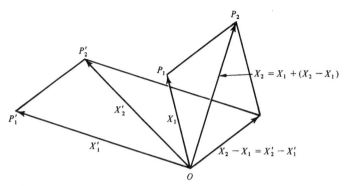

FIGURE 3-8. Difference of Two Vectors.

Note that by (3.4.6), these direction cosines are the components of a unit vector. Its direction and sense are the same as those of X. (Why?)

By what precedes, the **ordered differences** $y_1 - x_1$, $y_2 - x_2$, $y_3 - x_3$ determine uniquely the length, direction, and sense of a directed line segment. Also, two directed line segments are defined to be equal if and only if they have the same length, direction, and sense. The results of this section therefore show that *two directed line segments are equal if and only if their ordered differences are equal.*

A given set of ordered differences may be used as the components of a vector X. Also, a given set of ordered differences, that is, a given vector X, determines infinitely many equal directed segments $\overrightarrow{P_1 P_2}$, any particular one of which is determined as soon as its initial point P_1 or its terminal point P_2 is given. For example, if $X = [3, -1, 2]^{\mathsf{T}}$ and $P_1 = (2, 1, -1)$, then we must have $P_2 = (5, 0, 1)$. Figure 3-8 shows the relationship between segments equal to the segment $\overrightarrow{P_1 P_2}$ and the associated vector of ordered differences: $X = X_2 - X_1$, where X_2 is determined by P_2 and X_1 by P_1.

3.5 Orthogonality of Two Vectors

Let

$$V = \begin{bmatrix} v_1 \\ v_2 \\ v_3 \end{bmatrix} \quad \text{and} \quad W = \begin{bmatrix} w_1 \\ w_2 \\ w_3 \end{bmatrix}$$

be nonzero vectors in $\mathscr{E}^3$. Then the **angle** θ between V and W is defined to be the undirected angle θ, $0 \le \theta \le \pi$, between the directed segments

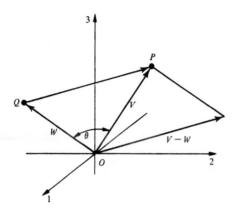

FIGURE 3-9. Angle Between Two Vectors.

$\overrightarrow{OP}$ and $\overrightarrow{OQ}$ which represent these vectors (Figure 3-9). The directed segment $\overrightarrow{QP}$ has the same length as the vector $V - W$. Hence, by the law of cosines,

$$|V - W|^2 = |V|^2 + |W|^2 - 2|V||W|\cos\theta$$

or

$$(v_1 - w_1)^2 + (v_2 - w_2)^2 + (v_3 - w_3)^2$$
$$= (v_1^2 + v_2^2 + v_3^2) + (w_1^2 + w_2^2 + w_3^2) - 2|V||W|\cos\theta,$$

from which we obtain

$$\cos\theta = \frac{v_1 w_1 + v_2 w_2 + v_3 w_3}{|V||W|}.$$

The expression in the numerator is the *scalar product* $V^{\mathsf{T}}W$ that we introduced in Chapter 1. Hence

(3.5.1) $$\cos\theta = \frac{V^{\mathsf{T}}W}{|V||W|}, \qquad 0 \le \theta \le \pi.$$

Now

$$\frac{|V^{\mathsf{T}}W|}{|V||W|} = |\cos\theta| \le 1,$$

so

(3.5.2) $$|V^\mathsf{T}W| \le |V||W|.$$

This last result is one form of the famous **Cauchy–Schwarz inequality,** which is frequently useful. Although $\cos \theta$ was defined only for nonzero V and W, (3.5.2) holds without exception. (Why?)

Nonzero vectors V and W are orthogonal (at right angles), that is, $\theta = \pi/2$, if and only if $\cos \theta = 0$. This holds if and only if the numerator in (3.5.1) vanishes:

(3.5.3) $V^\mathsf{T}W = 0,\quad V \ne 0,\quad W \ne 0$ \qquad (orthogonality condition).

It is often necessary to find a vector V orthogonal to each of two given vectors A and B; that is, one has to solve the equations

$$\begin{aligned} A^\mathsf{T}V &= 0 \\ B^\mathsf{T}V &= 0 \end{aligned} \quad \text{or} \quad \begin{aligned} a_1 v_1 + a_2 v_2 + a_3 v_3 &= 0 \\ b_1 v_1 + b_2 v_2 + b_3 v_3 &= 0 \end{aligned}$$

for the v's. For example, let

$$A = \begin{bmatrix} 1 \\ 1 \\ 0 \end{bmatrix}, \qquad B = \begin{bmatrix} 0 \\ 1 \\ 1 \end{bmatrix}.$$

Then

$$\begin{aligned} v_1 + v_2 \quad\;\; &= 0 \\ v_2 + v_3 &= 0. \end{aligned}$$

The complete solution of this system is

$$V = t \begin{bmatrix} 1 \\ -1 \\ 1 \end{bmatrix}.$$

Thus there are actually infinitely many solutions, all having the same direction since they are all multiples of a common vector, but differing in length or sense or both. The vectors A, B, and V, for the case $t = 1$, are shown in Figure 3-10.

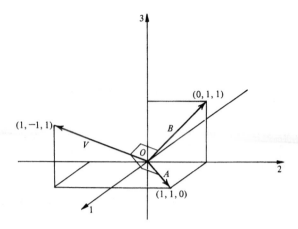

FIGURE 3-10. Vector Orthogonal to Two Given Vectors.

3.6   Exercises

1. Write the vector $X = X_2 - X_1$ determined by the directed segment $\overrightarrow{P_1P_2}$, where $P_1 = (-1, 0, 2)$ and $P_2 = (2, -1, 1)$. Then find the length and the direction cosines of this vector. Illustrate with a figure.

2. Find the vector which is the sum of the vectors determined by $\overrightarrow{P_1P_2}$ and $\overrightarrow{Q_1Q_2}$, where

$$P_1 = (1, 0, -1), \qquad P_2 = (4, 4, -2), \qquad Q_1 = (2, -1, 0), \qquad Q_2 = (3, 1, 2).$$

Illustrate with a figure.

3. If $\cos \alpha_1 = \frac{1}{3}$, $\cos \alpha_2 = \frac{2}{3}$, what is $\cos \alpha_3$? Illustrate with a figure (two cases).

4. Find a scalar multiple of the vector $[1, 1, 1]^{\mathsf{T}}$ which has length 1. Is there more than one such scalar multiple?

5. Show that there exist scalars t_1, t_2, and t_3, not all zero, such that

$$t_1 \begin{bmatrix} 1 \\ 2 \\ -3 \end{bmatrix} + t_2 \begin{bmatrix} -2 \\ -1 \\ 3 \end{bmatrix} + t_3 \begin{bmatrix} 4 \\ -1 \\ -3 \end{bmatrix} = \begin{bmatrix} 0 \\ 0 \\ 0 \end{bmatrix}.$$

Then illustrate the equation with a figure.

6. Find all sets of values of t_1, t_2, and t_3 such that

$$t_1 \begin{bmatrix} 1 \\ 2 \\ -3 \end{bmatrix} + t_2 \begin{bmatrix} -2 \\ -1 \\ 3 \end{bmatrix} + t_3 \begin{bmatrix} 4 \\ -1 \\ -3 \end{bmatrix} = \begin{bmatrix} 3 \\ 0 \\ -3 \end{bmatrix}.$$

*7. Show that, for a given set of direction cosines,

$$\begin{bmatrix} \cos \alpha_1 \\ \cos \alpha_2 \\ \cos \alpha_3 \end{bmatrix}$$

is a unit vector and illustrate with a figure.

*8. Show that if

$$X = \begin{bmatrix} x_1 \\ x_2 \\ x_3 \end{bmatrix},$$

then $X/|X|$, that is,

$$\begin{bmatrix} \dfrac{x_1}{|X|} \\[2ex] \dfrac{x_2}{|X|} \\[2ex] \dfrac{x_3}{|X|} \end{bmatrix},$$

is a unit vector with the same direction and sense as X. (Show that, in fact,

$$\frac{X}{|X|} = \begin{bmatrix} \cos \alpha_1 \\ \cos \alpha_2 \\ \cos \alpha_3 \end{bmatrix},$$

where α_1, α_2, and α_3 are direction angles of X.)

9. If the distance between the points $(-1, a, 2a)$ and $(2, -a, a)$ is $\sqrt{29}$, what are the possible values for a?

10. Illustrate the equation $U + V + W = 0$ with a figure. Do the same for $U + V + W + X = 0$.

*11. What is the equation of the locus of the endpoints of all vectors $\overrightarrow{OP}$ of length r in $\mathscr{E}^2$? In $\mathscr{E}^3$?

*12. In $\mathscr{E}^3$, show that for all real numbers α, β, the vector $\alpha X + \beta Y$ is coplanar with X and Y.

13. Prove that, for every real number a, the triangle with vertices $P_1 : (a, 1 - a, \sqrt{2})$, $P_2 : (2, a, -2a)$, and $P_3 : (3a - 1, -1, \sqrt{2})$ is a right triangle.

14. If the vectors

$$\begin{bmatrix} -1 \\ a \\ 2a \end{bmatrix} \quad \text{and} \quad \begin{bmatrix} 2 \\ -a \\ a \end{bmatrix}$$

are orthogonal, what are the possible values for a?

15. Interpret geometrically the equations

(a) $[x_1, x_2, x_3] \begin{bmatrix} -1 \\ 2 \\ 4 \end{bmatrix} = 0,$ (b) $[x_1, x_2, x_3] \begin{bmatrix} x_1 \\ x_2 \\ -x_3 \end{bmatrix} = 0.$

16. A *unit* vector is orthogonal to each of

$$\begin{bmatrix} 2 \\ -1 \\ 3 \end{bmatrix}, \quad \begin{bmatrix} -3 \\ 0 \\ 2 \end{bmatrix}.$$

Find all such vectors.

*17. Specialize the results of Sections 3.1 through 3.5 to $\mathscr{E}^2$.

3.7 The Vector Equation of a Line

Consider the line determined by two distinct points P_0 and P_1. Let the vectors determined by P_0 and P_1 be denoted by X_0 and X_1 respectively. Let X denote the vector determined by P where P is any point on the line. Then Figure 3-11 shows that X is X_0 plus some scalar multiple of $X_1 - X_0$, that is,

(3.7.1) $\qquad\qquad X = X_0 + t(X_1 - X_0) \qquad (t \text{ real}).$

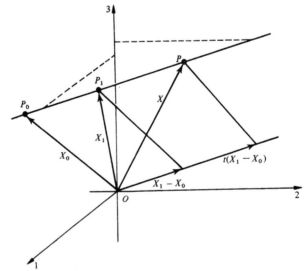

FIGURE 3.11. Vector Equation of a Line.

From Figure 3-11 one can see that the values of t are associated with points on the line as follows:

$t < 0$	points of the half-line starting at P_0 and not containing P_1
$t = 0$	point P_0
$0 < t < 1$	points between P_0 and P_1
$t = 1$	point P_1
$t > 1$	points of the half-line starting at P_1 and not containing P_0.

If P_0 has coordinates (x_{10}, x_{20}, x_{30}) and P_1 has coordinates (x_{11}, x_{21}, x_{31}), equation (3.7.1) is equivalent to the three scalar equations

$$(3.7.2) \quad \begin{aligned} x_1 &= x_{10} + t(x_{11} - x_{10}), \\ x_2 &= x_{20} + t(x_{21} - x_{20}), \\ x_3 &= x_{30} + t(x_{31} - x_{30}), \end{aligned}$$

which give the coordinates of the point P in terms of those of P_0 and P_1 since the coordinates of P are the same as the components of the vector X. These are called **parametric equations** for the line because the variable point P of the line is expressed in terms of the parameter t. In $\mathscr{E}^2$, one employs just the first two of these equations.

Given an equation (3.7.1), one can readily identify the line which it represents. For example, if

$$X = \begin{bmatrix} 2 \\ -1 \\ 4 \end{bmatrix} + t \begin{bmatrix} 5 \\ 0 \\ -3 \end{bmatrix},$$

then

$$X_0 = \begin{bmatrix} 2 \\ -1 \\ 4 \end{bmatrix}, \quad X_1 - X_0 = \begin{bmatrix} 5 \\ 0 \\ -3 \end{bmatrix}, \quad \text{and} \quad X_1 = \begin{bmatrix} 7 \\ -1 \\ 1 \end{bmatrix},$$

so the given vector equation represents the line on $P_0 : (2, -1, 4)$ and $P_1 : (7, -1, 1)$.

If P_0 is the origin and if P_1 is the point (a_1, a_2, a_3), the equations of the line reduce to

$$x_1 = ta_1$$
$$x_2 = ta_2$$
$$x_3 = ta_3$$

or, in vector form, if

$$A = \begin{bmatrix} a_1 \\ a_2 \\ a_3 \end{bmatrix},$$

(3.7.3) $$X = tA.$$

This equation shows that a line L lying on the origin carries all scalar multiples of a fixed vector A.

The set of all these vectors tA is **closed** with respect to addition and with respect to multiplication by a scalar; that is, for all real numbers t_1 and t_2,

$$(t_1 A) + (t_2 A) = (t_1 + t_2)A,$$

so the sum of any two vectors of the set is in the set and

$$t_1(t_2 A) = (t_1 t_2)A,$$

so any scalar multiple of any member of the set is in the set.

This illustrates an important mathematical concept. A set of vectors in $\mathscr{E}^2$ or $\mathscr{E}^3$ is called a **vector space** if and only if the sum of any two vectors in the set is also in the set and every real scalar multiple of every vector in the set is in the set. Examples of vector spaces are the set consisting of the zero vector alone and also the sets of all vectors in $\mathscr{E}^2$ or in $\mathscr{E}^3$. By what precedes, *the set of all vectors on the line whose equation is $X = tA$ is also a vector space.* Other examples will appear later.

The **direction of a line** is defined to be that of any segment $P_0 P_1$ on it. If we wish to give the line the same *sense* as that of $\overrightarrow{P_0 P_1}$, then **direction cosines of the line** are

$$\cos \alpha_1 = \frac{x_{11} - x_{10}}{|P_0 P_1|},$$

$$\cos \alpha_2 = \frac{x_{21} - x_{20}}{|P_0 P_1|},$$

$$\cos \alpha_3 = \frac{x_{31} - x_{30}}{|P_0 P_1|}.$$

From similar triangles, one can see that any other segment $\overrightarrow{Q_0 Q_1}$ on the line and with the same sense as $\overrightarrow{P_0 P_1}$ would determine the same direction cosines.

For every pair of points $P : (x_1, x_2, x_3)$, $Q : (y_1, y_2, y_3)$ on the line, the ordered differences $y_1 - x_1$, $y_2 - x_2$, and $y_3 - x_3$ are proportional to the direction cosines and are called **direction numbers** of the line.

3.8   The Vector Equation of a Plane

A plane in $\mathscr{E}^3$ is determined by any three noncollinear points on it. Let three such points be $P_0 : (x_{10}, x_{20}, x_{30})$, $P_1 : (x_{11}, x_{21}, x_{31})$, and $P_2 : (x_{12}, x_{22}, x_{32})$, and let the corresponding vectors be X_0, X_1, and X_2 respectively (Figure 3-12). Let $P : (x_1, x_2, x_3)$, with corresponding vector X, denote an arbitrary point of the plane determined by P_0, P_1, and P_2. Then, since P_0, P_1, P_2, and P are coplanar, the three vectors $X_1 - X_0$, $X_2 - X_0$, and $X - X_0$ are coplanar. Moreover, since $X_2 - X_0$ and $X_1 - X_0$ necessarily have distinct directions, $X - X_0$ may be written as the sum of appropriate multiples of these two vectors. That is, there exist scalars t_1 and t_2 such that

$$X - X_0 = t_1(X_1 - X_0) + t_2(X_2 - X_0)$$

or

(3.8.1) $$X = X_0 + t_1(X_1 - X_0) + t_2(X_2 - X_0).$$

This is the **parametric vector equation of a plane** on three given points.

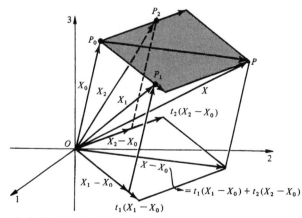

FIGURE 3-12. Plane on Three Given Points.

For example, consider the plane on the points $P_0:(0, 1, 1)$, $P_1:(1, 0, 1)$, and $P_2:(1, 1, 0)$. Substituting in (3.8.1), we find that its equation is

$$\begin{bmatrix} x_1 \\ x_2 \\ x_3 \end{bmatrix} = \begin{bmatrix} 0 \\ 1 \\ 1 \end{bmatrix} + t_1 \begin{bmatrix} 1 \\ -1 \\ 0 \end{bmatrix} + t_2 \begin{bmatrix} 1 \\ 0 \\ -1 \end{bmatrix}.$$

The equivalent scalar equations are

$$x_1 = t_1 + t_2$$
$$x_2 = 1 - t_1$$
$$x_3 = 1 - t_2.$$

Solving the second and third equations for t_1 and t_2, then substituting in the first and rearranging terms, we get the single, nonparametric equation of the plane:

$$x_1 + x_2 + x_3 = 2.$$

In the particular case when $X_0 = 0$, that is, when the origin is one of the three points determining the plane, (3.8.1) reduces to

(3.8.2) $$X = t_1 X_1 + t_2 X_2.$$

For example, the plane on $(0, 0, 0)$, $(1, 1, 0)$, and $(0, 1, 1)$ has the equation

$$\begin{bmatrix} x_1 \\ x_2 \\ x_3 \end{bmatrix} = t_1 \begin{bmatrix} 1 \\ 1 \\ 0 \end{bmatrix} + t_2 \begin{bmatrix} 0 \\ 1 \\ 1 \end{bmatrix}.$$

The equivalent scalar equations

$$x_1 = t_1$$
$$x_2 = t_1 + t_2$$
$$x_3 = \qquad t_2$$

are dependent and imply that

$$x_1 - x_2 + x_3 = 0.$$

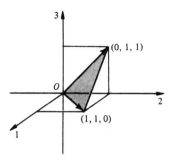

FIGURE 3-13. Plane on 0 Determined by Two Vectors.

If we consider this last equation as a system of equations, its complete solution is the vector equation of the plane from which the nonparametric equation was obtained. Thus the parametric and nonparametric representations of the plane are equivalent.

It is easy to sketch a triangular portion of the plane by sketching the vectors which determine it (Figure 3-13).

Now let X and Y be any two vectors of a plane with equation (3.8.2), where

$$X = t_1 X_1 + t_2 X_2, \qquad Y = s_1 X_1 + s_2 X_2.$$

Then

$$cX = (ct_1)X_1 + (ct_2)X_2$$

and

$$X + Y = (t_1 + s_1)X_1 + (t_2 + s_2)X_2,$$

so cX and $X + Y$ are also on the plane. Thus *the vectors of a plane on the origin constitute a vector space* (Section 3.7). We have now seen that the zero vector alone, the vectors on a line on the origin, the vectors on a plane on the origin, and $\mathscr{E}^3$ itself all constitute vector spaces. There are no other kinds of vector spaces in $\mathscr{E}^3$. Can you prove this?

We can now give a geometrical interpretation to the solution of a system of equations in three unknowns. The complete solution of $AX = 0$ always represents a vector space. It is the space consisting of the zero vector alone if the system has only the trivial solution, the vectors on a line on the origin if the solution has the form $X = t_1 X_1$, and the vectors on a plane on the origin if the solution has the form $X = t_1 X_1 + t_2 X_2$.

If $B \neq 0$, the solution of $AX = B$ never represents a vector space. It represents a single point if it has the form $X = X_0$, the solution is a line if

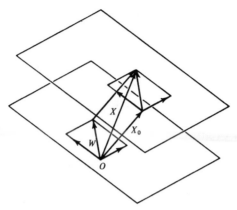

FIGURE 3-14. Translation of a Vector Space.

it has the form $X = X_0 + t_1 X_1$, and a plane if it has the form $X = X_0 + t_1 X_1 + t_2 X_2$. These spaces may be regarded as translations, by the addition of the vector X_0, of the vector spaces which are solutions of $AX = 0$ (Figure 3-14).

The single, nonparametric equation representing a plane in $\mathscr{E}^3$ may also be derived by vector considerations. Recall that there is a unique plane perpendicular to a given line L and containing a given point P_0. Also, a point P is on this plane if and only if the line L is perpendicular to the line on P and P_0. This is equivalent to the fact that P is on the plane if and only if a vector A having the same direction as L is orthogonal to the vector $X - X_0$ determined by $\overrightarrow{PP_0}$ (Figure 3-15). Hence for all points P of the plane and for no other points, we have

(3.8.3) $$A^{\mathsf{T}}(X - X_0) = 0.$$

If $A = [a_1, a_2, a_3]^{\mathsf{T}}$, this yields at once the familiar scalar form

(3.8.4) $$a_1(x_1 - c_1) + a_2(x_2 - c_2) + a_3(x_3 - c_3) = 0,$$

which may be rewritten in the form

(3.8.5) $$a_1 x_1 + a_2 x_2 + a_3 x_3 = b,$$

where

$$b = a_1 c_1 + a_2 c_2 + a_3 c_3.$$

The vector A is said to be **normal** to the plane.

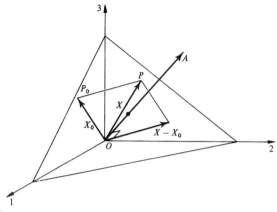

FIGURE 3-15. Plane Determined by a Point P_0 and a Vector A.

The vector equations (3.7.1), (3.8.1), and (3.8.3) are useful in the solution of problems about lines and planes. For example, in what point does the line on $(1, 1, 1)$ and $(0, -1, 2)$ intersect the plane on $(1, -1, 0)$ and with normal vector $[2, -3, 1]^T$? The line and the plane have respectively the equations

$$X = \begin{bmatrix} 1 \\ 1 \\ 1 \end{bmatrix} + t \begin{bmatrix} -1 \\ -2 \\ 1 \end{bmatrix}, \qquad [2, -3, 1]\left(X - \begin{bmatrix} 1 \\ -1 \\ 0 \end{bmatrix}\right) = 0.$$

Substituting from the first of these into the second, we get, for all points of the line that are also on the plane,

$$[2, -3, 1]\left(\begin{bmatrix} 0 \\ 2 \\ 1 \end{bmatrix} + t \begin{bmatrix} -1 \\ -2 \\ 1 \end{bmatrix}\right) = 0,$$

from which

$$-5 + 5t = 0, \qquad t = 1,$$

so the required point is $(0, -1, 2)$.

3.9 Exercises

1. Find parametric scalar equations for the line on
   (a) $P_1:(-1, 4, 3)$ and $P_2:(2, 0, 3)$,
   (b) $Q_1:(-4, 4, 3)$ and $Q_2:(2, 4, 3)$.
   What is special about each of these lines? Illustrate with figures.

2. Prove that the triangle with vertices P_1:$(-1, 4, 3)$, P_2:$(2, 0, 3)$, and P_3:$(-1, 4, 5)$ is a right triangle.

3. (a) Find the point where the line on P_1:$(1, 1, 1)$ and P_2:$(-1, 0, 3)$ meets the plane $2x_1 - 3x_2 + 7x_3 + 5 = 0$.

   (b) Show that the line on Q_1:$(2, -1, 4)$ and Q_2:$(-3, 5, 0)$ does not intersect, that is, is parallel to, the plane

$$2x_1 + x_2 - x_3 = 8.$$

4. For what value of x_3 will the line on $(2, 1, 4)$ and $(3, -1, 2)$ be parallel to the line on $(4, -2, 1)$ and $(2, 2, x_3)$?

5. Let the *unit* vector

$$A = \begin{bmatrix} a_1 \\ a_2 \\ a_3 \end{bmatrix}$$

define the direction of a line and let P_1:(b_1, b_2, b_3) be a point on it. Show that parametric equations for the line are

$$z_1 = b_1 + sa_1$$
$$z_2 = b_2 + sa_2$$
$$z_3 = b_3 + sa_3.$$

Then show that the parameter s may be used as the directed distance from point P_1 to point P:(z_1, z_2, z_3).

6. The two planes

$$x_1 + 2x_2 - x_3 + 4 = 0$$
$$2x_1 - 3x_2 + x_3 - 3 = 0$$

intersect in a line. Find parametric scalar equations for it.

7. (a) Find the equation of the plane on the three points $(1, 1, 2)$, $(1, 2, 1)$, and $(2, 1, 1)$. Illustrate with a figure.

   (b) Find the equation of the plane parallel to the x_3-axis and on the two points $(2, 0, 0)$ and $(0, 2, 0)$. (*Hint*: What are the direction numbers for a normal to the plane?)

*8. Prove in detail that the equation of the plane through the points $(c_1, 0, 0)$, $(0, c_2, 0)$, and $(0, 0, c_3)$, where $c_1 c_2 c_3 \neq 0$, can be written in the form

$$\frac{x_1}{c_1} + \frac{x_2}{c_2} + \frac{x_3}{c_3} = 1.$$

This is the **intercept form** of the equation of the plane.

9. Reduce the following equations of planes to intercept form. Then sketch the planes. In each case, find the point where the normal from the origin to the plane meets the plane. Show the normal on your sketch.

$$\text{(a)} \quad 3x_1 - 2x_2 + 6x_3 = 12,$$

$$\text{(b)} \quad 12x_1 + 3x_2 + 5x_3 = 4.$$

10. Interpret geometrically the complete solution of this system of equations and illustrate with a figure:

$$x_1 + x_2 + x_3 = 2$$

$$3x_1 + 3x_2 - 2x_3 = 6$$

$$x_1 + x_2 - 4x_3 = 2.$$

11. What is the geometrical interpretation of the system of equations

$$x_1 + \alpha x_2 + \alpha x_3 = 0$$

$$x_1 + x_2 - x_3 = 0$$

and of its complete solution? Illustrate with a figure.

3.10   Linear Combinations of Vectors in $\mathscr{E}^3$

Let $V_1, V_2, \ldots, V_k$ be vectors in $\mathscr{E}^3$. An expression of the form

$$t_1 V_1 + t_2 V_2 + \cdots + t_k V_k$$

is called a **linear combination** of the V's.

Consider the set of all linear combinations W of two fixed vectors V_1 and V_2 with distinct directions:

$$(3.10.1) \qquad\qquad W = t_1 V_1 + t_2 V_2.$$

The parallelogram law of addition shows that for all values of t_1 and t_2, the vector W is in the same plane with V_1 and V_2. Whenever two or more vectors lie in the same plane, we call them **coplanar**. Thus all linear combinations of two given vectors are coplanar.

On the other hand, if noncollinear vectors V_1 and V_2, and any vector W in the same plane are given, then we can use W as the diagonal of a parallelogram which is completed by drawing parallels to V_1 and V_2, through the terminal point of W, to intersect the lines carrying V_1 and V_2 respectively

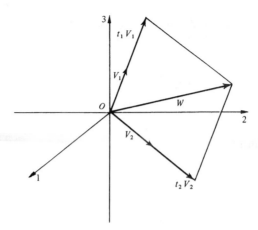

FIGURE 3-16. Coplanar Vectors.

(Figure 3-16). This determines scalar multiples $t_1 V_1$ and $t_2 V_2$ such that $W = t_1 V_1 + t_2 V_2$. That is, we have

Theorem 3.10.1: *The vector W is coplanar with noncollinear vectors V_1 and V_2 if and only if W can be written as a linear combination of V_1 and V_2.*

The t's are readily determined algebraically when V_1, V_2, and W are given. For example, if

$$V_1 = \begin{bmatrix} 1 \\ 2 \\ -1 \end{bmatrix}, \qquad V_2 = \begin{bmatrix} 3 \\ 0 \\ 2 \end{bmatrix}, \qquad W = \begin{bmatrix} 5 \\ 4 \\ 0 \end{bmatrix}$$

we have to solve

$$\begin{bmatrix} 5 \\ 4 \\ 0 \end{bmatrix} = t_1 \begin{bmatrix} 1 \\ 2 \\ -1 \end{bmatrix} + t_2 \begin{bmatrix} 3 \\ 0 \\ 2 \end{bmatrix}$$

for the t's. This equation is equivalent to the system

$$t_1 + 3t_2 = 5$$
$$2t_1 = 4$$
$$-t_1 + 2t_2 = 0.$$

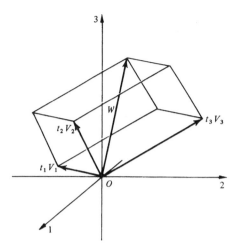

FIGURE 3-17. Linear Combination of Three Vectors.

Although there are more equations here than unknowns, the system is consistent and has the solution $t_1 = 2$, $t_2 = 1$, so

$$W = 2V_1 + V_2 .$$

If we do not know whether or not W is coplanar with V_1 and V_2, we simply attempt to solve equation (3.10.1) for the t's. If the corresponding system is consistent, W is coplanar with V_1 and V_2. If it is inconsistent, then W cannot be written as a linear combination of V_1 and V_2 and hence is not coplanar with them.

Consider now any linear combination

(3.10.2) $$W = t_1 V_1 + t_2 V_2 + t_3 V_3$$

where the V's are not coplanar. Then W is the diagonal of the parallelepiped determined by $t_1 V_1$, $t_2 V_2$, and $t_3 V_3$ (Figure 3-17).

On the other hand, if W, V_1, V_2, and V_3 are given (V_1, V_2, V_3 not coplanar), planes through the terminal point of W and parallel to those determined by pairs of V_1, V_2, and V_3 determine a parallelepiped and hence scalar multiples of the V's such that (3.10.2) holds. That is, we have

Theorem 3.10.2: *Every vector W in $\mathscr{E}^3$ is a linear combination of any three noncoplanar vectors.*

Again, for given V_1, V_2, V_3, W with V_1, V_2, and V_3 not coplanar, we can determine t_1, t_2, and t_3 algebraically. For example, if

$$V_1 = \begin{bmatrix} 1 \\ 0 \\ 0 \end{bmatrix}, \qquad V_2 = \begin{bmatrix} 1 \\ 1 \\ 0 \end{bmatrix}, \qquad V_3 = \begin{bmatrix} 1 \\ 1 \\ 1 \end{bmatrix},$$

then for any W, the equation (3.10.2) is equivalent to the system of equations

$$t_1 + t_2 + t_3 = w_1$$
$$t_2 + t_3 = w_2$$
$$t_3 = w_3,$$

which has the solution

$$t_1 = w_1 - w_2$$
$$t_2 = \quad w_2 - w_3$$
$$t_3 = \quad\quad\quad w_3.$$

That is, *any vector W in $\mathscr{E}^3$* can be expressed as a linear combination of the three given vectors, thus:

$$W = (w_1 - w_2)\begin{bmatrix} 1 \\ 0 \\ 0 \end{bmatrix} + (w_2 - w_3)\begin{bmatrix} 1 \\ 1 \\ 0 \end{bmatrix} + w_3\begin{bmatrix} 1 \\ 1 \\ 1 \end{bmatrix}.$$

3.11   Linear Dependence of Vectors; Bases

The vectors $V_1, V_2, \ldots, V_k$ are said to be **linearly dependent** if and only if there exist scalars $t_1, t_2, \ldots, t_k$, not all of which are zero, such that

(3.11.1) $$\qquad\qquad t_1 V_1 + t_2 V_2 + \cdots + t_k V_k = 0.$$

For example,

$$0\begin{bmatrix} 1 \\ -1 \\ 3 \end{bmatrix} + 2\begin{bmatrix} 1 \\ 2 \\ -1 \end{bmatrix} + (-1)\begin{bmatrix} 3 \\ -2 \\ 2 \end{bmatrix} + (-1)\begin{bmatrix} -1 \\ 6 \\ -4 \end{bmatrix} = \begin{bmatrix} 0 \\ 0 \\ 0 \end{bmatrix},$$

so the four vectors on the left are linearly dependent. The example illustrates the fact that the definition does *not* say that none of the t's may be zero.

If the equation (3.11.1) holds only when all the t's are zero, then we say that $V_1, V_2, \ldots, V_k$ are **linearly independent**.

For example, since

$$t_1 \begin{bmatrix} 1 \\ 0 \\ 0 \end{bmatrix} + t_2 \begin{bmatrix} 1 \\ 1 \\ 0 \end{bmatrix} + t_3 \begin{bmatrix} 1 \\ 1 \\ 1 \end{bmatrix} = \begin{bmatrix} 0 \\ 0 \\ 0 \end{bmatrix}$$

only if $t_1 = t_2 = t_3 = 0$, the three vectors $\begin{bmatrix} 1 \\ 0 \\ 0 \end{bmatrix}$, $\begin{bmatrix} 1 \\ 1 \\ 0 \end{bmatrix}$, and $\begin{bmatrix} 1 \\ 1 \\ 1 \end{bmatrix}$ are linearly independent.

Suppose now that the vectors V_1 and V_2 are linearly dependent. Then there exist t_1 and t_2, not both zero, such that

$$t_1 V_1 + t_2 V_2 = 0.$$

If $t_1 \neq 0$, then

$$V_1 = \left(-\frac{t_2}{t_1} \right) V_2,$$

so V_1 is a scalar multiple of V_2; that is, V_1 and V_2 are collinear. Similarly if $t_2 \neq 0$. Conversely, collinear (proportional) vectors are linearly dependent, for if

$$V_1 = kV_2 \qquad \text{or} \qquad V_2 = kV_1,$$

then

$$1 \cdot V_1 + (-k)V_2 = 0 \qquad \text{or} \qquad (-k)V_1 + 1 \cdot V_2 = 0$$

and at least one coefficient is not zero. Thus we have

Theorem 3.11.1: *Two vectors in $\mathscr{E}^3$ are linearly dependent if and only if they are collinear.*

If V_1, V_2, and V_3 are linearly dependent, then the equation

$$t_1 V_1 + t_2 V_2 + t_3 V_3 = 0$$

with not all the t's zero permits us to write at least one of the V's as a linear combination of the other two. For example, if $t_1 \neq 0$,

$$V_1 = \left(-\frac{t_2}{t_1} \right) V_2 + \left(-\frac{t_3}{t_1} \right) V_3,$$

so V_1 is coplanar with V_2 and V_3. Similarly if $t_2 \neq 0$ or $t_3 \neq 0$. Conversely, if V_1, V_2, and V_3 are coplanar, then an equation

$$V_i = \alpha_2 V_j + \alpha_3 V_k$$

holds for at least one permutation i, j, k of 1, 2, 3, so

$$1 \cdot V_i + (-\alpha_2)V_j + (-\alpha_3)V_k = 0$$

and the three vectors are linearly dependent. This proves

Theorem 3.11.2: *Three vectors in $\mathscr{E}^3$ are linearly dependent if and only if they are coplanar.*

Finally, consider four vectors V_1, V_2, V_3, and V_4 in $\mathscr{E}^3$. The equation

$$t_1 V_1 + t_2 V_2 + t_3 V_3 + t_4 V_4 = 0$$

is equivalent to three scalar equations in four unknowns:

$$v_{11}t_1 + v_{12}t_2 + v_{13}t_3 + v_{14}t_4 = 0$$
$$v_{21}t_1 + v_{22}t_2 + v_{23}t_3 + v_{24}t_4 = 0$$
$$v_{31}t_1 + v_{32}t_2 + v_{33}t_3 + v_{34}t_4 = 0,$$

where

$$V_j = \begin{bmatrix} v_{1j} \\ v_{2j} \\ v_{3j} \end{bmatrix}, \qquad j = 1, 2, 3.$$

By transforming this system to reduced echelon form, we can express one, two, or three of the variables, as the case may be, in terms of the remaining variables, of which there is at least one. Hence we can assign a nonzero value to at least one variable, so nontrivial solutions for the t's certainly exist. That is, we have

Theorem 3.11.3: *Any four vectors in $\mathscr{E}^3$ are necessarily linearly dependent.*

An important special case is this: Let V_1, V_2, and V_3 be any three linearly independent (noncoplanar) vectors in $\mathscr{E}^3$, and let W be any vector in $\mathscr{E}^3$. Then these four vectors are linearly dependent by the preceding theorem. Hence there exist scalars $t, t_1, t_2,$ and t_3 not all zero such that

$$tW + t_1 V_1 + t_2 V_2 + t_3 V_3 = 0.$$

In this equation, we must have $t \neq 0$, since otherwise V_1, V_2, and V_3 would be linearly dependent. (Why?) Hence we can solve for W:

$$W = \left(-\frac{t_1}{t}\right)V_1 + \left(-\frac{t_2}{t}\right)V_2 + \left(-\frac{t_3}{t}\right)V_3.$$

Thus we have proved

Theorem 3.11.4: *Every vector of $\mathscr{E}^3$ may be expressed as a linear combination of any three linearly independent (noncoplanar) vectors of $\mathscr{E}^3$.*

A set of three noncoplanar vectors, in terms of which every vector of $\mathscr{E}^3$ can necessarily be expressed as a linear combination, is called a **basis** for $\mathscr{E}^3$. The most obvious basis is the **standard basis** consisting of the set of elementary vectors,

$$E_1 = \begin{bmatrix} 1 \\ 0 \\ 0 \end{bmatrix}, \qquad E_2 = \begin{bmatrix} 0 \\ 1 \\ 0 \end{bmatrix}, \qquad E_3 = \begin{bmatrix} 0 \\ 0 \\ 1 \end{bmatrix},$$

for any vector X can be written thus:

$$X = x_1 E_1 + x_2 E_2 + x_3 E_3.$$

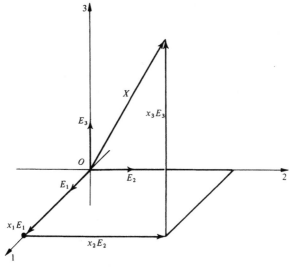

FIGURE 3-18. Standard Basis for $\mathscr{E}^3$.

Since in each case $E_i^T E_j = \delta_{ij}$, these three vectors are mutually orthogonal unit vectors. Any basis consisting of three mutually orthogonal unit vectors is called an **orthonormal basis** for $\mathscr{E}^3$. (The vectors E_1, E_2, E_3 are the same as the unit vectors *i*, *j*, *k* of vector analysis.) These ideas will be generalized in following chapters.

3.12   Exercises

1. (a) Express the vector

$$\begin{bmatrix} a_1 \\ a_2 \\ a_3 \end{bmatrix}$$

as a linear combination of the vectors

$$\begin{bmatrix} 0 \\ 1 \\ 1 \end{bmatrix}, \quad \begin{bmatrix} 1 \\ 0 \\ 1 \end{bmatrix}, \quad \begin{bmatrix} 1 \\ 1 \\ 0 \end{bmatrix}.$$

(b) Show that the vectors

$$\begin{bmatrix} 0 \\ 2 \\ -3 \end{bmatrix}, \quad \begin{bmatrix} 1 \\ 4 \\ 0 \end{bmatrix}, \quad \begin{bmatrix} 3 \\ 8 \\ 6 \end{bmatrix}$$

are linearly dependent.

*2. Show that if V_1, V_2, ..., V_k are vectors in $\mathscr{E}^3$ with $k \geq 4$, then the V's are necessarily linearly dependent.

3. Show that the three vectors E_1, $E_1 + E_2$, $E_1 + E_2 + E_3$ are linearly independent.

4. Are $E_1 - E_2$, $E_2 - E_3$, and $E_3 - E_1$ linearly independent? Why?

5. Given that X_1, X_2, and X_3 are linearly independent vectors in $\mathscr{E}^3$, prove that X_1, $X_1 + X_2$, and $X_1 + X_2 + X_3$ are linearly independent also.

*6. Show that the zero vector and any other vector V are linearly dependent.

*7. Show that if the vectors V_1, V_2, and V_3 are mutually orthogonal, then they are linearly independent.

8. Show that the vectors

$$\begin{bmatrix} 0 \\ \dfrac{1}{\sqrt{2}} \\ \dfrac{1}{\sqrt{2}} \end{bmatrix}, \quad \begin{bmatrix} 1 \\ 0 \\ 0 \end{bmatrix}, \quad \begin{bmatrix} 0 \\ \dfrac{-1}{\sqrt{2}} \\ \dfrac{1}{\sqrt{2}} \end{bmatrix},$$

constitute an orthonormal basis for $\mathscr{E}^3$.

9. Prove that

$$\begin{bmatrix} 1 \\ 2 \\ 2 \end{bmatrix}, \quad \begin{bmatrix} 0 \\ 1 \\ 2 \end{bmatrix}, \quad \begin{bmatrix} 0 \\ 0 \\ 3 \end{bmatrix}$$

constitute a basis for $\mathscr{E}^3$ by showing that the equation

$$\begin{bmatrix} w_1 \\ w_2 \\ w_3 \end{bmatrix} = t_1 \begin{bmatrix} 1 \\ 2 \\ 2 \end{bmatrix} + t_2 \begin{bmatrix} 0 \\ 1 \\ 2 \end{bmatrix} + t_3 \begin{bmatrix} 0 \\ 0 \\ 3 \end{bmatrix}$$

can be solved for the t's regardless of the values given to w_1, w_2, and w_3.

10. Find a linear combination of

$$\begin{bmatrix} 0 \\ 1 \\ 1 \end{bmatrix} \quad \text{and} \quad \begin{bmatrix} 1 \\ 1 \\ 0 \end{bmatrix}$$

which is orthogonal to

$$\begin{bmatrix} 0 \\ 1 \\ 1 \end{bmatrix}.$$

Illustrate with a figure.

*11. Prove that the set of all linear combinations of fixed vectors V_1, V_2, ..., V_k in $\mathscr{E}^3$ is a vector space. Discuss the various special cases.

*12. Show that the set of all vectors X in $\mathscr{E}^3$ which satisfy an equation $AX = 0$, where A is an $m \times 3$ matrix, is a vector space. Discuss geometrically.

*13. Prove that the set of all vectors X in $\mathscr{E}^3$ such that $AX = \lambda X$, where λ is a fixed scalar, is a vector space.

*14. Prove that the set of all vectors X in $\mathscr{E}^3$ such that $X^{\mathsf{T}}A = 0$, where A is a fixed vector in $\mathscr{E}^3$, is a vector space. Represent this space by an appropriate figure.

15. Given that U_1 and U_2 are orthogonal unit vectors in $\mathscr{E}^3$, show how to find U_3 such that U_1, U_2, and U_3 are an orthonormal set.

16. Perform the computation of Exercise 15 if

$$U_1 = \left[0, \frac{1}{\sqrt{2}}, -\frac{1}{\sqrt{2}}\right]^{\mathsf{T}}, \qquad U_2 = \left[\frac{1}{\sqrt{3}}, \frac{1}{\sqrt{3}}, \frac{1}{\sqrt{3}}\right]^{\mathsf{T}}.$$

CHAPTER
4

Vector Geometry in
n-Dimensional Space

4.1 The Real n-Space $\mathscr{R}^n$

The geometry of spaces of more than three dimensions is a purely mathematical creation since there are no corresponding, simple physical spaces to serve as guides to intuition. One way to develop such a geometry is to generalize, in the most natural way, various concepts and results of the analytic geometry of two and three dimensions. The reason for doing this is that many useful algebraic formulas and processes involving n variables are precisely analogous to formulas and processes involving two or three variables with which we have associated familiar geometric names. The development of n-dimensional geometry involves the association of a set of geometric names, similar to those used in two and three dimensions, with analogous arithmetic and algebraic objects and processes involving n variables. What happens in two and three dimensions helps us to know what to expect algebraically in higher-dimensional spaces, but we do not try to visualize, for example, four or more mutually perpendicular axes. When $n > 3$, "n-space," as we develop it, is nothing more or less than an arithmetic and algebraic construction. A synthetic development analogous to that of classical Euclidean geometry is also possible but is much less useful and is much more difficult.

We first define **real, arithmetic** n-**dimensional space**, $\mathscr{R}^n$, to be the set of all ordered n-tuples $(a_1, a_2, \ldots, a_n)$ of real numbers. An n-tuple $(a_1, a_2, \ldots, a_n)$ is called a **point** of $\mathscr{R}^n$. The numbers $a_1, a_2, \ldots, a_n$ are called the **coordinates** of the point. The point $O:(0, 0, \ldots, 0)$ is called the **origin** of $\mathscr{R}^n$. The set of all points $(0, \ldots, 0, a_i, 0, \ldots, 0)$ whose ith coordinate is an arbitrary real number is called the i-**axis of** $\mathscr{R}^n$.

A little later we shall introduce the concept of distance and then define what we mean by Euclidean n-space, $\mathscr{E}^n$, but first we examine some of the linear geometry that is independent of the definition of distance.

4.2 Vectors in $\mathscr{R}^n$

Let $P_0 : (a_1, a_2, \ldots, a_n)$ and $P_1 : (b_1, b_2, \ldots, b_n)$ be two points of $\mathscr{R}^n$. Then, as in $\mathscr{E}^3$, the set of points given by

$$(4.2.1) \quad (a_1 + t(b_1 - a_1), a_2 + t(b_2 - a_2), \ldots, a_n + t(b_n - a_n)), \qquad 0 \leq t \leq 1,$$

is called the **line segment** P_0P_1. [Recall (3.7.2) and the paragraph preceding those equations.] If we give to the points of this segment the order determined by increasing t continuously from 0 to 1 inclusive, we call the resulting ordered set the **directed line segment** $\overrightarrow{P_0P_1}$ of which P_0 is the **initial point** and P_1 is the **terminal point**.

With each real n-vector $X = [x_1, x_2, \ldots, x_n]^\mathsf{T}$, we associate a directed line segment $\overrightarrow{OP}$, where O is the origin and P is the point $(x_1, x_2, \ldots, x_n)$. As in $\mathscr{E}^3$, this directed segment will be called a **geometric vector**. In this chapter, we shall not distinguish verbally between X and $\overrightarrow{OP}$ and for brevity we shall call P "the terminal point of X." Since there is a one-to-one correspondence between points $P : (x_1, x_2, \ldots, x_n)$ and geometric vectors $\overrightarrow{OP}$, $\mathscr{R}^n$ may be considered to be either a space of points or a space of vectors.

In $\mathscr{E}^3$ there are familiar, purely geometric rules for adding vectors (the parallelogram law) and for multiplying them by scalars (with the aid of similar triangles). We saw in Chapter 3 that by describing geometric vectors by column matrices, we could replace the geometric operations by corresponding algebraic operations on the matrices representing the vectors. By contrast, in $\mathscr{R}^n$ there are no purely geometric rules for these same operations. Hence, in $\mathscr{R}^n$, we define the addition of vectors and the multiplication of vectors by scalars a priori in terms of these operations on the column matrices which represent the vectors. As a result, these operations have the same algebraic properties as do the corresponding operations on vectors in $\mathscr{E}^3$ or on matrices in general.

With each directed segment $\overrightarrow{P_0P_1}$ as defined above, we associate the n-vector of **ordered differences**

$$X_1 - X_0 = \begin{bmatrix} b_1 - a_1 \\ b_2 - a_2 \\ \vdots \\ b_n - a_n \end{bmatrix},$$

where X_1 and X_0 are the vectors with terminal points P_1 and P_0, respectively. Thus $\overrightarrow{P_0P_1}$ determines the geometric vector $\overrightarrow{OP}$, where P is the point $(b_1 - a_1, b_2 - a_2, \ldots, b_n - a_n)$. (Recall here Figure 3-6.)

Two directed segments are defined to be **equal** if and only if their associated n-vectors of ordered differences are equal. For example, the ordered pairs of points in $\mathscr{R}^4$, $P_0 : (2, 1, -3, 0)$ and $P_1 : (1, -1, 2, 1)$, $Q_0 : (1, 3, -1, 2)$ and $Q_1 : (0, 1, 4, 3)$, determine the equal directed segments $\overrightarrow{P_0P_1}$, $\overrightarrow{Q_0Q_1}$ and the associated vector

$$X = \begin{bmatrix} -1 \\ -2 \\ 5 \\ 1 \end{bmatrix}.$$

Recall here Figure 3-8.

Each given n-vector X in turn determines a class of infinitely many directed segments whose columns of ordered differences all equal X. Given the initial point of such a segment, the terminal point is easily found since the ordered differences are known. Thus, for the vector X in our example, if $(1, 1, 1, 1)$ is the initial point of one such segment, the terminal point must be $(0, -1, 6, 2)$.

4.3   Lines and Planes in $\mathscr{R}^n$

Given two distinct points $P_0 : (x_{10}, x_{20}, \ldots, x_{n0})$ and $P_1 : (x_{11}, x_{21}, \ldots, x_{n1})$, the **line** $P_0 P_1$ in $\mathscr{R}^n$ is defined to be the set of all points $P : (x_1, x_2, \ldots, x_n)$ determined by

(4.3.1) $$X = X_0 + t(X_1 - X_0) \qquad (t \text{ arbitrary}),$$

where the vectors X, X_0, and X_1 have terminal points P, P_0, and P_1 respectively. [This is the extension to $\mathscr{R}^n$ of the equation (3.7.1) of a line in 3-space.] It can be shown that the same set of points is determined by any two distinct points of the line. The reader should do this.

Note that, as in $\mathscr{E}^3$, P_1 and P_2 are on the same line on O if and only if all points of the directed segments $\overrightarrow{OP_1}$ and $\overrightarrow{OP_2}$ are on such a line. (Prove this.)

Three points are defined to be **collinear** if and only if they are not all distinct or two of them are distinct and the vector X corresponding to the third satisfies the equation (4.3.1) determined by the other two for some value of t. (Compare with Figure 3-11.)

For example, are the points $(1, 1, 2, -1)$, $(3, 0, 1, 4)$, and $(-5, 4, 5, -16)$ collinear? The problem is, in effect, to find a value for t, if one exists, such that

$$
\begin{aligned}
-5 &= 1 + (3 - 1)t & = 1 + 2t \\
4 &= 1 + (0 - 1)t & = 1 - t \\
5 &= 2 + (1 - 2)t & = 2 - t \\
-16 &= -1 + [4 - (-1)]t & = -1 + 5t.
\end{aligned}
$$

The solution of the first equation is $t = -3$ and this also satisfies the other three, so the three points are indeed collinear. This problem provides a good illustration of the usefulness of systems of linear equations having more equations than unknowns.

Given three noncollinear points P_0, P_1, and P_2, where $P_i = (x_{1i}, x_{2i}, \ldots, x_{ni})$, $i = 0, 1, 2$, the **plane** on P_0, P_1, P_2 is defined to be the set of all terminal points $(x_1, x_2, \ldots, x_n)$ of vectors X given by

$$(4.3.2) \qquad X = X_0 + t_1(X_1 - X_0) + t_2(X_2 - X_0),$$

where P_i is the terminal point of X_i, $i = 0, 1, 2$. (Compare Figure 3-12.) It can be shown that the same set of points is determined by any three noncollinear points of the plane.

For example, in $\mathscr{R}^4$, the plane on the three noncollinear points $(1, 1, 1, 1)$, $(1, 1, 0, 0)$, and $(0, 0, 2, 2)$ has the vector equation

$$(4.3.3) \qquad X = \begin{bmatrix} 1 \\ 1 \\ 1 \\ 1 \end{bmatrix} + t_1 \begin{bmatrix} 0 \\ 0 \\ -1 \\ -1 \end{bmatrix} + t_2 \begin{bmatrix} -1 \\ -1 \\ 1 \\ 1 \end{bmatrix}.$$

These ideas are readily generalized. A line in $\mathscr{R}^n$ is called a **1-flat** ; a plane in $\mathscr{R}^n$ is called a **2-flat**. If we assume that a $(k - 1)$-flat has been defined, then a **k-flat** is the set of terminal points P of the vectors X defined by

$$(4.3.4) \quad X = X_0 + t_1(X_1 - X_0) + t_2(X_2 - X_0) + \cdots + t_k(X_k - X_0),$$

where $X_0, X_1, \ldots, X_k$ correspond to $k + 1$ points $P_0, P_1, \ldots, P_k$ which are not all in the same $(k - 1)$-flat. (As in the case of lines and planes, the k-flat is determined by *any* $k + 1$ of it points, provided that they are not in the same $(k - 1)$-flat. This we can prove rigorously a little later.) The maximum value of k is n, in which case the n-flat is simply $\mathscr{R}^n$ itself, as our study of linear dependence in Chapter 5 will reveal.

A particularly important case for applications is the case $k = n - 1$. An $(n - 1)$-**flat**, or **hyperplane** as it is usually called, may also be defined by a single linear equation (as may a line in $\mathscr{E}^2$ and a plane in $\mathscr{E}^3$):

(4.3.5) $$a_1 x_1 + a_2 x_2 + \cdots + a_n x_n = b.$$

We shall be able to prove, using the results of Chapter 5, the equivalence of the representations (4.3.4) and (4.3.5) of a hyperplane.

In $\mathscr{E}^2$, a line ordinarily meets another line in a point, and in $\mathscr{E}^3$ a line ordinarily meets a plane in a point. Similarly, in $\mathscr{R}^n$ a line ordinarily meets a hyperplane in a point. To see this, rewrite (4.3.5) in matrix form:

(4.3.6) $$A^{\mathbf{T}} X = b$$

and substitute from (4.3.1) into (4.3.6):

$$A^{\mathbf{T}}(X_0 + t(X_1 - X_0)) = b,$$

so

(4.3.7) $$t A^{\mathbf{T}}(X_1 - X_0) = b - A^{\mathbf{T}} X_0$$

and hence, when $A^{\mathbf{T}}(X_1 - X_0) \neq 0$,

(4.3.8) $$t = \frac{b - A^{\mathbf{T}} X_0}{A^{\mathbf{T}}(X_1 - X_0)}.$$

Since there is, when $A^{\mathbf{T}}(X_1 - X_0) \neq 0$, a unique solution t, in this case the line meets the hyperplane in precisely one point. If $A^{\mathbf{T}}(X_1 - X_0) = 0$ and $b - A^{\mathbf{T}} X_0 \neq 0$, equation (4.3.7) is inconsistent, so no point of the line lies on the hyperplane. In this case the line and the hyperplane are said to be **parallel**. If $A^{\mathbf{T}}(X_1 - X_0) = 0$ and $b - A^{\mathbf{T}} X_0 = 0$, equation (4.3.7) is satisfied by all values of t, so the entire line is on the hyperplane.

For example, in $\mathscr{R}^4$, the line on $P_0 : (1, 0, 0, 0)$ and $P_1 : (0, 1, 0, 0)$ and the hyperplane

$$x_1 + x_2 + x_3 + x_4 = 4$$

may be represented respectively by the matrix equations

$$X = \begin{bmatrix} 1 \\ 0 \\ 0 \\ 0 \end{bmatrix} + t \begin{bmatrix} -1 \\ 1 \\ 0 \\ 0 \end{bmatrix} \quad \text{and} \quad [1, 1, 1, 1] X = 4.$$

From these, by substitution from the first equation into the second, we obtain

$$[1, 1, 1, 1]\left(\begin{bmatrix} 1 \\ 0 \\ 0 \\ 0 \end{bmatrix} + t \begin{bmatrix} -1 \\ 1 \\ 0 \\ 0 \end{bmatrix}\right) = 4$$

or

$$1 + 0 \cdot t = 4 \qquad \text{(inconsistent)}.$$

Since no point of the line $P_0 P_1$ is on the given hyperplane, the line is parallel to it.

On the other hand, the line on $Q_0:(1, 1, 1, 1)$ and $Q_1:(0, 2, 0, 2)$ has the equation

$$X = \begin{bmatrix} 1 \\ 1 \\ 1 \\ 1 \end{bmatrix} + t \begin{bmatrix} -1 \\ 1 \\ -1 \\ 1 \end{bmatrix}.$$

Its intersection with the same hyperplane is defined by the equation

$$4 + 0 \cdot t = 4,$$

which is satisfied by all values of t, so the entire line is on the given hyperplane. This is what one would expect since Q_0 and Q_1 are both on the hyperplane.

4.4   Linear Dependence and Independence in $\mathscr{R}^n$

Let $X_1, X_2, \ldots, X_k$ be vectors in $\mathscr{R}^n$. The expression $t_1 X_1 + t_2 X_2 + \cdots + t_k X_k$, where the t's are scalars, is called a **linear combination** of X_1, $X_2, \ldots, X_k$. If, for given vectors $X_1, X_2, \ldots, X_k$, there exist scalars t_1, $t_2, \ldots, t_k$, *not all zero*, such that

$$t_1 X_1 + t_2 X_2 + \cdots + t_k X_k = 0,$$

then we say that $X_1, X_2, \ldots, X_k$ are **linearly dependent**. If this equation holds true only when all the t's are zero, we say that $X_1, X_2, \ldots, X_k$ are **linearly independent**.

As in $\mathscr{E}^3$, linear dependence in $\mathscr{R}^n$ has a geometrical interpretation. For example, two vectors X_1 and X_2 in $\mathscr{R}^n$ are said to be **collinear vectors** if and only if their terminal points P_1 and P_2 lie on the same line on the origin.

The zero vector and any vector X are collinear. They are also linearly dependent since $1 \cdot 0 + 0 \cdot X = 0$. If $X_1 \neq 0$ and $X_2 \neq 0$ are collinear and if $X = tA$ is the equation of the corresponding line on the origin, there exist scalars $t_1 \neq 0$ and $t_2 \neq 0$ such that $X_1 = t_1 A$ and $X_2 = t_2 A$. Then $t_2 X_1 + (-t_1)X_2 = t_2 t_1 A - t_1 t_2 A = 0$, so that X_1 and X_2 are linearly dependent.

Suppose now that X_1 and X_2 are linearly dependent. If either is 0, the vectors are certainly collinear. If neither is 0 then there exist t_1 and t_2, neither zero, such that $t_1 X_1 + t_2 X_2 = 0$. Then $X_2 = -(t_1/t_2)X_1$, so the terminal point P_2 of X_2 is on the line on the origin determined by P_1. Thus we have proved

Theorem 4.4.1: *Two vectors in $\mathscr{R}^n$ are collinear if and only if they are linearly dependent.*

In general, suppose that $X_1, X_2, \ldots, X_k$ are linearly dependent. Then if $t_j \neq 0$ in

$$t_1 X_1 + t_2 X_2 + \cdots + t_j X_j + \cdots + t_k X_k = 0,$$

we can write

$$X_j = 0 + \left(-\frac{t_1}{t_j}\right)(X_1 - 0) + \cdots + \left(-\frac{t_{j-1}}{t_j}\right)(X_{j-1} - 0)$$

$$+ \left(-\frac{t_{j+1}}{t_j}\right)(X_{j+1} - 0) + \cdots + \left(-\frac{t_k}{t_j}\right)(X_k - 0),$$

so that all the vectors $X_1, X_2, \ldots, X_k$ belong to the same $(k-1)$-flat on the origin, or to some lower-dimensional flat space on O if $X_1, \ldots, X_{j-1}, X_{j+1}, \ldots, X_k$ are linearly dependent. The problem of dimension and its computation will be treated in detail in Chapter 5.

4.5   Vector Spaces in $\mathscr{R}^n$

A set $\mathscr{V}$ of vectors in $\mathscr{R}^n$ is called a **vector space** if and only if, whenever X and Y belong to $\mathscr{V}$, $X + Y$ and tX also belong to $\mathscr{V}$, that is, if and only if $\mathscr{V}$ is **closed** with respect to addition and with respect to multiplication by a scalar. The simplest vector spaces are the set consisting of the zero vector alone and the set consisting of all vectors in $\mathscr{R}^n$.

The set of all linear combinations of k given vectors $X_1, X_2, \ldots, X_k$ in $\mathscr{R}^n$ is a basic type of vector space. Indeed, if two such linear combinations are

$$X = t_1 X_1 + t_2 X_2 + \cdots + t_k X_k$$

and

$$Y = s_1 X_1 + s_2 X_2 + \cdots + s_k X_k,$$

then $X + Y$ and tX are also such linear combinations since

$$X + Y = (t_1 + s_1)X_1 + (t_2 + s_2)X_2 + \cdots + (t_k + s_k)X_k$$

and

$$tX = (tt_1)X_1 + (tt_2)X_2 + \cdots + (tt_k)X_k,$$

so the required closure properties hold.

In the case $k = 1$, if $X_1 \neq 0$, such a vector space is the set of all vectors X given by

$$X = tX_1 = 0 + t(X_1 - 0).$$

Hence this vector space is represented geometrically by a line on the origin. As in $\mathscr{E}^3$, the vector X corresponding to any point P on this line belongs to this vector space and every vector X of the space has its terminal point P on this line.

More generally, the vector space which is the set of all vectors X given by

$$X = t_1 X_1 + t_2 X_2 + \cdots + t_k X_k$$
$$= 0 + t_1(X_1 - 0) + t_2(X_2 - 0) + \cdots + t_k(X_k - 0)$$

is represented geometrically by a k-flat on the origin provided that X_1, $X_2, \ldots, X_k$ are not linearly dependent. These flat spaces will be treated more fully in the exercises of Chapters 5 and 6.

4.6 Exercises

1. Find the vector equation of the line on the points $(0, 1, 1, 0)$ and $(1, 0, 0, 1)$ in $\mathscr{R}^4$. Then eliminate the parameter from the corresponding scalar system of equations. How many hyperplanes must one intersect in order to define a line in $\mathscr{R}^4$? In $\mathscr{R}^n$?

2. Are the points $(0, 1, 1, 0)$, $(1, 0, 0, 1)$, and $(1, 1, 1, 1)$ collinear?

3. Write the vector equation of the line in $\mathscr{R}^n$ on O and P, where $P = (0, \ldots, 0, 1, 0, \ldots, 0)$, the 1 being the ith coordinate. Thus show that the i-axis is indeed a line.

4. Two points P_0 and P_1 of $\mathscr{R}^n$ are always collinear. Does it follow that the associated vectors X_0 and X_1 are always collinear? Illustrate with a figure in $\mathscr{E}^3$.

*5. Show that two nonzero vectors in $\mathscr{R}^n$ are collinear if and only if they are proportional.

*6. Show that two nonzero vectors in $\mathscr{R}^n$ are linearly dependent if and only if they are proportional.

7. Show that in $\mathscr{R}^n$ *any* two points of the line $X = X_0 + t(X_1 - X_0)$, where $X_1 \neq X_0$, determine the same line, that is, determine the same total set of points.

8. Show that

$$ X = \begin{bmatrix} 2 \\ 1 \\ -1 \\ 4 \end{bmatrix} + t \begin{bmatrix} 3 \\ 0 \\ 1 \\ -5 \end{bmatrix} \quad \text{and} \quad X = \begin{bmatrix} 5 \\ 1 \\ 0 \\ -1 \end{bmatrix} + t \begin{bmatrix} -6 \\ 0 \\ -2 \\ 10 \end{bmatrix} $$

represent the same line in $\mathscr{R}^4$.

9. Write the vector equation of the plane in $\mathscr{R}^4$ on the three points given in Exercise 2. Show that this plane lies on the origin.

*10. Three vectors in $\mathscr{R}^n$ are said to be **coplanar** if and only if their terminal points all lie on the same plane on O. Show that three vectors in $\mathscr{R}^n$ are coplanar if and only if they are linearly dependent.

11. Show that the four vectors in $\mathscr{R}^4$,

$$ \begin{bmatrix} 1 \\ -1 \\ 2 \\ 3 \end{bmatrix}, \quad \begin{bmatrix} 2 \\ 1 \\ -1 \\ 4 \end{bmatrix}, \quad \begin{bmatrix} -2 \\ 3 \\ 0 \\ 5 \end{bmatrix}, \quad \begin{bmatrix} -5 \\ 4 \\ 3 \\ 9 \end{bmatrix}, $$

are linearly dependent, whereas the first three are linearly independent. Show that, in consequence, these four vectors lie in a 3-flat on the origin and write its vector equation.

12. Use (4.3.4) to write the vector equations of the hyperplane (3-flat) in $\mathscr{R}^4$ determined by the points $(4, 0, 0, 0)$, $(0, 4, 0, 0)$, $(0, 0, 4, 0)$, and $(0, 0, 0, 4)$. Then eliminate the parameters from the corresponding set of scalar equations and show in this way that an equation for the 3-flat is $x_1 + x_2 + x_3 + x_4 = 4$.

*13. Generalize the intercept form (Section 3.9, Exercise 8) for the equation of a plane in $\mathscr{E}^3$ to an intercept form for the equation of a hyperplane in $\mathscr{R}^n$. What happens to this equation if the hyperplane fails to intersect a given axis? Can a hyperplane in $\mathscr{R}^n$ fail to intersect all n of the axes?

14. Find the point where the line OP, where $P = (1, 1, 1, 1)$, meets the hyperplane of Exercise 12.

*15. Show that if two distinct points of a line lie on a k-flat, then every point of the line lies on the k-flat. This property is the reason for calling a flat space *flat*. In $\mathscr{E}^3$, only a line and a plane contain all the points of *every* line which intersects them in two distinct points. A curved surface does not do this.

16. Show that if P_0 and P_1 with associated vectors X_0 and X_1 are such that $A^{\mathsf{T}}X_0 - b = A^{\mathsf{T}}X_1 - b$, then the line $P_0 P_1$ is parallel to the hyperplane with equation $A^{\mathsf{T}}X - b = 0$.

17. Show that every k-flat on the origin can be represented by an equation of the form

$$X = t_1 X_1 + t_2 X_2 + \cdots + t_k X_k,$$

where $X_1, X_2, \ldots, X_k$ are linearly independent.

18. Under what conditions will the equations $X = \sum_1^k t_i X_i$ and $X = \sum_1^k s_i Y_i$ represent the same vector space in $\mathscr{R}^n$?

19. Show that the vectors with terminal points on any k-flat *not* on the origin do *not* constitute a vector space. Illustrate with figures in $\mathscr{E}^2$ and $\mathscr{E}^3$.

20. Let a k-flat in $\mathscr{R}^n$ be defined by

$$X = X_0 + t_1(X_1 - X_0) + t_2(X_2 - X_0) + \cdots + t_k(X_k - X_0).$$

Show that the set of all vectors Y where

$$Y = X - X_0 = t_1(X_1 - X_0) + t_2(X_2 - X_0) + \cdots + t_k(X_k - X_0)$$

is a vector space. Interpret geometrically and sketch figures in $\mathscr{E}^2$ and $\mathscr{E}^3$.

21. Given a plane in $\mathscr{E}^3$, not on the origin, in how many ways can one derive a vector space in the manner of Exercise 20? What would you guess is the answer for a k-flat in $\mathscr{R}^n$? For a given k-flat, are these vector spaces distinct? Why?

4.7 Length and the Cauchy–Schwarz Inequality

The length of a vector in $\mathscr{R}^n$ can be defined in various sensible ways, each suitable to a particular purpose or application. On the basis of previous experience, it is natural to define the **length** $|X|$ of the vector X by

$$(4.7.1) \qquad |X| = (X^{\mathsf{T}}X)^{1/2} = \sqrt{x_1^2 + x_2^2 + \cdots + x_n^2}.$$

This function is also called the **norm** of X. (In other contexts, other functions of the components of X are also called norms.) Since a sum of squares of real numbers is zero if and only if each of the real numbers is zero, that is,

$$(4.7.2) \qquad x_1^2 + x_2^2 + \cdots + x_n^2 = 0 \quad \text{if and only if} \quad x_1 = x_2 = \cdots = x_n = 0,$$

it follows that a vector X in $\mathscr{R}^n$ has length zero if and only if it is the zero vector.

A vector is called a **unit vector** if and only if its length is 1. Some examples are

$$(4.7.3) \qquad E_1 = \begin{bmatrix} 1 \\ 0 \\ 0 \\ \vdots \\ 0 \end{bmatrix}, E_2 = \begin{bmatrix} 0 \\ 1 \\ 0 \\ \vdots \\ 0 \end{bmatrix}, \dots, E_n = \begin{bmatrix} 0 \\ 0 \\ \vdots \\ 0 \\ 1 \end{bmatrix}, \quad \text{and} \quad \begin{bmatrix} \dfrac{1}{\sqrt{n}} \\ \dfrac{1}{\sqrt{n}} \\ \vdots \\ \dfrac{1}{\sqrt{n}} \end{bmatrix}.$$

The first n of these will be called the **elementary unit vectors.**

If X is the vector associated with the directed segment $\overrightarrow{AB}$, then the **distance between A and B** is defined to be the length of X, which in this case is given by

$$(4.7.4) \qquad d(A, B) = \sqrt{(b_1 - a_1)^2 + (b_2 - a_2)^2 + \cdots + (b_n - a_n)^2}.$$

In view of this last definition, the length of a vector X is the distance between the origin O and the terminal point P of the vector X.

The space $\mathscr{R}^n$, with this definition of distance imposed on it, is called **Euclidean space of n dimensions** and is denoted by $\mathscr{E}^n$.

There are four basic properties that should be satisfied by any useful definition of distance. For all points A, B, and C of $\mathscr{R}^n$, the distance function $d(A, B)$ should be such that

$$(4.7.5)$$
    (a) $d(A, B) \geq 0$
    (b) $d(A, B) = 0$ if and only if $A = B$
    (c) $d(A, B) = d(B, A)$
    (d) $d(A, B) + d(B, C) \geq d(A, C)$ (the **triangle inequality**).

In the case of (4.7.4), properties (a) and (c) follow at once from the definition. Property (c) implies that distance is *not directed*. In the case of (b), one employs the result of (4.7.2). Property (d) is somewhat more difficult to prove. It rests on the following form of the famous Cauchy–Schwarz inequality:

$$(4.7.6) \qquad\qquad |X| \cdot |Y| \geq |X^{\mathsf{T}}Y|$$

or, in scalar form,

$$(4.7.7) \qquad\qquad \sqrt{\sum x_i^2} \cdot \sqrt{\sum y_i^2} \geq \left| \sum x_i y_i \right|,$$

where the summations run from 1 to n on i.

This last formula is proved by the following computation, where again summations on i and j run from 1 to n so far as allowed by any indicated restrictions:

$$
\begin{aligned}
\sum x_i^2 \cdot \sum y_i^2 - (\sum x_i y_i)^2 &= \sum_{i \neq j} x_i^2 y_j^2 + \sum x_i^2 y_i^2 - \sum x_i^2 y_i^2 - 2\sum_{i<j} x_i y_i x_j y_j \\
&= \sum_{i \neq j} x_i^2 y_j^2 - 2\sum_{i<j} x_i y_i x_j y_j \\
&= \sum_{i<j} x_i^2 y_j^2 - 2\sum_{i<j} x_i y_i x_j y_j + \sum_{i<j} x_j^2 y_i^2 \\
&= \sum_{i<j} (x_i^2 y_j^2 - 2x_i y_i x_j y_j + x_j^2 y_i^2) \\
&= \sum_{i<j} (x_i y_j - x_j y_i)^2.
\end{aligned}
$$

(If this is a little difficult to follow, try writing it out in full for $n = 3$.) Since the last member is necessarily non-negative, we have

$$
\sum x_i^2 \cdot \sum y_i^2 \geq (\sum x_i y_i)^2.
$$

The inequalities (4.7.7) and (4.7.6) now follow when we take the positive square root of each side.

From (4.7.7) we have

$$
\sqrt{\sum x_i^2} \cdot \sqrt{\sum y_i^2} \geq |\sum x_i y_i| \geq \sum x_i y_i.
$$

Multiplying by 2 and adding $\sum x_i^2$ and $\sum y_i^2$ to the extreme members, we get

$$
\sum x_i^2 + 2\sqrt{\sum x_i^2}\sqrt{\sum y_i^2} + \sum y_i^2 \geq \sum x_i^2 + 2\sum x_i y_i + \sum y_i^2
$$
$$
= \sum (x_i^2 + 2x_i y_i + y_i^2);
$$

that is,

$$
(\sqrt{\sum x_i^2} + \sqrt{\sum y_i^2})^2 \geq \sum (x_i + y_i)^2.
$$

Taking the positive square root on both sides, we have

$$
\sqrt{\sum x_i^2} + \sqrt{\sum y_i^2} \geq \sqrt{\sum (x_i + y_i)^2},
$$

which yields the **vector form of the triangle inequality**:

(4.7.8) $|X| + |Y| \geq |X + Y|.$

(See Figure 4-1 on the next page.)

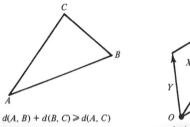

$$d(A, B) + d(B, C) \geqslant d(A, C)$$

$$|X| + |Y| \geqslant |X + Y|$$

FIGURE 4-1.   Triangle Inequality.

Now let

$$X = \begin{bmatrix} b_1 - a_1 \\ b_2 - a_2 \\ \vdots \\ b_n - a_n \end{bmatrix}, \qquad Y = \begin{bmatrix} c_1 - b_1 \\ c_2 - b_2 \\ \vdots \\ c_n - b_n \end{bmatrix},$$

so

$$X + Y = \begin{bmatrix} c_1 - a_1 \\ c_2 - a_2 \\ \vdots \\ c_n - a_n \end{bmatrix}.$$

Then $|X| = d(A, B)$, $|Y| = d(B, C)$, and $|X + Y| = d(A, C)$, so (4.7.8) becomes (4.7.5d) and the proof is complete. The inequalities (4.7.6) and (4.7.8) are frequently useful.

4.8   Angles and Orthogonality in $\mathscr{E}^n$

The **angle between two nonzero vectors** X and Y in $\mathscr{E}^n$ is defined to be the angle θ, $0 \leq \theta \leq \pi$, which is determined by

(4.8.1) $$\cos \theta = \frac{X^T Y}{|X| \cdot |Y|}, \qquad 0 \leq \theta \leq \pi.$$

Since, by the Cauchy–Schwarz inequality,

$$|X^T Y| \leq |X| \cdot |Y|,$$

we have from (4.8.1)

$$-1 \leq \cos \theta \leq 1,$$

so this is a reasonable definition.

The angles $\alpha_1, \alpha_2, \ldots, \alpha_n$ between a vector $X \neq 0$ and the elementary unit vectors $E_1, E_2, \ldots, E_n$ are called the **direction angles** of X and, by (4.8.1), are determined by

$$(4.8.2) \qquad \cos \alpha_i = \frac{X^\mathsf{T} E_i}{|X|} = \frac{x_i}{|X|}, \qquad i = 1, 2, \ldots, n.$$

These expressions are called the **direction cosines** of X. The direction angles determine both the sense and the direction of X. The vector $-X$ has direction cosines which are the negatives of those of X, so the direction angles of $-X$ are the supplements of those of X. We describe this by saying that X and $-X$ have **opposite senses** but the same direction. Squaring the direction cosines and adding, we find that

$$(4.8.3) \qquad \sum \cos^2 \alpha_i = 1.$$

The vector

$$(4.3.4) \qquad \frac{X}{|X|} = \begin{bmatrix} \dfrac{x_1}{|X|} \\[2mm] \dfrac{x_2}{|X|} \\[2mm] \vdots \\[2mm] \dfrac{x_n}{|X|} \end{bmatrix}$$

has the same direction cosines as X itself, so that it has the same direction and sense as X, but it has length 1. This unit vector is called the **normalized vector** associated with X. The process of dividing X by $|X|$ is called **normalization**.

The endpoints $(x_1, x_2, \ldots, x_n)$ of the members of the set of all *unit* vectors satisfy the equation

$$(4.8.5) \qquad x_1^2 + x_2^2 + \cdots + x_n^2 = 1.$$

The locus of these endpoints is called the **unit n-sphere with center at the origin.** For a unit vector $(x_1, x_2, \ldots, x_n)$, we have

$$(4.8.6) \qquad x_i = \cos \alpha_i, \qquad i = 1, 2, \ldots, n.$$

The **direction angles of a directed segment** $\overrightarrow{AB}$ are defined to be those of the corresponding vector and are therefore given by

$$(4.8.7) \qquad \cos \alpha_i = \frac{b_i - a_i}{d(A, B)}, \qquad i = 1, 2, \ldots, n.$$

We say that the vectors X and Y are **orthogonal** if and only if the angle between them is $\pi/2$. From (4.8.1) it follows that $\theta = \pi/2$ if and only if $X^{\mathsf{T}}Y = 0$; that is, X and Y are orthogonal if and only if $X^{\mathsf{T}}Y = 0$. For example, the vectors

$$X = \begin{bmatrix} 2 \\ -1 \\ 4 \\ 2 \end{bmatrix} \quad \text{and} \quad Y = \begin{bmatrix} -1 \\ 4 \\ 5 \\ -7 \end{bmatrix}$$

in $\mathscr{E}^4$ are orthogonal because $X^{\mathsf{T}}Y = 0$. Of particular importance is the fact that the elementary unit vectors are mutually orthogonal:

$$(4.8.8) \qquad E_i^{\mathsf{T}} E_j = 0, \qquad i \neq j.$$

Since in $\mathscr{E}^n$, just as in $\mathscr{E}^3$, we can write any vector X as a linear combination of the E_j's, thus:

$$\begin{bmatrix} x_1 \\ x_2 \\ \vdots \\ x_n \end{bmatrix} = x_1 \begin{bmatrix} 1 \\ 0 \\ 0 \\ \vdots \\ 0 \end{bmatrix} + x_2 \begin{bmatrix} 0 \\ 1 \\ 0 \\ \vdots \\ 0 \end{bmatrix} + \cdots + x_n \begin{bmatrix} 0 \\ 0 \\ \vdots \\ 0 \\ 1 \end{bmatrix},$$

and since all the E_j's are necessary for such a representation of an arbitrary vector X, we say that the E_j's form a **basis** for the vectors of $\mathscr{E}_n$. Since the E_j's are mutually orthogonal unit vectors, we call the basis **orthonormal**. We shall investigate such bases more extensively later.

4.9   Half-lines and Directed Distances

A point A on a line L divides the line into two **half-lines**. Let B be a point on L distinct from A and denote the corresponding vectors by X_B and X_A, respectively, so the line has the equation

$$(4.9.1) \qquad X = X_A + t(X_B - X_A).$$

Then the **positive half** of L, as determined by $\overrightarrow{AB}$, is defined to be all those points of L such that $t > 0$ and the **negative half** of L consists of all those points of L such that $t < 0$. These are **open half-lines**. If the conditions on t are replaced by $t \geq 0$ and $t \leq 0$ respectively, we get **closed half-lines**.

A simple but important example is the *j*-axis, for which we choose A as the origin and B as the **unit point** $(0, \ldots, 0, 1, 0, \ldots, 0)$ with a 1 as the *j*th coordinate, all other coordinates being zero. The equation of the *j*-axis is then

$$X = tE_j,$$

so the positive *j*-axis consists of those points $(0, \ldots, 0, t, 0, \ldots, 0)$ for which $t > 0$, as one would wish.

Two directed segments $\overrightarrow{AB}$ and $\overrightarrow{PQ}$ in $\mathscr{E}^n$, with associated nonzero vectors V and W, respectively, are said to have the **same direction** or to be **parallel** if and only if there exists a scalar k such that $W = kV$. They are further said to have the **same sense** if $k > 0$ and **opposite senses** if $k < 0$. The nonzero segment $\overrightarrow{PQ}$ on a line L whose positive half is determined by $\overrightarrow{AB}$ in the manner described above, will be said to be in the **positive sense** if its sense agrees with that of $\overrightarrow{AB}$ and in the **negative sense** if its sense is opposite to that of $\overrightarrow{AB}$.

Let P and Q on L have associated vectors X_P and X_Q respectively. Then, from (4.9.1),

$$X_P - X_Q = (t_P - t_Q)(X_B - X_A),$$

so that the sense of $\overrightarrow{PQ}$ is the same as or opposite to that of $\overrightarrow{AB}$ according as $t_P - t_Q$ is positive or negative. We may therefore define the **positive sense on the line** L to be the sense of increasing t. A line with a positive sense assigned to it is called a **directed line**.

We may now define the **directed distance** on L, from P to Q, denoted by $\overrightarrow{d}(P, Q)$, to be $d(P, Q)$ or $-d(P, Q)$ according as the sense of $\overrightarrow{PQ}$ is positive or negative.

For example, on the line in $\mathscr{E}^4$ for which $A = (0, 0, 0, 0)$ and $B = (1, 1, 1, 1)$, the *undirected* distance between the point $P:(t_1, t_1, t_1, t_1)$ and the point $Q:(t_2, t_2, t_2, t_2)$ is given by $\sqrt{4(t_2 - t_1)^2} = 2|t_2 - t_1|$. The *directed* distance from P to Q is then given by $2(t_2 - t_1)$, since this is positive when $t_2 > t_1$ and negative when $t_2 < t_1$.

4.10 Unitary *n*-Space

For certain physical applications, vectors with components from the complex field are useful, and definitions of length and orthogonality are needed. If X is such a vector and if we use $\sqrt{X^{\mathsf{T}}X}$ for the length of X, peculiarities arise. For example, if

$$X = \begin{bmatrix} 3 \\ 4 \\ 5i \end{bmatrix},$$

then $X^{\mathsf{T}}X = 3^2 + 4^2 + (5i)^2 = 0$, so a nonzero vector can have zero length, which is not ordinarily desirable. All such peculiarities are eliminated if we make this definition of length:

$$(4.10.1) \qquad\qquad |X| = \sqrt{X^*X},$$

where X^* denotes the tranjugate of X (see Section 1.12). Then length is a non-negative real number, for $\bar{x}_i x_i = |x_i|^2$, so that

$$|X| = \sqrt{\bar{x}_1 x_1 + \bar{x}_2 x_2 + \cdots + \bar{x}_n x_n} = \sqrt{\sum |x_i|^2} \geq 0.$$

In the case of our example, the length is now

$$\sqrt{3 \cdot 3 + 4 \cdot 4 + (-5i)(5i)} = \sqrt{50}$$

instead of 0. Since $\sum |x_i|^2 = 0$ if and only if each $x_i = 0$, the zero vector is the only vector whose length is zero according to this definition of length.

Similarly, if we were to use $X^{\mathsf{T}}Y = 0$ as a test for orthogonality, peculiarities would again appear. For example, the vector used in the example above would be orthogonal to itself since $X^{\mathsf{T}}X = 0$. Once again, we replace the transpose by the tranjugate in order to eliminate the peculiarity: We now define two vectors X and Y to be **orthogonal** if and only if neither is zero and

$$(4.10.2) \qquad\qquad X^*Y = 0 \qquad \text{(orthogonality condition)}.$$

The set of all n-vectors over the field of complex numbers, with the definitions of length and orthogonality given in (4.10.1) and (4.10.2), is called **unitary n-space** and we denote it by $\mathscr{U}^n$.

If X and Y are both real, both (4.10.1) and (4.10.2) reduce to the same rules we used before since then X^* and X^{T} are identical. Thus the geometry of $\mathscr{E}^n$ is a specialization of the geometry of $\mathscr{U}^n$. In $\mathscr{U}^n$, just as in $\mathscr{E}^n$, the elementary unit vectors $E_1, E_2, \ldots, E_n$ constitute an orthonormal basis since $E_i^* E_j = \delta_{ij}$, $i, j = 1, 2, \ldots, n$, and, for all X in $\mathscr{U}^n$,

$$X = x_1 E_1 + x_2 E_2 + \cdots + x_n E_n.$$

In $\mathscr{U}^n$, the **Cauchy–Schwarz inequality** takes the form

$$(4.10.3) \qquad\qquad |X| \cdot |Y| \geq |X^*Y|.$$

We give a different type of proof for the unitary case to illustrate another technique. If $X = 0$, the inequality is satisfied. Hence assume that $X \neq 0$ and consider the quadratic polynomial $p(t)$ defined as follows:

(4.10.4)
$$p(t) = \sum (|x_j|t - |y_j|)^2$$
$$= (\sum |x_j|^2)t^2 - 2(\sum |x_j| \cdot |y_j|)t + \sum |y_j|^2.$$

If $p(t_0) = 0$, where t_0 is real, then $\sum (|x_j|t_0 - |y_j|)^2$ is a vanishing sum of squares of real numbers, so that

$$|x_j|t_0 - |y_j| = 0, \qquad j = 1, 2, \ldots, n.$$

Solving this equation for $|y_j|$ and substituting in (4.10.4), we obtain

$$p(t) = \sum (|x_j|t - |x_j|t_0)^2 = (t - t_0)^2 \sum |x_j|^2 = (t - t_0)^2 |X|^2.$$

By hypothesis, $X \neq 0$, so that $|X| \neq 0$ also and therefore $p(t) = 0$ *only* when $t = t_0$. That is, if $p(t) = 0$ has a real root, that root is a *double* root. Hence the equation $p(t) = 0$ has either real and equal roots or conjugate complex roots, so that the discriminant of the quadratic polynomial $p(t)$ must be zero or negative:

$$4(\sum |x_j||y_j|)^2 - 4 \sum |x_j|^2 \cdot \sum |y_j|^2 \leq 0.$$

From this, since $|x_j| = |\bar{x}_j|$ and $|\bar{x}_j| \cdot |y_j| = |\bar{x}_j \cdot y_j|$, we obtain

(4.10.5)
$$\sum |x_j|^2 \cdot \sum |y_j|^2 \geq (\sum |x_j||y_j|)^2 = (\sum |\bar{x}_j y_j|)^2.$$

Using now the fact that the absolute value of a sum of terms is equal to or less than the sum of the absolute values of the separate terms, we have

$$\sum |x_j|^2 \cdot \sum |y_j|^2 \geq |\sum \bar{x}_j y_j|^2.$$

Taking square roots on both sides, we have (4.10.3), and the proof is complete.

Starting with the inequality of (4.10.5), we can obtain the triangle inequality for $\mathscr{U}^n$,

(4.10.6)
$$|X| + |Y| \geq |X + Y|,$$

in much the same manner as we obtained (4.7.8) from (4.7.7). The details are left as an exercise.

4.11   Exercises

1. The distance between the points $(a, 2a, -a, 3, 1)$ and $(3a, 2a, 2a, 1, 4)$ in $\mathscr{E}^5$ is $\sqrt{26}$. What are the possible values for a?

2. The vectors

$$\begin{bmatrix} a \\ 2a \\ 2a \\ 4 \end{bmatrix} \quad \text{and} \quad \begin{bmatrix} 4 \\ 2a \\ 2a \\ a \end{bmatrix}$$

   in $\mathscr{E}^4$ are orthogonal. What are the possible values for a?

3. The vector $[\frac{1}{2}, \frac{1}{2}, a, \sqrt{2}/2, b]^\mathsf{T}$ in $\mathscr{E}^5$ is a unit vector. What can you say about a and b?

4. Verify the Cauchy–Schwarz inequality for the vectors $[1, -2, 3, 4]^\mathsf{T}$ and $[4, 1, -2, -3]^\mathsf{T}$.

5. Find $\cos \theta$, where θ is the angle between the two vectors of Exercise 4.

6. Show that the points $(0, 0, 0, 0)$, $(1, -2, 3, 4)$, and $(4, -1, 2, -3)$ are the vertices of a right triangle in $\mathscr{E}^4$.

7. Show that the four vectors

$$V_1 = \begin{bmatrix} \dfrac{\sqrt{2}}{2} \\ \dfrac{\sqrt{2}}{2} \\ 0 \\ 0 \end{bmatrix}, \quad V_2 = \begin{bmatrix} \dfrac{\sqrt{2}}{2} \\ \dfrac{-\sqrt{2}}{2} \\ 0 \\ 0 \end{bmatrix}, \quad V_3 = \begin{bmatrix} 0 \\ 0 \\ \dfrac{\sqrt{2}}{2} \\ \dfrac{\sqrt{2}}{2} \end{bmatrix}, \quad V_4 = \begin{bmatrix} 0 \\ 0 \\ \dfrac{\sqrt{2}}{2} \\ \dfrac{-\sqrt{2}}{2} \end{bmatrix}$$

   constitute an orthonormal set. Show that they are a basis for $\mathscr{E}^4$ by solving the equation

$$X = t_1 V_1 + t_2 V_2 + t_3 V_3 + t_4 V_4$$

   for the t's in terms of the components of an arbitrary vector X.

8. Find a vector orthogonal to each of the vectors

$$\begin{bmatrix} 1 \\ -1 \\ 2 \end{bmatrix} \quad \text{and} \quad \begin{bmatrix} 2 \\ 0 \\ -1 \end{bmatrix}$$

   and then normalize all three. Is the result an orthonormal basis for the set of all real 3-vectors?

*9. Show that every pair of points P, Q of a hyperplane $A^{\mathsf{T}}X = b$ determines a vector orthogonal to the coefficient vector A of the equation. We describe this by saying that A is **orthogonal** or **normal** to the hyperplane.

*10. Show that the line defined by $X = X_0 + tA$ is normal to the hyperplane $A^{\mathsf{T}}X = b$. (Formulate the needed definition.)

*11. Given that A is a unit vector, determine the point where the line $X = tA$ meets the hyperplane $A^{\mathsf{T}}X = b$. Show that the distance between the origin and this point is $|b|$. Illustrate with figures in $\mathscr{E}^2$ and $\mathscr{E}^3$.

*12. Show that (a) the line with equation $X = X_0 + tB$ meets the hyperplane with equation $A^{\mathsf{T}}X = b$ in a unique point unless $A^{\mathsf{T}}B = 0$; (b) if $A^{\mathsf{T}}B = 0$ but $A^{\mathsf{T}}X_0 \neq b$, the line does not intersect the hyperplane, that is, is parallel to it; (c) if $A^{\mathsf{T}}B = 0$ and $A^{\mathsf{T}}X_0 = b$, all points of the line lie on the hyperplane.

*13. How does one define and compute the midpoint of a line segment AB in $\mathscr{E}^n$?

14. Find the length of the vector

$$X = \begin{bmatrix} 1 - i \\ 1 + i \\ 1 \\ 0 \end{bmatrix}$$

in $\mathscr{U}^4$, then normalize X.

15. Determine whether or not the vectors

$$\begin{bmatrix} i \\ -i \\ i \\ -i \\ 1 \end{bmatrix} \quad \text{and} \quad \begin{bmatrix} i \\ i \\ i \\ i \\ 1 \end{bmatrix}$$

in $\mathscr{U}^5$ are orthogonal.

*16. If X is a unit vector in $\mathscr{U}^n$, for what scalars α will αX still be a unit vector? (The set of scalars is now the field of complex numbers.)

17. The vectors

$$\begin{bmatrix} i \\ 1 \\ 0 \end{bmatrix} \quad \text{and} \quad \begin{bmatrix} 0 \\ 1 \\ i \end{bmatrix}$$

are each orthogonal to a *unit* vector

$$X = \begin{bmatrix} x_1 \\ x_2 \\ x_3 \end{bmatrix}$$

in $\mathscr{U}^3$. Find such a vector X. How many solutions are there?

*18. Prove that for all scalars α and all vectors X in $\mathscr{U}^n$,

$$|\alpha X| = |\alpha|\,|X|.$$

*19. Prove that $X*Y = \overline{Y*X}$.

*20. Show that if $A_1, A_2, \ldots, A_p$ are all orthogonal in $\mathscr{U}^n$ to a vector X, so is any linear combination $t_1 A_1 + t_2 A_2 + \cdots + t_p A_p$ of these vectors.

*21. Prove that if X and Y are vectors in $\mathscr{U}^n$ such that $|X| \leq 1$ and $|Y| \leq 1$, then $|X*Y| \leq 1$ also.

22. Prove that in $\mathscr{E}^n$, the vectors X_P and X_Q with terminal points P and Q respectively are orthogonal if and only if $d^2(P, Q) = d^2(O, P) + d^2(O, Q)$. Does the same theorem hold in $\mathscr{U}^n$?

*23. Prove that mutually orthogonal vectors in $\mathscr{E}^n$, and in $\mathscr{U}^n$, are linearly independent.

24. Show that in $\mathscr{U}^n$ the set of all vectors of the form

$$X = t_1 X_1 + t_2 X_2 + \cdots + t_k X_k,$$

where the X_j's are given and the t_j's are arbitrary complex numbers, constitutes a vector space.

*25. If $U_1, U_2, \ldots, U_n$ constitute an orthonormal basis for $\mathscr{U}^n$, and if B is any vector of $\mathscr{U}^n$, prove that

$$B = \sum_{i=1}^{n} (U_i^* B) U_i.$$

This amounts to solving

$$t_1 U_1 + t_2 U_2 + \cdots + t_n U_n = B$$

for the t's, which is particularly simple when the U's constitute an orthonormal basis for $\mathscr{U}^n$.

26. Show that in $\mathscr{E}^n$ the parameter s can be used as a directed distance from the terminal point of X_0 to the terminal point of X on the line with equation

$$X = X_0 + s \begin{bmatrix} \cos \alpha_1 \\ \cos \alpha_2 \\ \vdots \\ \cos \alpha_n \end{bmatrix}.$$

27. Let the positive sense on a line AB be that of $\overrightarrow{AB}$. Let P be a point on the line. Find formulas for

(a) $\dfrac{\overrightarrow{d}(A, P)}{\overrightarrow{d}(A, B)}$, (b) $\dfrac{\overrightarrow{d}(A, P)}{\overrightarrow{d}(P, B)}$.

Illustrate with figures in $\mathscr{E}^2$.

28. Show that the set of all points in $\mathscr{E}^n$ which are equidistant from P: $(a_1, a_2, \ldots, a_n)$ and $Q:(b_1, b_2, \ldots, b_n)$ is a hyperplane and find its equation. Find the point where this hyperplane intersects the line PQ. Specialize to the case $Q = (-a_1, -a_2, \ldots, -a_n)$. This hyperplane is called the **perpendicular bisector** of the line segment PQ.

29. Define what is meant by an **n-sphere** in $\mathscr{E}^n$. Then show that an equation of an n-sphere with center $P:(h_1, h_2, \ldots, h_n)$ and radius r is

$$(x_1 - h_1)^2 + (x_2 - h_2)^2 + \cdots + (x_n - h_n)^2 = r^2.$$

4.12   Linear Inequalities

In many applications, it is important to determine in what regions of $\mathscr{E}^n$ the values of a set of linear functions of the form

(4.12.1) $$f(X) = A^\mathsf{T}X - b$$

are simultaneously positive or negative. To see how the hyperplane determined by

(4.12.2) $$A^\mathsf{T}X - b = 0$$

is related to these regions for the corresponding function $f(X)$, we use familiar geometrical ideas from $\mathscr{E}^2$ and $\mathscr{E}^3$ as a guide.

We begin with a definition. In $\mathscr{E}^n$, distinct points P_0 and P_1 are said to be on **opposite sides of a hyperplane** if and only if the line $P_0 P_1$ intersects the hyperplane in a unique point R between P_0 and P_1. This is illustrated for $\mathscr{E}^2$ and $\mathscr{E}^3$ in Figure 4-2.

Let the hyperplane have the equation (4.12.2) and let P_0 and P_1 be

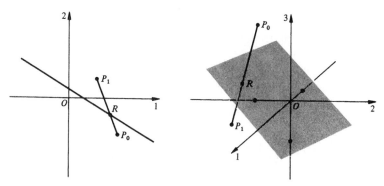

FIGURE 4-2. Opposite Sides of a Line and of a Plane.

terminal points of vectors X_0 and X_1 so that $P_0 P_1$ has the equation

$$X = X_0 + t(X_1 - X_0).$$

This line intersects the hyperplane in the point R for which

(4.12.3) $t = \dfrac{b - A^{\mathsf{T}} X_0}{(b - A^{\mathsf{T}} X_0) + (A^{\mathsf{T}} X_1 - b)}$,

provided that the denominator is not zero.

If now P_0 and P_1 are such that

$$A^{\mathsf{T}} X_0 - b < 0 \qquad \text{and} \qquad A^{\mathsf{T}} X_1 - b > 0,$$

we see from (4.12.3) that

$$0 < t < 1,$$

so R is between P_0 and P_1 and hence these two points are on opposite sides of the hyperplane. This implies that if P_0 is such that $A^{\mathsf{T}} X_0 - b < 0$, *no* point P_1 such that $A^{\mathsf{T}} X_1 - b > 0$ can be on the same side of the hyperplane as P_0. Similarly, if P_1 is such that $A^{\mathsf{T}} X_1 - b > 0$, *no* point P_0 such that $A^{\mathsf{T}} X_0 - b < 0$ can be on the same side of the hyperplane as P_1. We therefore have

Theorem 4.12.1: *In $\mathscr{E}^n$, all points for which $A^{\mathsf{T}} X - b > 0$ lie on one side of the hyperplane with equation $A^{\mathsf{T}} X - b = 0$ and all points for which $A^{\mathsf{T}} X - b < 0$ lie on the other side.*

This theorem justifies the following definition:

In $\mathscr{E}^n$, a linear function $f(X) = A^{\mathsf{T}} X - b$ determines two nonintersecting regions called **open half-spaces**: (1) a **positive half-space** consisting of all those points for which $A^{\mathsf{T}} X - b > 0$ and (2) a **negative half-space** consisting of all those points for which $A^{\mathsf{T}} X - b < 0$. The hyperplane with equation $A^{\mathsf{T}} X - b = 0$, that is, the set of all points such that $A^{\mathsf{T}} X - b = 0$, is said to constitute the **boundary** of each of these spaces.

The function $f(X)$ also determines two **closed half-spaces**: the region for which $A^{\mathsf{T}} X - b \geq 0$ and the region for which $A^{\mathsf{T}} X - b \leq 0$. Each of these half-spaces is the union of an open half-space and its boundary.

The theorem shows that two points are **on the same side of the hyperplane** with equation $A^{\mathsf{T}} X - b = 0$ if and only if they are in the same one of the two open half-spaces just defined.

Note that if the function involved is $g(X) = -A^{\mathsf{T}} X + b$, then $f(X) > 0$ if and only if $g(X) < 0$. Thus the half-space in which a function is positive is a property of the function, not of the associated hyperplane.

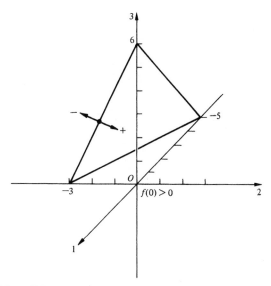

FIGURE 4-3. Half-Spaces Determined by the Function

$$f(X) = 6x_1 + 10x_2 - 5x_3 + 30.$$

Note also that the nature of an open half-space determined by $f(X)$ is identified by a single point in it. For example, let

$$f(X) = 6x_1 + 10x_2 - 5x_3 + 30.$$

Since $f(0) = 30$, the origin is in the positive half-space determined by this function (Figure 4-3).

Frequently one is concerned with the region determined by more than one inequality. Such a region is then the intersection of certain half-spaces. For example, what region in $\mathscr{E}^3$ is the intersection of the positive half-spaces associated with the two functions

$$f(X) = x_1 + x_2 - x_3 + 2$$

and

$$g(X) = 2x_1 + 2x_2 + x_3 + 4?$$

We determine the points where the associated planes cross the axes, sketch the two planes, and find that the origin is on the positive side of each. (Figure 4-4). The region in question is an infinite wedge formed by the two planes and opening toward the origin.

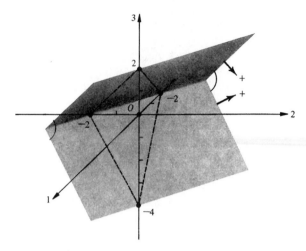

FIGURE 4-4. Intersection of Two Positive Half-Spaces.

In the same way, we can determine the intersection of the positive half-spaces associated with the functions

$$h(X) = -x_1 - x_2 - x_3 + 4$$

and

$$k(X) = x_1 + x_2 + x_3 + 4.$$

As Figure 4-5 shows, the required region is the infinite slab lying between the two planes $h(X) = 0$, $k(X) = 0$.

The intersection of the closed half-spaces determined by the three inequalities $x_1 \geq 0$, $x_2 \geq 0$, and $x_3 \geq 0$ is simply the first octant, inclusive of those portions of the coordinate planes that bound it.

The intersection of the half-spaces determined by the four inequalities

$$x_1 > 0$$
$$x_2 > 0$$
$$x_3 > 0$$
$$-x_1 - x_2 - x_3 + 2 > 0$$

is the interior of a tetrahedron with vertex at the origin and three edges along the coordinate axes (Figure 4-6).

One cannot enclose in $\mathscr{E}^3$ a region whose maximum diameter is finite with

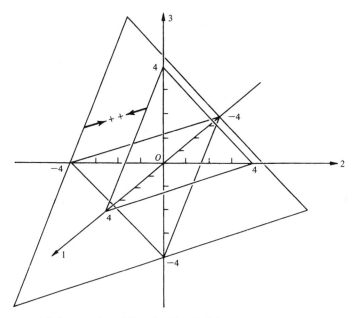

FIGURE 4-5. Intersection of Two Positive Half-Spaces.

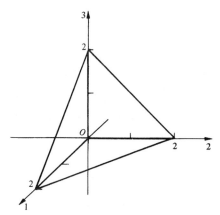

FIGURE 4-6. Tetrahedron Defined by

$$x_1 > 0, \; x_2 > 0, \; x_3 > 0, \; -x_1 - x_2 - x_3 + 2 > 0.$$

fewer than four linear inequalities. There is no upper limit on the number of inequalities one may use for the determination of either a finite or an infinite region in $\mathscr{E}^3$.

Consider now a k-flat represented by the parametric vector equation

$$X = X_0 + t_1(X_1 - X_0) + t_2(X_2 - X_0) + \cdots + t_k(X_k - X_0).$$

If X is the vector determined by a point P of the k-flat and if Y is determined by the point Q of the k-flat where

$$Y = X_0 + s_1(X_1 - X_0) + s_2(X_2 - X_0) + \cdots + s_k(X_k - X_0),$$

then, for all α, the terminal point of the vector $X + \alpha(Y - X)$ lies on the k-flat, since

$$X + \alpha(Y - X) = X_0 + (\alpha s_1 + (1 - \alpha)t_1)(X_1 - X_0) + \cdots$$
$$+ (\alpha s_k + (1 - \alpha)t_k)(X_k - X_0).$$

Thus we have proved

Theorem 4.12.2: *If the points P and Q lie on a k-flat, so does every point of the line PQ.*

Using this, we can easily prove a basic result showing how hyperplanes separate $\mathscr{E}^n$:

Theorem 4.12.3: *If a k-flat has no points in common with a hyperplane, then all its points are on the same side of the hyperplane.*

Let P and Q, with corresponding vectors X and Y, be on the k-flat. Let the hyperplane have the equation $A^T X - b = 0$ and suppose, contrary to what the theorem asserts, that $A^T X - b < 0$ and $A^T Y - b > 0$, so that P and Q lie on opposite sides of the hyperplane. Then some point of the line PQ will lie *on* the hyperplane if we can choose α so that

$$A^T(X + \alpha(Y - X)) - b = 0,$$

that is, so that

$$(A^T X - b) + \alpha((A^T Y - b) - (A^T X - b)) = 0.$$

Since $A^T Y - b > 0$ and $A^T X - b < 0$, the coefficient of α is a positive number. Hence we can solve for α:

$$\alpha = -\frac{A^T X - b}{A^T(Y - X)}.$$

Thus some point of PQ, which by Theorem 4.12.2 lies on the k-flat, also lies on the hyperplane, which contradicts the hypothesis of the theorem. Hence the k-flat cannot have points P and Q on opposite sides of the hyperplane and the proof is complete.

In linear programming problems, one has to study intersections of sets of half-spaces of the types described above. In particular, the set of points in $\mathscr{E}^n$ such that $x_1 > 0, x_2 > 0, \ldots, x_n > 0$ corresponds to the first quadrant in $\mathscr{E}^2$ and to the first octant in $\mathscr{E}^3$. It is called the **positive orthant** of $\mathscr{E}^n$; the **nonnegative orthant** of $\mathscr{E}^n$ is the intersection of the closed half-spaces defined by $x_1 \geq 0, x_2 \geq 0, \ldots, x_n \geq 0$.

4.13 Exercises

1. Identify and sketch the positive half-spaces in $\mathscr{E}^2$ determined by these functions:

   (a) $f(X) = x_1 + x_2 + 2,$ (b) $g(X) = 2 - x_1 - x_2,$

   (c) $h(X) = x_1 - 1,$ (d) $k(X) = 2 - x_2.$

2. Identify and sketch the intersection of the half-spaces in $\mathscr{E}^2$ determined by the inequalities

$$2x_1 - x_2 \geq 0$$
$$x_2 - 3 \leq 0$$
$$-x_1 + 2x_2 - 1 > 0.$$

3. Identify the positive half-space in $\mathscr{E}^3$ determined by the function $f(X) = x_1 + x_2 + 3x_3 + 3$ and illustrate with a sketch.

4. (a) Identify and sketch the intersection of the positive half-spaces in $\mathscr{E}^3$ determined by the functions

   (a) $\begin{cases} f(X) = x_1 + x_2 + x_3 + 3 \\ g(X) = x_1 + x_2 + x_3 - 3, \end{cases}$ (b) $\begin{cases} f(X) = x_1 + x_2 + x_3 + 3 \\ g(X) = x_1 - x_2 - x_3 - 3, \end{cases}$

   (c) $\begin{cases} f(X) = x_1 + x_2 - x_3 + 2 \\ g(X) = 2x_1 + 2x_2 + x_3 + 4, \end{cases}$ (d) $\begin{cases} f(X) = x_1 + x_2 + x_3 - 2 \\ g(X) = -x_1 - x_2 - x_3 + 4. \end{cases}$

5. Identify and sketch the region in $\mathscr{E}^3$ determined by the set of inequalities

$$-x_1 + 1 > 0, \quad -x_2 + 1 > 0, \quad -x_3 + 1 \geq 0,$$
$$x_1 + 1 > 0, \quad x_2 + 1 > 0, \quad x_3 + 1 \geq 0.$$

6. Two systems of inequalities are **equivalent** if and only if every solution of either system is also a solution of the other. Are the systems of inequalities

$$
\begin{array}{lll}
x_1 & \geq 0 & x_1 \geq 0 \\
x_1 + x_2 & \geq 0 \qquad \text{and} & x_2 \geq 0 \\
x_1 + x_2 + x_3 \geq 0 & & x_3 \geq 0
\end{array}
$$

equivalent? Why? Illustrate with figures in $\mathscr{E}^3$. What operations on a system of inequalities transform it into an equivalent system of inequalities?

7. For what sets of points in $\mathscr{E}^3$ do these systems of inequalities hold?

(a) $x_1^2 + x_2^2 + x_3^2 > 1$,

(b) $x_1^2 + x_2^2 + x_3^2 \leq 1$,

(c) $x_1^2 + x_2^2 + x_3^2 < 1, \; x_3 > 0$,

(d) $x_1^2 + x_2^2 + x_3^2 < 1, \; x_3 > 0, \; x_2 - x_3 > 0$.

*8. Given a k-flat, $k \leq n - 1$, show that there always exists a real linear function $f(X)$ such that all the points of the k-flat lie in the positive half-space determined by this function

9. Let P_A be the point where the normal line $X = tA$ meets the hyperplane $\pi: A^{\mathsf{T}}X - b = 0$. Define the **directed distance** $\overrightarrow{d(O, \pi)}$ from the origin O to π to be zero if $P_A = O$, $d(O, P_A)$ if $\overrightarrow{OP_A}$ and A have the same sense, and $-d(O, P_A)$ if $\overrightarrow{OP_A}$ and A have opposite senses. Show that $\overrightarrow{d(O, \pi)} = b/\|A\|$.

10. Two hyperplanes are said to be **parallel** if and only if their normals have the same direction. (This definition permits a hyperplane to be parallel to itself.) Define the **directed distance** $\overrightarrow{d(\pi, \pi')}$ from $\pi: A^{\mathsf{T}}X - b = 0$ to the parallel hyperplane $\pi': A^{\mathsf{T}}X - b' = 0$ to be $\overrightarrow{d(O, \pi')} - \overrightarrow{d(O, \pi)}$, so $\overrightarrow{d(\pi, \pi')} = (b' - b)/\|A\|$. Then define the **directed distance** $\overrightarrow{d(\pi, P_0)}$ from π to the terminal point P_0 of X_0 to be $\overrightarrow{d(\pi, \pi')}$ where $\pi': A^{\mathsf{T}}(X - X_0) = 0$ is the hyperplane parallel to π and on P_0. Show that $\overrightarrow{d(\pi, P_0)} = (A^{\mathsf{T}}X_0 - b)/\|A\|$. (Note that when $X_0 = 0$, this result is consistent with that of Exercise 9.)

11. How are the signs of the directed distances of Exercises 9 and 10 related to the half-spaces determined by the function $f(X) = A^{\mathsf{T}}X - b$? Construct a variety of examples in $\mathscr{E}^2$ and $\mathscr{E}^3$ and illustrate with figures.

CHAPTER
5

Vector Spaces

5.1 The General Definition of a Vector Space

In the preceding chapters we studied special sets of vectors in $\mathscr{E}^3$ and $\mathscr{E}^n$, sets we called "vector spaces." This geometrical concept may be generalized as follows. Assume first that we have a field $\mathscr{F}$ of "scalars" over which to work and a collection $\mathscr{V}$ of mathematical objects which we call "vectors" (even though they may not resemble geometric vectors). Equality of two vectors means they are identical. We call the set $\mathscr{V}$ a **vector space over** $\mathscr{F}$ if and only if its members satisfy the following requirements:

(A) There is an operation called "addition" such that, corresponding to any two members V_1 and V_2 of $\mathscr{V}$, there exists a unique "sum" $V_1 + V_2$. Moreover, addition of vectors from $\mathscr{V}$ obeys these rules:

(A$_1$) $V_1 + V_2$ is always a member of $\mathscr{V}$ (closure of $\mathscr{V}$ with respect to addition).

(A$_2$) For all V_1, V_2, V_3 in $\mathscr{V}$, $V_1 + (V_2 + V_3) = (V_1 + V_2) + V_3$ (the associative law of addition).

(A$_3$) For all V_1 and V_2 in $\mathscr{V}$, $V_1 + V_2 = V_2 + V_1$ (the commutative law of addition).

(A$_4$) There is a special member of $\mathcal{V}$, called the **zero vector** and denoted by 0, such that for each V of $\mathcal{V}$

$$V + 0 = V.$$

(A$_5$) To every member V of $\mathcal{V}$ there corresponds a **negative**, $-V$, which is also a member of $\mathcal{V}$, such that

$$V + (-V) = 0.$$

(S) There is an operation of multiplication by a scalar such that if α is any element of $\mathcal{F}$ and if V is any member of $\mathcal{V}$, there exists a uniquely defined product αV. Scalar multiplication is assumed to obey these rules, for all α, β in $\mathcal{F}$ and for all V, V_1, V_2 in $\mathcal{V}$:

(S$_1$) αV belongs to $\mathcal{V}$ (closure of $\mathcal{V}$ with respect to multiplication by scalars).

(S$_2$) $V\alpha = \alpha V$ (commutative law of multiplication by scalars).

(S$_3$) $\quad \alpha(V_1 + V_2) = \alpha V_1 + \alpha V_2$ ⎫distributive laws of

(S$_4$) $\quad\quad (\alpha + \beta)V = \alpha V + \beta V$ ⎬multiplication by scalars.

(S$_5$) $\quad\quad\quad \alpha(\beta V) = (\alpha\beta)V$ (associative law of multiplication by scalars).

(S$_6$) $\qquad\qquad\qquad\qquad 0 \cdot V = 0.$

(S$_7$) $\qquad\qquad\qquad\qquad 1 \cdot V = V.$

Note that in (S$_6$), the left 0 denotes the *scalar* zero, whereas the right zero denotes the *vector* zero.

The set consisting of the zero vector alone is a vector space, the **zero space** over $\mathcal{F}$. This space will be denoted by $\mathcal{Z}$.

The reader will recall from Chapter 4 that the vector spaces of $\mathcal{E}^n$, over the field of real numbers, exhibit all these properties. They provide the most familiar illustrations of this general definition. In most applications, $\mathcal{F}$ is $\mathcal{R}$ or $\mathcal{C}$. Examples we give are restricted to these two fields. The reader should *assume that the real field is intended in numerical exercises unless complex elements or coefficients imply that the complex field is required.*

The set of all n-vectors over a field $\mathcal{F}$ is the most basic type of vector space. This particular space will be denoted by the symbol $\mathcal{F}^n$. The symbol $\mathcal{R}^n$, used in Chapter 4 for real n-dimensional space, is an illustration of this notation. Complex n-dimensional space is denoted by $\mathcal{C}^n$. By the introduction of the distance function $|X| = \sqrt{X^*X}$, $\mathcal{C}^n$ is transformed into unitary n-space $\mathcal{U}^n$ just as the distance function $|X| = \sqrt{X^\mathsf{T}X}$ transforms $\mathcal{R}^n$ into $\mathcal{E}^n$.

There are other simple examples. To illustrate, consider the set $\mathcal{Q}$ of all quadratic polynomials

$$q = \alpha_1 x^2 + \alpha_2\, xy + \alpha_3\, y^2,$$

where α_1, α_2, and α_3 are arbitrary real numbers. Addition of two of these yields a third, addition is associative and commutative, the zero element is the zero polynomial $0x^2 + 0xy + 0y^2$, and the negative of $\alpha_1 x^2 + \alpha_2 xy + \alpha_3 y^2$ is $-\alpha_1 x^2 - \alpha_2\, xy - \alpha_3\, y^2$. If α is a real number, αq is $\alpha\alpha_1 x^2 + \alpha\alpha_2\, xy + \alpha\alpha_3\, y^2$, which is again a quadratic polynomial in the set. The required properties for multiplication by a scalar are easily verified. That is, $\mathscr{Q}$ is a vector space over the field of real numbers.

When we add these quadratic polynomials or multiply them by scalars, we operate only on the coefficients, not on the symbols x^2, xy, and y^2. In fact, if we establish a one-to-one correspondence

$$\alpha_1 x^2 + \alpha_2\, xy + \alpha_3\, y^2 \leftrightarrow \begin{bmatrix} \alpha_1 \\ \alpha_2 \\ \alpha_3 \end{bmatrix},$$

which assigns to each quadratic polynomial $\alpha_1 x^2 + \alpha_2\, xy + \alpha_3\, y^2$ the unique 3-vector $\begin{bmatrix} \alpha_1 \\ \alpha_2 \\ \alpha_3 \end{bmatrix}$ and to each 3-vector $\begin{bmatrix} \alpha_1 \\ \alpha_2 \\ \alpha_3 \end{bmatrix}$ the unique quadratic polynomial $\alpha_1 x^2 + \alpha_2\, xy + \alpha_3\, y^2$, then to the sum of polynomials

$$(\alpha_1 x^2 + \alpha_2\, xy + \alpha_3\, y^2) + (\beta_1 x^2 + \beta_2\, xy + \beta_3\, y^2)$$
$$= (\alpha_1 + \beta_1)x^2 + (\alpha_2 + \beta_2)xy + (\alpha_3 + \beta_3)y^2$$

there corresponds the sum of 3-vectors

$$\begin{bmatrix} \alpha_1 \\ \alpha_2 \\ \alpha_3 \end{bmatrix} + \begin{bmatrix} \beta_1 \\ \beta_2 \\ \beta_3 \end{bmatrix} = \begin{bmatrix} \alpha_1 + \beta_1 \\ \alpha_2 + \beta_2 \\ \alpha_3 + \beta_3 \end{bmatrix},$$

and to the scalar multiple

$$\alpha(\alpha_1 x^2 + \alpha_2\, xy + \alpha_3\, y^2) = \alpha\alpha_1 x^2 + \alpha\alpha_2\, xy + \alpha\alpha_3\, y^2$$

there corresponds the scalar multiple

$$\alpha \begin{bmatrix} \alpha_1 \\ \alpha_2 \\ \alpha_3 \end{bmatrix} = \begin{bmatrix} \alpha\alpha_1 \\ \alpha\alpha_2 \\ \alpha\alpha_3 \end{bmatrix}.$$

That is, instead of adding quadratic polynomials and multiplying them by scalars, we could just as well perform the same operations on the corresponding 3-vectors. (This is similar to omitting the variables in synthetic elimination.) Thus the distinction between the set of real quadratic polynomials and the set of real 3-vectors is one of notation only, as far as vector space operations are concerned.

Similarly, the set of all real linear polynomials

$$\alpha_1 x + \alpha_2 y + \alpha_3$$

is a vector space over the field of real numbers that differs only in notation from the space of all real 3-vectors. The concept of vector spaces that are fundamentally alike is a very important one, as we shall see.

An example of a vector space over the complex number field is the set of all polynomials

$$\alpha_n x^n + \alpha_{n-1} x^{n-1} + \cdots + \alpha_1 x + \alpha_0, \qquad n = 0, 1, 2, \ldots,$$

with complex numbers as coefficients. It is simple to verify that the properties of the definition all hold, for they are familiar properties of the algebra of polynomials.

Another kind of example is given by the set of all real functions $f(x)$ which are continuous on the closed interval $0 \le x \le 1$. Indeed, the sum of two such functions is also a real function continuous on $0 \le x \le 1$ and a scalar multiple of such a function is again a member of the set, as is the function which is 0 on $0 \le x \le 1$. The other properties are all familiar properties of algebra. Thus these functions constitute a vector space over the field of real numbers.

A most important kind of vector space for our purposes is given in

Theorem 5.1.1: *Over a given field $\mathscr{F}$, the set of all solutions of a system of homogeneous equations in n unknowns is a vector space.*

The set of homogeneous equations can be represented in matrix form by the equation

$$AX = 0,$$

where the coefficient matrix A is over $\mathscr{F}$. Now let Y_1 and Y_2 be vectors over $\mathscr{F}$ such that

$$AY_1 = 0 \qquad \text{and} \qquad AY_2 = 0.$$

Then

$$AY_1 + AY_2 = 0,$$

so

$$A(Y_1 + Y_2) = 0;$$

that is, $Y_1 + Y_2$ belongs to the set if Y_1 and Y_2 do (property A_1).

Also, for any scalar α,

$$\alpha A Y_1 = A(\alpha Y_1) = 0;$$

that is, αY_1 belongs to the set if Y_1 does (property S_1).

Moreover, the vector 0 is a solution (the trivial solution) and

$$-A Y_1 = A(-Y_1) = 0,$$

so (A_4) and (A_5) hold. The other properties of a vector space hold in this case because our vectors are matrices, and the proof is complete.

The preceding examples illustrate the fact that the vector-space structure is one of the basic structures of mathematics.

5.2 Linear Combinations and Linear Dependence

Let $V_1, V_2, \ldots, V_k$ be members of a vector space $\mathscr{V}$ over $\mathscr{F}$ and let $\alpha_1, \alpha_2, \ldots, \alpha_k$ be arbitrary scalars of $\mathscr{F}$. Then, by the properties (A_1), (A_2), and (S_1), repeatedly applied, the expression

$$(5.2.1) \qquad \alpha_1 V_1 + \alpha_2 V_2 + \cdots + \alpha_k V_k$$

represents a vector of $\mathscr{V}$. This vector is called a **linear combination** of V_1, $V_2, \ldots, V_k$.

An important result is

Theorem 5.2.1: *A subset $\mathscr{S}$ of a vector space $\mathscr{V}$ is itself a vector space if and only if $\mathscr{S}$ is closed with respect to addition and with respect to multiplication by scalars.*

Such a vector space $\mathscr{S}$ is called a **subspace** of $\mathscr{V}$.

First assume that $\mathscr{S}$ is a subset of $\mathscr{V}$ that is closed with respect to addition and with respect to multiplication by scalars. Let V belong to $\mathscr{S}$. Since $\mathscr{S}$ is closed with respect to multiplication by scalars, the vectors $0 \cdot V = 0$ and $(-1)V = -V$ also belong to $\mathscr{S}$. Then since $\mathscr{S}$ is closed with respect to addition and since $\mathscr{V}$ is a vector space, all the remaining properties hold at once.

The converse is immediate.

This theorem shows that the definitions given earlier of vector spaces in $\mathcal{E}^2$, $\mathcal{E}^3$, and $\mathcal{E}^n$ are consistent with the general definition given in the preceding section.

An application of this result is

Theorem 5.2.2: *The set of all linear combinations over $\mathscr{F}$ of the vectors V_1, $V_2, \ldots, V_k$ of a vector space $\mathscr{V}$ is a subspace of $\mathscr{V}$.*

By the preceding theorem, we need only show that (A_1) and (S_1) hold. If

$$\alpha_1 V_1 + \alpha_2 V_2 + \cdots + \alpha_k V_k$$

and

$$\beta_1 V_1 + \beta_2 V_2 + \cdots + \beta_k V_k$$

are arbitrary vectors of the set, then their sum is the linear combination

(5.2.2) $$(\alpha_1 + \beta_1)V_1 + (\alpha_2 + \beta_2)V_2 + \cdots + (\alpha_k + \beta_k)V_k,$$

which is also in the set, so (A_1) holds. Moreover,

(5.2.3) $$(\alpha\alpha_1)V_1 + (\alpha\alpha_2)V_2 + \cdots + (\alpha\alpha_k)V_k$$

is in the set, so (S_1) holds. Hence we have a vector space.

For example, the set of all real 2×2 matrices is a vector space over the set of real numbers. (Verify that all the properties hold.) Hence, by Theorem 5.2.2, the set of all matrices

$$\alpha_1 \begin{bmatrix} 1 & 0 \\ 0 & 1 \end{bmatrix} + \alpha_2 \begin{bmatrix} 0 & 1 \\ -1 & 0 \end{bmatrix} \qquad (\alpha_1, \alpha_2 \text{ real})$$

is also a vector space over the set of real numbers.

We say that the vectors $V_1, V_2, \ldots, V_k$ of a vector space $\mathscr{V}$ are **linearly dependent** over $\mathscr{F}$ if and only if there exist scalars $\alpha_1, \alpha_2, \ldots, \alpha_k$, *not all zero*, in $\mathscr{F}$ such that

(5.2.4) $$\alpha_1 V_1 + \alpha_2 V_2 + \cdots + \alpha_k V_k = 0.$$

In particular, *when $k = 1$, the zero vector is linearly dependent since $\alpha \cdot 0 = 0$* for all scalars α but *a nonzero vector V_1 is linearly independent since* $\alpha V_1 = 0$ if and only if $\alpha = 0$. Calling a single vector dependent or independent avoids the necessity of stating exceptional cases of many theorems.

For example, since

$$2\begin{bmatrix}1\\0\\0\end{bmatrix} + 3\begin{bmatrix}0\\1\\0\end{bmatrix} + (-5)\begin{bmatrix}0\\0\\1\end{bmatrix} + (-1)\begin{bmatrix}2\\3\\-5\end{bmatrix} = \begin{bmatrix}0\\0\\0\end{bmatrix},$$

the vectors

$$\begin{bmatrix}1\\0\\0\end{bmatrix}, \quad \begin{bmatrix}0\\1\\0\end{bmatrix}, \quad \begin{bmatrix}0\\0\\1\end{bmatrix}, \quad \begin{bmatrix}2\\3\\-5\end{bmatrix}$$

are linearly dependent. Similarly, the linear polynomials

$$2x_1 - 3x_2, \quad x_1 + 2x_2, \quad 2x_1 + 4x_2$$

are linearly dependent since

$$0 \cdot (2x_1 - 3x_2) + (-2)(x_1 + 2x_2) + 1 \cdot (2x_1 + 4x_2) \equiv 0,$$

that is, since this linear combination reduces identically to the zero polynomial.

This last example illustrates the fact that *some*, but *not all*, of the α's may be 0, as the definition says.

We say that $V_1, V_2, \ldots, V_k$ are **linearly independent** if and only if equation (5.2.4) holds only when (implies that) all the α's are 0. For example, we show that the three real linear polynomials

$$2x_1 + 3x_2 - x_3$$
$$x_1 - 2x_2 + x_3$$
$$3x_1 + x_2 - 4x_3$$

are linearly independent. We have

$$\alpha_1(2x_1 + 3x_2 - x_3) + \alpha_2(x_1 - 2x_2 + x_3) + \alpha_3(3x_1 + x_2 - 4x_3) \equiv 0$$

if and only if

$$(2\alpha_1 + \alpha_2 + 3\alpha_3)x_1 + (3\alpha_1 - 2\alpha_2 + \alpha_3)x_2 + (-\alpha_1 + \alpha_2 - 4\alpha_3)x_3 \equiv 0.$$

This identity is true if and only if the coefficients of the x's all vanish, that is, if and only if

$$2\alpha_1 + \alpha_2 + 3\alpha_3 = 0$$

$$3\alpha_1 - 2\alpha_2 + \alpha_3 = 0$$

$$-\alpha_1 + \alpha_2 - 4\alpha_3 = 0.$$

The computation

1	−1	4	(from equation 3)
0	1	−11	(from equation 2)
0	3	−5	(from equation 1)

1	0	−7
0	1	−11
0	0	28

shows that the only solution of these equations in the α's is the trival solution $(0, 0, 0)$, so the given expressions are indeed linearly independent.

Questions of linear dependence typically lead to systems of linear equations, as in the preceding example. To illustrate further, we determine whether or not the vectors in $\mathscr{R}^2$,

$$\begin{bmatrix} 2 \\ -1 \end{bmatrix}, \quad \begin{bmatrix} 1 \\ 2 \end{bmatrix}, \quad \begin{bmatrix} 7 \\ 4 \end{bmatrix},$$

are linearly dependent. The vector equation

$$\alpha_1 \begin{bmatrix} 2 \\ -1 \end{bmatrix} + \alpha_2 \begin{bmatrix} 1 \\ 2 \end{bmatrix} + \alpha_3 \begin{bmatrix} 7 \\ 4 \end{bmatrix} = \begin{bmatrix} 0 \\ 0 \end{bmatrix}$$

is equivalent to the system of scalar equations

$$2\alpha_1 + \alpha_2 + 7\alpha_3 = 0$$

$$-\alpha_1 + 2\alpha_2 + 4\alpha_3 = 0,$$

from which, by the usual methods,

$$\alpha_1 = -2\alpha_3$$

$$\alpha_2 = -3\alpha_3.$$

A convenient choice for α_3 is the value -1. Then a solution is $(2, 3, -1)$, so

$$2\begin{bmatrix} 2 \\ -1 \end{bmatrix} + 3\begin{bmatrix} 1 \\ 2 \end{bmatrix} - \begin{bmatrix} 7 \\ 4 \end{bmatrix} = \begin{bmatrix} 0 \\ 0 \end{bmatrix}.$$

A particularly useful special case is given in

Theorem 5.2.3: *Every nonempty subset of a set of vectors of the form*

$$(5.2.5) \quad \begin{bmatrix} a_{11} \\ a_{21} \\ a_{31} \\ a_{41} \\ \vdots \\ a_{n1} \end{bmatrix}, \begin{bmatrix} 0 \\ a_{22} \\ a_{32} \\ a_{42} \\ \vdots \\ a_{n2} \end{bmatrix}, \dots, \begin{bmatrix} 0 \\ \vdots \\ 0 \\ a_{kk} \\ a_{k+1,\,k} \\ \vdots \\ a_{nk} \end{bmatrix}, \quad a_{11}a_{22}\cdots a_{kk} \neq 0,$$

where the components belong to a field $\mathscr{F}$, is a linearly independent set over $\mathscr{F}$. A similar result holds for row vectors.

Indeed, if a linear combination of these vectors is to be zero, the coefficient of the first vector must be zero to eliminate the component a_{11}. Thereafter, the coefficient of the second vector must be zero to eliminate the component a_{22}. Continuing in this way, we see that all the coefficients must be zero, so the vectors are indeed linearly independent. A similar argument applies to any nonempty subset. (See also Corollary 5.4.2.)

A special case of this is the set of elementary n-vectors $E_1, E_2, \dots, E_n$, any subset of which is a linearly independent set.

We say that a vector V is a **linear combination** of the vectors $V_1, V_2, \dots, V_k$ if and only if there exist scalars $\alpha_1, \alpha_2, \dots, \alpha_k$ such that

$$(5.2.6) \quad V = \alpha_1 V_1 + \alpha_2 V_2 + \cdots + \alpha_k V_k.$$

Thus, since

$$\begin{bmatrix} 7 \\ 4 \end{bmatrix} = 2 \begin{bmatrix} 2 \\ -1 \end{bmatrix} + 3 \begin{bmatrix} 1 \\ 2 \end{bmatrix},$$

the vector $\begin{bmatrix} 7 \\ 4 \end{bmatrix}$ is a linear combination of $\begin{bmatrix} 2 \\ -1 \end{bmatrix}$ and $\begin{bmatrix} 1 \\ 2 \end{bmatrix}$. Indeed, we can write *any* 2-vector as a linear combination of these last two vectors, for the vector equation

$$\begin{bmatrix} x_1 \\ x_2 \end{bmatrix} = \alpha_1 \begin{bmatrix} 2 \\ -1 \end{bmatrix} + \alpha_2 \begin{bmatrix} 1 \\ 2 \end{bmatrix}$$

is equivalent to the pair of scalar equations

$$2\alpha_1 + \alpha_2 = x_1$$
$$-\alpha_1 + 2\alpha_2 = x_2,$$

the solution of which is

$$\alpha_1 = \tfrac{1}{5}(2x_1 - x_2)$$

$$\alpha_2 = \tfrac{1}{5}(x_1 + 2x_2),$$

so that as soon as $\begin{bmatrix} x_1 \\ x_2 \end{bmatrix}$ is given, we can compute the corresponding values of α_1 and α_2.

As another illustration, consider the complete solution of the system of equations

$$5x_1 - 8x_2 + 2x_3 - x_4 = 0$$

$$-7x_1 + 3x_2 + x_3 + x_4 = 0,$$

which we find by the methods of Chapter 2 to be, in parametric form,

$$x_1 = \phantom{2u+{}} 3u$$

$$x_2 = \phantom{2u+{}} 3v$$

$$x_3 = 2u + 5v$$

$$x_4 = 19u - 14v.$$

This can be written in vector form as

$$\begin{bmatrix} x_1 \\ x_2 \\ x_3 \\ x_4 \end{bmatrix} = u \begin{bmatrix} 3 \\ 0 \\ 2 \\ 19 \end{bmatrix} + v \begin{bmatrix} 0 \\ 3 \\ 5 \\ -14 \end{bmatrix}.$$

The two vectors on the right are linearly independent solutions of the given system. That is, the complete solution is the set of all linear combinations of a set of linearly independent particular solutions. We shall see in Chapter 6 that this holds true in general for systems of homogeneous linear equations.

We conclude this section with two simple theorems.

Theorem 5.2.4: *If the vectors $V_1, V_2, \ldots, V_n$ are linearly dependent, then at least one of them can be written as a linear combination of the rest.*

By hypothesis, the equation

$$\alpha_1 V_1 + \alpha_2 V_2 + \cdots + a_n V_n = 0$$

holds with at least one of the α's different from 0. Suppose that $\alpha_i \neq 0$. Then we can solve for V_i:

$$V_i = \left(-\frac{\alpha_1}{\alpha_i}\right)V_1 + \cdots + \left(-\frac{\alpha_{i-1}}{\alpha_i}\right)V_{i-1} + \left(-\frac{\alpha_{i+1}}{\alpha_i}\right)V_{i+1} + \cdots + \left(-\frac{\alpha_n}{\alpha_i}\right)V_n.$$

This proves the theorem. A converse result is this:

Theorem 5.2.5: *If V can be expressed as a linear combination of $V_1, V_2, \ldots, V_k$, then the vectors $V, V_1, V_2, \ldots, V_k$ are linearly dependent.*

By hypothesis, there exist $\alpha_1, \alpha_2, \ldots, \alpha_k$ such that

$$V = \alpha_1 V_1 + \alpha_2 V_2 + \cdots + \alpha_k V_k.$$

This equation can be written

$$(-1)V + \alpha_1 V_1 + \alpha_2 V_2 + \cdots + \alpha_k V_k = 0.$$

Since at least the coefficient of V is not zero, the desired conclusion follows.

5.3   Exercises

1. Show that the set of all real n-vectors X which simultaneously satisfy two systems of homogeneous equations, $AX = 0$ and $BX = 0$, with real coefficients, is a vector space over the field of real numbers.
*2. Show that every linear combination of solutions $Y_1, Y_2, \ldots, Y_k$ of an equation $AX = 0$ is also a solution.
3. Let A and B be fixed $n \times n$ real matrices. Show that the set of all $n \times n$ real matrices W such that $AWB = 0$ is a vector space, the scalars being the real numbers.
4. Show that the set of all real linear polynomials of the form

$$a_1 x_1 + a_2 x_2 + \cdots + a_n x_n,$$

   where n is fixed, is a vector space over the field of real numbers. Then do the same for the set of all real quadratic polynomials of the form

$$a_{11}x_1^2 + a_{12}x_1 x_2 + \cdots + a_{1n}x_1 x_n + a_{22}x_2^2 + a_{23}x_2 x_3 + \cdots + a_{nn}x_n^2,$$

   where again n is fixed.
5. Is the set of all **Hermitian forms** in n complex variables, that is, the set of all expressions of the form

$$\sum a_{ij}\bar{z}_i z_j \qquad (a_{ji} = \bar{a}_{ij})$$

   a vector space over the field of complex numbers? Why? Is it a vector space over the field of real numbers? Why?

6. Show that the set of all real $n \times n$ matrices A such that $AG = GA$, where G is a fixed real $n \times n$ matrix, is a vector space over the field of real numbers. Show that this space is also closed with respect to multiplication.

7. Prove that the set of all real 3-vectors X such that $X^T A = 0$, where A is a fixed real 3-vector, is a vector space over the field of real numbers. Interpret geometrically in $\mathscr{E}^3$.

8. For each of the following systems of equations, represent in matrix form the vector space of all solutions of the system:

(a) $\quad 2x_1 - x_2 + 4x_3 - 3x_4 = 0$
$\quad 5x_1 - 2x_2 - x_3 + x_4 = 0,$

(b) $\quad x_1 - x_2 + x_3 = 0,$

(c) $\quad x_1 + x_2 + x_3 + x_4 = 0$
$\quad x_1 - x_2 + x_3 + x_4 = 0$
$\quad x_1 + x_2 - x_3 + x_4 = 0$
$\quad x_1 + x_2 + x_3 - x_4 = 0.$

9. Find a linear combination of

$$\begin{bmatrix} 0 \\ 1 \\ 1 \end{bmatrix} \quad \text{and} \quad \begin{bmatrix} 1 \\ -1 \\ 1 \end{bmatrix}$$

that is orthogonal to

$$\begin{bmatrix} 1 \\ 2 \\ 0 \end{bmatrix}.$$

Illustrate with a figure.

10. Prove that two nonzero vectors A_1 and A_2 of a vector space $\mathscr{V}$ are linearly dependent if and only if there exists a scalar c such that $A_1 = cA_2$.

*11. Prove that k n-vectors $A_1, A_2, \ldots, A_k$ over a field $\mathscr{F}$ are linearly dependent if two of these vectors are the same or have proportional components. Give an example that shows the converse is not true.

*12. Prove that, if a set of vectors of a vector space $\mathscr{V}$ includes the zero vector, then the vectors of the set are linearly dependent.

13. Prove that the vectors

$$\begin{bmatrix} 1 \\ 0 \\ 0 \\ 0 \end{bmatrix}, \begin{bmatrix} 1 \\ 1 \\ 0 \\ 0 \end{bmatrix}, \begin{bmatrix} 1 \\ 1 \\ 1 \\ 0 \end{bmatrix}, \begin{bmatrix} 1 \\ 1 \\ 1 \\ 1 \end{bmatrix}$$

are linearly independent.

14. Prove that the vectors of Exercise 13 together with any real vector

$$\begin{bmatrix} a_1 \\ a_2 \\ a_3 \\ a_4 \end{bmatrix}$$

are linearly dependent.

15. Show that vectors V_1, V_2, ..., V_k are linearly dependent if and only if $\alpha_1 V_1$, $\alpha_2 V_2$, ..., $\alpha_k V_k$, where the α's are nonzero scalars, are linearly dependent.

16. Given that V_1, V_2, and V_3 are linearly independent vectors of an arbitrary vector space $\mathscr{V}$, prove that V_1, $V_1 + V_2$, and $V_1 + V_2 + V_3$ are also linearly independent vectors of $\mathscr{V}$. Generalize.

17. Prove that vectors

$$A = \begin{bmatrix} a_1 \\ a_2 \end{bmatrix} \qquad \text{and} \qquad B = \begin{bmatrix} b_1 \\ b_2 \end{bmatrix}$$

in $\mathscr{E}^2$ are linearly dependent if and only if $a_1 b_2 - a_2 b_1 = 0$.

18. Given that $a_1 b_2 - a_2 b_1 \neq 0$, show how to express an arbitrary vector

$$X = \begin{bmatrix} x_1 \\ x_2 \end{bmatrix}$$

in $\mathscr{E}^2$ as a linear combination of

$$A = \begin{bmatrix} a_1 \\ a_2 \end{bmatrix}, \qquad B = \begin{bmatrix} b_1 \\ b_2 \end{bmatrix};$$

that is, find the coefficients α and β in the equation

$$X = \alpha A + \beta B.$$

Then solve the problem for the special case

$$A = \begin{bmatrix} 1 \\ 2 \end{bmatrix}, \qquad B = \begin{bmatrix} 2 \\ 1 \end{bmatrix}.$$

19. Prove that the vectors

$$A = \begin{bmatrix} a_1 \\ a_2 \\ \vdots \\ a_n \end{bmatrix} \qquad \text{and} \qquad B = \begin{bmatrix} b_1 \\ b_2 \\ \vdots \\ b_n \end{bmatrix}$$

in $\mathscr{E}^n$ are linearly dependent if and only if $a_i b_j - a_j b_i = 0$ whenever $1 \leq i < j \leq n$.

20. Given that the vectors V_1, V_2, ..., V_k of a vector space $\mathscr{V}$ are linearly independent, under what conditions on the coefficients α_{ij} are the vectors $\alpha_{11}V_1$, $\alpha_{21}V_1 + \alpha_{22}V_2$, $\alpha_{31}V_1 + \alpha_{32}V_2 + \alpha_{33}V_3$, ..., $\alpha_{k1}V_1 + \alpha_{k2}V_2 + \cdots + \alpha_{kk}V_k$ also linearly independent?

21. Which columns of the **upper triangular matrix**

$$\begin{bmatrix} a_{11} & a_{12} & \cdots & a_{1n} \\ 0 & a_{22} & \cdots & a_{2n} \\ \vdots & & & \\ 0 & 0 & \cdots & a_{nn} \end{bmatrix}$$

are linearly independent if $\prod_{i=1}^{n} a_{ii} \neq 0$? If precisely one of the a_{ii}'s is zero?

*22. Prove that any p nonzero vectors V_1, V_2, ..., V_p of a vector space $\mathscr{V}$ are linearly dependent if and only if for some integer $k \geq 2$, V_k is a linear combination of V_1, V_2, ..., V_{k-1}.

*23. Prove that k given vectors A_1, A_2, ..., A_k in $\mathscr{R}^n$ are linearly dependent if and only if the equation

$$[A_1, A_2, ..., A_k]\begin{bmatrix} x_1 \\ x_2 \\ \vdots \\ x_k \end{bmatrix} = 0$$

has a nontrivial solution.

*24. Prove that k n-vectors A_1, A_2, ..., A_k over a number field $\mathscr{F}$ are linearly independent if and only if the equation $[A_1, A_2, ..., A_k]X_{k \times 1} = 0$ has only the trivial solution.

25. Are the matrices

$$\begin{bmatrix} 2 & -1 \\ 4 & 6 \end{bmatrix}, \quad \begin{bmatrix} 3 & 2 \\ 8 & 3 \end{bmatrix}, \quad \begin{bmatrix} -5 & -8 \\ -16 & 4 \end{bmatrix}, \quad \begin{bmatrix} 0 & -7 \\ -4 & 13 \end{bmatrix}$$

linearly dependent over the rational number field?

26. Prove that, if columns $j_1, j_2, ..., j_r$ of a matrix B are linearly dependent, then columns $j_1, j_2, ..., j_r$ of AB are also linearly dependent. (This implies that, if columns $j_1, j_2, ..., j_r$ of AB are independent, so are those of B.)

27. Prove that if $P_1, ..., P_k$ are linearly independent vectors in $\mathscr{R}^n$ and if $P = \sum_{i=1}^{k} a_i P_i$, then $P_1 - P$, $P_2 - P$, ..., $P_k - P$ are linearly independent if and only if $\sum \alpha_i \neq 1$.

*28. Prove that, over an arbitrary field $\mathscr{F}$, a necessary and sufficient condition that k linear combinations of k linearly independent n-vectors be also independent is that the matrix of the coefficients of combination be nonsingular.

*29. Show that, over an arbitrary field $\mathscr{F}$, the concept of the linear dependence of polynomials of a given degree in a given set of n variables is essentially the same as the concept of the linear dependence of vectors.

30. Show that, over an arbitrary field $\mathscr{F}$, every matrix of order 2 can be expressed as a linear combination of the matrices

$$\begin{bmatrix} 1 & 0 \\ 0 & 0 \end{bmatrix}, \quad \begin{bmatrix} 0 & 1 \\ 0 & 0 \end{bmatrix}, \quad \begin{bmatrix} 0 & 0 \\ 1 & 0 \end{bmatrix}, \quad \begin{bmatrix} 0 & 0 \\ 0 & 1 \end{bmatrix}.$$

Generalize this result.

31. Show that the Pauli spin matrices σ_x, σ_y, σ_z (Exercise 22, Section 1.9) and the identity matrix I_2 are a linearly independent set and then show how an arbitrary matrix $A = [\alpha_{ij}]$ of order 2 may be written as a linear combination of these four.

32. If $P = a_1\sigma_x + a_2\sigma_y + a_3\sigma_z + a_4 I_2$ and $Q = b_1\sigma_x + b_2\sigma_y + b_3\sigma_z + b_4 I_2$, where the a's and b's are scalars, express PQ as a linear combination of σ_x, σ_y, σ_z, and I_2.

*33. Show that a subset $\mathscr{S}$ of a vector space $\mathscr{V}$ over $\mathscr{F}$ is also a vector space if and only if, for all V_1 and V_2 in $\mathscr{S}$ and for all α_1 and α_2 in $\mathscr{F}$, $\alpha_1 V_1 + \alpha_2 V_2$ belongs to $\mathscr{S}$. Show also that $\mathscr{S}$ is a vector space if and only if, for all V_1 and V_2 in $\mathscr{S}$ and for all α in $\mathscr{F}$, $V_1 + \alpha V_2$ belongs to $\mathscr{S}$.

34. A homogeneous linear polynomial

$$a_1 x_1 + a_2 x_2 + \cdots + a_n x_n,$$

where the a_i's are arbitrary scalars, is called a **linear form** in the variables $x_1, x_2, \ldots, x_n$. Show that the set of all linear forms in $x_1, x_2, \ldots, x_n$ with coefficients belonging to a field $\mathscr{F}$ is a vector space over $\mathscr{F}$. Then investigate the linear dependence or independence of each of these sets of real linear forms in three variables:

(a)
$$\begin{aligned} 2x_1 + x_2 + x_3 \\ x_1 + 2x_2 + x_3 \\ x_1 + x_2 + 2x_3, \end{aligned}$$

(b)
$$\begin{aligned} 2x_1 + x_2 + x_3 \\ x_1 + 2x_2 + x_3 \\ x_1 + x_2 + 2x_3 \\ x_1 + x_2 + x_3. \end{aligned}$$

*35. Show that if $A_1, A_2, \ldots, A_m$ and X belong to $\mathscr{F}^n$, the linear forms $A_1^T X$, $A_2^T X, \ldots, A_m^T X$ are linearly dependent if and only if $A_1, A_2, \ldots, A_m$ are linearly dependent.

36. Show that m linear expressions over a field $\mathscr{F}$,

$$a_{i1} x_1 + a_{i2} x_2 + \cdots + a_{in} x_n, \qquad i = 1, 2, \ldots, m$$

are linearly independent if (a) each expression contains, with nonzero coefficient, a variable appearing in no other expression of the set, or (b) each expression contains, with nonzero coefficient, a variable appearing in no previous linear expression of the set.

37. A homogeneous quadratic polynomial

$$\sum_{i,j=1}^{n} a_{ij} x_i x_j,$$

where the a_{ij}'s are arbitrary scalars, is called a **quadratic form** in the variables $x_1, x_2, ..., x_n$. The set of all quadratic forms in $x_1, x_2, ..., x_n$ with coefficients in a field $\mathscr{F}$ is a vector space over $\mathscr{F}$. Investigate the linear dependence or independence of these sets of real quadratic forms:

(a) $2x_1^2 + x_1x_2 + x_2^2$

   $x_1^2 + 2x_1x_2 + x_2^2$

   $x_1^2 + x_1x_2 + 2x_2^2,$

(b) $2x_1^2 + x_1x_2 + x_2^2$

   $x_1^2 + 2x_1x_2 + x_2^2$

   $x_1^2 + x_1x_2 + 2x_2^2$

   $x_1^2 + x_1x_2 + x_2^2.$

Compare with Exercise 34.

38. A homogeneous polynomial

$$\sum_{j=1}^{m} \sum_{i=1}^{n} a_{ij} x_i y_j,$$

where the a_{ij}'s are arbitrary scalars, is called a **bilinear form** in the variables $x_1, x_2, ..., x_n$ and $y_1, y_2, ..., y_m$. The set of all bilinear forms with coefficients in a field $\mathscr{F}$ is a vector space over $\mathscr{F}$. Investigate the linear dependence or independence of the real bilinear forms in $x_1, x_2 ; y_1, y_2 :$

$$x_1 y_1 - x_1 y_2 + x_2 y_1 - x_2 y_2$$

$$-x_1 y_1 + x_1 y_2 - x_2 y_1 + x_2 y_2$$

$$2x_1 y_1 + x_1 y_2 - 2x_2 y_1 - 2x_2 y_2 .$$

Linear, quadratic, and bilinear forms are important in a wide variety of mathematical, physical, and statistical applications.

5.4   Basic Theorems on Linear Dependence

The following theorems make it easy to apply the concept of linear dependence. The results are independent of the field $\mathscr{F}$ over which the vector spaces are defined.

Theorem 5.4.1: *If some p ($p > 0$) vectors of the set of k vectors $V_1, V_2, ..., V_k$ are linearly dependent, then all k of the vectors are linearly dependent.*

Suppose that $V_1, V_2, ..., V_p$ are linearly dependent. Then there exist scalars $\alpha_1, \alpha_2, ..., \alpha_p$, not all 0, such that

$$\alpha_1 V_1 + \alpha_2 V_2 + \cdots + \alpha_p V_p = 0.$$

Hence

$$\alpha_1 V_1 + \alpha_2 V_2 + \cdots + \alpha_p V_p + 0 \cdot V_{p+1} + \cdots + 0 \cdot V_k = 0.$$

Since the coefficients in this latter equation are not all 0, the k vectors are linearly dependent. The same conclusion holds if any other subset of p vectors is linearly dependent.

We have at once

Corollary 5.4.2: *If the vectors V_1, V_2, ..., V_k are linearly independent, then no nonempty subset of these is linearly dependent.*

A special case of Theorem 5.4.1 is

Corollary 5.4.3: *If one of the vectors V_1, V_2, ..., V_k is the zero vector, the set is linearly dependent.*

This is because the zero vector alone is linearly dependent, so we can apply the theorem for the case $p = 1$.

Theorem 5.4.4: *If the vectors V_1, V_2, ..., V_k are linearly independent, but the vectors V_1, V_2, ..., V_k, W are linearly dependent, then W can be written as a linear combination of the V's.*

There exists, by hypothesis, a relation

$$\alpha W + \alpha_1 V_1 + \alpha_2 V_2 + \cdots + \alpha_k V_k = 0$$

with not all the α's zero. In particular, $\alpha \neq 0$, for if it were zero, the V's would be linearly dependent, contrary to hypothesis. Hence

$$W = \left(-\frac{\alpha_1}{\alpha}\right)V_1 + \left(-\frac{\alpha_2}{\alpha}\right)V_2 + \cdots + \left(-\frac{\alpha_k}{\alpha}\right)V_k.$$

Theorem 5.4.5: *Any $k + 1$ linear combinations of V_1, V_2, ..., V_k are linearly dependent.*

Let the linear combinations be

$$W_1 = \alpha_{11}V_1 + \alpha_{12}V_2 + \cdots + \alpha_{1k}V_k$$
$$W_2 = \alpha_{21}V_1 + \alpha_{22}V_2 + \cdots + \alpha_{2k}V_k$$
$$\vdots$$
$$W_{k+1} = \alpha_{k+1,\,1}V_1 + \alpha_{k+1,\,2}V_2 + \cdots + \alpha_{k+1,\,k}V_k.$$

We must show there exist scalars β_1, β_2, ..., β_{k+1}, not all zero, such that

$$\beta_1 W_1 + \beta_2 W_2 + \cdots + \beta_{k+1} W_{k+1} = 0.$$

By substituting for the W's, expanding, and collecting, we reduce this condition to the form

$$\sum_{i=1}^{k} (\beta_1 \alpha_{1i} + \beta_2 \alpha_{2i} + \cdots + \beta_{k+1} \alpha_{k+1, i}) V_i = 0.$$

This equation will certainly hold if all the coefficients of the V's are zero:

$$\beta_1 \alpha_{11} + \beta_2 \alpha_{21} + \cdots + \beta_{k+1} \alpha_{k+1, 1} = 0$$
$$\beta_1 \alpha_{12} + \beta_2 \alpha_{22} + \cdots + \beta_{k+1} \alpha_{k+1, 2} = 0$$
$$\vdots$$
$$\beta_1 \alpha_{1k} + \beta_2 \alpha_{2k} + \cdots + \beta_{k+1} \alpha_{k+1, k} = 0.$$

Here we have k homogeneous equations in $k + 1$ unknowns. By transforming this system to reduced echelon form, we are able to express some subset of p of these variables in terms of the remaining $k + 1 - p$ variables, $p \le k$. Thus there always exist nontrivial solutions for the β's, so the theorem is proved.

From this result and Theorem 5.4.1, we have

Corollary 5.4.6: *More than k linear combinations of $V_1, V_2, \ldots, V_k$ are always linearly dependent.*

A particularly useful special case is

Corollary 5.4.7: *More than n vectors of $\mathscr{F}^n$ are always linearly dependent.*

In fact, if $E_1, E_2, \ldots, E_n$ denote the elementary n-vectors, then any n-vector X of $\mathscr{F}^n$ can be written

$$X = x_1 E_1 + x_2 E_2 + \cdots + x_n E_n,$$

as we saw in Chapter 4. Since all n-vectors are linear combinations of the n n-vectors $E_1, E_2, \ldots, E_n$, more than n n-vectors are linearly dependent by the preceding corollary.

For example, we know at once by this corollary that

$$\begin{bmatrix} 2 \\ 1 \end{bmatrix}, \quad \begin{bmatrix} 1 \\ 2 \end{bmatrix}, \quad \begin{bmatrix} -4 \\ 5 \end{bmatrix}$$

are linearly dependent over the real field.

5.5 Dimension and Basis

We call a vector space $\mathscr{V}$ **finite dimensional** if and only if there exists a finite subset $\mathscr{S}$ of vectors of $\mathscr{V}$ such that every vector of $\mathscr{V}$ can be written as a linear combination of the vectors of $\mathscr{S}$. Such a finite subset $\mathscr{S}$ is said to **span** or to **generate** the vector space $\mathscr{V}$.

For example, the set of all n-vectors over a field $\mathscr{F}$ is finite dimensional, since for every vector X in $\mathscr{F}^n$,

$$X = x_1 E_1 + x_2 E_2 + \cdots + x_n E_n,$$

where the E_j's are the elementary n-vectors mentioned earlier.

The set of all real 2×2 matrices is finite dimensional, since

$$\begin{bmatrix} a_{11} & a_{12} \\ a_{21} & a_{22} \end{bmatrix} = a_{11}\begin{bmatrix} 1 & 0 \\ 0 & 0 \end{bmatrix} + a_{12}\begin{bmatrix} 0 & 1 \\ 0 & 0 \end{bmatrix} + a_{21}\begin{bmatrix} 0 & 0 \\ 1 & 0 \end{bmatrix} + a_{22}\begin{bmatrix} 0 & 0 \\ 0 & 1 \end{bmatrix}.$$

The **zero space** $\mathscr{Z}$, consisting of the zero vector alone, is spanned by the zero vector and hence is finite dimensional.

The set of all real quadratic polynomials

$$ax^2 + bxy + cy^2$$

is finite dimensional since each is a linear combination of the three quadratic polynomials x^2, xy, y^2.

On the other hand, the set of all real polynomials

$$a_n x^n + a_{n-1} x^{n-1} + \cdots + a_1 x + a_0, \qquad n = 0, 1, \ldots,$$

is not finite dimensional, for a linear combination of the polynomials of a finite set can have degree no higher than that of the highest-degree polynomial in the set. Hence we cannot represent all polynomials as linear combinations of the polynomials of any finite set.

The finite-dimensional vector spaces are the ones most commonly seen in applications, and from here on we treat them exclusively.

Consider next the equations

$$\begin{bmatrix} x_1 \\ x_2 \end{bmatrix} = x_1\begin{bmatrix} 1 \\ 2 \end{bmatrix} + x_2\begin{bmatrix} -3 \\ 0 \end{bmatrix} + x_2\begin{bmatrix} 3 \\ 3 \end{bmatrix} + (x_1 + x_2)\begin{bmatrix} 0 \\ -2 \end{bmatrix};$$

$$\begin{bmatrix} x_1 \\ x_2 \end{bmatrix} = x_1\begin{bmatrix} 1 \\ 0 \end{bmatrix} + x_2\begin{bmatrix} 0 \\ 1 \end{bmatrix}.$$

They show that each of the sets of vectors

$$\begin{bmatrix} 1 \\ 2 \end{bmatrix}, \quad \begin{bmatrix} -3 \\ 0 \end{bmatrix}, \quad \begin{bmatrix} 3 \\ 3 \end{bmatrix}, \quad \begin{bmatrix} 0 \\ -2 \end{bmatrix} \quad \text{and} \quad \begin{bmatrix} 1 \\ 0 \end{bmatrix}, \quad \begin{bmatrix} 0 \\ 1 \end{bmatrix}$$

spans $\mathscr{E}^2$, but the second set does so more economically. The vectors of the second (economical) set are linearly independent but those of the first (extravagant) set are not.

Moreover, if we omit either vector from the second set, we cannot get all vectors of $\mathscr{E}^2$, for multiples of one of these vectors always have at least one component zero. On the other hand, since

$$\begin{bmatrix} x_1 \\ x_2 \end{bmatrix} = \frac{x_2}{2} \begin{bmatrix} 1 \\ 2 \end{bmatrix} + \frac{x_2 - 2x_1}{6} \begin{bmatrix} -3 \\ 0 \end{bmatrix},$$

we could omit the last two vectors of the first set and still obtain all vectors of $\mathscr{E}^2$.

These two examples illustrate the following definition and theorem.

We define a set of nonzero vectors spanning a vector space $\mathscr{V}$ to be a **basis** for $\mathscr{V}$ if and only if no proper subset of the spanning set also spans $\mathscr{V}$. The simplest example of a basis is the set of elementary vectors $E_1, E_2, \ldots, E_n$ of $\mathscr{F}^n$. The first two vectors of the preceding numerical example,

$$\begin{bmatrix} 1 \\ 2 \end{bmatrix} \quad \text{and} \quad \begin{bmatrix} -3 \\ 0 \end{bmatrix},$$

are a basis for $\mathscr{E}^2$. The reader should show in each case that no proper subset spans the space in question.

Theorem 5.5.1: *Every basis for a finite-dimensional vector space $\mathscr{V}$ contains only linearly independent vectors.*

Indeed, if the vectors of a spanning set are not linearly independent, we can write one of them as a linear combination of the others. Then, substituting the resulting expression for this vector into any linear combination containing it, and rearranging a bit, we can express any vector of $\mathscr{V}$ as a linear combination of the remaining vectors of the spanning set. The spanning set is thus not a basis, since a proper subset of it also spans $\mathscr{V}$. Hence a basis contains only linearly independent vectors.

To illustrate, we return to the numerical example. Expressing the last two vectors of the spanning set of four vectors as linear combinations of the first two, we have

$$\begin{bmatrix} 3 \\ 3 \end{bmatrix} = \frac{3}{2} \begin{bmatrix} 1 \\ 2 \end{bmatrix} + \left(-\frac{1}{2} \right) \begin{bmatrix} -3 \\ 0 \end{bmatrix}$$

and

$$\begin{bmatrix} 0 \\ -2 \end{bmatrix} = (-1)\begin{bmatrix} 1 \\ 2 \end{bmatrix} + \left(-\frac{1}{3}\right)\begin{bmatrix} -3 \\ 0 \end{bmatrix}.$$

Hence, substituting and collecting, we find that

$$\alpha_1 \begin{bmatrix} 1 \\ 2 \end{bmatrix} + \alpha_2 \begin{bmatrix} -3 \\ 0 \end{bmatrix} + \alpha_3 \begin{bmatrix} 3 \\ 3 \end{bmatrix} + \alpha_4 \begin{bmatrix} 0 \\ -2 \end{bmatrix}$$

$$= \left(\alpha_1 + \frac{3}{2}\alpha_3 - \alpha_4\right)\begin{bmatrix} 1 \\ 2 \end{bmatrix} + \left(\alpha_2 - \frac{1}{2}\alpha_3 - \frac{1}{3}\alpha_4\right)\begin{bmatrix} -3 \\ 0 \end{bmatrix},$$

so any linear combination of the four vectors can be expressed as a linear combination of the first two. Hence the four are not a basis for $\mathscr{E}^2$. The first two, being linearly independent, *are* a basis for $\mathscr{E}^2$, as the next theorem reveals:

Theorem 5.5.2: *Every set of linearly independent vectors which spans a finite-dimensional vector space $\mathscr{V}$ is a basis for $\mathscr{V}$.*

Suppose that $V_1, V_2, \ldots, V_k$ are linearly independent and span $\mathscr{V}$. Then, in particular, $V_1, V_2, \ldots, V_k$ belong to $\mathscr{V}$. Now, if these vectors are not a basis for $\mathscr{V}$, then some proper subset thereof spans $\mathscr{V}$. Hence there must exist some V_i that is not in this proper subset. Since V_i is in $\mathscr{V}$, it must be a linear combination of the vectors in the spanning subset. This implies that $V_1, V_2, \ldots, V_k$ are linearly dependent, contrary to hypothesis. The contradiction shows that $V_1, V_2, \ldots, V_k$ must indeed constitute a basis for $\mathscr{V}$.

Theorem 5.5.3: *Every finite-dimensional vector space $\mathscr{V} \neq \mathscr{Z}$ has at least one basis.*

Since $\mathscr{V}$ is finite dimensional and does not consist of the zero vector alone, there exists a finite set $V_1, V_2, \ldots, V_k$ of nonzero vectors of $\mathscr{V}$ which span $\mathscr{V}$. If these are linearly independent, they are a basis for $\mathscr{V}$ by the previous theorem. If not, we can write one of them, say V_i, as a linear combination of the others, and replace it by this expression. Then every linear combination of $V_1, V_2, \ldots, V_k$ can be expressed as a linear combination of $V_1, V_2, \ldots, V_{i-1}, V_{i+1}, \ldots, V_k$, in the manner of the last preceding numerical example. If this remaining set of vectors is not linearly independent, we can eliminate another in the same way, and so on, as long as the reduced set is linearly dependent. Since we cannot eliminate *all* of $V_1, V_2, \ldots, V_k$ and still span $\mathscr{V}$, we must eventually arrive at a linearly independent set. By the preceding theorem, this is a basis for $\mathscr{V}$.

Theorem 5.5.4: *Any two bases of a finite-dimensional vector space $\mathscr{V}$ contain the same number of vectors.*

Let $\mathscr{B}_1$ and $\mathscr{B}_2$ denote any two bases of $\mathscr{V}$. Every vector of $\mathscr{B}_1$ can be written as a linear combination of the vectors of $\mathscr{B}_2$. If there were more vectors in $\mathscr{B}_1$ than in $\mathscr{B}_2$, they would then be linearly dependent, by Corollary 5.4.6. Since they are linearly independent by hypothesis, the number of vectors in $\mathscr{B}_1$ is no greater than the number of vectors in $\mathscr{B}_2$. In the same way, the number of vectors in $\mathscr{B}_2$ is no greater than the number of vectors in $\mathscr{B}_1$. Hence $\mathscr{B}_1$ and $\mathscr{B}_2$ contain the same number of vectors.

The unique number of vectors in a basis of $\mathscr{V}$ is called the **dimension** of $\mathscr{V}$. For example, the dimension of the set of all vectors in $\mathscr{F}^n$ is n, because this set of vectors has the basis $E_1, E_2, \ldots, E_n$. The space $\mathscr{Z}$ consisting of the zero vector alone is defined to have dimension 0. It has no basis.

The preceding theorems show that the dimension of a finite-dimensional vector space $\mathscr{V}$ is the *maximum* number of linearly independent vectors in $\mathscr{V}$ and is also the *minimum* number of vectors in a spanning set for $\mathscr{V}$.

Theorem 5.5.5: *If $V_1, V_2, \ldots, V_k$ is a basis for a vector space $\mathscr{V}$, if $V = \alpha_1 V_1 + \alpha_2 V_2 + \cdots + \alpha_k V_k$, and if $\alpha_i \neq 0$, then the set $V_1, V_2, \ldots, V_{i-1}, V, V_{i+1}, \ldots, V_k$ is also a basis for $\mathscr{V}$.*

In fact, since $\alpha_i \neq 0$,

$$V_i = \left(-\frac{\alpha_1}{\alpha_i}\right)V_1 + \cdots + \left(-\frac{\alpha_{i-1}}{\alpha_i}\right)V_{i-1} + \frac{1}{\alpha_i}V$$
$$+ \left(-\frac{\alpha_{i+1}}{\alpha_i}\right)V_{i+1} + \cdots + \left(-\frac{\alpha_k}{\alpha_i}\right)V_k.$$

Hence, if W is any vector of $\mathscr{V}$ and if

$$W = \beta_1 V_1 + \beta_2 V_2 + \cdots + \beta_k V_k,$$

by substituting for V_i and rearranging, we have

$$W = \left(\beta_1 - \frac{\alpha_1 \beta_i}{\alpha_i}\right)V_1 + \cdots + \left(\beta_{i-1} - \frac{\alpha_{i-1}\beta_i}{\alpha_i}\right)V_{i-1} + \frac{\beta_i}{\alpha_i}V$$
$$+ \left(\beta_{i+1} - \frac{\alpha_{i+1}\beta_i}{\alpha_i}\right)V_{i+1} + \cdots + \left(\beta_k - \frac{\alpha_k \beta_i}{\alpha_i}\right)V_k,$$

so that any vector W can be written as a linear combination of $V_1, \ldots, V_{i-1}$, $V, V_{i+1}, \ldots, V_k$. Since the dimension of $\mathscr{V}$ is k, no proper subset of these k vectors can span $\mathscr{V}$. Hence this set is a basis for $\mathscr{V}$.

Theorem 5.5.6: *If a vector space $\mathscr{V}$ has dimension k, then any k linearly independent vectors of $\mathscr{V}$ form a basis for $\mathscr{V}$.*

Since $\mathscr{V}$ has dimension k, it has a basis $V_1, V_2, \ldots, V_k$. Let $W_1, W_2, \ldots, W_k$ be any linearly independent set of k vectors. Since $W_1 \neq 0$, there exist scalars α not all zero such that

$$W_1 = \alpha_1 V_1 + \alpha_2 V_2 + \cdots + \alpha_k V_k.$$

If $\alpha_i \neq 0$, then, by the preceding theorem,

$$V_1, V_2, \ldots, V_{i-1}, W_1, V_{i+1}, \ldots, V_k$$

is a basis for $\mathscr{V}$. Since $W_2 \neq 0$, there exist scalars β not all zero such that

$$W_2 = \beta_1 V_1 + \beta_2 V_2 + \cdots + \beta_{i-1} V_{i-1} + \beta_i W_1 + \beta_{i+1} V_{i+1} + \cdots + \beta_k V_k.$$

Now some $\beta_j, j \neq i$, is not zero, since otherwise we would have $W_2 = \beta_i W_1$, which contradicts the fact that the W's are linearly independent. Then, by the preceding theorem, we can replace V_j by W_2, so that

$$V_1, \ldots, V_{i-1}, W_1, V_{i+1}, \ldots, V_{j-1}, W_2, V_{j+1}, \ldots, V_k$$

is a basis for $\mathscr{V}$.

We now show in the same way that W_3 can replace one of the remaining V's, and so on, until all the V's have been replaced by the W's, and the proof is complete.

The same sort of argument may be used to establish the **Steinitz replacement theorem**:

Theorem 5.5.7: *Let $V_1, V_2, \ldots, V_k$ be a basis for a vector space $\mathscr{V}$ and let $W_1, W_2, \ldots, W_p$ be any set of linearly independent vectors of $\mathscr{V}$. Then we can select $k - p$ vectors $V_{i_{p+1}}, \ldots, V_{i_k}$ from the given basis so that $W_1, \ldots, W_p, V_{i_{p+1}}, \ldots, V_{i_k}$ is a basis for $\mathscr{V}$.*

The reader should supply the details.

Theorem 5.5.8: *The representation of a vector V of a finite-dimensional vector space $\mathscr{V}$ as a linear combination of the vectors of a given basis for $\mathscr{V}$ is unique.*

Let $V_1, V_2, \ldots, V_k$ be a basis for $\mathscr{V}$. Then there exist scalars $\alpha_1, \alpha_2, \ldots, \alpha_k$ such that

$$V = \alpha_1 V_1 + \alpha_2 V_2 + \cdots + \alpha_k V_k.$$

If also

$$V = \beta_1 V_1 + \beta_2 V_2 + \cdots + \beta_k V_k,$$

then by subtraction we have

$$(\alpha_1 - \beta_1)V_1 + (\alpha_2 - \beta_2)V_2 + \cdots + (\alpha_k - \beta_k)V_k = 0.$$

Since $V_1, V_2, \ldots, V_k$ are linearly independent, the coefficients here are all zero, and hence

$$\alpha_1 = \beta_1, \alpha_2 = \beta_2, \ldots, \alpha_k = \beta_k.$$

This proves the theorem.

5.6 Computation of the Dimension of a Vector Space

Suppose that a vector space $\mathscr{V}$ is spanned by $V_1, V_2, \ldots, V_k$, so that it is the set of all linear combinations of the form

$$(5.6.1) \qquad \alpha_1 V_1 + \alpha_2 V_2 + \cdots + \alpha_i V_i + \cdots + \alpha_j V_j + \cdots + \alpha_k V_k.$$

By what precedes, to determine the dimension of $\mathscr{V}$, we must determine the maximum number of linearly independent vectors that exist in $\mathscr{V}$. To accomplish this, we can use the following three **elementary operations** on the spanning set:

1. Interchange any two of the V's. That is, the vectors $V_1, \ldots, V_j, \ldots, V_i, \ldots,$ V_k span the same space as do $V_1, \ldots, V_i, \ldots, V_j, \ldots, V_k$. Indeed, by the commutative law of addition, we can interchange $\alpha_i V_i$ and $\alpha_j V_j$ in the sum (5.6.1), which shows that the totality of linear combinations is unaffected by the interchange.

Observe that, by repeated interchange of two of the V's, we can rearrange the V's in any order we wish without altering the space they span.

2. Multiply any vector V_i of the set by an arbitrary nonzero scalar. That is, if $\alpha \neq 0$, the vectors $V_1, V_2, \ldots, \alpha V_i, \ldots, V_k$ span the same space as do $V_1, V_2, \ldots, V_i, \ldots, V_k$. In fact, $\alpha_1 V_1 + \cdots + \alpha_i V_i + \cdots + \alpha_k V_k = \alpha_1 V_1 + \cdots + (\alpha_i/\alpha)(\alpha V_i) + \cdots + \alpha_k V_k$, so that any linear combination of the vectors of one set is expressible as a linear combination of those of the other set. Hence from either set we obtain the same totality of linear combinations. By applying this operation repeatedly, we can multiply each of the V's by an arbitrary nonzero scalar and still obtain the same vector space $\mathscr{V}$.

3. *Add any scalar multiple of one of the V's, say* αV_j, *to any other vector,* V_i, *of the set.* Again, this does not alter the totality of linear combinations since

$$\alpha_1 V_1 + \cdots + \alpha_i V_i + \cdots + \alpha_j V_j + \cdots + \alpha_k V_k$$
$$= \alpha_1 V_1 + \cdots + \alpha_i(V_i + \alpha V_j) + \cdots + (\alpha_j - \alpha_i \alpha)V_j + \cdots + \alpha_k V_k,$$

Hence the space spanned by $V_1, \ldots, V_i + \alpha V_j, \ldots, V_j, \ldots, V_k$ is the same as that spanned by $V_1, \ldots, V_i, \ldots, V_j, \ldots, V_k$.

Repeated use of these three operations often makes it possible to transform a set of vectors to a set with a recognizably linearly independent subset. For example, let $V_1, V_2, \ldots, V_k$ be vectors in $\mathscr{F}^n$. We first arrange them for convenience as the columns of an $n \times k$ matrix:

$$\begin{bmatrix} a_{11} & a_{12} & \cdots & a_{1k} \\ a_{21} & a_{22} & \cdots & a_{2k} \\ \vdots & & & \\ a_{n1} & a_{n2} & \cdots & a_{nk} \end{bmatrix}.$$

We now apply the sweepout procedure to the *columns* of this matrix, using only the three operations outlined above. The object is to obtain at the left a maximal set of columns such that if $p > j$, the first nonzero element in column p is in a lower row than is the first nonzero element in column j. All remaining columns are zero vectors. The end result will be said to be in **column echelon form.** At the end of the process, we have a set of vectors of $\mathscr{F}^n$, the linearly independent members of which are easy to identify by Theorem 5.2.3.

Note that we do *not* work on rows. To do so would not necessarily leave invariant the vector space spanned by the columns of the resulting matrices. For example, the columns of

$$\begin{bmatrix} 0 & 0 \\ 1 & 0 \\ 0 & 1 \end{bmatrix}$$

span a space for all of whose vectors $x_1 = 0$. If we add the second row to the first, this is no longer true, so the operation has changed the column space.

We illustrate the procedure just described by finding the dimension of, and a basis for, the subspace of $\mathscr{E}^4$ spanned by the columns of the matrix

$$\begin{bmatrix} 1 & 2 & -1 & 0 \\ 2 & 4 & -2 & 0 \\ -1 & 4 & -5 & -2 \\ 3 & 3 & 0 & 1 \end{bmatrix},$$

namely, for the **column space** of the matrix. We use the upper left 1 to sweep the first row:

$$\begin{bmatrix} 1 & 0 & 0 & 0 \\ 2 & 0 & 0 & 0 \\ -1 & 6 & -6 & -2 \\ 3 & -3 & 3 & 1 \end{bmatrix}.$$

One now sees by inspection that columns 1 and 4 are independent and that columns 2 and 3 can be swept clean with column 4 so that the dimension is 2 and a basis is the pair of vectors

$$\begin{bmatrix} 1 \\ 2 \\ -1 \\ 3 \end{bmatrix} \quad \text{and} \quad \begin{bmatrix} 0 \\ 0 \\ -2 \\ 1 \end{bmatrix}.$$

A simple but important special case is that of the identity matrix, I_n, whose columns are the elementary vectors, so that its column space is the set of all n-vectors over whatever field $\mathscr{F}$ one may be using. When $\mathscr{F}$ is the real field, the column space of I_n is just the set of all vectors in $\mathscr{E}^n$ or $\mathscr{R}^n$.

If one only needs to know whether or not the n-vectors $A_1, A_2, \ldots, A_k$ are linearly dependent, one may apply this same computational procedure to the matrix $[A_1, A_2, \ldots, A_k]$. If a column of zeros can be obtained, the vectors are dependent, otherwise not. This is often the easiest way to determine whether linear dependence exists. Of course, if coefficients expressing the dependence are needed, a system of equations must be solved.

5.7 Exercises

1. Find a basis for the subspace of vectors in $\mathscr{R}^3$ spanned by

$$\begin{bmatrix} 2 \\ 1 \\ 3 \end{bmatrix}, \quad \begin{bmatrix} -1 \\ 4 \\ 0 \end{bmatrix}, \quad \begin{bmatrix} 4 \\ 2 \\ 6 \end{bmatrix}, \quad \begin{bmatrix} 2 \\ -8 \\ 0 \end{bmatrix}.$$

2. Prove that the vectors $E_1, E_1 + E_2, E_1 + E_2 + E_3, \ldots, E_1 + E_2 + \cdots + E_n$ constitute a basis for the vectors of $\mathscr{F}^n$.

3. Do the vectors $E_1 + E_2, E_2 + E_3, \ldots, E_{n-1} + E_n, E_n$ constitute a basis for the vectors of $\mathscr{F}^n$?

4. Do the vectors $E_1 + E_2, E_2 + E_3, \ldots, E_{n-1} + E_n, E_n + E_1$ constitute a basis for the vectors of $\mathscr{F}^n$? (Consider two cases: n even, n odd.)

5. Show that if $A_1, A_2, \ldots, A_k$ are linearly independent vectors of $\mathscr{R}^n$, then so are

$$A_1, A_1 + 2A_2, A_1 + 2A_2 + 3A_3, \ldots, A_1 + 2A_2 + 3A_3 + \cdots + kA_k.$$

6. Determine a basis for the column space of each matrix. Then express each of the columns as a linear combination of the vectors of the basis. Here $\mathscr{F}$ is the real field.

(a)
$$\begin{bmatrix} 1 & 3 & 0 & 0 \\ 2 & 2 & 4 & 0 \\ -1 & 4 & -7 & 0 \\ 3 & 0 & 9 & 1 \end{bmatrix},$$

(b)
$$\begin{bmatrix} 0 & 0 & 1 & 1 & 1 & 0 \\ 0 & 1 & 1 & 1 & 0 & 0 \\ 1 & 1 & 1 & 0 & 0 & 0 \\ 0 & 1 & 1 & 1 & 0 & 0 \\ 0 & 0 & 1 & 1 & 1 & 0 \end{bmatrix},$$

(c)
$$\begin{bmatrix} 1 & a & c & 0 \\ 0 & b & d & 1 \\ 1 & a & c & 0 \\ 0 & b & d & 1 \end{bmatrix}.$$

7. Show how to express any vector X in $\mathscr{E}^3$ as a linear combination of

$$\begin{bmatrix} 1 \\ 0 \\ 0 \end{bmatrix}, \quad \begin{bmatrix} 1 \\ 2 \\ 0 \end{bmatrix}, \quad \begin{bmatrix} 1 \\ 2 \\ 3 \end{bmatrix}.$$

Then prove without further computation that these three vectors constitute a basis for the vectors of $\mathscr{E}^3$.

*8. Prove that the vectors $X_1, X_2, \ldots, X_n$ in $\mathscr{F}^n$ constitute a basis for $\mathscr{F}^n$ if and only if every vector X of $\mathscr{F}^n$ can be written as a linear combination of these vectors. (This requires *no* computation.)

*9. Prove that n vectors $X_1, X_2, \ldots, X_n$ in $\mathscr{F}^n$ form a basis for the vectors of $\mathscr{F}^n$ if and only if each of the n equations

$$\alpha_{i1} X_1 + \alpha_{i2} X_2 + \cdots + \alpha_{in} X_n = E_i, \qquad i = 1, 2, \ldots, n$$

can be solved for the α's. This requires no computation.

10. Prove that a finite-dimensional vector space $\mathscr{V} \neq \mathscr{Z}$ over a number field $\mathscr{F}$ has infinitely many bases.

11. Denote an $n \times n$ matrix with a 1 in the ij-position and zeros elsewhere by E_{ij}. Use these matrices to show that the set of all $n \times n$ matrices over a given field $\mathscr{F}$ is a finite-dimensional vector space of dimension n^2.

12. Prove that the set of all real, upper-triangular matrices

$$\begin{bmatrix} a_{11} & a_{12} & \cdots & a_{1n} \\ 0 & a_{22} & \cdots & a_{2n} \\ \vdots & & & \\ 0 & 0 & \cdots & a_{nn} \end{bmatrix}$$

is a vector space over the real number field. Find the dimension of this space of matrices and give a basis for it.

13. Prove without any computation that

$$\begin{bmatrix} 1 \\ 2 \\ 2 \end{bmatrix}, \begin{bmatrix} 0 \\ 1 \\ 2 \end{bmatrix}, \begin{bmatrix} 0 \\ 0 \\ 3 \end{bmatrix}$$

constitute a basis for the vectors of $\mathscr{E}^3$. Then express $[3, 6, 9]^\mathsf{T}$ as a linear combination of the vectors of this basis. Which of the vectors of this basis may be replaced by $[3, 6, 9]^\mathsf{T}$, the resulting set still being a basis?

14. Which vectors of the basis

$$\begin{bmatrix} 1 \\ 2 \\ -1 \\ 3 \end{bmatrix}, \begin{bmatrix} 2 \\ 1 \\ 1 \\ 1 \end{bmatrix}, \begin{bmatrix} 3 \\ -2 \\ 1 \\ 4 \end{bmatrix}, \begin{bmatrix} 1 \\ 1 \\ 1 \\ 1 \end{bmatrix}$$

for the vectors of $\mathscr{E}^4$ can be replaced by the vector $[2, -3, 0, 3]^\mathsf{T}$, the result still being a basis for $\mathscr{E}^4$?

15. Prove that Theorem 5.5.6 follows from Theorems 5.4.5 and 5.4.4.

*16. Prove that if $V_1, V_2, \ldots, V_k$ are linearly independent vectors of an n-dimensional vector space $\mathscr{V}$, there always exist vectors $V_{k+1}, \ldots, V_n$ such that $V_1, V_2, \ldots, V_n$ is a basis for $\mathscr{V}$. In particular, if $\mathscr{V}$ is $\mathscr{F}^n$, $V_{k+1}, \ldots, V_n$ may be chosen to be $n - k$ of the elementary n-vectors.

17. Let $Y, X_1, \ldots, X_n$ belong to a vector space $\mathscr{V}$. Prove that if Y is a linear combination of $X_1, X_2, \ldots, X_n$, but not of $X_1, X_2, \ldots, X_{n-1}$, then X_n is a linear combination of $Y, X_1, \ldots, X_{n-1}$.

18. Given that $X_1, X_2, \ldots, X_k$ are linearly independent vectors of a vector space $\mathscr{V}$ and that $c_j \neq 0$, prove that $X_1, X_2, \ldots, X_{j-1}, \sum_{p=1}^k c_p X_p, X_{j+1}, \ldots, X_k$ are also linearly independent.

19. Let $A_1, \ldots, A_k$ be linearly independent vectors of a vector space $\mathscr{V}$ and let $A_1, \ldots, A_k, B_1, \ldots, B_p$ be linearly dependent. Show that at least one of the B's is dependent on the A's and the remaining B's. Which of the A's are linearly dependent on the remaining A's and B's?

20. The vector A_i of the set $A_1, A_2, \ldots, A_k$ is independent of the other vectors of the set if and only if whenever $c_1 A_1 + \cdots + c_i A_i + \cdots + c_k A_k = 0$ it follows that $c_i = 0$. Give an example of a set $\{A_1, A_2, A_3, A_4\}$ of linearly dependent vectors of $\mathscr{R}^4$ such that A_4 is independent of A_1, A_2, and A_3.

21. Use mathematical induction to prove that the points $P_0, P_1, \ldots, P_k$ of $\mathscr{R}^n$ lie on a k-flat and not on some lower-dimensional flat space if and only if $X_1 - X_0$, $X_2 - X_0, \ldots, X_k - X_0$ are linearly independent, where X_i is the vector determined by P_i.

22. Under what conditions do the real n-vectors $X_1, X_2, \ldots, X_k$ determine a k-flat on the origin?

23. Show that if $P_0, P_1, \ldots, P_k$ determine a k-flat in $\mathscr{R}^n$, if the terminal points $Q_0, Q_1, \ldots, Q_k$ of the vectors $Y_0, Y_1, \ldots, Y_k$ are on this k-flat, and if $Y_1 - Y_0$, $Y_2 - Y_0, \ldots, Y_k - Y_0$ are linearly independent, then $Q_0, Q_1, \ldots, Q_k$ determine the same k-flat.

5.8 Orthonormal Bases

In Chapter 4 we pointed out that the basis $E_1, E_2, \ldots, E_n$ for $\mathscr{E}^n$ consists of mutually orthogonal unit vectors. In general, a basis $V_1, V_2, \ldots, V_k$ of a subspace $\mathscr{V}$ of $\mathscr{E}^n$ or $\mathscr{U}^n$ is called an **orthonormal basis** for $\mathscr{V}$ if and only if it consists of mutually orthogonal unit vectors. Such bases are particularly useful.

When a basis $V_1, V_2, \ldots, V_k$ is known for a subspace $\mathscr{V}$ of $\mathscr{E}^n$, an orthonormal basis $U_1, U_2, \ldots, U_k$ is readily determined by what is known as the **Gram–Schmidt process**. In this process, the first vector of the orthonormal basis is chosen as the unit vector

$$U_1 = \frac{V_1}{|V_1|}.$$

To obtain the second vector, we first find a vector W_2 which is a special sort of linear combination of the *next* of the V_j's, namely V_2, and the *previously found* U_j's, namely U_1:

$$W_2 = V_2 - \alpha_{21} U_1.$$

The coefficient α_{21} is chosen so as to make U_1 and W_2 orthogonal:

$$U_1^\mathsf{T} W_2 = U_1^\mathsf{T} V_2 - \alpha_{21} U_1^\mathsf{T} U_1 = U_1^\mathsf{T} V_2 - \alpha_{21} = 0.$$

That is,

$$\alpha_{21} = U_1^\mathsf{T} V_2,$$

so

$$W_2 = V_2 - (U_1^\mathsf{T} V_2) U_1.$$

Note that $W_2 \neq 0$ since otherwise V_1 and V_2 would be linearly dependent. We now define U_2 as the unit vector

$$U_2 = \frac{W_2}{|W_2|},$$

and observe that $U_1^\mathsf{T} U_2 = 0$. Moreover, U_2 belongs to $\mathscr{V}$ since it is actually a linear combination of V_1 and V_2.

Next we construct W_3 as a special linear combination of V_3 and the previously found U_j's:

$$W_3 = V_3 - \alpha_{31} U_1 - \alpha_{32} U_2,$$

which is orthogonal to each of the unit vectors U_1 and U_2:

$$U_1^T W_3 = U_1^T V_3 - \alpha_{31} \quad\quad = 0$$
$$U_2^T W_3 = U_2^T V_3 \quad\quad - \alpha_{32} = 0.$$

Here we find

$$\alpha_{31} = U_1^T V_3, \qquad \alpha_{32} = U_2^T V_3,$$

so

$$W_3 = V_3 - (U_1^T V_3)U_1 - (U_2^T V_3)U_2 .$$

We have $W_3 \neq 0$ because V_1, V_2, and V_3 are linearly independent, so we may define

$$U_3 = \frac{W_3}{|W_3|}.$$

Now $U_1^T U_3 = 0$, $U_2^T U_3 = 0$, and since U_3 is actually a linear combination of V_1, V_2, and V_3, it too is in $\mathscr{V}$.

The general pattern is now clear. We define

$$(5.8.1a) \quad \begin{aligned} W_1 &= V_1, \\ W_j &= V_j - (U_1^T V_j)U_1 - (U_2^T V_j)U_2 - \cdots - (U_{j-1}^T V_j)U_{j-1}, \\ & \qquad\qquad\qquad\qquad\qquad j = 2, 3, \ldots, k, \end{aligned}$$

and, since $W_j \neq 0, j = 1, 2, \ldots, k,$

$$(5.8.1b) \quad U_j = \frac{W_j}{|W_j|}, \qquad j = 1, 2, 3, \ldots, k.$$

It is not hard to show that each such U_j is orthogonal to all previous U_j's, so that we have indeed an orthonormal set. Also, each U_j is in $\mathscr{V}$, since it is expressible as a linear combination of $V_1, V_2, \ldots, V_j$. Moreover, as we have seen before, mutually orthogonal unit vectors are linearly independent, so $U_1, U_2, \ldots, U_k$ constitute a basis for $\mathscr{V}$.

As an example, we find an orthonormal basis for the vector space in $\mathscr{E}^n$ spanned by

$$V_1 = \begin{bmatrix} 1 \\ 0 \\ 1 \end{bmatrix} \quad \text{and} \quad V_2 = \begin{bmatrix} 1 \\ 1 \\ 0 \end{bmatrix}.$$

We have first

$$U_1 = \begin{bmatrix} \dfrac{1}{\sqrt{2}} \\ 0 \\ \dfrac{1}{\sqrt{2}} \end{bmatrix}.$$

Then

$$W_2 = \begin{bmatrix} 1 \\ 1 \\ 0 \end{bmatrix} - \left(\left[\frac{1}{\sqrt{2}}, 0, \frac{1}{\sqrt{2}} \middle| \begin{bmatrix} 1 \\ 1 \\ 0 \end{bmatrix} \right] \right) \begin{bmatrix} \dfrac{1}{\sqrt{2}} \\ 0 \\ \dfrac{1}{\sqrt{2}} \end{bmatrix}$$

$$= \begin{bmatrix} 1 \\ 1 \\ 0 \end{bmatrix} - \frac{1}{\sqrt{2}} \begin{bmatrix} \dfrac{1}{\sqrt{2}} \\ 0 \\ \dfrac{1}{\sqrt{2}} \end{bmatrix} = \begin{bmatrix} 1 \\ 1 \\ 0 \end{bmatrix} - \begin{bmatrix} \dfrac{1}{2} \\ 0 \\ \dfrac{1}{2} \end{bmatrix} = \begin{bmatrix} \dfrac{1}{2} \\ 1 \\ -\dfrac{1}{2} \end{bmatrix}.$$

We have $|W_2| = \sqrt{\frac{3}{2}}$, so

$$U_2 = \begin{bmatrix} \dfrac{1}{\sqrt{6}} \\ \dfrac{2}{\sqrt{6}} \\ -\dfrac{1}{\sqrt{6}} \end{bmatrix}.$$

The computations begin to get messy in even this simple example. The procedure is, indeed, not adapted to efficient paper-and-pencil work, but it is readily programmed for a computer. For the purpose of deriving other results, if often suffices to know that an orthonormal basis exists.

An alternative procedure, sometimes useful for paper-and-pencil computations, is to determine just an *orthogonal* basis first, then normalize as the final step. The equations to be solved are

$$W_1 = A_1;$$

$$W_2 = A_2 - \alpha_{21} W_1, \quad W_1^\mathsf{T} W_2 = 0;$$

$$W_3 = A_3 - \alpha_{31} W_1 - \alpha_{32} W_2, \quad W_1^\mathsf{T} W_3 = 0, \ W_2^\mathsf{T} W_3 = 0;$$

$$\vdots$$

$$W_k = A_k - \alpha_{k1} W_1 - \cdots - \alpha_{k, k-1} W_{k-1}, \quad W_1^\mathsf{T} W_k = 0, \ldots, W_{k-1}^\mathsf{T} W_k = 0;$$

after which

$$U_j = \frac{W_j}{|W_j|}, \qquad j = 1, 2, \ldots, k.$$

This avoids the introduction of radicals until the very last step, but the arithmetic may still be no simpler. When $A_1, A_2, \ldots, A_k$ have integer components, it often is possible to simplify the computations by using

$$W_j = t_j(A_j - \alpha_{j1} W_1 - \cdots - \alpha_{j, j-1} W_{j-1}),$$

where the factor t_j is chosen so as to make the components of W_j as simple as possible.

The Gram–Schmidt process applies equally well in $\mathscr{U}^n$. One only needs to change the transpose to the tranjugate throughout.

An implication of these algorithms is

Theorem 5.8.1: *Every subspace of* $\mathscr{E}^n$ *or* $\mathscr{U}^n$ *has an orthonormal basis.*

5.9 Exercises

1. For what real values of a is the n-vector

$$\begin{bmatrix} a \\ \vdots \\ a \\ 0 \\ 0 \\ \vdots \\ 0 \end{bmatrix},$$

where there are k entries a and $n - k$ entries 0, a unit vector?

2. Find a linear combination of

$$\begin{bmatrix} -1 \\ 2 \\ 0 \end{bmatrix} \quad \text{and} \quad \begin{bmatrix} 0 \\ 2 \\ 1 \end{bmatrix}$$

that is orthogonal to

$$\begin{bmatrix} -1 \\ 2 \\ 0 \end{bmatrix}.$$

Then find an orthonormal basis for the subspace of vectors of $\mathscr{E}^3$ which is spanned by the first two given vectors. Illustrate with a figure.

3. Use the Gram–Schmidt process to construct orthonormal bases for the subspaces of $\mathscr{E}^4$ having the following bases:

(a) $\quad V_1 = \begin{bmatrix} 1 \\ 1 \\ 1 \\ 0 \end{bmatrix}, \quad V_2 = \begin{bmatrix} 0 \\ 1 \\ -1 \\ 0 \end{bmatrix}, \quad V_3 = \begin{bmatrix} 1 \\ 2 \\ 3 \\ 0 \end{bmatrix},$

(b) $\quad V_1 = \begin{bmatrix} 1 \\ 0 \\ -1 \\ 0 \end{bmatrix}, \quad V_2 = \begin{bmatrix} 1 \\ 0 \\ 1 \\ 1 \end{bmatrix}, \quad V_3 = \begin{bmatrix} -1 \\ 1 \\ 0 \\ 0 \end{bmatrix},$

(c) $\quad V_1 = \begin{bmatrix} 1 \\ 0 \\ 0 \\ 0 \end{bmatrix}, \quad V_2 = \begin{bmatrix} 1 \\ 1 \\ 0 \\ 0 \end{bmatrix}, \quad V_3 = \begin{bmatrix} 1 \\ 1 \\ 1 \\ 0 \end{bmatrix}.$

*4. Prove that if $U_1, U_2, \ldots, U_k$ are an orthonormal basis for a subspace of $\mathscr{E}^n$ and if

$$X = \alpha_1 U_1 + \alpha_2 U_2 + \cdots + \alpha_k U_k,$$

then

$$|X|^2 = \alpha_1^2 + \alpha_2^2 + \cdots + \alpha_k^2.$$

This reveals one of several reasons why an orthonormal basis is useful.

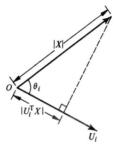

FIGURE 5-1. Projection of a Vector X on a Unit Vector.

*5. Prove that if U_1, U_2, ..., U_k are an orthonormal basis for a subspace $\mathscr{V}$ of $\mathscr{E}^n$, and if

$$X = \alpha_1 U_1 + \alpha_2 U_2 + \cdots + \alpha_k U_k,$$

then

$$U_i^\mathsf{T} X = \alpha_i, \qquad i = 1, 2, \ldots, k.$$

Hence, for each X in $\mathscr{V}$,

$$X = \sum_{i=1}^{k} (U_i^\mathsf{T} X) U_i.$$

Note that the associative law cannot be applied to the products in the terms of this sum. Why not?

The coefficient $U_i^\mathsf{T} X$ in this last equation has an important geometrical interpretation. If θ_i is the angle between X and U_i, we have, since $|U_i| = 1$,

$$\cos \theta = \frac{U_i^\mathsf{T} X}{|X|} \qquad \text{or} \qquad U_i^\mathsf{T} X = |X| \cos \theta_i.$$

Thus $U_i^\mathsf{T} X$ is the **projection of X on the unit vector** U_i. (See Figure 5-1.)

6. Show that if a vector X of $\mathscr{E}^n$ is orthogonal to each of the linearly independent vectors A_1, A_2, ..., A_k of $\mathscr{E}^n$, then X, A_1, A_2, ..., A_k are linearly independent.

*7. Prove that if U_1, U_2, ..., U_r are a set of orthonormal vectors in $\mathscr{E}^n$, then it is always possible to find vectors U_{r+1}, U_{r+2}, ..., U_n such that U_1, U_2, ..., U_n constitute an orthonormal basis for $\mathscr{E}^n$.

8. Show that Exercises 4, 5, 6, and 7 apply to $\mathscr{U}^n$ as well as to $\mathscr{E}^n$ if one only changes the transpose to the tranjugate throughout.

9. Show that the set of all real 4-vectors X orthogonal to each of the vectors

$$\begin{bmatrix} 1 \\ 1 \\ 1 \\ 1 \end{bmatrix}, \quad \begin{bmatrix} 1 \\ -1 \\ 1 \\ -1 \end{bmatrix}, \quad \begin{bmatrix} 1 \\ 0 \\ -1 \\ 0 \end{bmatrix}$$

constitutes a vector space in $\mathscr{E}^4$. Then find an orthonormal basis for this space.

5.10   Intersection and Sum of Two Vector Spaces

Let $\mathscr{U}$ and $\mathscr{V}$ be subspaces of a vector space $\mathscr{W}$ over a field $\mathscr{F}$. The set of all vectors which are in both $\mathscr{U}$ and $\mathscr{V}$ is called their **intersection** and is denoted by $\mathscr{U} \cap \mathscr{V}$. Since the zero vector is in every vector space, $\mathscr{U} \cap \mathscr{V}$ is never empty. We have

Theorem 5.10.1: *The intersection of two subspaces $\mathscr{U}$ and $\mathscr{V}$ of a vector space $\mathscr{W}$ is again a vector space.*

Let V_1 and V_2 belong to each of $\mathscr{U}$ and $\mathscr{V}$, hence to $\mathscr{U} \cap \mathscr{V}$. Then $V_1 + V_2$ and αV_1 are in each of $\mathscr{U}$ and $\mathscr{V}$, that is, in $\mathscr{U} \cap \mathscr{V}$, because $\mathscr{U}$ and $\mathscr{V}$ are both vector spaces. This proves that the subset $\mathscr{U} \cap \mathscr{V}$ of $\mathscr{W}$ is a vector space. It may be that $\mathscr{U} \cap \mathscr{V} = \mathscr{Z}$.

Consider now the subset of $\mathscr{W}$ consisting of all vectors of the form $\alpha U + \beta V$, where U is in $\mathscr{U}$ and V is in $\mathscr{V}$. We call this the **sum** of $\mathscr{U}$ and $\mathscr{V}$ and denote it by $\mathscr{U} + \mathscr{V}$. Note that U and V are not fixed in this definition and that α and β are arbitrary scalars of $\mathscr{F}$.

Theorem 5.10.2: *The sum of two subspaces $\mathscr{U}$ and $\mathscr{V}$ of a vector space $\mathscr{W}$ is again a vector space.*

If $\alpha_1 U_1 + \beta_1 V_1$ and $\alpha_2 U_2 + \beta_2 V_2$ are two vectors of $\mathscr{U} + \mathscr{V}$, so are $(\alpha_1 U_1 + \alpha_2 U_2) + (\beta_1 V_1 + \beta_2 V_2)$ and $(\alpha\alpha_1)U_1 + (\alpha\beta_1)V_1$, so the sum is indeed a vector space.

Now let $\mathscr{U}$ and $\mathscr{V}$ be finite dimensional with dimensions u and v respectively, let $\mathscr{U} \cap \mathscr{V}$ have dimension r, and let $\mathscr{U} + \mathscr{V}$ have dimension s. Let a basis for $\mathscr{U} \cap \mathscr{V}$ be the linearly independent vectors

$$X_1, X_2, \ldots, X_r.$$

(If $\mathscr{U} \cap \mathscr{V} = \mathscr{Z}$, then there are no X_j's.) Starting with these, let a basis for $\mathscr{U}$ be

$$X_1, X_2, \ldots, X_r, U_{r+1}, \ldots, U_u$$

and let a basis for $\mathscr{V}$ be

$$X_1, X_2, \ldots, X_r, V_{r+1}, \ldots, V_v.$$

(How do we know such bases exist?) Then the $r + (u - r) + (v - r) = u + v - r$ vectors

(5.10.1) $X_1, X_2, \ldots, X_r, U_{r+1}, \ldots, U_u, V_{r+1}, \ldots, V_v$

are linearly independent, for if

$$\sum_{i=1}^{r} \alpha_i X_i + \sum_{j=r+1}^{u} \beta_j U_j + \sum_{k=r+1}^{v} \gamma_k V_k = 0,$$

then

$$\sum \alpha_i X_i + \sum \beta_j U_j = \sum (-\gamma_k)V_k,$$

so that if some γ_k is not 0, the vector of $\mathscr{V}$ on the right is equal to a vector of $\mathscr{U}$ on the left. This means that $\sum (-\gamma_k)V_k$, being in both $\mathscr{U}$ and $\mathscr{V}$, must be a linear combination of $X_1, X_2, \ldots, X_r$, so that the vectors X_1, $X_2, \ldots, X_r, V_{r+1}, \ldots, V_v$ are linearly dependent, a contradiction. Therefore every γ_k must be 0 and hence, since $X_1, X_2, \ldots, X_r, U_{r+1}, \ldots, U_u$ are linearly independent, all the α's and β's must also be 0. This proves that the total set is independent, as claimed.

Now every U in $\mathscr{U}$, and every V in $\mathscr{V}$, is a linear combination of vectors in this set of $u + v - r$ independent vectors, and therefore so is every vector $\alpha U + \beta V$ of $\mathscr{U} + \mathscr{V}$. That is, these vectors are an independent set which span $\mathscr{U} + \mathscr{V}$ and hence are a basis for it. Thus the dimension of $\mathscr{U} + \mathscr{V}$ is $s = u + v - r$. We summarize in

Theorem 5.10.3: *If $\mathscr{U}$ of dimension u and $\mathscr{V}$ of dimension v are subspaces of a vector space $\mathscr{W}$, and if $\mathscr{U} \cap \mathscr{V}$ and $\mathscr{U} + \mathscr{V}$ have dimension r and s respectively, then*

(5.10.2) $r + s = u + v.$

This is often written

(5.10.3) $\dim(\mathscr{U} + \mathscr{V}) = \dim \mathscr{U} + \dim \mathscr{V} - \dim(\mathscr{U} \cap \mathscr{V})$

These results have easy geometrical interpretations. For example, let $\mathscr{U}$ and $\mathscr{V}$ be distinct two-dimensional vector spaces in $\mathscr{E}^3$, so that they can be represented by distinct planes on the origin. These two planes intersect in a line, which represents the space $\mathscr{U} \cap \mathscr{V}$ (Figure 5-2). The space $\mathscr{U} + \mathscr{V}$, spanned by X_1 in $\mathscr{U} \cap \mathscr{V}$, U_2 in $\mathscr{U}$, and V_2 in $\mathscr{V}$, is simply $\mathscr{E}^3$. Here, (5.10.2) becomes $1 + 3 = 2 + 2$.

If $\mathscr{U}$ is a one-dimensional and $\mathscr{V}$ is a two-dimensional vector space in $\mathscr{E}^3$, $\mathscr{U}$ not a subspace of $\mathscr{V}$, then we can represent these as a line and a plane on O, the line not lying in the plane. In this case $\mathscr{U} \cap \mathscr{V}$ is just the zero vector and a basis U_1 for $\mathscr{U}$, together with a basis V_1, V_2 for $\mathscr{V}$, is a basis for the set of all vectors in $\mathscr{E}^3$. Here (5.10.2) becomes $0 + 3 = 1 + 2$.

When, as in this example, $\mathscr{U} \cap \mathscr{V}$ is the zero vector and $\mathscr{U} + \mathscr{V}$ is the

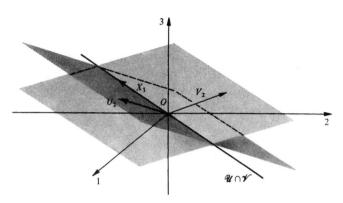

FIGURE 5-2. Intersection and Sum of Two Vector Spaces.

set of all vectors in $\mathcal{W}$, we call $\mathcal{U}$ and $\mathcal{V}$ **complementary subspaces** of $\mathcal{W}$ (Figure 5-3).

Theorem 5.10.4: *If $\mathcal{U}$ and $\mathcal{V}$ are complementary subspaces of $\mathcal{W}$, then every vector X of $\mathcal{W}$ has a unique representation as the sum of a vector of $\mathcal{U}$ and a vector of $\mathcal{V}$.*

In this case, $\dim(\mathcal{U} \cap \mathcal{V}) = 0$, so $\dim \mathcal{W} = \dim \mathcal{U} + \dim \mathcal{V}$ and, from (5.10.1), a basis for $\mathcal{W}$ has the form

$$U_1, U_2, \ldots, U_u, V_1, \ldots, V_v.$$

Hence for each X in $\mathcal{W}$ there exist scalars λ_i, μ_j such that

$$X = \sum_{i=1}^{u} \lambda_i U_i + \sum_{j=1}^{v} \mu_j V_j.$$

By Theorem 5.5.8, this representation is unique and the proof is complete.

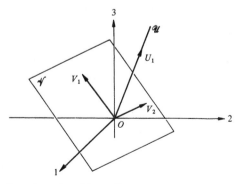

FIGURE 5-3. Complementary Vector Spaces.

5.11 Exercises

1. Determine the dimensions of the sum and of the intersection of the vector spaces defined by the columns of these matrices:

$$\begin{bmatrix} 1 & 1 & 1 & 4 \\ 0 & 1 & 1 & 3 \\ 0 & 0 & 1 & 2 \\ 0 & 0 & -1 & -2 \end{bmatrix}, \quad \begin{bmatrix} 0 & 0 & -1 & 2 \\ 0 & 0 & -1 & 2 \\ 0 & 1 & -1 & 3 \\ 1 & 1 & 1 & 0 \end{bmatrix}.$$

2. Find the dimension of the sum and of the intersection of the column spaces of the matrices

$$\begin{bmatrix} 1 & 3 & 1 \\ 2 & 2 & 14 \\ -1 & 1 & -13 \\ 4 & 0 & 40 \end{bmatrix} \quad \text{and} \quad \begin{bmatrix} 2 & 1 & 4 \\ 0 & -1 & 2 \\ 2 & 2 & 2 \\ -4 & -5 & -2 \end{bmatrix}.$$

What do the results imply?

3. Prove that the intersection of subspaces $\mathscr{V}_1, \mathscr{V}_2, \ldots, \mathscr{V}_k$ of a vector space $\mathscr{V}$ is again a vector space.

4. The points whose coordinates satisfy a nontrivial real linear equation

$$a_1 x_1 + a_2 x_2 + \cdots + a_n x_n = 0$$

constitute a hyperplane on the origin, while the corresponding vectors constitute a vector space. How, then, may one interpret geometrically the set of all solutions of a system of m such linear equations in n unknowns?

5. Let $\mathscr{U}$ and $\mathscr{V}$ be subspaces of an n-dimensional vector space $\mathscr{N}$ over a field $\mathscr{F}$. Let $\mathscr{U}$ and $\mathscr{V}$ have dimensions u and v respectively. Prove that $\dim(\mathscr{U} \cap \mathscr{V})$ is at least $u + v - n$. Give an example for which it is exactly $u + v - n$. Under what conditions on u and v is $\dim(\mathscr{U} \cap \mathscr{V})$ greater than 0?

6. Given that $\mathscr{U}$ and $\mathscr{V}$ are vector spaces in $\mathscr{F}^n$, of dimensions u and v respectively, and that $\mathscr{U} + \mathscr{V}$ is the set of all vectors of $\mathscr{F}^n$, prove that $u + v \geq n$ and that, if $u + v = n$, then $\mathscr{U} \cap \mathscr{V}$ is just the zero vector.

7. Show that if, in an n-dimensional vector space $\mathscr{W}$ over a field $\mathscr{F}$, *one* vector has a unique representation as the sum of a vector from a subspace $\mathscr{U}$ of dimension u and a vector from a subspace $\mathscr{V}$ of dimension v, then $\mathscr{U} + \mathscr{V}$ has dimension $u + v$ and $\mathscr{U} \cap \mathscr{V}$ has dimension 0.

8. Prove by induction that if $\mathscr{U}_1, \mathscr{U}_2, \ldots, \mathscr{U}_k$ in $\mathscr{F}^n$ each have dimension $n - 1$, then $\mathscr{U}_1 \cap \mathscr{U}_2 \cap \cdots \cap \mathscr{U}_k$ has dimension at least $n - k$. In particular, if $k < n$, $\mathscr{U}_1 \cap \mathscr{U}_2 \cap \cdots \cap \mathscr{U}_k$ does not consist of the zero vector alone. Illustrate the various possible cases in $\mathscr{E}^3$ with figures.

9. Show that any two of the subspaces spanned by $X_1, X_2, \ldots, X_r$, by $U_{r+1}, \ldots, U_u$, and by $V_{r+1}, \ldots, V_v$, as defined in the proof of Theorem 5.10.2, have only the zero space as their intersection.

10. If $\mathcal{V}_1$ and $\mathcal{V}_2$ are subspaces of $\mathcal{V}$ such that $\mathcal{V}_1 \cap \mathcal{V}_2 = \mathcal{Z}$, then $\mathcal{V}_1 + \mathcal{V}_2$ is called the **direct sum** of these two spaces. (Such sums are of particular importance in more advanced work in linear algebra.) Show that the set $\mathcal{M}_s$ of all $n \times n$ symmetric matrices over a field $\mathcal{F}$ and the set $\mathcal{M}_{ss}$ of all $n \times n$ skew-symmetric matrices over $\mathcal{F}$ are two such vector spaces. Show that if M denotes any $n \times n$ matrix over $\mathcal{F}$, there exist unique matrices M_s in $\mathcal{M}_s$ and M_{ss} in $\mathcal{M}_{ss}$ such that

$$M = M_s + M_{ss}.$$

Hence show that if $\mathcal{M}$ is the space of all $n \times n$ matrices over $\mathcal{F}$,

$$\mathcal{M} = \mathcal{M}_s + \mathcal{M}_{ss}.$$

11. Show that $\mathscr{E}^n$ is the direct sum of the vector spaces $\{\alpha E_1\}, \{\alpha E_2\}, \ldots, \{\alpha E_n\}$ determined by the n elementary vectors.

12. Prove that the intersection of a subspace $\mathcal{V}_1$ of $\mathscr{F}_1^n$ and a subspace $\mathcal{V}_2$ of $\mathscr{F}_2^n$ is a vector space over the intersection of $\mathscr{F}_1$ and $\mathscr{F}_2$, where $\mathscr{F}_1$ and $\mathscr{F}_2$ are subfields of a field $\mathscr{F}$.

5.12 Isomorphic Vector Spaces

Recall that the set of all real quadratic polynomials of the form $ax^2 + bxy + cy^2$ is a three-dimensional vector space over the field of real numbers. By arraying the coefficients in the form of a 3-vector, we can put these quadratic forms, as they are called, in one-to-one correspondence with the vectors of $\mathscr{E}^3$:

$$ax^2 + bxy + cy^2 \leftrightarrow \begin{bmatrix} a \\ b \\ c \end{bmatrix}.$$

The double-headed arrow means that to each quadratic form there corresponds a unique 3-vector and to each 3-vector there corresponds a unique quadratic form. Corresponding members of the two spaces are called **images** of each other under the correspondence.

If α is any scalar, then

$$\alpha(ax^2 + bxy + cy^2) = \alpha ax^2 + \alpha bxy + \alpha cy^2 \leftrightarrow \begin{bmatrix} \alpha a \\ \alpha b \\ \alpha c \end{bmatrix} = \alpha \begin{bmatrix} a \\ b \\ c \end{bmatrix};$$

that is, to a scalar multiple of a vector of either of the spaces, there corresponds the same scalar multiple of the image vector. We say that the correspondence "preserves" the operation of scalar multiplication.

Now let

$$q_i = a_i x^2 + b_i xy + c_i y^2, \qquad i = 1, 2.$$

Then

$$q_1 + q_2 = (a_1 + a_2)x^2 + (b_1 + b_2)xy + (c_1 + c_2)y^2$$

$$\leftrightarrow \begin{bmatrix} a_1 + a_2 \\ b_1 + b_2 \\ c_1 + c_2 \end{bmatrix} = \begin{bmatrix} a_1 \\ b_1 \\ c_1 \end{bmatrix} + \begin{bmatrix} a_2 \\ b_2 \\ c_2 \end{bmatrix},$$

so that the image of the sum of two vectors of either space is the sum of their images. Thus the correspondence also preserves the operation of addition.

Since these two operations are preserved by the correspondence, it follows by induction that to any linear combination of vectors of the one space there corresponds the same linear combination of the image vectors, so the linear algebraic properties of the two spaces are the same. We describe this complex of ideas by saying that the spaces are **isomorphic.**

The two vector spaces are distinct concrete representations of the same abstract entity: the three-dimensional vector space over the field of real numbers. This space has, of course, infinitely many concrete representations, all mutually isomorphic and all isomorphic to the space of real 3-vectors. (The relation of this abstract space to its concrete representations is analogous to the relation of a positive integer to the sets of which it is the cardinal number.) Thus, when we study the linear algebra of $\mathscr{R}^3$, we really study the linear algebra of all three-dimensional vector spaces over the real field.

The preceding ideas are generalized in the following definition. Let $\mathscr{U}$ and $\mathscr{V}$ be finite-dimensional vector spaces over a field $\mathscr{F}$ whose vectors are in one-to-one correspondence, that is, such that each U in $\mathscr{U}$ has a unique image V in $\mathscr{V}$ and such that U is in turn the unique image of $V: U \leftrightarrow V$. Then $\mathscr{U}$ and $\mathscr{V}$ are said to be **isomorphic** under this correspondence if and only if, whenever $U_1 \leftrightarrow V_1$ and $U_2 \leftrightarrow V_2$,

(5.12.1) $\alpha U_1 \leftrightarrow \alpha V_1$ for all scalars α in $\mathscr{F}$,

and

(5.12.2) $U_1 + U_2 \leftrightarrow V_1 + V_2,$

that is, if and only if the correspondence preserves the operation of multiplication by a scalar and the operation of addition.

Concerning isomorphic vector spaces, we have several basic theorems.

Theorem 5.12.1: *If $\mathscr{U}$ and $\mathscr{V}$ are isomorphic, then their zero vectors correspond.*

This follows from (5.12.1) in the case when $\alpha = 0$.

Theorem 5.12.2: *If $\mathscr{U}$ and $\mathscr{V}$ are isomorphic and if $U_1, U_2, \ldots, U_k$ correspond respectively to $V_1, V_2, \ldots, V_k$, then for all scalars $\alpha_1, \alpha_2, \ldots, \alpha_k$,*

$$(5.12.3) \quad \alpha_1 U_1 + \alpha_2 U_2 + \cdots + \alpha_k U_k \leftrightarrow \alpha_1 V_1 + \alpha_2 V_2 + \cdots + \alpha_k V_k.$$

This follows from (5.12.1) and (5.12.2) by induction.

Theorem 5.12.3: *If $\mathscr{U}$ and $\mathscr{V}$ are isomorphic and if $U_1, U_2, \ldots, U_k$ correspond respectively to $V_1, V_2, \ldots, V_k$, then $U_1, U_2, \ldots, U_k$ are linearly dependent (independent) if and only if $V_1, V_2, \ldots, V_k$ are linearly dependent (independent).*

The relation (5.12.3) holds. If either combination is zero with not all the α's zero, then the image combination must be zero also by Theorem 5.12.1. By this same theorem, if either combination cannot be 0 unless all the α's are 0, then neither can the image combination be 0 unless all the α's are 0. The theorem follows.

Theorem 5.12.4: *Isomorphic finite-dimensional vector spaces have the same dimension.*

Let $\mathscr{U}$ and $\mathscr{V}$ be isomorphic and finite dimensional. Let $U_1, U_2, \ldots, U_k$ be a basis for $\mathscr{U}$. Then $U_1, U_2, \ldots, U_k$ are independent and hence, by Theorem 5.12.3, so are their images $V_1, V_2, \ldots, V_k$ in $\mathscr{V}$, so the dimension of $\mathscr{V}$ is at least as great as that of $\mathscr{U}$. Similarly, the dimension of $\mathscr{U}$ is at least as great as that of $\mathscr{V}$. Hence the dimensions are the same.

We have finally

Theorem 5.12.5: *Every vector space $\mathscr{V}$ of dimension n over $\mathscr{F}$ is isomorphic to the space $\mathscr{F}^n$ of all n-vectors over $\mathscr{F}$.*

Choose a basis $V_1, V_2, \ldots, V_n$ for $\mathscr{V}$ and define the one-to-one correspondence

$$\alpha_1 V_1 + \alpha_2 V_2 + \cdots + \alpha_n V_n \leftrightarrow \begin{bmatrix} \alpha_1 \\ \alpha_2 \\ \vdots \\ \alpha_n \end{bmatrix}.$$

It requires only a simple computation to show that this correspondence preserves both multiplication by a scalar and addition. The theorem follows.

In view of Theorem 5.12.5, to study the linear algebra of $\mathscr{V}$, we need only study the linear algebra of the set of all n-vectors over $\mathscr{F}$. (This is somewhat analogous to our use of synthetic elimination to solve a system of linear equations.) Thus our almost exclusive attention to such n-tuples is fully justified.

5.13   Exercises

1. Show that in $\mathscr{E}^3$ the vector spaces defined by

$$X = t_1 \begin{bmatrix} 1 \\ 0 \\ 0 \end{bmatrix} + t_2 \begin{bmatrix} 1 \\ -1 \\ 1 \end{bmatrix}, \qquad Y = s_1 \begin{bmatrix} 0 \\ 0 \\ 1 \end{bmatrix} + s_2 \begin{bmatrix} 1 \\ -1 \\ 1 \end{bmatrix},$$

are isomorphic. Define a convenient one-to-one correspondence between the vectors of the two spaces such that the usual operations are preserved. Illustrate with a figure showing several pairs of corresponding vectors in the two spaces. Sketch it all on one coordinate system. Then generalize.

2. By defining a one-to-one correspondence such that the operations of addition and multiplication are preserved, show that the field of complex numbers $a + bi$ is isomorphic to the set of real matrices

$$\begin{bmatrix} a & -b \\ b & a \end{bmatrix},$$

so that this set of matrices is also a field. What matrices correspond to 1 and to i respectively? What matrices correspond to $c(a + bi)$, where c is real, to $-(a + bi)$, to $(a + bi)^{-1}$, to $a - bi$? What additional operations are therefore preserved? What is the matrix equivalent of the equation $\overline{(a + bi)(c + di)} = \overline{a + bi} \cdot \overline{c + di}$? What special properties are exhibited by this set of matrices that do not hold for 2×2 matrices in general?

3. Prove that isomorphism of vector spaces is an equivalence relation, that is, that it is reflexive, symmetric, and transitive.

4. Prove that any two finite-dimensional vector spaces which are over the same field $\mathscr{F}$ and which have the same dimension are isomorphic.

CHAPTER
6

The Rank of a Matrix

6.1 The Rank of a Matrix

The **column rank** of an $m \times n$ matrix A over a field $\mathscr{F}$ is defined to be the dimension of its column space; that is, it is the maximum number of linearly independent columns in A. A simple but important example is I_n, whose columns are linearly independent, so that the column rank is n.

We have already seen in Section 5.6 how this dimension and a basis for the column space may be simultaneously computed in the case of a matrix with constant entries: We simply reduce to column echelon form or to a sufficiently close approximation thereof, using elementary column transformations only. If the entries are not constants, the column rank may vary with the values of those entries. Thus the real matrix

$$\begin{bmatrix} 2 & x \\ -1 & 3 \end{bmatrix}$$

has column rank 2 unless $x = -6$. In the latter case, the column rank is 1 since the two columns are proportional, so only one is linearly independent.

The rows of A also span a vector space, the **row space** of A. Its dimension is called the **row rank** of A. The row rank of A and a basis for the row space may be determined by reducing A to echelon or near-echelon form in the usual way, operating on rows only. The remaining nonzero rows are linearly independent [the transpose of the result in (5.2.5)] and their number is the row rank. For example, the row rank of

$$\begin{bmatrix} 1 & 3 & 0 & 1 & -1 \\ 0 & 0 & 1 & 0 & 2 \\ 0 & 0 & 0 & 0 & 0 \end{bmatrix}$$

is two.

When we reduce A to column echelon form, *using elementary column transformations only*, we find both the column rank and a basis for the column space of A. If we want only to know the column rank, and do not need a basis for the column space, we can use row operations also, for we have

Theorem 6.1.1: *Elementary row transformations leave the column rank of a matrix invariant and elementary column transformations leave the row rank invariant.*

Let

$$A_{n \times p} = [A_1, A_2, \ldots, A_p].$$

Suppose that we interchange rows i and k of $A_{n \times p}$ and denote the resulting matrix by

$$B = [B_1, B_2, \ldots, B_p].$$

Then an equation

(6.1.1) $$\sum_{j=1}^{p} \alpha_j A_j = 0$$

holds if and only if the scalar equations

(6.1.2) $$\sum_{j=1}^{p} \alpha_j a_{qj} = 0, \quad q = 1, 2, \ldots, n,$$

hold.

But by interchanging equations i and k in (6.1.2), we obtain an equivalent system of equations which is, in fact,

$$\sum_{j=1}^{p} \alpha_j b_{qj} = 0, \quad q = 1, 2, \ldots, n,$$

so (6.1.1) holds if and only if

(6.1.3)
$$\sum_{j=1}^{p} \alpha_j B_j = 0$$

also holds. This implies that if any subset of columns of B is a linearly dependent set, then the corresponding subset of columns of A must be a linearly dependent set and vice versa. [Columns that do not appear in the subset have zero coefficients in equations (6.1.1) and (6.1.3).] Hence linearly *independent* sets must also correspond, so that the column rank is the same for A and B.

Next suppose that we multiply the ith row of A by a nonzero constant α and call the resulting matrix

$$C_{n \times p} = [C_1, C_2, \ldots, C_p].$$

If we now multiply the ith equation of (6.1.2) by α, we obtain an equivalent system of equations which is, in fact,

$$\sum_{j=1}^{p} \alpha_j c_{qj} = 0, \qquad q = 1, 2, \ldots, n,$$

so (6.1.1) holds if and only if

(6.1.4)
$$\sum_{j=1}^{p} \alpha_j C_j = 0.$$

By the same reasoning used in the case of the previous operation, if any columns of C are dependent (independent), the corresponding columns of A must also be dependent (independent) and conversely, so that A and C have the same column rank.

Finally, suppose we add α times the ith row of A to the kth row of A, $i \neq k$, and denote the resulting matrix by

$$G_{n \times p} = [G_1, G_2, \ldots, G_p].$$

Now if we add α times the ith equation of (6.1.2) to the kth equation, we obtain an equivalent system which is, in fact,

$$\sum_{j=1}^{p} \alpha_j g_{qj} = 0, \qquad q = 1, 2, \ldots, n,$$

so (6.1.1) holds if and only if

(6.1.5)
$$\sum_{j=1}^{p} \alpha_j G_j = 0.$$

Thus, as before, we conclude that columns of G are linearly dependent (independent) if and only if the corresponding columns of A are dependent (independent), so the column rank is once again invariant.

In the same way, we can show that the elementary column operations do not change the row rank, and the proof is complete.

If we now apply elementary row transformations appropriately to the column echelon form, we can sweep the nonzero columns with their leading 1's, then rearrange the rows, if necessary, so that the remaining 1's all appear on the main diagonal. The result is a matrix of the form

(6.1.6)
$$\begin{bmatrix} I_r & \vdots & 0 \\ \hdashline 0 & \vdots & 0 \end{bmatrix}$$

in which the 0's denote zero matrices, and in which the number of linearly independent rows and the number of linearly independent columns are the same. Since none of the transformations used change either the row or the column rank, we have proved

Theorem 6.1.2: *The column rank and the row rank of a matrix A over a field $\mathscr{F}$ are the same.*

This result makes it proper to refer to the common value of the row and the column rank of a matrix A as the **rank** of A. We denote the rank of A by the symbol $r(A)$. The matrix (6.1.6) to which A is reduced is called the **rank normal form** of A. At times some or all of the zero matrices bordering the identity matrix I_r are absent and we have one of the following instead:

(6.1.7) $$[I_r \quad 0], \qquad \begin{bmatrix} I_r \\ 0 \end{bmatrix}, \qquad I_r.$$

Let the symbol " $\sim$ " mean "has the same rank and order as." Then an example illustrating the above theorems is the following:

$$\begin{bmatrix} 2 & -1 & 3 \\ 1 & 2 & 0 \end{bmatrix} \sim \begin{bmatrix} 1 & 2 & 0 \\ 2 & -1 & 3 \end{bmatrix} \quad \text{(interchange rows)}$$

$$\sim \begin{bmatrix} 1 & 0 & 0 \\ 2 & -5 & 3 \end{bmatrix} \quad \begin{array}{l}\text{(subtract twice first} \\ \text{column from second)}\end{array}$$

$$\sim \begin{bmatrix} 1 & 0 & 0 \\ 0 & -5 & 3 \end{bmatrix} \quad \begin{array}{l}\text{(subtract twice first} \\ \text{row from second)}\end{array}$$

$$\sim \begin{bmatrix} 1 & 0 & 0 \\ 0 & 1 & 3 \end{bmatrix} \quad \text{(divide second column by } -5)$$

$$\sim \begin{bmatrix} 1 & 0 & 0 \\ 0 & 1 & 0 \end{bmatrix} \quad \begin{array}{l}\text{(subtract three times} \\ \text{second column from third).}\end{array}$$

The rank is 2 and the final result is the rank normal form.

6.2 Basic Theorems About the Rank of a Matrix

Suppose an $n \times n$ matrix A over a field $\mathscr{F}$ has rank n so that the columns of A, as well as the rows of A, are linearly independent. Since the columns of A then form a basis for the set of all n-vectors over $\mathscr{F}$, each vector E_j is a unique linear combination of the columns of A, that is, each equation $AX_j = E_j$ has a unique solution for X_j. Then, since corresponding columns are equal,

$$[AX_1, AX_2, \ldots, AX_n] = [E_1, E_2, \ldots, E_n]$$

or, if

$$X = [X_1, X_2, \ldots, X_n],$$
(6.2.1)
$$AX = I_n.$$

The rows of A are the columns of A^T and are independent, so that each vector E_j is also a unique linear combination of the columns of A^T. That is, each equation $A^\mathsf{T}Y_j = E_j$ has a unique solution for Y_j. Then

$$[A^\mathsf{T}Y_1, A^\mathsf{T}Y_2, \ldots, A^\mathsf{T}Y_n] = [E_1, E_2, \ldots, E_n]$$

or, if

$$Y = [Y_1, Y_2, \ldots, Y_n],$$
$$A^\mathsf{T}Y = I_n,$$

so that

(6.2.2)
$$Y^\mathsf{T}A = I_n.$$

Then, from (6.2.1) and (6.2.2), we have

$$Y^\mathsf{T} = Y^\mathsf{T}(AX) = (Y^\mathsf{T}A)X = X,$$

so (6.2.1) and (6.2.2) combine to yield

$$AX = XA = I,$$

from which $X = A^{-1}$.

Suppose, conversely, that A^{-1} exists and put

$$A^{-1} = [C_1, C_2, \ldots, C_n].$$

From the equation

$$A[C_1, C_2, \ldots, C_n] = I_n = [E_1, E_2, \ldots, E_n]$$

we have then

$$AC_1 = E_1, AC_2 = E_2, \ldots, AC_n = E_n.$$

Thus each of the n linearly independent vectors $E_1, E_2, \ldots, E_n$ is a linear combination of the columns of A, which are therefore independent also, so that A has rank n. We have thus proved

Theorem 6.2.1: *An $n \times n$ matrix A over a field $\mathscr{F}$ has rank n if and only if A^{-1} exists, that is, if and only if A is nonsingular.*

Now let

$$A_{m \times n} B_{n \times p} = C_{m \times p}.$$

If we write A as a matrix of columns $[A_1, A_2, \ldots, A_n]$, then

$$AB = [(b_{11}A_1 + b_{21}A_2 + \cdots + b_{n1}A_n), \ldots, (b_{1p}A_1 + b_{2p}A_2 + \cdots + b_{np}A_n)],$$

so that the columns of C are linear combinations of the columns of A. Hence the columns of C all belong to the column space of A, and the maximum number of independent columns in C therefore does not exceed the maximum number of independent columns of A. That is, the rank of C cannot exceed the rank of A. Similarly, if we write B as a matrix of rows,

$$\begin{bmatrix} B^{(1)} \\ B^{(2)} \\ \vdots \\ B^{(n)} \end{bmatrix},$$

then

$$AB = \begin{bmatrix} a_{11}B^{(1)} + a_{12}B^{(2)} + \cdots + a_{1n}B^{(n)} \\ \vdots \\ a_{m1}B^{(1)} + a_{m2}B^{(2)} + \cdots + a_{mn}B^{(n)} \end{bmatrix},$$

so the rows of the product C are in the row space of B. Hence the rank of C cannot exceed the rank of B. Combining these results, we have

Theorem 6.2.2: *The rank of the product of two matrices cannot exceed the rank of either factor.*

For example, if A and B both have rank 2, we could have

$$\begin{bmatrix} 1 & 0 & 0 & 0 \\ 0 & 0 & 1 & 0 \end{bmatrix} \begin{bmatrix} 0 & 0 \\ 1 & 0 \\ 0 & 0 \\ 0 & 1 \end{bmatrix} = \begin{bmatrix} 0 & 0 \\ 0 & 0 \end{bmatrix},$$

or

$$\begin{bmatrix} 1 & 0 & 0 \\ 0 & 1 & 0 \end{bmatrix} \begin{bmatrix} 1 & 0 \\ 0 & 0 \\ 0 & 1 \end{bmatrix} = \begin{bmatrix} 1 & 0 \\ 0 & 0 \end{bmatrix},$$

or

$$\begin{bmatrix} 1 & 0 & 0 \\ 0 & 1 & 0 \end{bmatrix} \begin{bmatrix} 1 & 0 \\ 0 & 1 \\ 0 & 0 \end{bmatrix} = \begin{bmatrix} 1 & 0 \\ 0 & 1 \end{bmatrix},$$

so the rank of the product may be 0, 1, or 2.

Now suppose that A^{-1} exists and that

$$AB = M, \qquad \text{so} \qquad B = A^{-1}M.$$

Then, by the preceding theorem,

$$r(M) \le r(B) \qquad \text{and} \qquad r(B) \le r(M).$$

A similar argument applies when A is the second factor, so that we have

Theorem 6.2.3: *If A is nonsingular, the rank of the product AB is the same as the rank of B, and the rank of CA is the same as the rank of C.*

Next, suppose that

$$(6.2.3) \qquad\qquad AB = I_n,$$

where A and B are both $n \times n$. By Theorem 6.2.2, since $r(I_n) = n$, we have $r(A) = r(B) = n$. Then, by Theorem 6.2.1, A^{-1} exists. Multiplying (6.2.3) by A^{-1}, we get $B = A^{-1}$. Similarly, B^{-1} exists and $A = B^{-1}$. In summary, we have

Theorem 6.2.4: *If $A_{n \times n} B_{n \times n} = I_n$, then A and B are nonsingular and $A^{-1} = B$, $B^{-1} = A$.*

We defined B to be an inverse of A if and only if $AB = BA = I$. This theorem says that it suffices to show that $AB = I$ in order to show that B is the inverse of A.

Concerning the rank of the sum of two matrices, we have

Theorem 6.2.5: *If A and B are both m × n matrices, then*

$$(6.2.3) \qquad\qquad |r(A) - r(B)| \le r(A + B) \le r(A) + r(B).$$

Indeed, every column of $A + B$ is a linear combination of the set of columns of both A and B. In this set, at most $r(A) + r(B)$ columns are linearly independent. Hence at most this many columns of $A + B$ are linearly independent. That is, $r(A + B) \le r(A) + (B)$.

Now let $A + B = C$, so that $A = C + (-B)$. Then, since $r(-B) = r(B)$, what was just proved implies that $r(A) \le r(C) + r(B)$, so that

$$r(A) - r(B) \le r(C) = r(A + B).$$

Also, $B = C + (-A)$. Now we have $r(B) \le r(C) + r(A)$, so that

$$r(B) - r(A) \le r(C) = r(A + B).$$

Combining the two inequalities just established, we have

$$|r(A) - r(B)| \le r(A + B),$$

and the proof of the theorem is complete.

6.3   Matrix Representation of Elementary Transformations

It is interesting and useful to represent the elementary row and column transformations in matrix form.

Consider the facts that

$$(6.3.1) \qquad\qquad E_i^{\mathsf{T}} A = A^{(i)}, \qquad A E_j = A_j,$$

where E_i is the ith elementary unit vector, where $A^{(i)}$ denotes the ith row of A, and where A_j denotes its jth column (write out the details in full). From these facts, we see that to interchange two rows (columns) of a matrix A, we may simply interchange the corresponding rows (columns) of a conformable identity matrix and premultiply (postmultiply) A by the results. For example,

$$\begin{bmatrix} E_2^T \\ E_1^T \\ E_3^T \end{bmatrix} \begin{bmatrix} a_1 & a_2 \\ b_1 & b_2 \\ c_1 & c_2 \end{bmatrix} = \begin{bmatrix} 0 & 1 & 0 \\ 1 & 0 & 0 \\ 0 & 0 & 1 \end{bmatrix} \begin{bmatrix} a_1 & a_2 \\ b_1 & b_2 \\ c_1 & c_2 \end{bmatrix} = \begin{bmatrix} b_1 & b_2 \\ a_1 & a_2 \\ c_1 & c_2 \end{bmatrix}$$

and

$$\begin{bmatrix} a_1 & a_2 & a_3 \\ b_1 & b_2 & b_3 \end{bmatrix} [E_1, E_3, E_2] = \begin{bmatrix} a_1 & a_2 & a_3 \\ b_1 & b_2 & b_3 \end{bmatrix} \begin{bmatrix} 1 & 0 & 0 \\ 0 & 0 & 1 \\ 0 & 1 & 0 \end{bmatrix} = \begin{bmatrix} a_1 & a_3 & a_2 \\ b_1 & b_3 & b_2 \end{bmatrix}.$$

From the facts

(6.3.2) $$(\alpha E_i^T)A = \alpha A^{(i)}, \qquad A(\alpha E_i) = \alpha A_i,$$

or from the multiplicative behavior of a diagonal matrix, we conclude that to multiply the ith row (column) of A by the scalar α, we may multiply the ith row (column) of a conformable identity matrix by α and premultiply (postmultiply) A by the result. For example,

$$\begin{bmatrix} E_1^T \\ \alpha E_2^T \\ E_3^T \end{bmatrix} \begin{bmatrix} a_1 & a_2 \\ b_1 & b_2 \\ c_1 & c_2 \end{bmatrix} = \begin{bmatrix} 1 & 0 & 0 \\ 0 & \alpha & 0 \\ 0 & 0 & 1 \end{bmatrix} \begin{bmatrix} a_1 & a_2 \\ b_1 & b_2 \\ c_1 & c_2 \end{bmatrix} = \begin{bmatrix} a_1 & a_2 \\ \alpha b_1 & \alpha b_2 \\ c_1 & c_2 \end{bmatrix},$$

$$\begin{bmatrix} a_1 & a_2 \\ b_1 & b_2 \\ c_1 & c_2 \end{bmatrix} [E_1, \alpha E_2] = \begin{bmatrix} a_1 & a_2 \\ b_1 & b_2 \\ c_1 & c_2 \end{bmatrix} \begin{bmatrix} 1 & 0 \\ 0 & \alpha \end{bmatrix} = \begin{bmatrix} a_1 & \alpha a_2 \\ b_1 & \alpha b_2 \\ c_1 & \alpha c_2 \end{bmatrix}.$$

Finally, we observe that

(6.3.3) $$(E_i^T + \alpha E_j^T)A = A^{(i)} + \alpha A^{(j)} \qquad \text{and} \qquad A(E_i + \alpha E_j) = A_i + \alpha A_j,$$

from which it follows that to add α times the jth row (column) of A to the ith row (column) of A, we may first do the same to an identity matrix and then premultiply (postmultiply) A by the result. For example,

$$\begin{bmatrix} E_1^T + \alpha E_2^T \\ E_2^T \\ E_3^T \end{bmatrix} \begin{bmatrix} a_1 & a_2 \\ b_1 & b_2 \\ c_1 & c_2 \end{bmatrix} = \begin{bmatrix} 1 & \alpha & 0 \\ 0 & 1 & 0 \\ 0 & 0 & 1 \end{bmatrix} \begin{bmatrix} a_1 & a_2 \\ b_1 & b_2 \\ c_1 & c_2 \end{bmatrix} = \begin{bmatrix} a_1 + \alpha b_1 & a_2 + \alpha b_2 \\ b_1 & b_2 \\ c_1 & c_2 \end{bmatrix}$$

and

$$\begin{bmatrix} a_1 & a_2 \\ b_1 & b_2 \\ c_1 & c_2 \end{bmatrix} [E_1, E_2 + \alpha E_1] = \begin{bmatrix} a_1 & a_2 \\ b_1 & b_2 \\ c_1 & c_2 \end{bmatrix} \begin{bmatrix} 1 & \alpha \\ 0 & 1 \end{bmatrix} = \begin{bmatrix} a_1 & \alpha a_1 + a_2 \\ b_1 & \alpha b_1 + b_2 \\ c_1 & \alpha c_1 + c_2 \end{bmatrix}.$$

In summary, we have

Theorem 6.3.1: *To effect an elementary transformation of a matrix A, first apply the same transformation to a conformable identity matrix, then pre-multiply A by the result if the operation is on rows, postmultiply if it is on columns.*

The matrices that effect elementary transformations are called **elementary matrices**.

Every elementary transformation has an inverse; that is, there exists a second elementary transformation which precisely undoes the first. Thus if we interchange two parallel lines of A and then interchange them again, the end result is simply A. Hence an elementary transformation of this type is its own inverse. If we multiply a line of A by the scalar $\alpha \neq 0$, we can then return to A by multiplying the same line by $1/\alpha$. If we add α times a line j of A to another, parallel line i of A, we can restore A by adding $-\alpha$ times line j to line i. Thus we have

Theorem 6.3.2: *Every elementary transformation has an inverse which is an elementary transformation of the same kind.*

To get the matrix of an elementary transformation, we perform the transformation on an identity matrix. If we now write the matrix of the *inverse* transformation and apply it appropriately to the preceding elementary matrix, we must get the identity matrix as the product. That is, we have

Theorem 6.3.3: *Every elementary matrix has an inverse which is the matrix of the inverse elementary transformation.*

In view of this result, the inverse of an elementary matrix may be written by inspection. For example,

$$\begin{bmatrix} 0 & 1 & 0 \\ 1 & 0 & 0 \\ 0 & 0 & 1 \end{bmatrix}^{-1} = \begin{bmatrix} 0 & 1 & 0 \\ 1 & 0 & 0 \\ 0 & 0 & 1 \end{bmatrix}, \qquad \begin{bmatrix} 1 & 0 & 0 \\ 0 & \frac{1}{2} & 0 \\ 0 & 0 & 1 \end{bmatrix}^{-1} = \begin{bmatrix} 1 & 0 & 0 \\ 0 & 2 & 0 \\ 0 & 0 & 1 \end{bmatrix},$$

$$\begin{bmatrix} 1 & 0 & -3 \\ 0 & 1 & 0 \\ 0 & 0 & 1 \end{bmatrix}^{-1} = \begin{bmatrix} 1 & 0 & 3 \\ 0 & 1 & 0 \\ 0 & 0 & 1 \end{bmatrix}.$$

Now recall the example of reduction to rank normal form given in Section 6.1. We can represent each of the elementary transformations used there in matrix form and represent the reduction as matrix multiplication.

If we number the elementary matrices to show the order in which they are used, we have

$$
\begin{bmatrix} 1 & 0 \\ -2 & 1 \end{bmatrix} \begin{bmatrix} 0 & 1 \\ 1 & 0 \end{bmatrix} \begin{bmatrix} 2 & -1 & 3 \\ 1 & 2 & 0 \end{bmatrix}
$$
$$
\quad (3) \qquad\quad (1)
$$

$$
\times \begin{bmatrix} 1 & -2 & 0 \\ 0 & 1 & 0 \\ 0 & 0 & 1 \end{bmatrix} \begin{bmatrix} 1 & 0 & 0 \\ 0 & -\frac{1}{5} & 0 \\ 0 & 0 & 1 \end{bmatrix} \begin{bmatrix} 1 & 0 & 0 \\ 0 & 1 & -3 \\ 0 & 0 & 1 \end{bmatrix} = \begin{bmatrix} 1 & 0 & 0 \\ 0 & 1 & 0 \end{bmatrix}.
$$
$$
\qquad (2) \qquad\qquad (4) \qquad\qquad (5)
$$

We can combine the first two factors here, and the last three, to obtain

$$
\begin{bmatrix} 0 & 1 \\ 1 & -2 \end{bmatrix} \begin{bmatrix} 2 & -1 & 3 \\ 1 & 2 & 0 \end{bmatrix} \begin{bmatrix} 1 & \frac{2}{5} & -\frac{6}{5} \\ 0 & -\frac{1}{5} & \frac{3}{5} \\ 0 & 0 & 1 \end{bmatrix} = \begin{bmatrix} 1 & 0 & 0 \\ 0 & 1 & 0 \end{bmatrix},
$$

in which the first and third factors of the left member, being products of nonsingular matrices, are also nonsingular.

The procedure of the example is fully general. Given a matrix A over a field $\mathscr{F}$, we can reduce A to rank normal form by row and column transformations using only elements of $\mathscr{F}$. Next we represent these transformations in matrix form and express the reduced form of A as a matrix product. Finally, by multiplying them together in the correct order, we combine the matrices of the row operations into a single matrix and those of the column operations into another. These products are nonsingular since their factors are all nonsingular. Thus we may conclude

Theorem 6.3.4: *Given a matrix A over a field $\mathscr{F}$, there exist nonsingular matrices B and C, also over $\mathscr{F}$, such that the product BAC is the rank normal form of A.*

When A is nonsingular, its rank normal form is just the identity matrix of the same order. In this case, there exist elementary matrices $R_1, \ldots, R_h$ and $C_1, \ldots, C_k$ such that

$$
R_h \cdots R_1 A C_1 \cdots C_k = I,
$$

so

(6.3.4) $$A = R_1^{-1} \cdots R_h^{-1} C_k^{-1} \cdots C_1^{-1}.$$

The matrices on the right in this equation are also elementary matrices. Hence we have

Theorem 6.3.5: *A nonsingular matrix over a field $\mathscr{F}$ can be expressed as a product of elementary matrices over $\mathscr{F}$.*

For example:

$$\begin{bmatrix} 1 & 3 & -2 \\ 0 & 1 & -1 \\ 0 & -1 & 0 \end{bmatrix} \sim \begin{bmatrix} 1 & 0 & -2 \\ 0 & 1 & -1 \\ 0 & -1 & 0 \end{bmatrix}$$ (subtract three times first column from second)

$$\sim \begin{bmatrix} 1 & 0 & 0 \\ 0 & 1 & -1 \\ 0 & -1 & 0 \end{bmatrix}$$ (add twice first column to third)

$$\sim \begin{bmatrix} 1 & 0 & 0 \\ 0 & 0 & -1 \\ 0 & -1 & 0 \end{bmatrix}$$ (add third row to second)

$$\sim \begin{bmatrix} 1 & 0 & 0 \\ 0 & -1 & 0 \\ 0 & 0 & -1 \end{bmatrix}$$ (exchange second and third rows)

$$\sim \begin{bmatrix} 1 & 0 & 0 \\ 0 & 1 & 0 \\ 0 & 0 & 1 \end{bmatrix}$$ (multiply second column by -1, then multiply third column by -1).

Hence

$$\begin{bmatrix} 1 & 0 & 0 \\ 0 & 0 & 1 \\ 0 & 1 & 0 \end{bmatrix}\begin{bmatrix} 1 & 0 & 0 \\ 0 & 1 & 1 \\ 0 & 0 & 1 \end{bmatrix}\begin{bmatrix} 1 & 3 & -2 \\ 0 & 1 & -1 \\ 0 & -1 & 0 \end{bmatrix}\begin{bmatrix} 1 & -3 & 0 \\ 0 & 1 & 0 \\ 0 & 0 & 1 \end{bmatrix}$$

$$\times \begin{bmatrix} 1 & 0 & 2 \\ 0 & 1 & 0 \\ 0 & 0 & 1 \end{bmatrix}\begin{bmatrix} 1 & 0 & 0 \\ 0 & -1 & 0 \\ 0 & 0 & 1 \end{bmatrix}\begin{bmatrix} 1 & 0 & 0 \\ 0 & 1 & 0 \\ 0 & 0 & -1 \end{bmatrix} = I.$$

Writing inverses of the elementary matrices by inspection and paying careful attention to the order in which factors appear, we obtain the factorization

$$\begin{bmatrix} 1 & 3 & -2 \\ 0 & 1 & -1 \\ 0 & -1 & 0 \end{bmatrix} = \begin{bmatrix} 1 & 0 & 0 \\ 0 & 1 & -1 \\ 0 & 0 & 1 \end{bmatrix}\begin{bmatrix} 1 & 0 & 0 \\ 0 & 0 & 1 \\ 0 & 1 & 0 \end{bmatrix}\begin{bmatrix} 1 & 0 & 0 \\ 0 & 1 & 0 \\ 0 & 0 & -1 \end{bmatrix}$$

$$\times \begin{bmatrix} 1 & 0 & 0 \\ 0 & -1 & 0 \\ 0 & 0 & 1 \end{bmatrix}\begin{bmatrix} 1 & 0 & -2 \\ 0 & 1 & 0 \\ 0 & 0 & 1 \end{bmatrix}\begin{bmatrix} 1 & 3 & 0 \\ 0 & 1 & 0 \\ 0 & 0 & 1 \end{bmatrix}.$$

In certain applications, it is essential to be able to express a given nonsingular matrix as a product of elementary matrices. The factorization is not unique, for there is always an unlimited number of ways in which to reduce A to rank normal form.

This procedure affords an often useful means of inverting a matrix. In fact, from (6.3.4) we have at once that

$$(6.3.5) \qquad A^{-1} = C_1 \cdots C_k R_h \cdots R_1.$$

For example, inspection yields the factorization

$$\begin{bmatrix} 1 & 2 & 3 \\ 0 & 1 & 2 \\ 0 & 0 & 1 \end{bmatrix} = \begin{bmatrix} 1 & 0 & -1 \\ 0 & 1 & 0 \\ 0 & 0 & 1 \end{bmatrix} \begin{bmatrix} 1 & 2 & 0 \\ 0 & 1 & 0 \\ 0 & 0 & 1 \end{bmatrix} \begin{bmatrix} 1 & 0 & 0 \\ 0 & 1 & 2 \\ 0 & 0 & 1 \end{bmatrix},$$

so

$$\begin{bmatrix} 1 & 2 & 3 \\ 0 & 1 & 2 \\ 0 & 0 & 1 \end{bmatrix}^{-1} = \begin{bmatrix} 1 & 0 & 0 \\ 0 & 1 & -2 \\ 0 & 0 & 1 \end{bmatrix} \begin{bmatrix} 1 & -2 & 0 \\ 0 & 1 & 0 \\ 0 & 0 & 1 \end{bmatrix} \begin{bmatrix} 1 & 0 & 1 \\ 0 & 1 & 0 \\ 0 & 0 & 1 \end{bmatrix}$$

$$= \begin{bmatrix} 1 & -2 & 1 \\ 0 & 1 & -2 \\ 0 & 0 & 1 \end{bmatrix}.$$

We close this section with a definition and a theorem. Two matrices over a field $\mathscr{F}$ are said to be **equivalent** if and only if they have the same rank normal form, that is, if and only if they have the same order and the same rank. If A and B are equivalent with normal form N, then there exist nonsingular matrices C, D, F, and G such that

$$CAD = N = FBG.$$

We can solve this equation for either A or B, and since C, D, F, and G are products of elementary matrices, we have

Theorem 6.3.6: *If two matrices A and B over the same field $\mathscr{F}$ are equivalent, either may be transformed into the other by a sequence of elementary transformations; that is, there exist nonsingular matrices P and Q such that*

$$PAQ = B, \qquad P^{-1}BQ^{-1} = A.$$

6.4   Exercises

1. Use elementary transformations to reduce

(i) $\begin{bmatrix} 1 & 0 & 1 \\ 0 & 2 & 0 \\ 1 & 0 & 1 \end{bmatrix}$,

(ii) $\begin{bmatrix} 2 & -1 & 3 & 4 & 0 \\ 1 & 3 & 1 & -2 & 2 \\ 4 & 1 & 0 & 5 & -3 \\ 1 & -1 & -4 & 3 & -5 \end{bmatrix}$,

(iii) $\begin{bmatrix} 1 \\ 2 \\ -1 \\ 4 \\ 3 \end{bmatrix}$,

(iv) $\begin{bmatrix} 1 & a & b & 0 \\ 0 & c & d & 1 \\ 1 & a & b & 0 \\ 0 & c & d & 1 \end{bmatrix}$,

to (a) column echelon form, (b) row echelon form, and (c) rank normal form.

2. How does the rank of each of these matrices vary with the value of t?

(a) $\begin{bmatrix} 1 & 1 & t \\ 1 & t & 1 \\ t & 1 & 1 \end{bmatrix}$,

(b) $\begin{bmatrix} 0 & 1 & t \\ 1 & t & -1 \\ t & -1 & 0 \end{bmatrix}$,

(c) $\begin{bmatrix} t & 2 & -1 \\ 2 & 4 & -2 \\ -1 & -2 & t \end{bmatrix}$.

3. Under what conditions, if any, will the rank of the matrix

$$\begin{bmatrix} 1 & 0 & 0 \\ 0 & h-2 & 2 \\ 0 & k-1 & h+2 \\ 0 & 0 & 3 \end{bmatrix}$$

be less than 3, and what will that rank be?

4. (a) Give examples to show that the rank of the sum of two matrices of ranks 2 and 3, respectively, can be 1, 2, 3, 4, or 5.
   (b) Write the rank normal forms for matrices of rank 3 and orders (4, 3), (3, 6), (4, 5), and (3, 3) respectively.

5. Show that αA has the same rank as A for all nonzero scalars α.

6. Give an example that shows that the rank normal form of the product of two matrices is not necessarily the product of their rank normal forms.

7. Let E and F denote the matrices of arbitrary elementary transformations. Give verbal interpretations to the equations

$$E(AB) = (EA)B$$

and

$$(AB)F = A(BF).$$

8. Invent simple examples of matrices that illustrate these relationships:

(a) $r(A + B) = r(A) + r(B)$,   (d) $r(HK) < r(H)$ and $< r(K)$,
(b) $r(C + D) = r(C) - r(D)$,   (e) $r(LM) = r(L) = r(M)$,
(c) $r(FG) = r(F)$ but $< r(G)$,   (f) $r(NP) < r(N)$ but $= r(P)$.

9. Factor the matrix

$$\begin{bmatrix} 2 & 5 \\ 3 & 1 \end{bmatrix}$$

into a product of elementary matrices.

10. Find nonsingular matrices A and B such that

$$A \begin{bmatrix} 2 & -1 & 4 \\ 3 & 0 & 1 \end{bmatrix} B$$

is in rank normal form.

11. Find the inverse of the product

$$\begin{bmatrix} 0 & 0 & 1 \\ 0 & 1 & 0 \\ 1 & 0 & 0 \end{bmatrix} \begin{bmatrix} 1 & 0 & -2 \\ 0 & 1 & 0 \\ 0 & 0 & 1 \end{bmatrix} \begin{bmatrix} 1 & 0 & 0 \\ 0 & -2 & 0 \\ 0 & 0 & 1 \end{bmatrix}$$

without first computing the product.

12. Factor into a product of elementary matrices, then find the inverse:

(a) $\begin{bmatrix} 1 & a & b & c \\ 0 & 1 & 0 & 0 \\ 0 & 0 & 1 & 0 \\ 0 & 0 & 0 & 1 \end{bmatrix}$,   (b) $\begin{bmatrix} a & 1 & 0 \\ 1 & 0 & 0 \\ b & c & 2 \end{bmatrix}$.

13. Determine matrices P and Q such that

$$P \begin{bmatrix} 2 & -1 & 0 \\ 3 & 0 & 1 \end{bmatrix} Q = \begin{bmatrix} 1 & -1 & 1 \\ 0 & 1 & 2 \end{bmatrix}.$$

14. Determine the ranks of the $n \times n$ matrices

(a) $\begin{bmatrix} n-1 & 1 & \cdots & 1 \\ 1 & n-1 & \cdots & 1 \\ \vdots & & & \\ 1 & 1 & \cdots & n-1 \end{bmatrix}$,   (b) $\begin{bmatrix} 1-n & 1 & \cdots & 1 \\ 1 & 1-n & \cdots & 1 \\ \vdots & & & \\ 1 & 1 & \cdots & 1-n \end{bmatrix}$.

15. Show that three points (x_1, x_2), (y_1, y_2), and (z_1, z_2) in the plane are collinear if and only if the rank of the matrix

$$\begin{bmatrix} x_1 & x_2 & 1 \\ y_1 & y_2 & 1 \\ z_1 & z_2 & 1 \end{bmatrix}$$

is less than 3.

16. Show that the locus of points (x, y, z) in $\mathscr{E}^3$, such that the matrix

$$\begin{bmatrix} x & y & z \\ 1 & x & y \end{bmatrix}$$

has rank 1, is the cubic space curve with parametric equations $x = t$, $y = t^2$, $z = t^3$.

17. Given that A is a 5×5 matrix, write the elementary matrices that accomplish each of these elementary transformations on A:
  (a) adds 2 times row 3 of A to row 5 of A,
  (b) multiplies row 2 of A by $\frac{1}{3}$,
  (c) interchanges rows 1 and 4 of A.
  Then write
  (d) the single matrix that does all three operations in one step and in the order stated,
  (e) the inverse of the matrix found in (d).

18. Under what conditions on x will these matrices:

  (a) $\begin{bmatrix} x & \sqrt{2} & 0 \\ \sqrt{2} & x & \sqrt{2} \\ 0 & \sqrt{2} & x \end{bmatrix}$,   (b) $\begin{bmatrix} 4-x & 2\sqrt{5} & 0 \\ 2\sqrt{5} & 4-x & \sqrt{5} \\ 0 & \sqrt{5} & 4-x \end{bmatrix}$

  fail to have an inverse? (Examine the rank.)

19. Given the matrix equation

$$AWB + C = DWB,$$

in which all matrices are $n \times n$ and C is nonsingular, what other matrices must have inverses and what is then the solution for W?

*20. Prove, using the concept of rank, that if $AB = I$, where all three matrices are of order n, then A^{-1} and B^{-1} exist.

21. Which columns of a matrix in row echelon form are certainly independent?

22. Given the vectors

$$\begin{bmatrix} 1 \\ 1 \\ 1 \\ 0 \end{bmatrix}, \quad \begin{bmatrix} 0 \\ -1 \\ 1 \\ 1 \end{bmatrix}, \quad \begin{bmatrix} 2 \\ 1 \\ 0 \\ 0 \end{bmatrix},$$

find a fourth vector such that the resulting set of four vectors constitutes a basis for $\mathscr{E}^4$.

*23. Name as many different ways as you can to prove that a given set of n real n-vectors, $A_1, A_2, \ldots, A_n$, constitutes a basis for $\mathscr{E}^n$.

24. Compute the rank of

$$\begin{bmatrix} 0 & t & 2t & 1 \\ t & 0 & 1 & t \\ t-1 & t & t+1 & t \\ t & -1 & -1 & 0 \end{bmatrix}$$

as a function of t. Here it pays to be clever rather than systematic in reducing to diagonal form by sweepout.

25. Use elementary transformations to reduce the matrix

$$\begin{bmatrix} 1 & a & b \\ -a & 1 & c \\ -b & -c & 1 \end{bmatrix}$$

to pseudo-diagonal form; then find its inverse.

26. Given that $a \neq 0$, compute

$$\begin{bmatrix} a & a_{12} \\ 0 & a \end{bmatrix}^{-1} \quad \text{and} \quad \begin{bmatrix} a & a_{12} & a_{13} \\ 0 & a & a_{23} \\ 0 & 0 & a \end{bmatrix}^{-1}$$

with the aid of elementary transformations and then generalize.

27. Given that

$$M = \begin{bmatrix} 1 & 0 & \cdots & 0 & 0 \\ 0 & 1 & \cdots & 0 & 0 \\ \vdots & & & & \\ a_1 & a_2 & \cdots & a_{n-1} & 1 \end{bmatrix},$$

find, by thinking of M as a product of elementary matrices, a formula for M^k where k is any integer.

28. Given that

$$A_{n \times m} B_{m \times p} C_{p \times n} = I_n,$$

(a) what can you say about the ranks of A, B, and C and
(b) what can you say about the sizes of the integers m and p?

29. Let E denote the elementary matrix that effects a certain elementary column transformation. Show that the matrix that effects the corresponding row transformation is E^{T}.

30. Reduce the symmetric matrix

$$A = \begin{bmatrix} 1 & 2 & 0 \\ 2 & 8 & 6 \\ 0 & 6 & 9 \end{bmatrix}$$

to rank normal form by performing *symmetrical* pairs of elementary operations, that is, by employing exactly the same elementary transformations on rows as on columns. Then represent the result as a product BAC. How are the matrices B and C related?

At times, though not in this case, complex numbers must be used to reduce a real symmetric matrix to rank normal form by symmetrical pairs of elementary operations.

31. Prove that if A is a real symmetric matrix, then there exists a nonsingular matrix P such that PAP^T is the rank normal form of A, but that P is not necessarily real. Prove that if A is also nonsingular, then $A^{-1} = P^\mathsf{T}P$. Give an example in which P is real, another in which it is not.

*32. Prove that, if a matrix R of order n effects a given elementary transformation on the rows (columns) of every $n \times n$ matrix A, then R must be the corresponding elementary matrix defined in Theorem 6.3.1.

33. Prove that if a matrix has rank r, then it has at least one $r \times r$ submatrix which is nonsingular.

34. Show that any finite sequence of elementary row transformations of the first r rows of an $m \times p$ matrix A, $r \leq m$, can be effected by premultiplication of A by a nonsingular matrix of the form

$$\begin{bmatrix} R & 0 \\ \hline 0 & I \end{bmatrix},$$

where R is of order r. (Similarly for column transformations.)

35. Use the preceding exercise to prove that, if A_r is nonsingular and of order r, then the matrix

$$\begin{bmatrix} A_r & I_r \\ \hline I_r & 0_r \end{bmatrix}$$

may be reduced to the form

$$\begin{bmatrix} I_r & B_r \\ \hline C_r & 0_r \end{bmatrix}$$

by operations on the first r rows and first r columns only, and that $CB = A^{-1}$.

36. Consider the partitioned matrix

$$A = \begin{bmatrix} A_{11} & A_{12} & \cdots & A_{1q} \\ A_{21} & A_{22} & \cdots & A_{2q} \\ \vdots & & & \\ A_{p1} & A_{p2} & \cdots & A_{pq} \end{bmatrix},$$

where the A_{ij}'s are submatrices of orders (m_i, n_j) respectively. Let

$$\begin{bmatrix} I_{m_1} & & & \\ & I_{m_2} & 0 & \\ & 0 & \ddots & \\ & & & I_{m_p} \end{bmatrix}, \quad \begin{bmatrix} I_{n_1} & & & \\ & I_{n_2} & 0 & \\ & 0 & \ddots & \\ & & & I_{n_q} \end{bmatrix},$$

be partitioned identity matrices of orders $\sum m_i$ and $\sum n_j$ respectively. Show that elementary transformations of the *rows or columns of submatrices* of A may be effected by first applying these transformations to the rows or columns of submatrices of the partitioned identity matrices given above, then pre- or post-multiplying as before. Illustrate with examples.

37. Prove that the elementary transformations of partitioned matrices described in Exercise 36 do not alter rank.

38. Prove that if A_1 and B_1, A_2 and B_2, ..., A_k and B_k are equivalent pairs of matrices, then the partitioned matrices

$$M_1 = \begin{bmatrix} A_1 & 0 & \cdots & 0 \\ 0 & A_2 & \cdots & 0 \\ \vdots & & & \\ 0 & 0 & \cdots & A_k \end{bmatrix} \quad \text{and} \quad M_2 = \begin{bmatrix} B_1 & 0 & \cdots & 0 \\ 0 & B_2 & \cdots & 0 \\ \vdots & & & \\ 0 & 0 & \cdots & B_k \end{bmatrix}$$

are also equivalent. If you know matrices that transform A_1 into B_1, ..., A_k into B_k, what matrices will transform M_1 into M_2?

39. Show that the rank of the matrix M_1 in Exercise 38 is the sum of the ranks of the A's.

40. Prove that the elementary transformation which interchanges two parallel lines of a matrix can be effected by a suitable succession of elementary transformations of the other two kinds.

41. Show that to reduce a matrix to rank normal form it suffices to use transformations of the types
(a) add one line to any other parallel line,
(b) multiply a line by a nonzero scalar α,
so that every square matrix over a given field $\mathscr{F}$ may be written as a product of matrices of the form $I + \alpha E_{ij}$, α a scalar, $i \neq j$, and diagonal matrices over $\mathscr{F}$, of which at most one is singular.

*42. Show that it is possible, by using only row transformations, to reduce a matrix A over $\mathscr{F}$ to an equivalent matrix $[\alpha_{ij}]$ such that $\alpha_{ij} = 0$ if $i > j$, and, by using only column transformations, to an equivalent matrix $[\beta_{ij}]$ such that $\beta_{ij} = 0$ if $i < j$. Moreover, if $\alpha_{jj} \neq 0$, row transformations will also yield $\alpha_{ij} = 0$ for $i < j$ and if $\beta_{ii} \neq 0$, column transformations will also yield $\beta_{ij} = 0$ for $i > j$. Finally, if A has rank r, then no more than r of the elements α_{jj} and no more than r of the elements β_{ii} may differ from zero.

*43. Prove that the m linear forms in n variables

$$\alpha_{i1} x_1 + \alpha_{i2} x_2 + \cdots + \alpha_{in} x_n, \qquad i = 1, 2, \ldots, m,$$

are linearly dependent if and only if the rank of the matrix $[\alpha_{ij}]$ is less than m. In particular, if $m = n$, they are linearly dependent if and only if $[\alpha_{ij}]$ is singular, and if $m > n$, they are always linearly dependent.

44. Investigate the linear dependence or independence of the three linear forms

$$x_1 + 2x_2 - 3x_3$$

$$x_1 + x_2 - x_3$$

$$x_2 - 2x_3 .$$

*45. Let $V_1, V_2, \ldots, V_k$ be a basis for a subspace of $\mathscr{V}$ of $\mathscr{F}^n$. Show that a vector X of $\mathscr{F}^n$ belongs to $\mathscr{V}$ if and only if $[X, V_1, V_2, \ldots, V_k]$ has rank k.

*46. Prove that every system of linear equations equivalent under the elementary operations to the system represented by $A_{m \times n} X = B$ may be represented by an equation

$$CAX = CB,$$

where C is a nonsingular matrix of order m. Conversely, if C is any nonsingular matrix of order m, then the system represented by $CAX = CB$ is equivalent to that represented by $AX = B$.

*47. Show that, if A has rank r, then there exist nonsingular matrices R and C such that RA and AC respectively have the forms

$$\begin{bmatrix} G_1 \\ G_2 \\ \vdots \\ G_r \\ 0 \\ \vdots \\ 0 \end{bmatrix} \quad \text{and} \quad [F_1, F_2, \ldots, F_r, 0, \ldots, 0],$$

where $G_1, \ldots, G_r$ are independent rows and $F_1, \ldots, F_r$ are independent columns. Use these results to prove again that the rank of a product $AB = P$ cannot exceed the rank of either factor.

*48. Show that if an $n \times n$ matrix A over a field $\mathscr{F}$ satisfies a polynomial equation

$$\alpha_p A^p + \alpha_{p-1} A^{p-1} + \cdots + \alpha_1 A + \alpha_0 I_n = 0,$$

where $\alpha_0 \neq 0$, then A is nonsingular. What formula for A^{-1} as a polynomial function of A is implied by this equation?

6.5   Homogeneous Systems of Linear Equations

Consider the homogeneous system of m linear equations in n unknowns, over a field $\mathscr{F}$, represented by the equation

(6.5.1) $$AX = 0.$$

Let the rank of A be r. Then, when we transform the system to reduced echelon form, we get an equivalent system of r nontrivial equations

$$(6.5.2) \qquad\qquad\qquad BX = 0.$$

If $r = n$, $B = I_n$ and the solution is $X = 0$. If $r < n$, the system can be solved for r of the variables in terms of the remaining $n - r$.

Substituting parameters $t_1, t_2, \ldots, t_{n-r}$ for these $n - r$ variables and rearranging in vector notation, as we have done before, we obtain the complete solution in the form

$$(6.5.3) \qquad\qquad X = t_1 X_1 + t_2 X_2 + \cdots + t_{n-r} X_{n-r},$$

where the vectors $X_1, X_2, \ldots, X_{n-r}$ are particular solutions of (6.5.1). The particular solution X_i is obtainable from (6.5.3) by assigning all the t's but t_i the value 0 and assigning t_i the value 1.

In view of how the X_i's are obtained, they are linearly independent, for when we combine the equations

$$x_{i_1} = t_1$$
$$x_{i_2} = t_2$$
$$\vdots$$
$$x_{i_{n-r}} = t_{n-r}$$

which introduce the parameters with the equations obtained from (6.5.2) in order to accomplish the rearrangement (6.5.3), the result is that X_k has a 1 in row i_k and all other X_j's have 0's in row i_k, $k = 1, 2, \ldots, n - r$. By a familiar argument, these isolated 1's guarantee the independence of the X_j's. The next example will help to make this clear.

Combining all these observations, we obtain

Theorem 6.5.1: *The set of all solutions of a system of m homogeneous linear equations in n unknowns,*

$$AX = 0,$$

where A has rank r < n, is the set of all linear combinations of n − r linearly independent particular solutions and thus is an (n − r)-dimensional vector space. If r = n, the solution set is the 0-dimensional space consisting of the zero vector alone, so AX = 0 has nontrivial solutions if and only if the rank of A is less than the number of unknowns.

The vector space of all solutions of $AX = 0$ is called the **null space** of the matrix A.

For example, consider

$$x_1 - 2x_2 + x_3 + 4x_4 - x_5 = 0$$
$$2x_1 + x_2 - x_3 + 5x_4 + x_5 = 0$$
$$x_1 + 13x_2 - 8x_3 - 5x_4 + 8x_5 = 0.$$

Here the reduced echelon form is

$$x_1 \quad - \tfrac{1}{5}x_3 + \tfrac{14}{5}x_4 + \tfrac{1}{5}x_5 = 0$$
$$x_2 - \tfrac{3}{5}x_3 - \tfrac{3}{5}x_4 + \tfrac{3}{5}x_5 = 0.$$

Now we put

$$x_3 = 5t_1$$
$$x_4 = 5t_2$$
$$x_5 = 5t_3$$

(the coefficients 5 simplify matters) and get, after rearrangement of the equations,

$$\begin{bmatrix} x_1 \\ x_2 \\ x_3 \\ x_4 \\ x_5 \end{bmatrix} = t_1 \begin{bmatrix} 1 \\ 3 \\ 5 \\ 0 \\ 0 \end{bmatrix} + t_2 \begin{bmatrix} -14 \\ 3 \\ 0 \\ 5 \\ 0 \end{bmatrix} + t_3 \begin{bmatrix} -1 \\ -3 \\ 0 \\ 0 \\ 5 \end{bmatrix}.$$

It is easily checked that we have three particular solutions on the right. The last three components of each of these vectors make clear their linear independence. The complete solution represents a three-dimensional subspace of $\mathscr{R}^5$.

Now let $V_1, V_2, \ldots, V_{n-r}$ denote *any* $n - r$ linearly independent solutions of (6.5.1). Then, by Theorem 5.5.6, these vectors also constitute a basis for the null space of A, so that every solution of (6.5.1) can be written in the form

(6.5.4) $X = \alpha_1 V_1 + \alpha_2 V_2 + \cdots + \alpha_{n-r} V_{n-r}$

and every such combination is a solution. That is, we have

Theorem 6.5.2: *The complete solution of a system of homogeneous linear equations $AX = 0$ in n unknowns, where A has rank $r < n$, can be represented as the* *set of all linear combinations of any $n - r$ linearly independent particular solutions.*

To illustrate the use of this theorem, consider the system

$$x_1 + x_2 - 4x_3 - 2x_4 = 0$$
$$2x_1 - 3x_2 + 2x_3 + 2x_4 = 0.$$

Note that $r = 2$, $n - r = 2$, so that we need two independent particular solutions. If we put $x_3 = 1$, $x_4 = 0$, we get the equations

$$x_1 + x_2 = 4$$
$$2x_1 - 3x_2 = -2,$$

from which $x_1 = x_2 = 2$, so one solution is

$$\begin{bmatrix} 2 \\ 2 \\ 1 \\ 0 \end{bmatrix}.$$

Alternatively, if we put $x_1 = 0$, $x_2 = 2$, we get

$$2x_3 + x_4 = 1$$
$$x_3 + x_4 = 3,$$

from which $x_3 = -2$, $x_4 = 5$, so another solution is

$$\begin{bmatrix} 0 \\ 2 \\ -2 \\ 5 \end{bmatrix}.$$

The complete solution is then

$$X = t_1 \begin{bmatrix} 2 \\ 2 \\ 1 \\ 0 \end{bmatrix} + t_2 \begin{bmatrix} 0 \\ 2 \\ -2 \\ 5 \end{bmatrix},$$

since the two particular solutions are linearly independent. By Theorem 6.5.2, this representation of the complete solution is not unique.

The preceding theorem shows that the set of all solutions of an equation $AX = 0$, where A has rank r, is a vector space of dimension $n - r$. The converse of this result is also true. Suppose that we have given a subspace $\mathcal{V}$ of $\mathcal{F}^n$, where $\mathcal{V}$ has dimension $n - r$, $0 \leq r \leq n$. If $r = 0$, then $\mathcal{V} = \mathcal{F}^n$ and every vector X of $\mathcal{F}^n$ satisfies the equation $AX = 0$, where $A = 0_{n \times n}$. If $r = n$, then $\mathcal{V} = \mathcal{Z}$ and the vector 0 is the complete solution of the equation $AX = 0$, where $A = I_n$. If $0 < r < n$, then $\mathcal{V}$ has a basis $A_1, A_2, \ldots, A_{n-r}$. Since the A_j's are linearly independent, by Theorem 5.4.4 the matrix

$$[X, A_1, A_2, \ldots, A_{n-r}]$$

has rank $n - r$ if and only if X is a linear combination over $\mathcal{F}$ of the A_j's, that is, if and only if X belongs to $\mathcal{V}$.

By examining the A's, we can rearrange the rows of the matrix so that the *last* $n - r$ rows are independent:

(6.5.5)
$$\begin{bmatrix} R_{i_1} \\ \vdots \\ R_{i_r} \\ R_{i_{r+1}} \\ \vdots \\ R_{i_n} \end{bmatrix}.$$

Then, for precisely the vectors X of $\mathcal{V}$, each of the first r rows of this matrix is a linear combination of the last $n - r$ rows. That is, for such vectors X and no others there exist scalars β_{jm} in $\mathcal{F}$ such that

(6.5.6) $$R_{i_j} = \beta_{j1} R_{i_{r+1}} + \beta_{j2} R_{i_{r+2}} + \cdots + B_{j,n-r} R_{i_n}, \qquad j = 1, 2, \ldots, r.$$

These equations imply in particular that, for the first column entries,

(6.5.7)
$$x_{i_1} = \beta_{11} x_{i_{r+1}} + \beta_{12} x_{i_{r+2}} + \cdots + \beta_{1,n-r} x_{i_n}$$
$$\vdots$$
$$x_{i_r} = \beta_{r1} x_{i_{r+1}} + \beta_{r2} x_{i_{r+2}} + \cdots + \beta_{r,n-r} x_{i_n}.$$

Since this system of r equations in $x_1, x_2, \ldots, x_n$ has rank r because of the distinct left members and since its solutions are precisely the vectors of $\mathcal{V}$, we have proved

Theorem 6.5.3: *Every* $(n - r)$-*dimensional subspace of* $\mathcal{F}^n$, *where* $0 \leq r \leq n$, *may be defined as the set of all solutions of a system of homogeneous linear equations of rank* r *over* $\mathcal{F}$.

For example, a real vector X belongs to the subspace of $\mathcal{R}^4$ with the basis

$$\begin{bmatrix} 1 \\ 1 \\ 1 \\ 1 \end{bmatrix}, \quad \begin{bmatrix} 1 \\ 2 \\ 0 \\ -1 \end{bmatrix} \quad \text{if and only if} \quad \begin{bmatrix} x_1 & 1 & 1 \\ x_2 & 1 & 2 \\ x_3 & 1 & 0 \\ x_4 & 1 & -1 \end{bmatrix}$$

has rank 2. The last two rows are independent. Sweeping the last two columns of the first two rows, we obtain the equivalent matrix

$$\begin{bmatrix} x_1 - 2x_3 + x_4 & 0 & 0 \\ x_2 - 3x_3 + 2x_4 & 0 & 0 \\ x_3 & 1 & 0 \\ x_4 - x_3 & 0 & -1 \end{bmatrix}.$$

The rank of this matrix will be 2, that is, X belongs to the subspace in question, if and only if X satisfies

$$x_1 = 2x_3 - x_4$$
$$x_2 = 3x_3 - 2x_4,$$

so this system of equations defines the subspace.

The first two theorems of this section are useful in studying the rank of the product of two matrices. Recall first that the vector space of solutions of an equation $A_{m \times n} X = 0$ is called the null space of A. The dimension of the null space is often called the **nullity** or **deficiency** of A and is denoted by $N(A)$. Since A has n columns,

(6.5.8) $$N(A) = n - r(A).$$

In the case of a product $A_{m \times n} B_{n \times p}$, we have at once

(6.5.9) $$N(AB) = p - r(AB) \geq p - r(B) = N(B).$$

However, the nullity of the product AB bears no general relation to the nullity of A, as the following examples show:

$$A = \begin{bmatrix} 1 & 0 & 0 & 0 \\ 0 & 1 & 0 & 0 \end{bmatrix}, \quad B = \begin{bmatrix} 1 & 0 \\ 0 & 1 \\ 0 & 0 \\ 0 & 0 \end{bmatrix}, \quad AB = \begin{bmatrix} 1 & 0 \\ 0 & 1 \end{bmatrix},$$

$$\begin{cases} N(AB) = 0, \\ N(A) = 2, \\ N(AB) < N(A). \end{cases}$$

$$A = \begin{bmatrix} 0 & 0 & 1 & 0 \\ 0 & 0 & 0 & 1 \end{bmatrix}, \qquad B = \begin{bmatrix} 1 & 0 \\ 0 & 1 \\ 0 & 0 \\ 0 & 0 \end{bmatrix}, \qquad AB = \begin{bmatrix} 0 & 0 \\ 0 & 0 \end{bmatrix},$$

$$\begin{cases} N(AB) = 2, \\ N(A) = 2, \\ N(AB) = N(A). \end{cases}$$

$$A = \begin{bmatrix} 0 & 0 & 1 & 0 \\ 0 & 0 & 0 & 1 \end{bmatrix}, \qquad B = \begin{bmatrix} 1 & 0 & 0 \\ 0 & 1 & 0 \\ 0 & 0 & 0 \\ 0 & 0 & 0 \end{bmatrix}, \qquad AB = \begin{bmatrix} 0 & 0 & 0 \\ 0 & 0 & 0 \end{bmatrix},$$

$$\begin{cases} N(AB) = 3, \\ N(A) = 2, \\ N(AB) > N(A). \end{cases}$$

On the other hand, if A and B are both square and of order n, so that A, B, and AB all have the same number of columns, we have, from $r(AB) \le \min(r(A), r(B))$,

(6.5.10) $N(AB) \ge \max(N(A), N(B))$ (A and B square).

In summary, we have

Theorem 6.5.4: *The nullity of the product AB is at least as great as the nullity of B. When A and B are square, the nullity of the product AB is at least as great as the nullity of either factor.*

Again let A and B be arbitrary $m \times n$ and $n \times p$ matrices respectively. Let $\mathcal{N}_A$ denote the null space of A, $\mathcal{N}_B$ the null space of B, and $\mathcal{N}_{AB}$ the null space of AB. Then if $BX = 0$, it follows that $ABX = 0$ also, so $\mathcal{N}_B$ is a subspace of $\mathcal{N}_{AB}$. We may therefore begin with any basis X_1, $X_2, \ldots, X_{N(B)}$ for $\mathcal{N}_B$ and augment it with vectors $X_{N(B)+1}, \ldots, X_{N(AB)}$ to form a basis for $\mathcal{N}_{AB}$.

We next observe that the vectors $BX_{N(B)+1}, \ldots, BX_{N(AB)}$ all belong to the null space of A. Moreover, they are linearly independent, for if

$$\sum_{j=N(B)+1}^{N(AB)} \lambda_j BX_j = 0, \quad \text{that is, if} \quad B \sum_{j=N(B)+1}^{N(AB)} \lambda_j X_j = 0,$$

then the vector

$$\sum_{j=N(B)+1}^{N(AB)} \lambda_j X_j$$

belongs to $\mathcal{N}_B$. Hence, for suitable scalars μ_i,

$$\sum_{j=N(B)+1}^{N(AB)} \lambda_j X_j = \sum_{i=1}^{N(B)} \mu_i X_i.$$

Because the X's are linearly independent, all the λ's (and all the μ's as well) must be 0, so $BX_{N(B)+1}, \ldots, BX_{N(AB)}$ are independent as claimed and therefore span a subspace of dimension $N(AB) - N(B)$ of $\mathcal{N}_A$. This implies that $N(A) \geq N(AB) - N(B)$. We have thus

Theorem 6.5.5: *The nullity of the product of two matrices $A_{m \times n}$ and $B_{n \times p}$ is not greater than the sum of their nullities:*

(6.5.11) $$N(AB) \leq N(A) + N(B).$$

When A and B are both square and of order n, we can combine (6.5.10) and (6.5.11) to obtain **Sylvester's law of nullity:**

(6.5.12) $$\max(N(A), N(B)) \leq N(AB) \leq N(A) + N(B)$$
$$(A \text{ and } B \text{ square}).$$

In the general case, since $N(A) = n - r(A)$, $N(B) = p - r(B)$, and $N(AB) = p - r(AB)$, (6.5.11) and Theorem 6.2.2 provide upper and lower bounds for the rank of a product:

Theorem 6.5.6: *For $A_{m \times n}$ and $B_{n \times p}$,*

(6.5.13) $$r(A) + r(B) - n \leq r(AB) \leq \min(r(A), r(B)).$$

6.6 Nonhomogeneous Systems of Linear Equations

Consider now a nonhomogeneous system of linear equations represented by

(6.6.1) $$AX = B,$$

where $A_{m \times n}$ and $B_{m \times 1} \neq 0$ are matrices over a given field $\mathscr{F}$. We saw in Chapter 1 how to solve such a system by what amounted to transforming the matrix

$$[A, \quad B],$$

which is called the **augmented matrix** of the system, to reduced echelon form:

$$\begin{bmatrix}
1 & \alpha_{12} & \cdots & 0 & \alpha_{1,i_1+1} & \cdots & 0 & \alpha_{1k} & \cdots & \alpha_{1n} & \vdots & \beta_1 \\
0 & \cdots & 0 & 1 & \alpha_{2,i_1+1} & \cdots & 0 & \alpha_{2k} & \cdots & \alpha_{2n} & \vdots & \beta_2 \\
\vdots & & & & & & & & & & \vdots & \\
0 & \cdots\cdots\cdots\cdots\cdots\cdots & & & 0 & 1 & \alpha_{rk} & \cdots & \alpha_{rn} & \vdots & \beta_r \\
0 & 0 & \cdots\cdots\cdots\cdots\cdots & & 0 & 0 & 0 & \cdots & 0 & \vdots & \beta_{r+1} \\
\vdots & & & & & & & & & \vdots & \\
0 & 0 & \cdots\cdots\cdots\cdots\cdots & & 0 & 0 & 0 & \cdots & 0 & \vdots & \beta_m
\end{bmatrix}.$$

The system is consistent if and only if $\beta_{r+1} = \cdots = \beta_m = 0$, in which case we can solve for r unknowns in terms of the remaining $n - r$ unknowns.

Since only row operations are used to obtain the reduced echelon form, so that the submatrix consisting of the first n columns is a reduced form of A, it follows that A has the rank r, where r is the number of rows of this submatrix with leading 1's.

The rank of the entire reduced matrix, including the column of β's, is the same as that of $[A, B]$. Hence the rank of the augmented matrix is also r if all of $\beta_{r+1}, \ldots, \beta_m$ are 0. Moreover, it will be r *only* if this is true, for if at least one of this set of β's is unequal to zero, we have *exactly one additional linearly independent row*, so that the rank of the augmented matrix must then be $r + 1$. This observation proves

Theorem 6.6.1: *A system of nonhomogeneous linear equations $AX = B$ is consistent if and only if the coefficient and augmented matrices, A and $[A, B]$, have the same rank.*

When the system is consistent, the common rank of A and $[A, B]$ is called the **rank of the system**.

Consider the system

(6.6.2)
$$\begin{aligned}
x_1 + x_2 - 3x_3 + x_4 &= 0 \\
2x_1 - x_2 + x_3 - x_4 &= 1 \\
x_1 + 4x_2 - 10x_3 + 4x_4 &= -1.
\end{aligned}$$

Here the reduced echelon form of the system is

$$\begin{aligned}
x_1 \qquad - \tfrac{2}{3}x_3 \qquad &= \tfrac{1}{3} \\
x_2 - \tfrac{7}{3}x_3 + x_4 &= -\tfrac{1}{3}.
\end{aligned}$$

Thus $r = 2$, $n - r = 2$. We have, if we put $x_3 = t_1$, $x_4 = t_2$:

$$
\begin{aligned}
x_1 &= \quad \tfrac{1}{3} + \tfrac{2}{3}t_1 \\
x_2 &= -\tfrac{1}{3} + \tfrac{7}{3}t_1 - t_2 \\
x_3 &= \qquad\quad t_1 \\
x_4 &= \qquad\qquad\quad t_2
\end{aligned}
$$

or, in vector notation

$$
(6.6.3) \qquad X =
\begin{bmatrix} \tfrac{1}{3} \\ -\tfrac{1}{3} \\ 0 \\ 0 \end{bmatrix}
+ t_1
\begin{bmatrix} \tfrac{2}{3} \\ \tfrac{7}{3} \\ 1 \\ 0 \end{bmatrix}
+ t_2
\begin{bmatrix} 0 \\ -1 \\ 0 \\ 1 \end{bmatrix}.
$$

Here the first column on the right provides a particular solution of the original system, obtainable by putting $t_1 = t_2 = 0$.

If we now replace the constant terms of the given system of equations by zeros, we get the *corresponding homogeneous system.* Solving this system in the same way, we simply drop the constant terms throughout and get

$$
(6.6.4) \qquad X = t_1
\begin{bmatrix} \tfrac{2}{3} \\ \tfrac{7}{3} \\ 1 \\ 0 \end{bmatrix}
+ t_2
\begin{bmatrix} 0 \\ -1 \\ 0 \\ 1 \end{bmatrix},
$$

which shows that the complete solution (6.6.3) is the sum of a particular solution of the nonhomogeneous system (6.6.2) and the complete solution (6.6.4) of the corresponding homogeneous system. This fact is completely general, as is shown by

Theorem 6.6.2: *The complete solution of a nonhomogeneous system $AX = B$ is the sum of any particular solution of the given system and the complete solution of the corresponding homogeneous system $AX = 0$.*

To prove this, let X_0 denote any particular solution of $AX = B$ and let Y denote an arbitrary solution so that $AY = B$ and $AX_0 = B$ are both true equations. Subtracting, we have $A(Y - X_0) = 0$, so that $Y - X_0$ is a solution of the corresponding homogeneous system $AX = 0$. Put $Y - X_0 = W$ so that

$$
Y = X_0 + W;
$$

that is, any solution Y of $AX = B$ is the sum of the particular solution X_0 of $AX = B$ and some solution W of $AX = 0$.

Now let X_0 denote the same particular solution of $AX = B$, let W denote *any* solution of $AX = 0$, and let

$$Y = X_0 + W.$$

Then, since $AW = 0$,

$$AY = AX_0 + AW = AX_0 = B;$$

that is, Y is a solution of $AX = B$.

The argument shows that if we write

(6.6.5) $Y = X_0 + W,$

where X_0 is any particular solution of $AX = B$ and W is now the *complete* solution of $AX = 0$, we get *every solution and only solutions* of $AX = B$; that is, we have the complete solution of this equation, as the theorem asserts.

We illustrate with an example. Consider the system

$$x_1 + x_2 - x_3 - x_4 = 2$$
$$x_1 - x_2 + x_3 - x_4 = 0.$$

Here, if we put $x_3 = x_4 = 0$, we get the equations

$$x_1 + x_2 = 2$$
$$x_1 - x_2 = 0,$$

so $x_1 = x_2 = 1$. Thus a particular solution is

$$\begin{bmatrix} 1 \\ 1 \\ 0 \\ 0 \end{bmatrix}.$$

When we turn to the corresponding homogeneous system, we can see by inspection that two linearly independent solutions are

$$\begin{bmatrix} 1 \\ 0 \\ 0 \\ 1 \end{bmatrix} \quad \text{and} \quad \begin{bmatrix} 0 \\ 1 \\ 1 \\ 0 \end{bmatrix}.$$

Hence the complete solution of the given system is, by the theorem,

(6.6.6)
$$X = \begin{bmatrix} 1 \\ 1 \\ 0 \\ 0 \end{bmatrix} + t_1 \begin{bmatrix} 1 \\ 0 \\ 0 \\ 1 \end{bmatrix} + t_2 \begin{bmatrix} 0 \\ 1 \\ 1 \\ 0 \end{bmatrix}.$$

The geometrical interpretation here is this: The complete solution of the homogeneous system,

(6.6.7)
$$W = t_1 \begin{bmatrix} 1 \\ 0 \\ 0 \\ 1 \end{bmatrix} + t_2 \begin{bmatrix} 0 \\ 1 \\ 1 \\ 0 \end{bmatrix},$$

is a two-dimensional vector space representable as a plane on the origin in $\mathscr{E}^4$. The complete solution (6.6.6) of the nonhomogeneous system is representable as a plane parallel to the one on the origin and lying on the point $(1, 1, 0, 0)$. Figure 3-14 gives an intuitive idea of the situation.

It is important here to realize that the vectors given by (6.6.6), that is, those vectors X whose terminal points lie on the plane on $(1, 1, 0, 0)$, *do not* constitute a vector space, whereas those given by (6.6.7) *do* constitute a vector space. The terminal points of these latter vectors lie on the plane on the origin. The first plane is said to be obtained from the second by **translation**. The translation is accomplished by adding the vector X_0 to each vector W of the space represented by the plane on the origin.

6.7 Exercises

1. Solve these systems of equations by the method of Theorem 6.6.2:

(a) $\quad x_1 + x_2 - x_3 = 1$
$\quad\;\; 3x_1 - x_2 + 2x_3 = 0,$

(b) $\quad 2x_1 - x_2 + x_3 = 4$
$\quad\;\; x_1 + 2x_2 - x_3 = -1$
$\quad\;\; 7x_1 - 16x_2 + 11x_3 = 29,$

(c) $\quad x_1 + x_2 + x_3 + x_4 = 1,$

(d) $\quad -3x_1 + x_2 + x_3 + 5x_4 = 4$
$\qquad\qquad\;\; 2x_2 - 3x_3 + 5x_4 = 4$
$\qquad\qquad\quad\; - x_3 + 5x_4 = 4.$

*2. Given that $A_{n \times n}$ has rank n, use Theorems 6.5.1 and 6.6.2 to show that the system of equations represented by $AX = B$, where X and B denote n-vectors, has a *unique* solution for X.

3. Find a homogeneous system of linear equations with the complete solution

$$t_1 \begin{bmatrix} 1 \\ 1 \\ 0 \\ 0 \end{bmatrix} + t_2 \begin{bmatrix} 0 \\ 1 \\ 1 \\ 1 \end{bmatrix}.$$

4. Find a nonhomogeneous system of linear equations with the complete solution

$$\begin{bmatrix} 1 \\ 1 \\ 1 \\ 1 \end{bmatrix} + t_1 \begin{bmatrix} 1 \\ 1 \\ 0 \\ 0 \end{bmatrix} + t_2 \begin{bmatrix} 0 \\ 1 \\ 1 \\ 1 \end{bmatrix}.$$

5. In each case, illustrate with a figure the locus of the terminal points of the vectors of the set:

   (a) $X = t_1 \begin{bmatrix} 1 \\ 1 \\ 0 \end{bmatrix} + t_2 \begin{bmatrix} 0 \\ 1 \\ 1 \end{bmatrix}$,

   (b) $X = t_1 \begin{bmatrix} 1 \\ 1 \\ 0 \end{bmatrix} + t_2 \begin{bmatrix} 0 \\ 1 \\ 1 \end{bmatrix}$,   where $t_1 + t_2 = 1$,

   (c) $X = \begin{bmatrix} 1 \\ 1 \\ 0 \end{bmatrix} + t \begin{bmatrix} -1 \\ 0 \\ 1 \end{bmatrix}$.

6. Using considerations of rank, show that if $AX = 0$ for all n-vectors X, then $A = 0$.

7. Show that $AX = 0$ has nontrivial solutions for the vector X if and only if the columns of A are linearly dependent. Similarly, show that $YB = 0$, where Y denotes a row vector, has nontrivial solutions for Y if and only if the rows of B are linearly dependent.

*8. Show that if $AB = 0$, where A and B are of order n and where neither A nor B is the zero matrix, then both A and B are singular.

*9. Show that if $A \neq 0$ is a singular matrix of order n, there always exist nonzero but necessarily singular matrices B and C of order n such that $AB = CA = 0$. Determine all such matrices B and C.

*10. Given an $m \times n$ matrix A, shown that there exist nonzero $n \times p$ matrices B such that $AB = 0$ if and only if the columns of A are linearly dependent. Similarly, given an $n \times p$ matrix B, there exist nonzero matrices A such that $AB = 0$ if and only if the rows of B are linearly dependent. Determine all such matrices in each case.

11. Prove that a necessary and sufficient condition that a homogeneous system of n

linear equations in n unknowns over a given field $\mathscr{F}$ have nontrivial solutions is that the coefficient matrix be singular.

*12. Prove that a necessary and sufficient condition that k linear combinations of k linearly independent n-vectors be also independent is that the matrix of the coefficients of combination be nonsingular.

*13. Prove that if A is an $m \times n$ matrix with rank n and if B is $n \times p$, then $r(AB) = r(B)$, and if $A_{m \times n}$ has rank m and C is $p \times m$, then $r(CA) = r(C)$.

14. Let A denote any skew-symmetric matrix of order 3:

$$A = \begin{bmatrix} 0 & c & -b \\ -c & 0 & a \\ b & -a & 0 \end{bmatrix}.$$

Obtain in parametric form the solutions of the equation $AX = 0$ and then show without computing the product that $AB = 0$, where

$$B = \begin{bmatrix} a^2 & ab & ac \\ ab & b^2 & bc \\ ac & bc & c^2 \end{bmatrix}.$$

15. Given n linear forms $A_1^T X, A_2^T X, \ldots, A_n^T X$, all of which vanish at the same nonzero vector B, that is, $A_1^T B = 0$, $A_2^T B = 0, \ldots, A_n^T B = 0$. Prove that the n forms are linearly dependent.

16. Prove that if A has rank r, then every set of $r + 1$ columns of A is linearly dependent.

17. Show that if a matrix A of order n has rank $r < n$, there are exactly $n - r$ linearly independent equations relating the columns (rows) of A. (*Hint:* If $\sum \alpha_j A_j = 0$, where A_j are the columns of A, then $[\alpha_1, \alpha_2, \ldots, \alpha_n]^T$ is a solution of the equation $AX = 0$.) Illustrate with an example. What is the corresponding result when A is of order (m, n)?

*18. (a) Show that if a matrix $A_{m \times n}$ has rank 1, then it can be written as the product of two matrices of rank 1:

$$A = \begin{bmatrix} \alpha_1 \\ \alpha_2 \\ \vdots \\ \alpha_m \end{bmatrix} [\beta_1, \beta_2, \ldots, \beta_n].$$

(b) Show that if A has rank 2, then it can be written as the product of two matrices of rank 2:

$$A = \begin{bmatrix} \alpha_{11} & \alpha_{12} \\ \alpha_{21} & \alpha_{22} \\ \vdots \\ \alpha_{m1} & \alpha_{m2} \end{bmatrix} \begin{bmatrix} \beta_{11} & \beta_{12} & \cdots & \beta_{1n} \\ \beta_{21} & \beta_{22} & \cdots & \beta_{2n} \end{bmatrix}.$$

(c) Now generalize.

19. In each case, under what conditions on t will the system of equations be consistent?

(a) $3x_1 + x_2 + x_3 = t$
$x_1 - x_2 + 2x_3 = 1 - t$
$x_1 + 3x_2 - 3x_3 = 1 + t$,

(b) $x_1 + x_2 + x_3 + x_4 = t$
$x_1 - x_2 + x_3 - x_4 = t - 4$
$x_1 + x_2 - x_3 - x_4 = t + 1$
$3x_1 + x_2 + x_3 - x_4 = 0$.

20. For what values of t will the system

(a) $x + 2y - z = 0$
$2x - 3y + z = 0$
$tx + 7y + z = 0$,

(b) $tx_1 + x_2 \quad\quad = 0$
$x_1 + tx_2 - x_3 = 0$
$\quad - x_2 + tx_3 = 0$

have a nontrivial solution? Why?

*21. (a) Show that $AX = B$ is consistent if and only if B is a linear combination of the columns of A.

(b) Show that $AX = B$ is consistent if and only if every solution of $A^{\mathsf{T}}Y = 0$ is also a solution of $B^{\mathsf{T}}Y = 0$.

22. Given that $A_{m \times n} X_{n \times 1} = B_{m \times 1}$ is consistent for every choice of the vector B, what can you say about the dimensions and rank of A?

23. Prove that precisely one of the following is true:

(a) $A_{m \times n} X_{n \times 1} = B$ is consistent;

(b) there exists a vector $Y_{m \times 1}$ such that $Y^{\mathsf{T}}A = 0$ and $Y^{\mathsf{T}}B = 1$. (*Hint:* Use the results of Exercise 21.)

24. Prove that if A is a matrix over the complex field, then A and A^*A have the same null space and therefore have the same rank. (*Hint:* If $A^*AX = 0$, then $X^*A^*AX = 0$, etc.) Prove similarly that A and AA^* have the same rank.

25. Suppose that the system of m equations in n unknowns $AX = B$ over the complex field is consistent and has rank m. Show that it has a unique solution X of the form

$$X = \sum_{j=1}^{m} c_j A^{(j)*},$$

where $A^{(j)}$ denotes the jth row of A.

26. Prove that the columns of AB, where B is *nonsingular*, span the same vector space as do the columns of A. Given that $X = \sum c_j A_j$ is any member of this vector space, determine an expression for X as a linear combination of the columns of AB.

27. (a) Under what conditions will k planes

$$a_j x + b_j y + c_j z + d_j = 0, \quad j = 1, 2, \ldots, k,$$

intersect in exactly one point P? Under what further conditions will each set of three of these planes intersect in P and in no other point?

(b) Show that the three hyperplanes in $\mathscr{E}^4$ with equations

$$x_1 - x_2 + x_3 - x_4 = 0$$
$$x_1 + x_2 - x_3 + 2x_4 = 3$$
$$3x_1 - x_2 + x_3 - 2x_4 = 1$$

intersect in a line. Find two points on the line and write a parametric vector equation for it.

28. Given a system of m equations in four unknowns, discuss in terms of rank and consistency the possible kinds of intersection of the corresponding hyperplanes in $\mathscr{E}^4$.

29. Show that the set of all vectors X in $\mathscr{E}^n$ which are orthogonal to each of the vectors $A_1, A_2, \ldots, A_k$ of $\mathscr{E}^n$ is a vector space. Assuming that the vectors $A_1, A_2, \ldots, A_k$ are linearly independent, what is the dimension of this space?

30. An orthogonal basis for $\mathscr{E}^4$ is to include the mutually orthogonal vectors

$$\begin{bmatrix} 2 \\ 1 \\ 0 \\ 0 \end{bmatrix}, \quad \begin{bmatrix} 1 \\ -2 \\ -2 \\ 1 \end{bmatrix}, \quad \begin{bmatrix} 0 \\ 0 \\ 1 \\ 2 \end{bmatrix}.$$

Find a fourth vector for the basis.

31. Given that $AX = B$ is a consistent system of m real equations in n unknowns, the rank of the system being r, interpret the system and its solution in geometrical language.

32. Given a k-flat in $\mathscr{E}^n$, show how to find a set of hyperplanes whose intersection is precisely the given k-flat.

33. Show that if X_1 is a solution of $AX = B_1$ and X_2 is a solution of $AX = B_2$, then $\alpha_1 X_1 + \alpha_2 X_2$ is a solution of $AX = \alpha_1 B_1 + \alpha_2 B_2$. Show how one may use this fact to write the complete solution of this last equation, given those of the first two. (The first part of this problem is the algebraic form of the **principle of superposition** of solutions of linear equations.)

34. Given that A is $m \times n$ and that $X_1, X_2, \ldots, X_m$ are solutions of the equations $AX = E_j, j = 1, 2, \ldots, m$ respectively, what is a solution of $AX = B$? In the particular case when $m = n$ and $r(A) = n$, how is the matrix $[X_1, X_2, \ldots, X_n]$ related to A?

35. The solution of a certain system of n equations in n unknowns $AX = B$ is of the form $x_j = a_j b/c$, where a_j, b, and c are integers such that b and c have no common factors other than ± 1. How may one alter the elements of A and B so as to get a system whose solution is $x_j = a_j, j = 1, 2, \ldots, n$?

36. Let X denote a $1 \times n$ row matrix, let A be an $n \times m$ matrix, and let B denote a $1 \times m$ row matrix. Show how the results of preceding sections may be adapted to the problem of solving the systems of equations $XA + B = 0$ and $XA = 0$, where X and B denote row vectors.

37. Prove that if A has rank n and M has rank 1, then $A - M$ has rank $\geq n - 1$. (Here A and M are both $n \times n$.) Then show that

$$B = A - \frac{XX^{\mathsf{T}}}{X^{\mathsf{T}}A^{-1}X}$$

has rank exactly $n - 1$ for every X such that $X^{\mathsf{T}}A^{-1}X \neq 0$. (*Hint:* If $Y = A^{-1}X$, $BY = 0$.)

38. Use diagonal matrices to provide examples of the extreme cases of Sylvester's law of nullity and of the inequalities (6.5.13).

39. Prove that if $AB = 0$, A and B both of order n, then $r(A) + r(B) \leq n$.

40. If $A_{m \times p}B_{p \times m} = I_m$, then A is called a **left inverse** of B and B is called a **right inverse** of A.

    (a) Show that if B^{-1} exists, then B^{-1} is the only left (right) inverse of B.

    (b) Show that if $A_{m \times p}B_{p \times m} = I_m$ and $B_{p \times m}C_{m \times p} = I_p$, then $m = p$ and $A = C = B^{-1}$.

    (c) Show that if $A_{m \times p}B_{p \times m} = I_m$, then $m \leq p$.

    (d) Show that if $r(B_{p \times m}) = m$ and $m < p$, then B has infinitely many left inverses.

41. If $AB = B$, then A is called a **left identity** for B. If $BC = B$, then C is called a **right identity** for B.

    (a) Show that a nonsingular matrix has only one right and one left identity element, namely I.

    (b) Show that the matrix

$$B = \begin{bmatrix} 0 & 0 & 0 \\ 2 & -1 & 3 \\ 4 & 1 & 6 \end{bmatrix}$$

does not have an inverse and that every matrix

$$A = \begin{bmatrix} a & 0 & 0 \\ b & 1 & 0 \\ c & 0 & 1 \end{bmatrix}$$

is a left identity for B.

    (c) Find a right identity element for B that is different from I_3.

42. In this exercise, all matrices are over a common number field $\mathscr{F}$.

    (a) Show that the set of all skew-symmetric matrices S of order n is a vector space of dimension $[n(n - 1)]/2$ and find a basis for it.

    (b) Show that for every n-vector B and every skew-symmetric matrix S, $B^{\mathsf{T}}SB = 0$.

    (c) Show that every vector SB, where S is skew-symmetric and B is a fixed nonzero n-vector, is a solution of the equation $BB^{\mathsf{T}}X = 0$.

    (d) Show that the set of all vectors of the form SB, where again S is skew-symmetric and B is a fixed nonzero n-vector, is a vector space of dimension $n - 1$ and find a basis for it.

    (e) Show that the vector space defined in (d) is the space of all solutions of the equation $BB^{\mathsf{T}}X = 0$.

(f) Show that, if $B \neq 0$, the equation $BB^{\mathsf{T}}X = B$ is consistent and has the complete solution

$$X = KB + \frac{1}{B^{\mathsf{T}}B} B$$

where K is the most general $n \times n$ skew-symmetric matrix.

(g) Explain the fact that there are $[n(n - 1)]/2$ arbitrary parameters in the matrix K but that the solution space of $BB^{\mathsf{T}}X = B$ has dimension only $n - 1$. (For $n > 2$, $[n(n - 1)]/2 > n - 1$.)

43. Under what conditions will the equation

$$AW = B,$$

where A and B are fixed matrices of order n over a field $\mathscr{F}$ and where W is an unknown matrix of order n, have

(a) a unique solution for W,

(b) more than one solution for W,

(c) no solutions for W?

Apply your conclusions in particular to the cases $B = 0$ and $B = A$.

44. Suppose that $A_{n \times n}$ is of rank r. Moreover, suppose that the first r rows and the first r columns of A are independent so that A can be written in the form

$$A = \begin{bmatrix} A_{11} & A_{12} \\ \hline A_{21} & A_{22} \end{bmatrix},$$

where A_{11} is $r \times r$ and A_{11}^{-1} exists. Prove there exist matrices B and C such that

$$A = \begin{bmatrix} A_{11} & A_{12} \\ \hline BA_{11} & BA_{12} \end{bmatrix} = \begin{bmatrix} A_{11} & A_{11}C \\ \hline BA_{11} & BA_{11}C \end{bmatrix}.$$

45. Let A be the matrix of Exercise 44. Show that the system of linear equations

$$AX = D$$

or

$$\begin{bmatrix} A_{11} & A_{12} \\ \hline A_{21} & A_{22} \end{bmatrix} \begin{bmatrix} X_1 \\ \hline X_2 \end{bmatrix} = \begin{bmatrix} D_1 \\ \hline D_2 \end{bmatrix},$$

has the complete solution

$$X_1 = A_{11}^{-1}D_1 - CX_2,$$

where X_1 and D_1 each have r components, provided that $D_2 = BD_1$.

46. Use Exercise 45 to solve the system

$$
\begin{bmatrix} 2 & 1 & -2 \\ 1 & 2 & 2 \\ 3 & -3 & 0 \end{bmatrix} \begin{bmatrix} x_1 \\ x_2 \\ x_3 \end{bmatrix} = \begin{bmatrix} 4 \\ -1 \\ 5 \end{bmatrix}.
$$

47. Given that $A_{m \times n}$ has rank n and that $B_{n \times p} = [B'_{n \times r}, B''_{n \times (p-r)}]$, where the columns of B' are independent and B has rank r, what can be said about the columns of $AB = [AB', AB'']$? Similarly, if

$$
A_{m \times n} = \begin{bmatrix} A'_{r \times n} \\ A''_{(m-r) \times n} \end{bmatrix}
$$

has rank r with the rows of A' independent and if $B_{n \times p}$ has rank n, what can be said about the rows of

$$
AB = \begin{bmatrix} A'B \\ A''B \end{bmatrix}?
$$

48. In the study of electric circuits, it is at times necessary to solve systems of equations of the form

$$
\begin{bmatrix} a_{11} & a_{12} & \cdots & a_{1n} \\ a_{21} & a_{22} & \cdots & a_{2n} \\ \vdots & & & \\ a_{n1} & a_{n2} & \cdots & a_{nn} \end{bmatrix} \begin{bmatrix} I_1 \\ I_2 \\ \vdots \\ I_n \end{bmatrix} = \begin{bmatrix} E \\ 0 \\ 0 \\ \vdots \\ 0 \end{bmatrix},
$$

where $[a_{ij}]_n$ is nonsingular, the I's are the "mesh currents," and E is a "source voltage." The "branch currents" are then expressions of the form

$$
i_p = \sum_{j=1}^{n} \alpha_{pj} I_j, \qquad p = 1, 2, \ldots, b,
$$

where each α is $+1$, 0, or -1. Show that the branch currents are proportional to E, that is,

$$
\begin{bmatrix} i_1 \\ i_2 \\ \vdots \\ i_b \end{bmatrix} = E \begin{bmatrix} \beta_1 \\ \beta_2 \\ \vdots \\ \beta_b \end{bmatrix},
$$

where the β's are constants depending on the a's and on the α's.

49. Show that a Hermitian matrix of rank 1 cannot have all its diagonal entries equal to zero.

50. Show that every Hermitian matrix of order n and rank 1 can be written in the form XX^*, where X is a suitably chosen n-vector.

51. Given that $\mathscr{F}$ is the field of rational functions of x with real coefficients, show that the rank of this matrix is 2:

$$\begin{bmatrix} 2x^2 + x - 1 & 3x^2 + 2 & 3x - 7 \\ 2x + 1 & 2x^2 - x + 1 & -6x^2 - 7x - 1 \\ x^2 - x - 1 & 1 & 4x^2 - 2x - 5 \end{bmatrix}.$$

6.8   Another Look at Nonhomogeneous Systems

The complete solution in the nonhomogeneous case can be described in another way that for certain purposes is more useful than that of Theorem 6.6.2. Suppose again that $AX = B$ is consistent with rank r, so the complete solution may be written

$$(6.8.1) \qquad X = X_0 + t_1 X_1 + \cdots + t_{n-r} X_{n-r}.$$

Here X_0 is a particular solution of $AX = B$ and $t_1 X_1 + \cdots + t_{n-r} X_{n-r}$ is the complete solution of $AX = 0$, so

$$(6.8.2) \qquad AX_0 = B; \quad AX_j = 0, \quad j = 1, 2, \ldots, n - r.$$

We can rearrange (6.8.1) in the form

$$(6.8.3) \quad X = \left(1 - \sum_{i=1}^{n-r} t_i\right) X_0 + t_1 (X_1 + X_0) + \cdots + t_{n-r}(X_{n-r} + X_0).$$

Note that the vectors $X_0, X_1 + X_0, \ldots, X_{n-r} + X_0$ are all solutions of $AX = B$ and that the sum of their coefficients in (6.8.3) is 1. Moreover, these vectors are linearly independent, for if

$$\alpha_0 X_0 + \alpha_1 (X_1 + X_0) + \cdots + \alpha_{n-r}(X_{n-r} + X_0) = 0,$$

then

$$(6.8.4) \qquad \left(\sum_{j=0}^{n-r} \alpha_j\right) X_0 + \alpha_1 X_1 + \cdots + \alpha_{n-r} X_{n-r} = 0.$$

Multiplying by A, we find by (6.8.2) that

$$\left(\sum_{j=0}^{n-r} \alpha_j\right) B = 0.$$

Since $B \neq 0$, it follows that

(6.8.5)
$$\sum_{j=0}^{n-r} \alpha_j = 0.$$

Hence (6.8.4) reduces to

(6.8.6)
$$\alpha_1 X_1 + \cdots + \alpha_{n-r} X_{n-r} = 0.$$

Because $X_1, X_2, \ldots, X_{n-r}$ are linearly independent, it follows from (6.8.6) and (6.8.5) that $\alpha_1 = 0, \ldots, \alpha_{n-r} = 0$, and $\alpha_0 = 0$, so the vectors X_0, $X_1 + X_0, \ldots, X_{n-r} + X_0$ are indeed linearly independent. From these observations we have

Theorem 6.8.1: *If the system of linear equations represented by $AX = B$ is consistent and has rank r, then the complete solution is expressible as a linear combination of $n - r + 1$ linearly independent solutions such that the sum of the coefficients in the combination is always 1.*

Now let $Y_0, Y_1, \ldots, Y_{n-r}$ be any $n - r + 1$ linearly independent solutions of $AX = B$ and let Y be any solution whatever of that equation. Then from (6.8.1) we see that $Y, Y_0, Y_1, \ldots, Y_{n-r}$ are $n - r + 2$ linear combinations of the $n - r + 1$ vectors $X_0, X_1, \ldots, X_{n-r}$ and hence are linearly dependent. Because $Y_0, Y_1, \ldots, Y_{n-r}$ are linearly independent, it follows from Theorem 5.4.4 that there exist scalars $t_0, t_1, \ldots, t_{n-r}$ such that

(6.8.7)
$$Y = t_0 Y_0 + t_1 Y_1 + \cdots + t_{n-r} Y_{n-r}.$$

Since every Y in this equation is a solution of $AX = B$, multiplication of (6.8.7) by A yields

$$B = \sum_{j=0}^{n-r} t_j B$$

from which, since $B \neq 0$,

(6.8.8)
$$\sum_{j=0}^{n-r} t_j = 1.$$

Thus every solution Y of $AX = B$ is given by (6.8.7) subject to the constraint (6.8.8). Moreover, if (6.8.8) is satisfied, every linear combination Y given by (6.8.7) is a solution since then

$$AY = \sum_{j=0}^{n-r} t_j (AY_j) = \left(\sum_{j=0}^{n-r} t_j \right) B = B.$$

We therefore have

Theorem 6.8.2: *If* $AX = B$ *is consistent and has rank* r, *and if* $Y_0, Y_1, \ldots, Y_{n-r}$ *are any* $n - r + 1$ *linearly independent solutions, then the complete solution is the set of all linear combinations* Y *given by*

$$Y = t_0 Y_0 + t_1 Y_1 + \cdots + t_{n-r} Y_{n-r},$$

where

$$t_0 + t_1 + \cdots + t_{n-r} = 1.$$

By way of illustration, consider the system

$$2x_1 + x_2 + x_3 + x_4 = 2$$
$$3x_1 - x_2 + x_3 - x_4 = 2$$
$$x_1 + 2x_2 - x_3 + x_4 = 1$$
$$6x_1 + 2x_2 + x_3 + x_4 = 5.$$

The system is consistent, the ranks of the coefficient and augmented matrices both being 3. Here $n - r + 1 = 2$ and the first three equations may be used to obtain the two independent solutions required.

Put $x_1 = 0$. Then from the first three equations we find $x_2 = 3$, $x_3 = 2$, $x_4 = -3$, so that a particular solution is $[0, 3, 2, -3]^T$. Next, putting $x_1 = 2$, we obtain the particular solution $[2, -5, -3, 6]^T$. These two solutions are linearly independent, so the complete solution of the system is

$$\begin{bmatrix} x_1 \\ x_2 \\ x_3 \\ x_4 \end{bmatrix} = t_1 \begin{bmatrix} 0 \\ 3 \\ 2 \\ -3 \end{bmatrix} + t_2 \begin{bmatrix} 2 \\ -5 \\ -3 \\ 6 \end{bmatrix}, \qquad t_1 + t_2 = 1.$$

6.9 The Variables one Can Solve for

For the discussion in Chapter 1, it was useful to write a system of linear equations in the form

(6.9.1)
$$f_1 = 0$$
$$f_2 = 0$$
$$\vdots$$
$$f_m = 0.$$

With the equations in this form, we say that any subset of the given set of m equations is **linearly dependent (independent)** if and only if the left members of the equations of the subset are linearly dependent (independent).

Thus the equations

$$
\begin{array}{ll}
2x - y + 3 = 0 & \qquad 2x - y = -3 \\
x + 3y - 2 = 0 \quad \text{or} & \qquad x + 3y = 2 \\
x + 10y - 9 = 0 & \qquad x + 10y = 9
\end{array}
$$

(6.9.2)

are linearly dependent equations because

$$
1 \cdot (2x - y + 3) + (-3)(x + 3y - 2) + 1 \cdot (x + 10y - 9) \equiv 0.
$$

From this identity we see that if we add the first of equations (6.9.2) to the third and subtract three times the second from their sum, we get a new third equation which is just $0 = 0$ and may be dropped. The given system of three equations is thus equivalent to a system containing just two equations. This illustrates

Theorem 6.9.1: *If one equation of a linear system of equations is a linear combination of other equations of the system, it may be deleted from the system without altering the set of solutions.*

Indeed, if

$$
f_i \equiv \alpha_1 f_{j_1} + \alpha_2 f_{j_2} + \cdots + \alpha_k f_{j_k},
$$

then

$$
f_i - \alpha_1 f_{j_1} - \alpha_2 f_{j_2} - \cdots - \alpha_k f_{j_k} \equiv 0.
$$

Thus the ith equation can be reduced to the form $0 = 0$ by transformations which lead to an equivalent system so that the solution set is not altered.

As the preceding example illustrates, the procedure and the outcome are the same whether the equations are written with the constant terms on the left as in (6.9.1) or on the right as in a system represented by

(6.9.3) $AX = B.$

Suppose that (6.9.3) is obtained from (6.9.1) by transposition of constant terms and that (6.9.3) is consistent, with rank r. Then there is at least one set of r linearly independent rows in $[A, B]$, which necessarily corresponds to a set of r linearly independent equations in (6.9.1). All rows are linear combinations of the r linearly independent rows. We can therefore reduce all rows other than the r independent ones to zeros; that is, we can eliminate all but the r linearly independent equations by subtracting from each of the dependent ones suitable multiples of the independent ones. This implies, in virtue of the preceding theorem,

Theorem 6.9.2: *If the system of m linear equations in n unknowns represented by $AX = B$ is consistent and of rank r, then any r linearly independent equations of this system form an equivalent system.*

Suppose that we want to solve for $x_{i_1}, x_{i_2}, \ldots, x_{i_r}$. Then we can rearrange the reduced system mentioned in the theorem in the form

$$c_{11}x_{i_1} + c_{12}x_{i_2} + \cdots + c_{1r}x_{i_r} = d_{11}x_{i_{r+1}} + d_{12}x_{i_{r+2}} + \cdots + d_{1,n-r}x_{i_n} + g_1$$

$$c_{21}x_{i_1} + c_{22}x_{i_2} + \cdots + c_{2r}x_{i_r} = d_{21}x_{i_{r+1}} + d_{22}x_{i_{r+2}} + \cdots + d_{2,n-r}x_{i_n} + g_2$$

$$\vdots$$

$$c_{r1}x_{i_1} + c_{r2}x_{i_2} + \cdots + c_{rr}x_{i_r} = d_{r1}x_{i_{r+1}} + d_{r2}x_{i_{r+2}} + \cdots + d_{r,n-r}x_{i_n} + g_r$$

or

(6.9.4) $$CX_1 = DX_2 + G,$$

where

$$X_1 = \begin{bmatrix} x_{i_1} \\ x_{i_2} \\ \vdots \\ x_{i_r} \end{bmatrix} \quad \text{and} \quad X_2 = \begin{bmatrix} x_{i_{r+1}} \\ x_{i_{r+2}} \\ \vdots \\ x_{i_n} \end{bmatrix}.$$

There exists a unique solution of (6.9.4) for $x_{i_1}, x_{i_2}, \ldots, x_{i_r}$ in terms of the remaining variables if and only if C^{-1} exists, that is, if and only if C has rank r. Now the r rows of C are rows of the $m \times r$ submatrix of A containing the r columns of coefficients of $x_{i_1}, x_{i_2}, \ldots, x_{i_r}$. This submatrix therefore has rank r if C does. Conversely, if the submatrix of coefficients of these variables has rank r, it contains r linearly independent rows. These constitute a nonsingular matrix C, which leads to a solvable equation of the form (6.9.4). We therefore have

Theorem 6.9.3: *If $AX = B$ is consistent and of rank r, we can solve for $x_{i_1}, x_{i_2}, \ldots, x_{i_r}$ in terms of the remaining variables if and only if the $m \times r$ submatrix of A which contains the coefficients of $x_{i_1}, x_{i_2}, \ldots, x_{i_r}$ has rank r. This observation applies whether or not $B = 0$.*

For example, we can solve

$$x_1 - 2x_2 + 3x_3 - x_4 = 1$$
$$2x_1 - 4x_2 + 5x_3 - 2x_4 = 0$$

for x_1 and x_3, x_2 and x_3, or x_3 and x_4, but *not* for x_1 and x_2, x_1 and x_4, or x_2 and x_4.

6.10   Basic Solutions

Assume that the system of r equations

$$AX = B$$

with real coefficients is consistent with rank r. Transforming first to the reduced echelon form, we can then write the solution in the form

$$x_{i_1} = L_1(x_{i_{r+1}}, \ldots, x_{i_n}) + \beta_1$$
$$x_{i_2} = L_2(x_{i_{r+1}}, \ldots, x_{i_n}) + \beta_2$$
$$\vdots$$
$$x_{i_r} = L_r(x_{i_{r+1}}, \ldots, x_{i_n}) + \beta_r,$$

where $L_1, L_2, \ldots, L_r$ are linear functions of the indicated variables. If in these r equations we put

$$x_{i_{r+1}} = x_{i_{r+2}} = \cdots = x_{i_n} = 0,$$

we get a solution

$$x_{i_1} = \beta_1$$
$$x_{i_2} = \beta_2$$
$$\vdots$$
$$x_{i_r} = \beta_r$$
$$x_{i_{r+1}} = 0$$
$$\vdots$$
$$x_{i_n} = 0.$$

A solution obtained by this procedure is called a **basic solution** of the given system. Such solutions are of importance in linear programming.

We can obtain as many basic solutions as there are sets of r variables for which we can solve in terms of the remaining $n - r$ variables. If we can solve for *every* set of r variables, then there will be

$$C(n, r) = \frac{n!}{r!(n - r)!}$$

basic solutions altogether. Frequently there are not this many basic solutions.

Consider, for example,

$$3x_1 - 2x_2 - x_3 - 4x_4 = -2$$
$$2x_1 + 5x_2 - x_3 - 4x_4 = \quad 5 \qquad (r = 2, n - r = 2).$$

Here we can solve for x_1 and x_2, for x_1 and x_3, for x_1 and x_4, for x_2 and x_3, for x_2 and x_4, but *not* for x_3 and x_4. To get the basic solution corresponding to x_1 and x_2, it is convenient first to put $x_3 = 0$ and $x_4 = 0$ and then solve the resulting equations,

$$3x_1 - 2x_2 = -2$$
$$2x_1 + 5x_2 = \quad 5.$$

This yields the basic solution $(0, 1, 0, 0)$. The other basic solutions are respectively $(-7, 0, -19, 0)$, $(-7, 0, 0, -\frac{19}{4})$, $(0, 1, 0, 0)$, and $(0, 1, 0, 0)$. Notice that a basic solution may have more zero values than those that were assigned. This explains why the basic solutions associated with x_1 and x_2, x_2 and x_3, and x_2 and x_4 are the same. Since there are three ways in which a pair of zeros can be picked from $(0, 1, 0, 0)$, we have three identical basic solutions.

When more than the $n - r$ variables assigned the value 0 assume the value 0 in a basic solution, the solution is called **degenerate**. Otherwise it is **nondegenerate**.

This example suggests a question: If $AX = B$ is a set of r consistent equations and if the rank is r, which sets of r variables $x_{i_1}, x_{i_2}, \ldots, x_{i_r}$ lead to nondegenerate basic solutions?

If such a solution exists, we must first be able to solve for these variables, so there must exist an equation (6.9.4) in which C is nonsingular. Then the columns of C are linearly independent and hence are a basis for $\mathscr{E}^r$. Since $AX = B$ has r equations, the matrix G of (6.9.4) is simply B and $C = [A_{i_1}, A_{i_2}, \ldots, A_{i_r}]$. If we now put $x_{i_{r+1}} = x_{i_{r+2}} = \cdots = x_{i_n} = 0$ in (6.9.4), we get

(6.10.1) $$CX_1 = B,$$

which says that

(6.10.2) $x_{i_1} A_{i_1} + x_{i_2} A_{i_2} + \cdots + x_{i_r} A_{i_r} = B$

or, in words, that B is a linear combination of the columns of C. If the
solution of (6.10.1) is a nondegenerate basic solution, no x_{i_j}, $j = 1, 2, \ldots, r$,
is zero and hence B can replace any one A_{i_j}, $j = 1, 2, \ldots, r$, and a basis for
$\mathscr{E}^r$ still results. That is, if a nondegenerate basic solution exists, every set of
r of the columns $A_{i_1}, A_{i_2}, \ldots, A_{i_r}, B$, is a linearly independent set.

Conversely, suppose that every such set of r columns is linearly independent. Then, since the A_{i_j}'s are independent but the $r + 1$ r-vectors
$A_{i_1}, A_{i_2}, \ldots, A_{i_r}, B$ are necessarily a dependent set, B is a unique linear
combination of the A_{i_j}'s; that is, there exist unique values $x_{i_1}, x_{i_2}, \ldots, x_{i_r}$
such that (6.10.2), and therefore (6.10.1), holds. Since any r of A_{i_1},
$A_{i_2}, \ldots, A_{i_r}, B$ are linearly independent, B can replace any of these A_{i_j}'s in
the basis they determine for $\mathscr{E}^r$ and hence no x_{i_j}, $j = 1, 2, \ldots, r$, can be 0.
That is, we have a nondegenerate basic solution. In summary, we have

Theorem 6.10.1: *The system of r real equations in n unknowns $AX = B$ has a
nondegenerate basic solution for $x_{i_1}, x_{i_2}, \ldots, x_{i_r}$ if and only if every set of r
of the vectors $A_{i_1}, A_{i_2}, \ldots, A_{i_r}, B$ is a linearly independent set.*

We return to the example given earlier in this section, for which $r = 2$.
The basic solution for x_1 and x_2 exists and is degenerate because, although
the first two of the columns

$$\begin{bmatrix} 3 \\ 2 \end{bmatrix}, \qquad \begin{bmatrix} -2 \\ 5 \end{bmatrix}, \qquad \begin{bmatrix} -2 \\ 5 \end{bmatrix}$$

are linearly independent, the last two are not. The basic solution for x_1
and x_3 exists and is nondegenerate, since every pair of the columns

$$\begin{bmatrix} 3 \\ 2 \end{bmatrix}, \qquad \begin{bmatrix} -1 \\ -1 \end{bmatrix}, \qquad \begin{bmatrix} -2 \\ 5 \end{bmatrix}$$

is linearly independent. A basic solution for x_3 and x_4 does not exist,
since the first two of the columns

$$\begin{bmatrix} -1 \\ -1 \end{bmatrix}, \qquad \begin{bmatrix} -4 \\ -4 \end{bmatrix}, \qquad \begin{bmatrix} -2 \\ 5 \end{bmatrix}$$

are not linearly independent.

6.11 Exercises

1. Determine by inspection how many basic solutions actually exist for the system

(a) $x_1 - 2x_2 - 3x_3 + x_4 = 2$ (b) $x_1 - x_2 + x_3 - x_4 = 1$
    $2x_1 + x_2 - 6x_3 - 2x_4 = 4,$ $x_1 + x_2 + x_3 - x_4 = 1,$

and state which of these are degenerate without finding the solutions.

2. Construct an example of two equations in four unknowns which has no nondegenerate basic solutions.

3. Find all basic solutions of

$$x_1 + x_2 - x_3 - 5x_4 = 10$$
$$x_1 - x_2 + x_3 - 2x_4 = 4.$$

4. For which sets of variables can one solve the following system? Which sets will lead to nondegenerate basic solutions?

$$x_1 + 2x_2 - x_3 - 2x_4 = 1$$
$$2x_1 - x_2 + 3x_3 + x_4 = -3$$
$$3x_1 + 2x_2 + 2x_3 - 2x_4 = -2.$$

5. Compute the basic solutions for the system of Exercise 4.

6. If you are seeking *only* nondegenerate basic solutions but obtain k additional zeros when you set some $n - r$ variables to zero, which additional combinations of variables, and how many, will you not need to set to zero?

*7. Show that m linear equations in n unknowns are linearly independent if either (a) each equation contains, with nonzero coefficient, an unknown appearing in no other equation, or (b) each equation contains, with nonzero coefficient, an unknown appearing in no previous equation. These tests are often used in network theory.

8. Reduce to an equivalent system of independent equations:

$$\frac{x_1 - 1}{1} = \frac{x_2 - 2}{2} = \frac{x_3 - 3}{3} = \frac{x_4 - 4}{4}.$$

*9. Show that a system of m linear equations in n uhknowns can contain at most $n + 1$ independent equations. When there are $n + 1$ independent equations, how many solutions does the system have?

10. How many independent quadratic equations of the form

$$Ax^2 + Bxy + Cy^2 + Dx + Ey + F = 0$$

may a simultaneous system of such equations contain? Could a system with this many independent equations have a simultaneous solution for x and y? (*Hint:* Write the system in question in matrix form.)

11. Show by examples that the equations of a system may be (a) dependent and consistent, (b) dependent and inconsistent, (c) independent and consistent, or (d) independent and inconsistent.

12. Determine in each case the locus in three-dimensional Euclidean space of the endpoints of the vectors of the set:

(a) $X = t_1 A_1 + t_2 A_2,$ $t_1 + t_2 = 1,$ $t_1, t_2 > 0.$

(b) $X = t_1 A_1 + t_2 A_2 + t_3 A_3,$ $t_1 + t_2 + t_3 = 1,$ $t_1, t_2, t_3 > 0.$

CHAPTER
7

Determinants

7.1 The Definition of a Determinant

We saw in Chapter 2 that the inverse of the matrix

$$A = \begin{bmatrix} a_{11} & a_{12} \\ a_{21} & a_{22} \end{bmatrix}$$

exists if and only if $a_{11}a_{22} - a_{12}a_{21} \neq 0$. When the inverse exists, it is given by

(7.1.1)
$$A^{-1} = \frac{1}{a_{11}a_{22} - a_{12}a_{21}} \begin{bmatrix} a_{22} & -a_{12} \\ -a_{21} & a_{11} \end{bmatrix}.$$

Thus the existence of the inverse and the formula for the inverse depend on the value of a certain scalar-valued function of the matrix. The value of this function is called the **determinant** of A, abbreviated "det A." It is often denoted by vertical bars instead of brackets on the array:

(7.1.2)
$$\det A = \begin{vmatrix} a_{11} & a_{12} \\ a_{21} & a_{22} \end{vmatrix} = a_{11}a_{22} - a_{12}a_{21}.$$

This same function arises in many other connections. For example, the complete solution of a rank 2 system

$$a_{11}x_1 + a_{12}x_2 + a_{13}x_3 = 0$$

$$a_{21}x_1 + a_{22}x_2 + a_{23}x_3 = 0$$

involves three such determinants, each associated with a 2×2 submatrix of the coefficient matrix:

$$(7.1.3) \quad \begin{bmatrix} x_1 \\ x_2 \\ x_3 \end{bmatrix} = t \begin{bmatrix} a_{12}a_{23} - a_{13}a_{22} \\ a_{13}a_{21} - a_{11}a_{23} \\ a_{11}a_{22} - a_{12}a_{21} \end{bmatrix} = t \begin{bmatrix} \begin{vmatrix} a_{12} & a_{13} \\ a_{22} & a_{23} \end{vmatrix} \\ - \begin{vmatrix} a_{11} & a_{13} \\ a_{21} & a_{23} \end{vmatrix} \\ \begin{vmatrix} a_{11} & a_{12} \\ a_{21} & a_{22} \end{vmatrix} \end{bmatrix}.$$

There is a simple pattern here which the reader should observe, for it generalizes to $n - 1$ independent equations in n unknowns, as we shall see later.

As another example, let us find the area of the triangle determined by the vectors

$$\begin{bmatrix} x_1 \\ x_2 \end{bmatrix} \quad \text{and} \quad \begin{bmatrix} y_1 \\ y_2 \end{bmatrix}$$

of $\mathscr{E}^2$ which terminate respectively in the points $P(x_1, x_2)$ and $Q(y_1, y_2)$, thus forming the triangle OPQ. In Figure 7-1, the area K of the triangle OPQ is the area of the triangle OSQ, plus the area of the trapezoid $SRPQ$, minus the area of triangle ORP; that is,

$$K = \tfrac{1}{2}y_1y_2 + \tfrac{1}{2}(x_2 + y_2)(x_1 - y_1) - \tfrac{1}{2}x_1x_2.$$

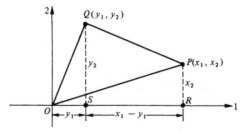

FIGURE 7-1. Area of a Triangle.

Expanding and collecting, we have

$$(7.1.4) \qquad K = \frac{1}{2}(x_1 y_2 - x_2 y_1) = \frac{1}{2}\begin{vmatrix} x_1 & y_1 \\ x_2 & y_2 \end{vmatrix}.$$

Here we have assumed in the figure that the directed angle θ from OP to OQ is positive. If it is negative, then the value of K computed from the formula just given will also be negative. (Can you show this?) The formula is valid no matter where P and Q are located in the plane. For most purposes, one uses the absolute value of the result.

These applications of determinants generalize readily to matrices of higher order, to more variables, and to higher dimensions. To accomplish such generalizations, we begin by recalling that a **permutation** of the integers $1, 2, \dots, n$ is simply a linear arrangement of these integers in some order: $j_1, j_2, \dots, j_{n-1}, j_n$. We can choose the first integer j_1 of such a permutation in n ways, after which we can choose the second, j_2, as any of the $n - 1$ remaining integers, then j_3 in $n - 2$ ways, $\dots$, the integer j_{n-1} in 2 ways, and finally j_n in 1 way, since it must be the one integer that remains. Thus there are $n(n - 1)(n - 2) \cdots 2 \cdot 1 = n!$ such permutations altogether. For example, the $3! = 6$ permutations of 1, 2, 3 are

$$1, 2, 3 \qquad 2, 3, 1 \qquad 3, 1, 2$$
$$1, 3, 2 \qquad 2, 1, 3 \qquad 3, 2, 1.$$

When no confusion can result, one often omits the commas in writing a permutation. Thus 2, 1, 3 may be written 213 and $j_1, j_2, \dots, j_n$ may be written $j_1 j_2 \dots j_n$.

Given a permutation π of the integers $1, 2, \dots, n$, we associate with it a measure, denoted by $\mu(\pi)$, of its departure from the normal order, as follows: For each of the integers $j_1, j_2, \dots, j_n$, count the number of smaller integers following it in π and add the results. The total is the **parity index** $\mu(\pi)$ of π. If $\mu(\pi)$ is even, we call π an **even permutation**. If $\mu(\pi)$ is odd, π is called an **odd permutation**. For example, if $n = 6$, we have

$$\mu(1, 2, 3, 4, 5, 6) = 0 + 0 + 0 + 0 + 0 + 0 = 0$$
$$\mu(1, 2, 3, 4, 6, 5) = 0 + 0 + 0 + 0 + 1 + 0 = 1$$
$$\mu(1, 4, 3, 2, 6, 5) = 0 + 2 + 1 + 0 + 1 + 0 = 4$$
$$\mu(6, 5, 4, 3, 2, 1) = 5 + 4 + 3 + 2 + 1 + 0 = 15.$$

The normal order is always an even permutation. Is the reverse order always odd?

When two permutations are both even or both odd, they are said to have the **same parity**. When one is even and the other is odd, they are said to have **opposite parity**.

In a permutation π, each instance of a larger integer preceding a smaller one is called an **inversion**, so $\mu(\pi)$ is the total number of inversions in π.

Since every time one integer is followed by a smaller one, the smaller integer is preceded by the larger one, it follows that *we can also compute $\mu(\pi)$ by counting the number of larger integers preceding each of $j_1, j_2, \ldots, j_n$ and totaling the results*. This should be checked in the preceding examples.

Let A be a square matrix with entries in an arbitrary field $\mathscr{F}$. We define the **determinant** of A, denoted by $\det A$ or by a square array with bars rather than brackets, as follows:

$$(7.1.5) \quad \det A = \begin{vmatrix} a_{11} & a_{12} & \cdots & a_{1n} \\ a_{21} & a_{22} & \cdots & a_{2n} \\ \vdots & & & \\ a_{n1} & a_{n2} & \cdots & a_{nn} \end{vmatrix} = \sum_{\pi} (-1)^{\mu(\pi)} a_{1j_1} a_{2j_2} \cdots a_{nj_n},$$

where $j_1, j_2, \ldots, j_n$ is a permutation π of $1, 2, \ldots, n$ and the summation extends over all $n!$ permutations π. Note that each of the $n!$ terms in this sum contains a special kind of product of n factors: The product contains one factor from each row of A since each row index appears exactly once and one factor from each column of A since each column index also appears exactly once. Moreover, every product of exactly n factors a_{ij} in which every row and every column is represented exactly once appears in this sum. Indeed, if we arrange the factors so that the row indices are in the natural order, the column indices appear in some permutation π of the natural order, so the product appears in some term of the sum (7.1.5).

The sign prefixed to a term in this sum is positive if π is an even permutation, negative if π is odd.

Because the row indices are kept in the normal order in each term, we call this the **row expansion** of $\det A$.

We have, for example, when $n = 3$,

$$\begin{vmatrix} a_{11} & a_{12} & a_{13} \\ a_{21} & a_{22} & a_{23} \\ a_{31} & a_{32} & a_{33} \end{vmatrix} = \sum_{\pi} (-1)^{\mu(\pi)} a_{1j_1} a_{2j_2} a_{3j_3}$$

$$= a_{11}a_{22}a_{33} + a_{12}a_{23}a_{31} + a_{13}a_{21}a_{32}$$

$$- a_{11}a_{23}a_{32} - a_{12}a_{21}a_{33} - a_{13}a_{22}a_{31}.$$

A particular case is

$$\begin{vmatrix} 1 & -1 & 2 \\ -1 & 0 & 3 \\ 1 & 4 & -3 \end{vmatrix} = 1 \cdot 0 \cdot (-3) + (-1) \cdot 3 \cdot 1 + 2 \cdot (-1) \cdot 4$$
$$- 1 \cdot 3 \cdot 4 - (-1)(-1)(-3) - 2 \cdot 0 \cdot 1$$
$$= 0 - 3 - 8 - 12 + 3 + 0$$
$$= -20.$$

Note that the sign factor $(-1)^{\mu(\pi)}$ appears in each term in addition to the signs carried by the individual entries of the matrix.

7.2 Some Basic Theorems

Since every row and every column of A has an entry in each term $(-1)^{\mu(\pi)}a_{1j_1}a_{2j_2} \cdots a_{nj_n}$ of det A, we have

Theorem 7.2.1: *If all the elements of one line of a square matrix A are zero, then det $A = 0$.*

(Recall that "line" means "row or column.")
There are other useful formulas for det A. One appears in

Theorem 7.2.2: *If A is an $n \times n$ matrix and π is $i_1, i_2, \ldots, i_n$,*

(7.2.1) $$\det A = \sum_{\pi} (-1)^{\mu(\pi)} a_{i_1 1} a_{i_2 2} \cdots a_{i_n n}.$$

That is, if we keep the column subscripts in the normal order in each term and extend the sum over all $n!$ permutations π of the row indices, we also obtain det A. This is called the **column expansion** of det A.

To prove this, note first that we have precisely all $n!$ products of n factors a_{ij} such that each row and each column is represented exactly once in each product. (Why?) The only question that remains is whether or not a given term has the same sign in both row and column expansions. To see that it does, consider the product $a_{1j_1}a_{2j_2} \cdots a_{nj_n}$. If $j_k = n$, move the factor a_{kj_k} to the far right so that we have

$$a_{1j_1} \cdots a_{k-1, j_{k-1}} a_{k+1, j_{k+1}} \cdots a_{nj_n} a_{kj_k}.$$

Then, among the rearranged column indices, since $j_k = n$, there are precisely $n - k$ smaller integers $j_{k+1}, \ldots, j_n$ that no longer follow n, so the parity index of the permutation of column indices is reduced by $n - k$. However, among row indices, the larger integers $k + 1, \ldots, n$ now all precede k, so the parity index of the permutation of row indices is increased by $n - k$.

If now j_m is $n - 1$ and if we move $a_{m j_m}$ to the next-to-the-last position on the right, we again reduce the parity index of the permutation of column indices and increase the parity index of the permutation of row indices by the same amount in each case. Continuing thus, we eventually get a product of the form $a_{i_1 1} a_{i_2 2} \cdots a_{i_n n}$ in which the parity index of $i_1, i_2, \ldots, i_n$ is the same as that of the permutation $j_1, j_2, \ldots, j_n$ with which we began. Thus the equal products $a_{i_1 1} a_{i_2 2} \cdots a_{i_n n}$ and $a_{1 j_1} a_{2 j_2} \cdots a_{n j_n}$ will have the same sign factor $(-1)^{\mu(\pi)}$ in the sums (7.2.1) and (7.1.5) respectively. This completes the proof.

The preceding theorem enables us to prove

Theorem 7.2.3: $\det A^{\mathsf{T}} = \det A$.

Since

$$A^{\mathsf{T}} = \begin{bmatrix} a_{11} & a_{21} & \cdots & a_{n1} \\ a_{12} & a_{22} & \cdots & a_{n2} \\ \vdots & & & \\ a_{1n} & a_{2n} & \cdots & a_{nn} \end{bmatrix},$$

the *row* expansion of $\det A^{\mathsf{T}}$ is $\sum_{\pi} (-1)^{\mu(\pi)} a_{i_1 1} a_{i_2 2} \cdots a_{i_n n}$, which is precisely the *column* expansion of $\det A$, so $\det A^{\mathsf{T}} = \det A$, as claimed.

This theorem is important because it shows that whatever we prove about rows and/or columns of the determinant of a matrix holds equally well for columns and/or rows since transposing does not change the value of the determinant. Thus, when we prove one theorem, we have really proved two. We may state this as

Theorem 7.2.4: *In every theorem about determinants, it is legitimate to interchange the words "row" and "column" throughout.*

A frequently useful result is

Theorem 7.2.5: *The interchange of any two integers of a permutation results in a permutation of opposite parity.*

Consider first the effect of the interchange of two adjacent integers, j_k and j_{k+1}. If $j_k < j_{k+1}$, the value of μ is increased by 1 by the interchange but if $j_k > j_{k+1}$, the value of μ is decreased by 1. In either case the parity is changed from odd to even or from even to odd.

Consider now the interchange of any two integers j_i and j_k, where $i < k$, in the permutation

$$j_1 \cdots j_{i-1} j_i j_{i+1} \cdots j_{k-1} j_k j_{k+1} \cdots j_n.$$

We can put j_i between j_k and j_{k+1} by $k - i$ successive interchanges of adjacent integers. Thereafter, we can put j_k in the slot vacated by j_i by $(k - 1) - i$ exchanges of adjacent integers. The total number of such interchanges is $2(k - i) - 1$, which is odd. Thus the net effect on μ is to change it from odd to even or from even to odd, so the theorem is proved. We can now prove

Theorem 7.2.6: *If two parallel lines of a square matrix A are interchanged, the determinant of the resulting matrix B is* $- \det A$.

Suppose we interchange rows i and k of A, where $i < k$, to get B.
The products $a_{1j_1} \cdots a_{ij_i} \cdots a_{kj_k} \cdots a_{nj_n}$ of det A are in one-to-one correspondence with equal products $a_{1j_1} \cdots a_{kj_k} \cdots a_{ij_i} \cdots a_{nj_n}$ in det B. By Theorem 7.2.5, we then have

$$
\begin{aligned}
\det B &= \sum (-1)^{\mu(j_1, \, \ldots, \, j_k, \, \ldots, \, j_i, \, \ldots, \, j_n)} a_{1j_1} \cdots a_{kj_k} \cdots a_{ij_i} \cdots a_{nj_n} \\
&= \sum - (-1)^{\mu(j_1, \, \ldots, \, j_i, \, \ldots, \, j_k, \, \ldots, \, j_n)} a_{1j_1} \cdots a_{ij_i} \cdots a_{kj_k} \cdots a_{nj_n} \\
&= - \det A.
\end{aligned}
$$

The same result applies to columns by Theorem 7.2.4.
For example, interchanging columns 1 and 3, we have

$$\det \begin{bmatrix} 1 & -1 & 2 \\ 0 & 1 & 3 \\ 0 & 0 & 1 \end{bmatrix} = -\det \begin{bmatrix} 2 & -1 & 1 \\ 3 & 1 & 0 \\ 1 & 0 & 0 \end{bmatrix} = 1.$$

This theorem yields next

Theorem 7.2.7: *If two parallel lines of a square matrix A are equal, det $A = 0$.*

For if we interchange the two equal lines, we obviously do not change the determinant. However, by Theorem 7.2.6, we change the sign of the determinant. Thus det $A = -\det A$, so det $A = 0$.
For example, it is easy to verify that

$$\det \begin{bmatrix} 1 & 2 & 1 \\ 0 & 5 & 0 \\ 1 & 2 & 1 \end{bmatrix} = 0.$$

Theorem 7.2.8: *If a matrix B results from a square matrix A by multiplying all entries in one line of A by k, then det B = k det A.*

Suppose that we multiply the *i*th row of *A* by *k* to get *B*. Then we have

$$\det B = \sum_{\pi} (-1)^{\mu(\pi)} a_{1j_1} a_{2j_2} \cdots (ka_{ij_i}) a_{i+1, j_{i+1}} \cdots a_{nj_n}.$$

Factoring out the common factor *k* from each term of this sum, we have

$$\det B = k\left(\sum_{\pi} (-1)^{\mu(\pi)} a_{1j_1} a_{2j_2} \cdots a_{ij_i} \cdots a_{nj_n} \right),$$

so

$$\det B = k \det A.$$

This is often a useful result, as the following examples show.

$$\begin{vmatrix} 12 & -24 & 36 \\ 2 & 5 & 9 \\ 4 & -2 & -6 \end{vmatrix} = 12 \begin{vmatrix} 1 & -2 & 3 \\ 2 & 5 & 9 \\ 4 & -2 & -6 \end{vmatrix}$$

$$= 12 \cdot 3 \begin{vmatrix} 1 & -2 & 1 \\ 2 & 5 & 3 \\ 4 & -2 & -2 \end{vmatrix} = 12 \cdot 3 \cdot 2 \begin{vmatrix} 1 & -2 & 1 \\ 2 & 5 & 3 \\ 2 & -1 & -1 \end{vmatrix}.$$

When we remove more than one factor, these factors are multiplied together, as is illustrated by this example.

We can also use the theorem to introduce nonzero factors:

$$\begin{vmatrix} \frac{1}{6} & \frac{1}{3} \\ \frac{1}{2} & 2 \end{vmatrix} = \frac{1}{6} \begin{vmatrix} 6 \cdot \frac{1}{6} & 6 \cdot \frac{1}{3} \\ \frac{1}{2} & 2 \end{vmatrix} = \frac{1}{6} \begin{vmatrix} 1 & 2 \\ \frac{1}{2} & 2 \end{vmatrix}$$

$$= \frac{1}{6} \cdot \frac{1}{2} \begin{vmatrix} 1 & 2 \\ 2 \cdot \frac{1}{2} & 2 \cdot 2 \end{vmatrix} = \frac{1}{12} \begin{vmatrix} 1 & 2 \\ 1 & 4 \end{vmatrix}.$$

That is, we can multiply a line by a factor $k \neq 0$ provided that we pay for it by multiplying the determinant by $1/k$.

Theorem 7.2.9: *If two parallel lines of a square matrix A are proportional, then det A = 0.*

Prove this, using Theorems 7.2.8 and 7.2.7.

7.3 The Cofactor in det *A* of an Element of *A*

In the expansion (7.1.5) of det *A*, we can gather all terms containing a given element a_{ij} of *A* and extract the common factor a_{ij}. The remaining factor is the **cofactor** A_{ij} of a_{ij}, that is, of the element in the *ij*-position. For example, when $n = 3$,

$$\begin{vmatrix} a_{11} & a_{12} & a_{13} \\ a_{21} & a_{22} & a_{23} \\ a_{31} & a_{32} & a_{33} \end{vmatrix} = \begin{aligned} & a_{11}(a_{22}a_{33} - a_{23}a_{32}) \\ & + a_{12}(a_{23}a_{31} - a_{21}a_{33}) \\ & + a_{13}(a_{21}a_{32} - a_{22}a_{31}). \end{aligned}$$

The expressions in parentheses are the cofactors of a_{11}, a_{12}, and a_{13} respectively.

Since every line of *A* is represented exactly once by a factor in each term in the expansion of det *A*, the cofactor of a_{ij} contains no other element of row *i* of *A* and no other element of column *j*. The preceding example illustrates this. Because of this fact, we can arrange the terms of det *A* in disjoint sets, each having one of $a_{i1}, a_{i2}, \ldots, a_{in}$ as a common factor. Extracting these common factors, we obtain

(7.3.1) $$\det A = a_{i1}A_{i1} + a_{i2}A_{i2} + \cdots + a_{in}A_{in}.$$

This is called **the expansion of det *A* in terms of the elements of the *i*th row**. Operating the same way with the elements of the *j*th column, we get **the expansion of det *A* in terms of the elements of the *j*th column**:

(7.3.2) $$\det A = a_{1j}A_{1j} + a_{2j}A_{2j} + \cdots + a_{nj}A_{nj}.$$

These are particularly useful formulas for evaluating determinants, as we shall see. In words, det *A* is equal to the sum of the products of the elements of any line of *A* by their cofactors.

Now suppose, for example, that we have a 4×4 matrix *B* such that

$$\det B = 4A_{31} + 5A_{32} - 6A_{33} + A_{34},$$

where the A_{ij}'s are the cofactors of the elements of the third row of *A*. This will hold if the third row of *B* is $[4, 5, -6, 1]$, while the other three rows of *B* are the same as those of *A*:

$$B = \begin{bmatrix} a_{11} & a_{12} & a_{13} & a_{14} \\ a_{21} & a_{22} & a_{23} & a_{24} \\ 4 & 5 & -6 & 1 \\ a_{41} & a_{42} & a_{43} & a_{44} \end{bmatrix}.$$

More generally, by the same principle, we have

Theorem 7.3.1: *The expansions*

$$\det A = a_{i1}A_{i1} + a_{i2}A_{i2} + \cdots + a_{in}A_{in}$$

and

$$\det A = a_{1j}A_{1j} + a_{2j}A_{2j} + \cdots + a_{nj}A_{nj}$$

imply that

$$c_1 A_{i1} + c_2 A_{i2} + \cdots + c_n A_{in}$$

is the determinant of a matrix the same as A except that the ith row of A has been replaced by the c's and

$$c_1 A_{1j} + c_2 A_{2j} + \cdots + c_n A_{nj}$$

is the determinant of a matrix the same as A except that the jth column has been replaced by the c's.

For example, if $n = 3$,

$$c_1 A_{13} + c_2 A_{23} + c_3 A_{33} = \det \begin{bmatrix} a_{11} & a_{12} & c_1 \\ a_{21} & a_{22} & c_2 \\ a_{31} & a_{32} & c_3 \end{bmatrix}.$$

This theorem enables us to prove

Theorem 7.3.2: *If $i \neq k$, $j \neq k$, then*

$$(7.3.3) \qquad a_{i1}A_{k1} + a_{i2}A_{k2} + \cdots + a_{in}A_{kn} = 0$$

and

$$(7.3.4) \qquad a_{1j}A_{1k} + a_{2j}A_{2k} + \cdots + a_{nj}A_{nk} = 0.$$

In words, *the sum of the products of the elements of one line by the cofactors of the elements of a different parallel line is zero.*

To prove this, note that by Theorem 7.3.1, the expression (7.3.3) is the expansion, in terms of the elements of the kth row, of the determinant of a matrix the same as A except that the kth row has been replaced by the ith row of A, which also appears in its normal place, of course. This matrix has two identical rows and its determinant is therefore zero. A similar argument holds for (7.3.4). For example, if $i = 1$, $k = 2$, $n = 3$, we have

$$a_{11}A_{21} + a_{12}A_{22} + a_{13}A_{23} = \det \begin{bmatrix} a_{11} & a_{12} & a_{13} \\ a_{11} & a_{12} & a_{13} \\ a_{31} & a_{32} & a_{33} \end{bmatrix} = 0.$$

We can use the **Kronecker delta** symbol, defined by

$$\delta_{rs} = \begin{cases} 1 & \text{if } r = s \\ 0 & \text{if } r \neq s \end{cases}$$

to combine (7.3.1) and (7.3.3), (7.3.2) and (7.3.4), as follows:

(7.3.5) $$a_{i1}A_{k1} + a_{i2}A_{k2} + \cdots + a_{in}A_{kn} = \delta_{ik} \det A$$

(7.3.6) $$a_{1j}A_{1k} + a_{2j}A_{2k} + \cdots + a_{nj}A_{nk} = \delta_{jk} \det A.$$

Consider now this example:

$$\begin{vmatrix} b_{11} + b_{12} & a_{12} & a_{13} \\ b_{21} + b_{22} & a_{22} & a_{23} \\ b_{31} + b_{32} & a_{32} & a_{33} \end{vmatrix}$$

$$= (b_{11} + b_{12})A_{11} + (b_{21} + b_{22})A_{21} + (b_{31} + b_{32})A_{31}$$

$$= (b_{11}A_{11} + b_{21}A_{21} + b_{31}A_{31}) + (b_{12}A_{11} + b_{22}A_{21} + b_{32}A_{31})$$

$$= \begin{vmatrix} b_{11} & a_{12} & a_{13} \\ b_{21} & a_{22} & a_{23} \\ b_{31} & a_{32} & a_{33} \end{vmatrix} + \begin{vmatrix} b_{12} & a_{12} & a_{13} \\ b_{22} & a_{22} & a_{23} \\ b_{32} & a_{32} & a_{33} \end{vmatrix}$$

or, in vector notation,

$$\det [B_1 + B_2, A_2, A_3] = \det [B_1, A_2, A_3] + \det [B_2, A_2, A_3].$$

This last equation suggests the following generalization, which is proved in exactly the manner suggested by the example:

Theorem 7.3.3: If, in a square matrix A, the column A_i is a sum of columns:

$$A_i = \sum_{k=1}^{p} B_k,$$

then

$$\det A = \sum_{k=1}^{p} \det [A_1, A_2, \ldots, A_{i-1}, B_k, A_{i+1}, \ldots, A_n].$$

A similar theorem holds for rows. For example,

$$\begin{vmatrix} a_1 + 2b_1 & a_2 + 2b_2 & a_3 + 2b_3 \\ b_1 & b_2 & b_3 \\ c_1 & c_2 & c_3 \end{vmatrix} = \begin{vmatrix} a_1 & a_2 & a_3 \\ b_1 & b_2 & b_3 \\ c_1 & c_2 & c_3 \end{vmatrix} + \begin{vmatrix} 2b_1 & 2b_2 & 2b_3 \\ b_1 & b_2 & b_3 \\ c_1 & c_2 & c_3 \end{vmatrix}$$

$$= \begin{vmatrix} a_1 & a_2 & a_3 \\ b_1 & b_2 & b_3 \\ c_1 & c_2 & c_3 \end{vmatrix}.$$

This example also illustrates

Theorem 7.3.4: *If a matrix B results from adding any multiple of one line of a square matrix A to another parallel line of A, then det B = det A.*

By the preceding theorem and Theorem 7.2.8,

$$\det [A_1, \ldots, A_{i-1}, A_i + kA_j, A_{i+1}, \ldots, A_n]$$

$$= \det [A_1, \ldots, A_{i-1}, A_i, A_{i+1}, \ldots, A_n]$$

$$+ k \det [A_1, \ldots, A_{i-1}, A_j, A_{i+1}, \ldots, A_n] = \det A,$$

since the second determinant of the sum is 0 because of the presence of two columns A_j. Similarly for rows.

For example:

$$\begin{vmatrix} 1 & 4 & -2 \\ 2 & 5 & -4 \\ -3 & 1 & 3 \end{vmatrix} = \begin{vmatrix} 1 & 0 & 0 \\ 2 & -3 & 0 \\ -3 & 13 & -3 \end{vmatrix},$$

where the 1 in the 1,1-position has been used to sweep the first row.

7.4 Cofactors and the Computation of Determinants

The concept of a cofactor has yielded a number of useful theorems. It is now time to obtain an explicit formula for the cofactor of an element. First we determine A_{nn}. If we group all terms in det A which contain a_{nn} and factor out a_{nn}, we get

$$A_{nn} a_{nn} = \left(\sum (-1)^{\mu(j_1, \ldots, j_{n-1}, n)} a_{1j_1} a_{2j_2} \cdots a_{n-1, j_{n-1}}\right) a_{nn},$$

where the summation extends over all permutations $j_1, j_2, \ldots, j_{n-1}$ of $1, 2, \ldots, n-1$. Since $\mu(j_1, \ldots, j_{n-1}, n) = \mu(j_1, \ldots, j_{n-1})$, the sum in parentheses is the expansion of

$$\begin{vmatrix} a_{11} & a_{12} & \cdots & a_{1, n-1} \\ a_{21} & a_{22} & \cdots & a_{2, n-1} \\ \vdots & & & \\ a_{n-1, 1} & a_{n-1, 2} & \cdots & a_{n-1, n-1} \end{vmatrix},$$

and this is the formula for A_{nn}.

Now let us partition A so as to isolate a_{ij} and then move a_{ij} to the n,n-position by $n - j$ interchanges of adjacent columns and then by $n - i$ interchanges of adjacent rows, thus:

$$\det A = \begin{vmatrix} B & C & D \\ E & a_{ij} & F \\ G & H & K \end{vmatrix} = (-1)^{n-j} \begin{vmatrix} B & D & C \\ E & F & a_{ij} \\ G & K & H \end{vmatrix}$$

$$= (-1)^{n-i}(-1)^{n-j} \begin{vmatrix} B & D & C \\ G & K & H \\ E & F & a_{ij} \end{vmatrix}.$$

Now $(-1)^{n-i}(-1)^{n-j} = (-1)^{i+j}$. Also, in the last determinant written, the cofactor of a_{ij}, since it is in the n,n-position, is

$$\begin{vmatrix} B & D \\ G & K \end{vmatrix}$$

by the preceding formula for a_{nn}. Hence in $\det A$ the cofactor of a_{ij} is

$$(-1)^{i+j} \begin{vmatrix} B & D \\ G & K \end{vmatrix}.$$

Now this last determinant is the determinant of a matrix the same as A except that the ith row and the jth column have been deleted. We therefore have

Theorem 7.4.1: *The cofactor A_{ij} of a_{ij} in the expansion of $\det A$ is $(-1)^{i+j}$ times the determinant of the submatrix of A obtained by deleting the ith row and the jth column of A.*

For example, the cofactor of the element k in

$$\begin{vmatrix} 3 & 4 & -6 \\ 1 & -1 & k \\ a & 1 & 1 \end{vmatrix} \quad \text{is} \quad (-1)^{2+3}\begin{vmatrix} 3 & 4 \\ a & 1 \end{vmatrix}$$

since the element k is in row 2 and column 3. The cofactor of a_{41} in

$$\begin{vmatrix} 0 & 0 & 0 & a_{14} \\ 0 & 0 & a_{23} & 0 \\ 0 & a_{32} & 0 & 0 \\ a_{41} & 0 & 0 & 0 \end{vmatrix} \quad \text{is} \quad (-1)^{4+1}\begin{vmatrix} 0 & 0 & a_{14} \\ 0 & a_{23} & 0 \\ a_{32} & 0 & 0 \end{vmatrix}.$$

Now suppose all elements but a_{ij} in the ith row of A are zero:

$$\begin{vmatrix} a_{11} & \cdots & a_{1j} & \cdots & a_{1n} \\ \vdots & & & & \\ 0 & \cdots & 0 \;\; a_{ij} \;\; 0 & \cdots & 0 \\ \vdots & & & & \\ a_{n1} & \cdots & a_{nj} & \cdots & a_{nn} \end{vmatrix}$$

Then, expanding along the ith row, we have just

$$\det A = a_{ij} A_{ij}$$

since all the other terms of the expansion are zero. We can often use this observation together with a sweepout process to compute determinants. For example, a useful result concerns triangular matrices:

$$\begin{vmatrix} a_{11} & a_{12} & \cdots & a_{1n} \\ 0 & a_{22} & \cdots & a_{2n} \\ \vdots & & & \\ 0 & 0 & \cdots & a_{nn} \end{vmatrix} = a_{11}\begin{vmatrix} a_{22} & a_{23} & \cdots & a_{2n} \\ 0 & a_{33} & \cdots & a_{3n} \\ \vdots & & & \\ 0 & 0 & & a_{nn} \end{vmatrix} = \cdots = a_{11}a_{22}\cdots a_{nn}.$$

If the zeros are not there initially, we can often create them. If we apply Theorem 7.3.4 and sweep the first row, we have

$$\begin{vmatrix} 1 & 2 & -1 \\ -2 & 1 & 4 \\ 3 & 0 & 5 \end{vmatrix} = \begin{vmatrix} 1 & 0 & 0 \\ -2 & 5 & 2 \\ 3 & -6 & 8 \end{vmatrix} = 1\cdot(-1)^{1+1}\begin{vmatrix} 5 & 2 \\ -6 & 8 \end{vmatrix} = 52.$$

In the next example, by sweeping the first column, we get

$$\begin{vmatrix} 1 & a & a^2 \\ 1 & b & b^2 \\ 1 & c & c^2 \end{vmatrix} = \begin{vmatrix} 0 & a - c & a^2 - c^2 \\ 0 & b - c & b^2 - c^2 \\ 1 & c & c^2 \end{vmatrix} = 1 \cdot (-1)^{3+1} \begin{vmatrix} a - c & a^2 - c^2 \\ b - c & b^2 - c^2 \end{vmatrix}$$

$$= (a - c)(b - c) \begin{vmatrix} 1 & a + c \\ 1 & b + c \end{vmatrix} = (a - c)(b - c)(b - a).$$

The idea is to use a sweepout procedure to reduce the computation of the determinant of a matrix of order n to the computation of the determinant of a matrix of order $n - 1$, and so on, until the order becomes low enough that the determinant can be computed conveniently.

This reduction can be made fully systematic so that it can be programmed. This is illustrated by the following example.

$$\begin{vmatrix} 4 & 2 & 3 \\ 5 & -2 & 4 \\ -3 & 6 & 8 \end{vmatrix} = 4 \begin{vmatrix} 1 & \frac{1}{2} & \frac{3}{4} \\ 5 & -2 & 4 \\ -3 & 6 & 8 \end{vmatrix} \qquad \text{Create a 1 in the 1,1-position.}$$

$$= 4 \begin{vmatrix} 1 & \frac{1}{2} & \frac{3}{4} \\ 0 & -\frac{9}{2} & \frac{1}{4} \\ 0 & \frac{15}{2} & \frac{41}{4} \end{vmatrix} \qquad \text{Sweep column 1.}$$

$$= 4 \cdot \left(-\frac{9}{2}\right) \begin{vmatrix} 1 & \frac{1}{2} & \frac{3}{4} \\ 0 & 1 & -\frac{1}{18} \\ 0 & \frac{15}{2} & \frac{41}{4} \end{vmatrix} \qquad \text{Create a 1 in the 2,2-position.}$$

$$= 4 \cdot \left(-\frac{9}{2}\right) \begin{vmatrix} 1 & \frac{1}{2} & \frac{3}{4} \\ 0 & 1 & -\frac{1}{18} \\ 0 & 0 & \frac{32}{3} \end{vmatrix} \qquad \text{Sweep column 2 below the 2,2-position.}$$

$$= 4 \cdot \left(-\frac{9}{2}\right) \cdot \frac{32}{3}$$

$$= -192.$$

The reader should give a general description of this systematic process.

7.5   Exercises

1. Compute

    (a) all the $\mu(\pi)$ for $n = 4$,

    (b) $\mu(2, 1, 4, 3, 6, 5, 8, 7, \ldots, 2n, 2n - 1)$,

    (c) $\mu(n, n - 1, n - 2, \ldots, 3, 2, 1)$.

2. Compute:

(a) $\begin{vmatrix} a-b & a+b \\ a+b & a-b \end{vmatrix},$

(b) $\begin{vmatrix} 1 & a & bc \\ 1 & b & ca \\ 1 & c & ab \end{vmatrix},$

(c) $\begin{vmatrix} 1 & 3 & -2 & 4 \\ 1 & 4 & 3 & 0 \\ 1 & 5 & 5 & 4 \\ 1 & 6 & -2 & -3 \end{vmatrix},$

(d) $\begin{vmatrix} 2 & 1 & -3 & -1 \\ 1 & 4 & -8 & 0 \\ -1 & -1 & 4 & 2 \\ 3 & 2 & 1 & 5 \end{vmatrix},$

(e) $\begin{vmatrix} 2 & -1 & 1 & 0 \\ -1 & 4 & 0 & 1 \\ 1 & 0 & 2 & -2 \\ 0 & 1 & -2 & 1 \end{vmatrix},$

(f) $\begin{vmatrix} 1 & 1 & 1 & 1 \\ 2 & 4 & 8 & 16 \\ 3 & 9 & 27 & 81 \\ 4 & 16 & 64 & 256 \end{vmatrix},$

(g) $\begin{vmatrix} 0 & i & -i & 1+i \\ i & 0 & i & -1 \\ -i & i & 0 & i \\ 1+i & -1 & i & 0 \end{vmatrix},$

(h) $\begin{vmatrix} a_1 & a_2 & 0 & 0 \\ 0 & a_1 & a_2 & 0 \\ 0 & 0 & a_1 & a_2 \\ a_2 & 0 & 0 & a_1 \end{vmatrix}.$

*3. Compute the determinant of the lower triangular matrix:

$$\begin{bmatrix} a_{11} & 0 & 0 & \cdots & 0 \\ a_{21} & a_{22} & 0 & \cdots & 0 \\ a_{31} & a_{32} & a_{33} & \cdots & 0 \\ \vdots & & & & \\ a_{n1} & a_{n2} & a_{n3} & \cdots & a_{nn} \end{bmatrix}.$$

A similar result holds for upper triangular matrices.

4. Compute by first reducing to triangular form by the "sweepout" process:

(a) $\begin{vmatrix} 1 & 2 & 3 & 4 \\ -1 & 1 & 2 & 3 \\ 1 & -1 & 1 & 2 \\ -1 & 1 & -1 & 1 \end{vmatrix},$

(b) $\begin{vmatrix} 1 & -2 & 0 & 3 & 6 \\ 2 & -4 & 1 & 5 & 11 \\ -1 & 2 & 3 & -2 & -5 \\ 3 & -6 & 4 & 1 & 10 \\ -3 & 7 & 2 & -7 & -16 \end{vmatrix},$

(c) $\begin{vmatrix} 1 & 1 & 0 & 1 \\ 1 & 0 & 1 & 1 \\ 1 & 1 & 1 & 1 \\ 1 & x & y & z \end{vmatrix},$

(d) $\begin{vmatrix} 1 & 1 & 1 & 1 \\ a & 1 & 1 & 1 \\ b & a & 1 & 1 \\ c & b & a & 1 \end{vmatrix}.$

5. Evaluate:

$$\begin{vmatrix} 0 & 0 & 0 & a_{1n} \\ 0 & 0 & a_{2,n-1} & 0 \\ \vdots & & & \\ 0 & a_{n-1,2} & 0 & 0 \\ a_{n1} & 0 & 0 & 0 \end{vmatrix}.$$

Don't jump at a wrong conclusion here!

*6. Prove that if the ith row of A is a linear combination of the other rows, then det A is zero. Prove also that if $r(A_{n \times n}) < n$, then det $A = 0$.

*7. Prove that $\det(\lambda A) = \lambda^n \det A$, where A is $n \times n$ and where λ is a scalar.

*8, Prove that, if A and B are both of order n,

  (a) det $A^{\mathsf{T}}B$ = det AB^{T} = det $A^{\mathsf{T}}B^{\mathsf{T}}$ = det AB,

  (b) det A^*B^* = $\overline{\det AB}$.

*9. Show that, if A is a square matrix, det A^*A is a non-negative real number and that, if A is Hermitian, det A is real.

10. (a) Given that

$$A = \begin{bmatrix} 1 & 2 & -1 \\ 4 & 0 & 3 \\ 1 & 2 & 5 \end{bmatrix},$$

compute $[A_{ij}]_{3 \times 3}^{\mathsf{T}}$, where A_{ij} denotes the cofactor of a_{ij}. Then compute the product $A \cdot [A_{ij}]^{\mathsf{T}}$.

  (b) If A is any $n \times n$ matrix such that det $A \neq 0$, compute

$$A \cdot \frac{1}{\det A} [A_{ij}]^{\mathsf{T}}.$$

11. (a) Show that if $A = [A_1, A_2, \ldots, A_n]$ is an n-square matrix such that for a fixed column A_j,

$$A_j = \sum_{k=1}^{p} \lambda_k B_k, \quad \text{then det } A = \sum_{k=1}^{p} \lambda_k \det[A_1, A_2, \ldots, A_{j-1}, B_k, A_{j+1}, \ldots, A_n].$$

  (b) Given that

$$A_j = \begin{bmatrix} a_j & d & g \\ b_j & e & h \\ c_j & f & k \end{bmatrix}, \quad j = 1, 2, \ldots, p$$

prove that

$$\det \sum_{j=1}^{p} A_j = p^2 \sum_{j=1}^{p} (\det A_j).$$

What is the corresponding result for $n \times n$ matrices?

12. Given that $A = [a_{ij}]_{n \times n}$, form the matrix

$$
B = \begin{bmatrix}
a_{11} & a_{12} & \cdots & a_{1n} & -\sum_{(j)} a_{1j} \\
a_{21} & a_{22} & \cdots & a_{2n} & -\sum_{(j)} a_{2j} \\
\vdots & & & & \\
a_{n1} & a_{n2} & \cdots & a_{nn} & -\sum_{(j)} a_{nj} \\
-\sum_{(i)} a_{i1} & -\sum_{(i)} a_{i2} & \cdots & -\sum_{(i)} a_{in} & -\sum_{(i)}\sum_{(j)} a_{ij}
\end{bmatrix}.
$$

Let B_{ii} denote the cofactor of the ith main diagonal entry of B. Prove that for $i = 1, 2, \ldots, n+1$, $B_{ii} = \det A$.

13. Show that $\det [\delta_{ij} a_{ij}]_{n \times n} = \prod_{j=1}^{n} a_{jj}$, where δ_{ij} is the Kronecker delta.

14. Prove that

$$
\begin{vmatrix}
1-n & 1 & 1 & \cdots & 1 \\
1 & 1-n & 1 & \cdots & 1 \\
1 & 1 & 1-n & \cdots & 1 \\
\vdots & & & & \\
1 & 1 & 1 & \cdots & 1-n
\end{vmatrix}_{n \times n} = 0.
$$

(This is useful in statistical theory, as is the next one.)

15. Prove that

$$
\begin{vmatrix}
x+\lambda & x & x & \cdots & x \\
x & x+\lambda & x & \cdots & x \\
\vdots & & & & \\
x & x & x & \cdots & x+\lambda
\end{vmatrix}_{n \times n} = \lambda^{n-1}(nx + \lambda).
$$

16. Expand and interpret geometrically:

$$
\begin{vmatrix}
x_1 & x_2 & x_3 & 1 \\
2 & 0 & 0 & 1 \\
0 & 3 & 0 & 1 \\
0 & 0 & -4 & 1
\end{vmatrix} = 0.
$$

17. Expand by successive splitting of columns and simplify:

$$
\begin{vmatrix}
a_1 + a_2 & a_2 + a_3 & a_3 + a_1 \\
b_1 + b_2 & b_2 + b_3 & b_3 + b_1 \\
c_1 + c_2 & c_2 + c_3 & c_3 + c_1
\end{vmatrix}.
$$

18. Show that elementary transformations can help you to write the expanded form of this determinant in factored form:

$$\begin{vmatrix} 1 & 1 & 1 \\ a & b & c \\ a^3 & b^3 & c^3 \end{vmatrix}.$$

19. Compute, in factored form,

$$\begin{vmatrix} a & b & 0 & 0 \\ c & d & 0 & 0 \\ 0 & 0 & \alpha & \beta \\ 0 & 0 & \gamma & \delta \end{vmatrix}.$$

Can you state the general rule suggested by this example?

20. Compute this determinant, which appears in the study of fluid mechanics:

$$\det \begin{bmatrix} \sin \alpha \cos \theta & \sin \alpha \sin \theta & \cos \alpha \\ r \cos \alpha \cos \theta & r \cos \alpha \sin \theta & -r \sin \alpha \\ -r \sin \alpha \sin \theta & r \sin \alpha \cos \theta & 0 \end{bmatrix}.$$

21. By using the properties developed in Section 7.2, show without expanding that

$$\begin{vmatrix} a-b & 1 & a \\ b-c & 1 & b \\ c-a & 1 & c \end{vmatrix} = \begin{vmatrix} a & 1 & b \\ b & 1 & c \\ c & 1 & a \end{vmatrix}.$$

22. Show without expanding that

$$\begin{vmatrix} 1 & \alpha & \beta\gamma \\ 1 & \beta & \gamma\alpha \\ 1 & \gamma & \alpha\beta \end{vmatrix} = \begin{vmatrix} 1 & \alpha & \alpha^2 \\ 1 & \beta & \beta^2 \\ 1 & \gamma & \gamma^2 \end{vmatrix}.$$

*23. (a) Prove, using mathematical induction, that if

$$V = \begin{bmatrix} x_1^{n-1} & x_1^{n-2} & \cdots & x_1 & 1 \\ x_2^{n-1} & x_2^{n-2} & \cdots & x_2 & 1 \\ \vdots & & & & \\ x_n^{n-1} & x_n^{n-2} & \cdots & x_n & 1 \end{bmatrix},$$

then

$$\det V = \prod_{1 \le i < j \le n} (x_i - x_j).$$

The matrix V is called a **Vandermonde matrix**. From the expansion it follows that V is nonsingular if and only if all the x_i's are distinct. This fact is often useful.

(b) If the x_i are not all distinct, what is the rank of V?

*24. Prove that, if A is skew-symmetric and of odd order, then $\det A = 0$.

25. Show that, if in a square matrix A all the elements for which the sum of the subscripts is odd are multiplied by -1, the determinant of the new matrix equals $\det A$.

26. Given that A is a square matrix over the complex field, show that

$$\det A^* = \det \overline{A} = \overline{\det A}.$$

27. Use the fact that every row and every column is represented exactly once in each term of the expansion

$$\det A = \sum_{\pi} (-1)^{\mu(\pi)} a_{1j_1} a_{2j_2} \cdots a_{nj_n}$$

to prove that when the a_{ij}'s are indeterminates, $\det A$ is not a factorable polynomial. (*Hint:* If $\det A = fg$ and if f contains an element from row (column) j, then, by the first observation, g cannot contain any element from row (column) j, etc.)

28. Let A be an $n \times n$ partitioned matrix:

$$A = [A_1, A_2, A_3],$$

where A_1 and A_2 each have r columns. Let B be an arbitrary $r \times r$ matrix. Prove that

$$\det [A_1 + A_2 B, A_2, A_3] = \det A.$$

State a similar result for rows. Generalize.

29. Given $A_{n \times p}, B_{n \times q}, C_{n \times r}$, where $p + q + r = 2n$, prove that

$$\det \begin{bmatrix} A & B & 0 \\ A & 0 & C \end{bmatrix} = \pm \det \begin{bmatrix} B & A & 0 \\ B & 0 & C \end{bmatrix}.$$

When should the negative sign be used? Generalize this result.

30. Let A be a partitioned square matrix whose blocks A_{ij} are matrices of order p:

$$A = [A_{ij}], \qquad i, j = 1, 2, \ldots, m.$$

Let B be a matrix the same as A except that the jth column of submatrices, where j is fixed, is replaced by a linear combination of the columns of the partitioned form of A:

$$B_{ij} = \sum_{k=1}^{m} c_k A_{ik}, \qquad i = 1, 2, \ldots, m.$$

Prove that

$$\det B = c_j^p \det A.$$

31. (a) Show by an example that $\det (A + B) \not\equiv \det A + \det B$.
   (b) Given that

$$A = \begin{bmatrix} a_{11} & a_{12} \\ a_{21} & a_{22} \end{bmatrix}, \qquad B = \begin{bmatrix} b_{11} & b_{12} \\ b_{21} & b_{22} \end{bmatrix},$$

   show that $\det (A + B) = \det A + \det B$ if and only if

$$\det \begin{bmatrix} a_{11} & b_{12} \\ a_{21} & b_{22} \end{bmatrix} + \det \begin{bmatrix} b_{11} & a_{12} \\ b_{21} & a_{22} \end{bmatrix} = 0.$$

32. Prove that two matrices

$$\begin{bmatrix} a & b \\ 0 & c \end{bmatrix} \qquad \text{and} \qquad \begin{bmatrix} \alpha & \beta \\ 0 & \gamma \end{bmatrix}$$

commute if and only if

$$\det \begin{bmatrix} b & a - c \\ \beta & \alpha - \gamma \end{bmatrix} = 0.$$

33. Compute

$$\det \begin{bmatrix} a & a & \cdots & a & a \\ b & a & \cdots & a & a \\ 0 & b & \cdots & a & a \\ \vdots & & & & \\ 0 & 0 & \cdots & a & a \\ 0 & 0 & \cdots & b & a \end{bmatrix}_n.$$

34. Compute the determinant of this tridiagonal matrix:

$$\begin{bmatrix} a & a & & & & \\ b & a & a & & \mathbf{0} & \\ & b & & & & \\ & & & & a & a \\ & \mathbf{0} & & & b & a \\ & & & b & a \end{bmatrix}_n.$$

(A **tridiagonal matrix** is one in which nonzero entries may appear only on the main diagonal and the two adjacent diagonals. Tridiagonal matrices are important in linear computations and in certain applications.)

35. Let

$$d_n = \det \begin{bmatrix} 0 & k & k & \cdots & k & k \\ k & 0 & k & \cdots & k & k \\ k & k & 0 & \cdots & k & k \\ \vdots & & & & & \\ k & k & k & \cdots & 0 & k \\ k & k & k & \cdots & k & r \end{bmatrix}_n.$$

Take $d_1 = r$, and prove by mathematical induction that $d_n = (-k)^{n-1}[(n-1)k - (n-2)r]$.

36. Prove that, if

$$d_n = \det \begin{bmatrix} 2\cos\theta & 1 & 0 & 0 & \cdots & 0 & 0 & 0 \\ 1 & 2\cos\theta & 1 & 0 & \cdots & 0 & 0 & 0 \\ 0 & 1 & 2\cos\theta & 1 & \cdots & 0 & 0 & 0 \\ \vdots & & & & & & & \\ 0 & 0 & 0 & 0 & \cdots & 1 & 2\cos\theta & 1 \\ 0 & 0 & 0 & 0 & \cdots & 0 & 1 & 2\cos\theta \end{bmatrix},$$

then

$$d_n = 2d_{n-1}\cos\theta - d_{n-2} \qquad (n \geq 3),$$

and hence, by induction, that if $\theta \neq k\pi$, k any integer,

$$d_n = \csc\theta \sin(n+1)\theta.$$

Finally show that, if $\theta = k\pi$, then

$$d_n = (-1)^{k(n+1)}(n+1).$$

37. Compute

$$\det \begin{bmatrix} 2\cos\alpha & -1 & 0 & 0 & \cdots & 0 & 0 & 0 \\ -1 & 2\cos\alpha & -1 & 0 & \cdots & 0 & 0 & 0 \\ \vdots & & & & & & & \\ 0 & 0 & 0 & 0 & \cdots & -1 & 2\cos\alpha & -1 \\ 0 & 0 & 0 & 0 & \cdots & 0 & -1 & 2\cos\alpha \end{bmatrix}_{n \times n}$$

by reducing this problem to the previous one.

38. Use the result of the preceding exercise to show that

$$\det \begin{bmatrix} 2 & -1 & 0 & 0 & \cdots & 0 & 0 & 0 \\ -1 & 2 & -1 & 0 & \cdots & 0 & 0 & 0 \\ \vdots & & & & & & & \\ 0 & 0 & 0 & 0 & \cdots & -1 & 2 & -1 \\ 0 & 0 & 0 & 0 & \cdots & 0 & -1 & 2 \end{bmatrix}_{n \times n} = n + 1.$$

39. This result appeared in a research paper on economics:

$$\det \begin{bmatrix} \dfrac{1+a_1}{a_1} & 1 & 1 & \cdots & 1 \\ 1 & \dfrac{1+a_2}{a_2} & 1 & \cdots & 1 \\ \vdots & & & & \\ 1 & 1 & 1 & \cdots & \dfrac{1+a_n}{a_n} \end{bmatrix} = \dfrac{1+\sum a_i}{\prod a_i}.$$

Prove that the result is correct.

40. Show by inspection that the equation

$$\begin{vmatrix} 0 & \alpha - x & \beta - x \\ -\alpha - x & 0 & \gamma - x \\ -\beta - x & -\gamma - x & 0 \end{vmatrix} = 0$$

has at least one real root. (See Exercise 24.) When will it have a multiple root? (Many other interesting problems concerning determinants and matrices may be found in the problem sections of the *American Mathematical Monthly* and *The Mathematics Magazine*.)

41. Two quadratic equations.

$$a_0 x^2 + a_1 x + a_2 = 0, \qquad a_0 \neq 0,$$
$$b_0 x^2 + b_1 x + b_2 = 0, \qquad b_0 \neq 0,$$

have a common root if and only if the determinant

$$\begin{vmatrix} a_0 & a_1 & a_2 & 0 \\ 0 & a_0 & a_1 & a_2 \\ b_0 & b_1 & b_2 & 0 \\ 0 & b_0 & b_1 & b_2 \end{vmatrix}$$

vanishes. Use this fact to show that, for any values of α and β, the equations

$$\alpha x^2 + x + (1 - \alpha) = 0$$
$$(1 - \beta)x^2 + x + \beta = 0,$$

will have a common root. (The determinant above is a special case of the resultant of two polynomials, the vanishing of which is a necessary and sufficient condition that the polynomials have at least one root in common, provided we assume the polynomials have nonvanishing, leading coefficients. See M. Bôcher, *Introduction to Higher Algebra*, Macmillan, New York, 1949, p. 195, where resultants are discussed in detail.)

42. In the study of the stability of oscillating electrical or vibrating dynamical systems, the following definition is important. An algebraic equation with real coefficients is said to be a **stable equation** if and only if all its roots have negative real parts. The following tests are due to Hermite.

The equations $(a_0 > 0$ in each case)

(α) $\qquad\qquad\qquad a_0 + a_1 x + a_2 x^2 = 0,$

(β) $\qquad\qquad\qquad a_0 + a_1 x + a_2 x^2 + a_3 x^3 = 0,$

(γ) $\qquad\qquad\qquad a_0 + a_1 x + a_2 x^2 + a_3 x^3 + a_4 x^4 = 0,$

are respectively stable if and only if

(α') $\qquad a_1 > 0, \qquad a_2 > 0,$

(β') $\qquad a_1 > 0, \qquad \begin{vmatrix} a_1 & a_3 \\ a_0 & a_2 \end{vmatrix} > 0, \qquad a_3 > 0,$

(γ') $\qquad a_1 > 0, \qquad \begin{vmatrix} a_1 & a_3 \\ a_0 & a_2 \end{vmatrix} > 0, \qquad \begin{vmatrix} a_1 & a_3 & 0 \\ a_0 & a_2 & a_4 \\ 0 & a_1 & a_3 \end{vmatrix} > 0, \qquad a_4 > 0,$

with similar tests for higher-order equations. Using these tests, investigate the stability of the equations:

(a) $\qquad\qquad\qquad 1 + 2x + 3x^2 = 0,$

(b) $\qquad\qquad\qquad 1 + 2x + 3x^2 + x^3 = 0,$

(c) $\qquad\qquad\qquad 1 + 2x + 3x^2 + x^3 + 2x^4 = 0.$

Also show that, if (γ) is stable, (β) is stable, and if (β) is stable, (α) is stable, but not conversely.

43. In problems in vibrations, it becomes necessary to solve equations similar to the following for λ:

$$\begin{vmatrix} 4 - \lambda & -1 & 0 \\ -1 & 4 - \lambda & -1 \\ 0 & -1 & 4 - \lambda \end{vmatrix} = 0.$$

Expand the determinant and solve the resulting equation. The values of λ found in this way are used to obtain the "normal modes of vibration."

44. Show that the number of multiplications and divisions required in computing the determinant of an $n \times n$ matrix by sweepout is

$$\frac{n-1}{3}(n^2 + n + 3),$$

while the number required to apply the definition of det A directly is $(n-1)n!$ Make a table comparing these two numbers for $n = 1, 2, 3, 4, 5, 6$.

45. Show that $\det [a + (i-1)n + j]_n = 0$ if $n > 2$ and i and j are row and column indices respectively.

*46. Given that A and C are square, prove that

$$\det \left[\begin{array}{c|c} A & 0 \\ \hline B & C \end{array} \right] = \det A \cdot \det C$$

by using the definition of a determinant and noting which terms are necessarily zero.

*47. Generalize Exercise 46 and prove the result by induction.

*48. Show that no permutation can be reduced to natural order by both an even and an odd number of interchanges of adjacent integers.

49. Prove that if a permutation has a total of k inversions, then it can be reduced to the natural order by exactly k interchanges of adjacent integers, but not by fewer. Thus k is the *minimum number* of such interchanges that will accomplish this result.

*50. Prove that a permutation is even (odd) if and only if it may be reduced to the natural order by an even (odd) number of interchanges of adjacent integers.

51. Prove that at most $n(n-1)/2$ interchanges of adjacent integers are required to restore a given permutation of $1, 2, \ldots, n$ to the natural order.

*52. Prove that for $n \geq 2$, there are equal numbers of even and odd permutations of the integers from 1 to n.

*53. Prove that a permutation $j_1 j_2 \ldots j_n$ is even or odd depending upon whether the following product is positive or negative:

$$\prod_{n \geq k > p \geq 1} (j_k - j_p) = (j_n - j_{n-1})(j_n - j_{n-2}) \cdots (j_n - j_2)(j_n - j_1)$$

$$(j_{n-1} - j_{n-2}) \cdots (j_{n-1} - j_2)(j_{n-1} - j_1)$$

$$\vdots$$

$$(j_3 - j_2)(j_3 - j_1)$$

$$(j_2 - j_1).$$

54. If π denotes a permutation $j_1 j_2 \ldots j_n$ of $1, 2, \ldots, n$, then an often useful notation is the **epsilon symbol** defined by

$$\varepsilon_{j_1 j_2 \cdots j_n} = (-1)^{\mu(\pi)},$$

that is, $\varepsilon_{j_1 j_2 \cdots j_n}$ is 1 or -1 according as $j_1, j_2, \ldots, j_n$ is even or odd. Prove that, if the E_j are a permutation of the elementary n-vectors, then

$$\det [E_{j_1}, E_{j_2}, \ldots, E_{j_n}] = \varepsilon_{j_1 j_2 \cdots j_n}.$$

55. Prove that if $k_1, k_2, \ldots, k_n$ and $j_1, j_2, \ldots, j_n$ are permutations of $1, 2, \ldots, n$, then

$$\varepsilon_{k_1 k_2 \cdots k_n} \varepsilon_{j_{k_1} j_{k_2} \cdots j_{k_n}} = \varepsilon_{j_1 j_2 \cdots j_n}.$$

56. Show that, if $i_1, i_2, \ldots, i_n$ and $j_1, j_2, \ldots, j_n$ are any permutations of $1, 2, \ldots, n$, and if $A = [a_{ij}]_n$, then

$$
\det \begin{bmatrix}
a_{i_1 j_1} & a_{i_1 j_2} & \cdots & a_{i_1 j_n} \\
a_{i_2 j_1} & a_{i_2 j_2} & \cdots & a_{i_2 j_n} \\
\vdots & & & \\
a_{i_n j_1} & a_{i_n j_2} & \cdots & a_{i_n j_n}
\end{bmatrix} = \varepsilon_{i_1 i_2 \cdots i_n} \varepsilon_{j_1 j_2 \cdots j_n} \det A.
$$

*57. Prove the following theorem. If the elements a_{ij} of a matrix A of order n are differentiable functions of a parameter t, than $d(\det A)/(dt)$ is the sum of the determinants of n matrices, each the same as A except that, in the first, the elements of the first row have been replaced by their derivatives with respect to t, in the second the elements of the second row have been replaced by their derivatives with respect to t, etc. (A similar result holds for columns by Theorem 7.2.4.) Then show that

$$
\frac{d}{dt}(\det A) = \sum_{i,\, j=1}^{n} A_{ij} \frac{da_{ij}}{dt}.
$$

58. Compute the derivative of the determinant in each case:

$$
\text{(a)} \quad \begin{vmatrix}
1+t & 1-t & 1 \\
1-t & 1+t & 0 \\
1 & 0 & t
\end{vmatrix}, \qquad \text{(b)} \quad \det[a - \delta_{ij} t]_{n \times n}
$$

according to the rule given in Exercise 57.

59. Let $A = [a_{ij} + \delta_{ij} x]$, where the a_{ij} are constant and δ_{ij} is the Kronecker delta. Compute $d(\det A)/dx$.

60. Treating the elements of $A = [a_{ij}]_n$ as n^2 independent variables, evaluate

$$
\frac{\partial}{\partial a_{ij}}(\det A).
$$

7.6   The Determinant of the Product of Two Matrices

Let

$$
D = \mathrm{diag}\,[d_1, d_2, \ldots, d_n]
$$

and

$$
G = \mathrm{diag}\,[g_1, g_2, \ldots, g_n].
$$

Then

$$DG = \text{diag}\,[d_1 g_1, d_2 g_2, \ldots, d_n g_n],$$

so

$$\det DG = d_1 g_1 d_2 g_2 \cdots d_n g_n = d_1 d_2 \cdots d_n \cdot g_1 g_2 \cdots g_n;$$

that is,

(7.6.1) $$\det DG = \det D \det G.$$

This formula actually holds for the product of any two matrices of order n. To prove this for A and B nonsingular, recall that adding any multiple of one line of a matrix to any other parallel line of the matrix does not alter the determinant. Also, one can reduce a nonsingular matrix to a diagonal matrix with the same determinant by using row operations of this kind only, or by column operations of this kind only. For example:

$$\begin{bmatrix} 0 & 1 & 2 \\ 3 & -1 & 4 \\ 1 & 2 & 0 \end{bmatrix} \sim \begin{bmatrix} 1 & 3 & 2 \\ 0 & -7 & 4 \\ 1 & 2 & 0 \end{bmatrix}$$ Add row 3 to row 1; subtract three times row 3 from row 2.

$$\sim \begin{bmatrix} 1 & 3 & 2 \\ 0 & -7 & 4 \\ 0 & -1 & -2 \end{bmatrix}$$ Subtract row 1 from row 3.

$$\sim \begin{bmatrix} 1 & 0 & \frac{26}{7} \\ 0 & -7 & 4 \\ 0 & 0 & -\frac{18}{7} \end{bmatrix}$$ Add $\frac{3}{7}$ times row 2 to row 1; subtract $\frac{1}{7}$ times row 2 from row 3.

$$\sim \begin{bmatrix} 1 & 0 & 0 \\ 0 & -7 & 0 \\ 0 & 0 & -\frac{18}{7} \end{bmatrix}$$ Add $\frac{26}{18}$ times row 3 to row 1; add $\frac{28}{18}$ times row 3 to row 2.

The reduction is thus accomplished, and because of the type of operation used,

$$\det \begin{bmatrix} 0 & 1 & 2 \\ 3 & -1 & 4 \\ 1 & 2 & 0 \end{bmatrix} = \det \begin{bmatrix} 1 & 0 & 0 \\ 0 & -7 & 0 \\ 0 & 0 & -\frac{18}{7} \end{bmatrix} = 18.$$

A similar procedure is possible using the analogous column operations. In no case is the procedure necessarily unique.

This process can be described in a general way. It is left to the reader to do this.

Given two nonsingular matrices A and B, there exist, in view of the preceding observations, matrices $R_1, R_2, \ldots, R_k$ representing row transformations of the type named and matrices $C_1, C_2, \ldots, C_p$ representing column transformations of the type named such that

$$R_k R_{k-1} \cdots R_2 R_1 A = D_A \qquad \text{or} \qquad A = R_1^{-1} R_2^{-1} \cdots R_k^{-1} D_A$$

and

$$B C_1 C_2 \cdots C_p = D_B \qquad \text{or} \qquad B = D_B \, C_p^{-1} \cdots C_2^{-1} C_1^{-1},$$

where D_A and D_B are diagonal matrices such that

(7.6.2) $\qquad \det A = \det D_A \qquad \text{and} \qquad \det B = \det D_B$.

Hence

$$AB = R_1^{-1} \cdots R_k^{-1} D_A D_B C_p^{-1} \cdots C_1^{-1}$$

and, since the inverse of an elementary transformation is an elementary transformation of the same kind, the matrices R_i^{-1} and C_j^{-1} here do not alter the determinant either, so

$$\det(AB) = \det(D_A D_B)$$
$$= \det D_A \det D_B \qquad \text{by (7.6.1)}$$

or

(7.6.3) $\qquad \det(AB) = \det A \det B \qquad \text{by (7.6.2)}.$

If A or B is singular, that is, if $r(A) < n$ or $r(B) < n$, then $r(AB) < n$ also, so that $\det A = 0$ or $\det B = 0$ and $\det(AB) = 0$ by Exercise 6, Section 7.5. Hence, in this case also, $\det(AB) = \det A \det B$.

We have thus proved

Theorem 7.6.1: *The determinant of the product of two square matrices is the product of their determinants.*

This same technique of reduction to diagonal form by transformations that leave the determinant invariant allows us to conclude

Theorem 7.6.2: *If $A_1, A_2, \ldots, A_n$ are square matrices, then*

$$(7.6.4) \qquad \det \begin{bmatrix} A_1 & 0 & \cdots & 0 \\ 0 & A_2 & \cdots & 0 \\ \vdots & & & \\ 0 & 0 & \cdots & A_n \end{bmatrix} = \det A_1 \det A_2 \cdots \det A_n.$$

In fact, for each $i, i = 1, 2, \ldots, n$, there are elementary transformations that reduce A_i to diagonal form without affecting any other A_j and without affecting the determinant. The effect of all these is to reduce the given block matrix to diagonal form. The determinant is just the product of all these diagonal elements, but the factors of this product can be grouped so that they yield the product $\det A_1 \det A_2 \cdots \det A_n$.

The result of Theorem 7.6.1 may be generalized to the case in which A is $m \times n$ and B is $n \times m$. Suppose first that $m > n$. We can apply Theorem 7.6.1 if we first adjoin $m - n$ columns of zeros and $m - n$ rows of zeros to A and B respectively. The product is still AB, so we have

$$\det AB = \det \begin{bmatrix} a_{11} & \cdots & a_{1n} \\ a_{21} & \cdots & a_{2n} \\ \vdots & & \\ a_{m1} & \cdots & a_{mn} \end{bmatrix} \begin{bmatrix} b_{11} & b_{12} & \cdots & b_{1m} \\ \vdots & & & \\ b_{n1} & b_{n2} & \cdots & b_{nm} \end{bmatrix}$$

$$= \det \begin{bmatrix} a_{11} & \cdots & a_{1n} & 0 & \cdots & 0 \\ a_{21} & \cdots & a_{2n} & 0 & \cdots & 0 \\ \vdots & & & & & \\ a_{m1} & \cdots & a_{mn} & \underbrace{0 \quad \cdots \quad 0}_{m - n \text{ cols.}} \end{bmatrix} \left. \begin{bmatrix} b_{11} & b_{12} & \cdots & b_{1m} \\ \vdots & & & \\ b_{n1} & b_{n2} & \cdots & b_{nm} \\ 0 & 0 & \cdots & 0 \\ \vdots & & & \\ 0 & 0 & \cdots & 0 \end{bmatrix} \right\} m - n \text{ rows} = 0.$$

This proves

Theorem 7.6.3: *If A is $m \times n$ and B is $n \times m$ and if $m > n$, then $\det AB = 0$.*

The case $m < n$ is more difficult and includes Theorem 7.6.1 as a special case. We begin with a definition: A **major determinant** of a $p \times q$ matrix is the determinant of any square submatrix of maximum order. For example,

$$\det \begin{bmatrix} a_{11} & a_{13} \\ a_{21} & a_{23} \end{bmatrix}$$

is one of the three major determinants of

$$\begin{bmatrix} a_{11} & a_{12} & a_{13} \\ a_{21} & a_{22} & a_{23} \end{bmatrix}$$

but not of the matrix

$$\begin{bmatrix} a_{11} & a_{12} & a_{13} \\ a_{21} & a_{22} & a_{23} \\ a_{31} & a_{32} & a_{33} \end{bmatrix}.$$

In fact, *the* determinant of a square matrix is its only major determinant.

Suppose now that A is $m \times n$ and B is $n \times m$, and that $m \le n$. Then a major determinant of A and a major determinant of B are said to be **corresponding majors** of A and B if and only if the columns of A used to form the major of A have the same indices as do the rows of B used to form the major of B. For example, in

$$\begin{bmatrix} \alpha_1 & \alpha_2 & \alpha_3 \\ \beta_1 & \beta_2 & \beta_3 \end{bmatrix} \quad \text{and} \quad \begin{bmatrix} a_1 & b_1 \\ a_2 & b_2 \\ a_3 & b_3 \end{bmatrix},$$

$$\det \begin{bmatrix} \alpha_1 & \alpha_3 \\ \beta_1 & \beta_3 \end{bmatrix} \ (columns\ 1\ and\ 3) \quad \text{and} \quad \det \begin{bmatrix} a_1 & b_1 \\ a_3 & b_3 \end{bmatrix} \ (rows\ 1\ and\ 3)$$

are corresponding majors. What are the other two pairs of corresponding majors?

We shall now establish

Theorem 7.6.4: *If A is an $m \times n$ matrix and B is an $n \times m$ matrix, and if $m \le n$, then $\det AB$ is equal to the sum of the products of the corresponding majors of A and B.*

For example,

$$\det \left(\begin{bmatrix} x_1 & x_2 & x_3 \\ y_1 & y_2 & y_3 \end{bmatrix} \cdot \begin{bmatrix} x_1 & y_1 \\ x_2 & y_2 \\ x_3 & y_3 \end{bmatrix} \right) = \begin{vmatrix} x_1 & x_2 \\ y_1 & y_2 \end{vmatrix} \cdot \begin{vmatrix} x_1 & y_1 \\ x_2 & y_2 \end{vmatrix}$$

$$+ \begin{vmatrix} x_1 & x_3 \\ y_1 & y_3 \end{vmatrix} \cdot \begin{vmatrix} x_1 & y_1 \\ x_3 & y_3 \end{vmatrix} + \begin{vmatrix} x_2 & x_3 \\ y_2 & y_3 \end{vmatrix} \cdot \begin{vmatrix} x_2 & y_2 \\ x_3 & y_3 \end{vmatrix}$$

$$= \begin{vmatrix} x_1 & x_2 \\ y_1 & y_2 \end{vmatrix}^2 + \begin{vmatrix} x_1 & x_3 \\ y_1 & y_3 \end{vmatrix}^2 + \begin{vmatrix} x_2 & x_3 \\ y_2 & y_3 \end{vmatrix}^2.$$

To prove the theorem, we first write A as a matrix of n m-vectors, and then form the product AB:

$$AB = [A_1, A_2, \ldots, A_n] \cdot \begin{bmatrix} b_{11} & b_{12} & \cdots & b_{1m} \\ b_{21} & b_{22} & \cdots & b_{2m} \\ \vdots & & & \\ b_{n1} & b_{n2} & \cdots & b_{nm} \end{bmatrix}$$

$$= [(b_{11}A_1 + b_{21}A_2 + \cdots + b_{n1}A_n),$$

$$(b_{12}A_1 + b_{22}A_2 + \cdots + b_{n2}A_n), \ldots, (b_{1m}A_1 + b_{2m}A_2 + \cdots + b_{nm}A_n)].$$

If we now proceed to expand det AB by repeated application of Theorem 7.3.3, we note that, by decomposing the first column, we obtain n determinants. From *each* of these, by decomposing the second column, we obtain n more determinants, etc. When the process is complete, we will have obtained n^m determinants in which all possible combinations, including repetitions, of the columns $A_1, A_2, \ldots, A_n$ appear:

$$\det AB = \sum_{j_1, j_2, \ldots, j_m = 1}^{n} \det [(b_{j_1 1}A_{j_1}), (b_{j_2 2}A_{j_2}), \ldots, (b_{j_m m}A_{j_m})].$$

Evidently, every term in the above sum in which $j_1, j_2, \ldots, j_m$ are not all distinct will vanish because of proportional columns. Thus the summation needs to be extended only over the $P_m^n = n!/(n-m)!$ permutations of the integers $1, 2, \ldots, n$ taken m at a time. Designating such a sum by $\sum_{(j)}'$ and factoring out the b's, we have

$$\det AB = \sum_{(j)}' (b_{j_1 1}b_{j_2 2} \cdots b_{j_m m}) \det [A_{j_1}, A_{j_2}, \ldots, A_{j_m}].$$

The terms of this sum may be grouped into sets of $m!$ terms, each set involving only permutations $j_1 j_2 \ldots j_m$ *of the same m digits.* Let $k_1 k_2 \ldots k_m$ denote such a set of m digits arranged in the natural order, and let

$$\varepsilon_{j_1 j_2 \cdots j_m}^{k_1 k_2 \cdots k_m} = \pm 1,$$

depending on whether $j_1 j_2 \ldots j_m$ is an even or odd permutation of $k_1 k_2 \ldots k_m$. Then

$$\det [A_{j_1}, A_{j_2}, \ldots, A_{j_m}] = \varepsilon_{j_1 j_2 \cdots j_m}^{k_1 k_2 \cdots k_m} \det [A_{k_1}, A_{k_2}, \ldots, A_{k_m}],$$

so that we can write

$$\det AB = \sum_{(k)} \left(\sum_{(j)} \varepsilon_{j_1 j_2 \cdots j_m}^{k_1 k_2 \cdots k_m} b_{j_1 1} b_{j_2 2} \cdots b_{j_m m} \right) \det [A_{k_1}, A_{k_2}, \ldots, A_{k_m}].$$

Here the inner sum is now extended over all permutations $j_1 \ldots j_m$ of $k_1 \ldots k_m$, and the outer sum is extended over all C_m^n natural-order combinations of m of the digits $1, 2, \ldots, n$. Examination of the sum in parentheses reveals that it is the major determinant of B formed from the rows with indices $k_1, k_2, \ldots, k_m$. This is precisely the corresponding major of the major, $\det [A_{k_1}, A_{k_2}, \ldots, A_{k_m}]$, of A. Since the summation $\sum_{(k)}$ includes all possible pairs of such majors, the theorem has been proved.

In the case $m = n$, each matrix has just one major, so we have in this case a second proof of Theorem 7.6.1.

7.7 A Formula for A^{-1}

We begin with

Theorem 7.7.1: A^{-1} exists if and only if $\det A \neq 0$.

Suppose that A^{-1} exists. Then

$$AA^{-1} = I,$$

so, by Theorem 7.6.1,

$$\det A \cdot \det A^{-1} = \det I = 1.$$

Hence

$$\det A \neq 0,$$

as the theorem asserts. Of course we also have

$$\det A^{-1} \neq 0.$$

In fact,

(7.7.1) $$\det (A^{-1}) = \frac{1}{\det A} = (\det A)^{-1}.$$

Note the two uses of the superscript -1: The first denotes the inverse of a matrix, the second the multiplicative inverse (reciprocal) of a scalar.

Now suppose that det $A \neq 0$ and define

$$(7.7.2) \qquad B = \frac{1}{\det A} \begin{bmatrix} A_{11} & A_{21} & \cdots & A_{n1} \\ A_{12} & A_{22} & \cdots & A_{n2} \\ \vdots & \vdots & & \vdots \\ A_{1n} & A_{2n} & \cdots & A_{nn} \end{bmatrix} = \frac{1}{\det A} [A_{ij}]^{\mathsf{T}},$$

where the A_{ij}'s are the cofactors of the a_{ij}'s in det A. Then, with the aid of the identities

$$a_{i1}A_{k1} + a_{i2}A_{k2} + \cdots + a_{in}A_{kn} = \delta_{ik} \det A$$

$$a_{1j}A_{1k} + a_{2j}A_{2k} + \cdots + a_{nj}A_{nk} = \delta_{jk} \det A,$$

it is not hard to prove that

$$AB = BA = I_n.$$

The reader should execute the computation. Thus, when det $A \neq 0$, the matrix B is the unique inverse of A, and the proof of the theorem is complete.

The formula for A^{-1} that we have just derived is recorded in

Theorem 7.7.2: *If* det $A \neq 0$, *then*

$$(7.7.3) \qquad A^{-1} = \frac{1}{\det A} [A_{ij}]^{\mathsf{T}}.$$

The matrix $[A_{ij}]^{\mathsf{T}}$ is called the **adjoint matrix** of A and is denoted by adj A or by $\mathscr{A}$. We have

$$(7.7.4) \qquad A \cdot [A_{ij}]^{\mathsf{T}} = [A_{ij}]^{\mathsf{T}} \cdot A = (\det A)I.$$

Applying Theorem 7.6.1 to (7.7.4) we have

$$\det A \cdot \det(\text{adj } A) = (\det A)^n.$$

This is an identity in the n^2 variables a_{ij} and hence it is legitimate to cancel the not identically zero factor det A. This yields

Theorem 7.7.3: *For every square matrix* A,

$$\det(\text{adj } A) = (\det A)^{n-1}.$$

7.8   Determinants and the Rank of a Matrix

Given a matrix $A_{m \times n}$ over an arbitrary field if $r(A) = k$, then there exists at least one set of k linearly independent rows, but no set of $k + 1$ or more linearly independent rows. In a submatrix consisting of k linearly independent rows of A, there must exist at least one set of k linearly independent columns. These columns then form a $k \times k$ submatrix of A which has an inverse, since it has rank k. Hence its determinant is not zero. Moreover, since no set of $k + 1$ or more rows is linearly independent, no square submatrix of order $k + 1$ or more has a nonzero determinant.

Conversely, if there exists in A a $k \times k$ submatrix M with determinant not zero, but no higher-order square submatrix with determinant not zero, then the rows of M are linearly independent and hence so are the corresponding rows of A. Thus the rank of A is at least k. By the preceding paragraph, it cannot be more than k, since no square submatrix of order higher than k has a nonzero determinant.

We have thus proved

Theorem 7.8.1: *The rank of a matrix A is the order of the highest-order square submatrix of A whose determinant is not zero.*

The determinant of a square submatrix of a matrix A is called a **minor determinant,** or simply a **minor,** of A. If the submatrix is of order r, its determinant is called a **minor of order r** of A. Thus the rank of a matrix A is the order of the highest-order nonzero minor of A.

The sweepout procedure is usually easier than the determinant method of finding the rank of a matrix. However, at times the determinant rule gives a quick answer. For example, regardless of the values of a, b, c, and d, the rank of

$$\begin{bmatrix} 1 & a & b & 0 \\ 0 & c & d & 1 \\ 1 & a & b & 0 \\ 0 & c & d & 1 \end{bmatrix}$$

is 2 because

$$\det \begin{bmatrix} 1 & 0 \\ 0 & 1 \end{bmatrix} \neq 0,$$

but the determinant of any third- or higher-order submatrix is zero, since it has at least two identical rows.

In many cases, prior use of appropriate elementary transformations makes it easier to apply Theorem 7.8.1.

Consider now the case of n homogeneous equations in n unknowns, represented by $AX = 0$. We have seen in Chapter 6 that there are nontrivial solutions if and only if the rank of A is less than n. By the preceding theorem, the rank of A is less than n if and only if det $A = 0$. We may therefore conclude

Theorem 7.8.2: *A system of n homogeneous equations in n unknowns, $AX = 0$ in matrix form, has nontrivial solutions if and only if det $A = 0$.*

For example, a particular system of the family of systems

$$tx_1 + x_2 + x_3 = 0$$
$$x_1 + tx_2 + x_3 = 0$$
$$x_1 + x_2 + tx_3 = 0$$

has nontrivial solutions if and only if

$$\begin{vmatrix} t & 1 & 1 \\ 1 & t & 1 \\ 1 & 1 & t \end{vmatrix} = 0,$$

that is, if and only if

$$\begin{vmatrix} t-1 & 0 & 1-t \\ 0 & t-1 & 1-t \\ 1 & 1 & t \end{vmatrix} = 0.$$

Factoring $(t - 1)$ from each of the first two rows, we have the equation

$$(t-1)^2 \begin{vmatrix} 1 & 0 & -1 \\ 0 & 1 & -1 \\ 1 & 1 & t \end{vmatrix} = 0,$$

that is, the equation

$$(t - 1)^2(t + 2) = 0,$$

so nontrivial solutions exist if and only if $t = 1$ or $t = -2$. When $t = 1$, the system has rank 1 so that the solution space is a two-dimensional vector space, but when $t = -2$, the rank is 2, so that the solution space is a one-dimensional vector space. The reader should solve the system in each of these two cases.

The theorem implies that if $AX = 0$ has only the trivial solution, then $\det A \neq 0$, and conversely. Hence, putting together a number of facts and definitions that we have observed at one time or another, we see that all of the following are equivalent for a square matrix of order n:

$$\det A \neq 0,$$

$$A^{-1} \text{ exists.}$$

$$A \text{ is nonsingular,}$$

$$r(A) = n,$$

(7.8.1)   the rows of A are linearly independent,

the columns of A are linearly independent,

$$AX = 0 \text{ has only the trivial solution,}$$

$$AX = Y \text{ has a unique solution for each } Y.$$

7.9   Solution of Systems of Equations by Using Determinants

Often one can derive a determinantal formula for the solutions of a system of equations. Such formulas are attractive and are useful in theoretical derivations. In numerical cases, they involve more computation than does the sweep out process except in special instances, such as a small number of variables or coefficient matrices with many elements zero (**sparse matrices**).

As a first example, consider $n - 1$ homogeneous equations in n unknowns, represented by $AX = 0$, where the rank of A is $n - 1$. Then we have $n - (n - 1) = 1$ linearly independent solution and all others are multiples of this one. To the $n - 1$ equations add an nth equation with coefficients all zero. Then the enlarged system is equivalent to the original one and has a square coefficient matrix whose determinant is zero. Hence, if $\tilde{A}$ denotes the new coefficient matrix, we have

$$a_{i1}\tilde{A}_{n1} + a_{i2}\tilde{A}_{n2} + \cdots + a_{in}\tilde{A}_{nn} = \delta_{in} \det \tilde{A} = 0, \qquad i = 1, 2, \ldots, n,$$

so $(\tilde{A}_{n1}, \tilde{A}_{n2}, \ldots, \tilde{A}_{nn})$ is a solution. Moreover, since $\tilde{A}$ has rank $n - 1$, *not all* of the cofactors $\tilde{A}_{n1}, \tilde{A}_{n2}, \ldots, \tilde{A}_{nn}$, which are formed from the elements of the first $n - 1$ rows, are zero. That is, the complete solution is

$$(7.9.1) \qquad \begin{bmatrix} x_1 \\ x_2 \\ \vdots \\ x_n \end{bmatrix} = t \begin{bmatrix} \tilde{A}_{n1} \\ \tilde{A}_{n2} \\ \vdots \\ \tilde{A}_{nn} \end{bmatrix}.$$

For example, suppose we need a unit vector in $\mathscr{E}^4$ which is orthogonal to each of the three independent vectors

$$\begin{bmatrix} 1 \\ 1 \\ 0 \\ 1 \end{bmatrix}, \quad \begin{bmatrix} 1 \\ -1 \\ 1 \\ 0 \end{bmatrix}, \quad \begin{bmatrix} 0 \\ 1 \\ 1 \\ -1 \end{bmatrix}.$$

Then what we seek first is a solution to the system of three equations in four unknowns:

$$x_1 + x_2 \qquad + x_4 = 0$$
$$x_1 - x_2 + x_3 \qquad = 0$$
$$x_2 + x_3 - x_4 = 0.$$

By the formula (7.9.1) we determine the four cofactors of a hypothetical fourth row. These are obtained by deleting successive columns of the coefficient matrix, computing the determinant, and attaching the proper sign:

$$- \begin{vmatrix} 1 & 0 & 1 \\ -1 & 1 & 0 \\ 1 & 1 & -1 \end{vmatrix}, \quad \begin{vmatrix} 1 & 0 & 1 \\ 1 & 1 & 0 \\ 0 & 1 & -1 \end{vmatrix},$$

$$- \begin{vmatrix} 1 & 1 & 1 \\ 1 & -1 & 0 \\ 0 & 1 & -1 \end{vmatrix}, \quad \begin{vmatrix} 1 & 1 & 0 \\ 1 & -1 & 1 \\ 0 & 1 & 1 \end{vmatrix},$$

so the complete solution is

$$\begin{bmatrix} x_1 \\ x_2 \\ x_3 \\ x_4 \end{bmatrix} = t \begin{bmatrix} 3 \\ 0 \\ -3 \\ -3 \end{bmatrix}.$$

The corresponding normalized vector is one of these two:

$$\pm \frac{1}{\sqrt{3}} \begin{bmatrix} 1 \\ 0 \\ -1 \\ -1 \end{bmatrix}.$$

Next consider n equations in n unknowns represented by

$$AX = B,$$

where A^{-1} exists. Then $X = A^{-1}B$, or

$$X = \frac{1}{\det A} \begin{bmatrix} A_{11} & A_{21} & \cdots & A_{n1} \\ A_{12} & A_{22} & \cdots & A_{n2} \\ \vdots & & & \\ A_{1n} & A_{2n} & \cdots & A_{nn} \end{bmatrix} \begin{bmatrix} b_1 \\ b_2 \\ \vdots \\ b_n \end{bmatrix}$$

$$= \frac{1}{\det A} \begin{bmatrix} (b_1 A_{11} + b_2 A_{21} + \cdots + b_n A_{n1}) \\ (b_1 A_{12} + b_2 A_{22} + \cdots + b_n A_{n2}) \\ \vdots \\ (b_1 A_{1n} + b_2 A_{2n} + \cdots + b_n A_{nn}) \end{bmatrix},$$

is the unique (Ex. 2, Sect. 7.5) solution of the system. In scalar form,

$$x_j = \frac{1}{\det A}(b_1 A_{1j} + b_2 A_{2j} + \cdots + b_n A_{nj}), \qquad j = 1, 2, \ldots, n.$$

The expression in parentheses is the determinant of a matrix the same as A except that the jth column has been replaced by the b's.

We summarize all this in **Cramer's rule**:

Theorem 7.9.1: *Given a system of n equations in n unknowns,*

$$AX = B,$$

such that A^{-1} exists, then the unique solution is given in determinant form by

$$(7.9.2) \quad x_j = \frac{\det [A_1, \ldots, A_{j-1}, B, A_{j+1}, \ldots, A_n]}{\det A}, \qquad j = 1, 2, \ldots, n.$$

For example, to solve

$$(k - 1)x_1 + \qquad kx_2 = \quad 1$$
$$kx_1 + (k - 1)x_2 = -1,$$

we first compute the determinant of the coefficient matrix:

$$\begin{vmatrix} k-1 & k \\ k & k-1 \end{vmatrix} = (k-1)^2 - k^2 = 1 - 2k.$$

Then, if $1 - 2k \neq 0$, that is, if $k \neq \frac{1}{2}$,

$$x_1 = \frac{\begin{vmatrix} 1 & k \\ -1 & k-1 \end{vmatrix}}{1 - 2k} = \frac{2k-1}{1-2k} = -1,$$

$$x_2 = \frac{\begin{vmatrix} k-1 & 1 \\ k & -1 \end{vmatrix}}{1 - 2k} = \frac{1 - 2k}{1 - 2k} = 1.$$

Thus every system of this one-parameter family of systems of equations, except the one for which $k = \frac{1}{2}$, has the unique solution $(-1, 1)$.

As another example, consider the system

$$\begin{aligned} x_1 - a_{12} x_2 \qquad\qquad &= b_1 \\ x_2 - a_{23} x_3 &= b_2 \\ -a_{31} x_1 \qquad\qquad + x_3 &= b_3. \end{aligned}$$

Here

$$\det A = \begin{vmatrix} 1 & -a_{12} & 0 \\ 0 & 1 & -a_{23} \\ -a_{31} & 0 & 1 \end{vmatrix} = 1 - a_{12} a_{23} a_{31},$$

$$x_1 = \frac{1}{\det A} \begin{vmatrix} b_1 & -a_{12} & 0 \\ b_2 & 1 & -a_{23} \\ b_3 & 0 & 1 \end{vmatrix} = \frac{1}{\det A} (b_1 + b_2 a_{12} + b_3 a_{12} a_{23}),$$

$$x_2 = \frac{1}{\det A} \begin{vmatrix} 1 & b_1 & 0 \\ 0 & b_2 & -a_{23} \\ -a_{31} & b_3 & 1 \end{vmatrix} = \frac{1}{\det A} (b_2 + b_3 a_{23} + b_1 a_{23} a_{31}).$$

$$x_3 = \frac{1}{\det A} \begin{vmatrix} 1 & -a_{12} & b_1 \\ 0 & 1 & b_2 \\ -a_{31} & 0 & b_3 \end{vmatrix} = \frac{1}{\det A} (b_3 + b_1 a_{31} + b_2 a_{31} a_{12}).$$

There is a fairly simple pattern here, one which suggests that there is a simple solution to a similar system of n equations in n unknowns. Can you establish the general rule?

7.10   A Geometrical Application of Determinants

Recall Theorem 7.8.2, which says that a homogeneous system of n linear equations in n unknowns has nontrivial solutions if and only if the determinant of the coefficients is zero.

This theorem is often used in geometrical applications. For example, suppose that $(a_1, a_2), (b_1, b_2)$ are distinct points so that they determine a line in $\mathscr{E}^2$. Then (x_1, x_2) is a point on this line if and only if the three equations

$$Ax_1 + Bx_2 + C = 0$$
$$Aa_1 + Ba_2 + C = 0$$
$$Ab_1 + Bb_2 + C = 0$$

have a nontrivial solution for the coefficients A, B, C, that is, by Theorem 7.8.2, if and only if

$$\begin{vmatrix} x_1 & x_2 & 1 \\ a_1 & a_2 & 1 \\ b_1 & b_2 & 1 \end{vmatrix} = 0.$$

This linear equation in x_1 and x_2 is satisfied by (a_1, a_2) and by (b_1, b_2), so it is an equation of the line on these two points.

We saw in Chapter 4 that n points not in the same $(n-2)$-flat determine a hyperplane. The equation of such a hyperplane is readily written down in determinant form. Indeed, generalizing from the preceding example, we write the equation

$$(7.10.1) \qquad \begin{vmatrix} x_1 & x_2 & \cdots & x_n & 1 \\ a_{11} & a_{12} & \cdots & a_{1n} & 1 \\ a_{21} & a_{22} & \cdots & a_{2n} & 1 \\ \vdots & & & & \\ a_{n1} & a_{n2} & \cdots & a_{nn} & 1 \end{vmatrix} = 0,$$

which is satisfied by the coordinates of each of the n points

$$(a_{11}, a_{12}, \ldots, a_{1n}), (a_{21}, a_{22}, \ldots, a_{2n}), \ldots, (a_{n1}, a_{n2}, \ldots, a_{nn}),$$

since substitution results in two equal rows in each case. Thus the locus lies on the points in question. If we expand along the first row, we see that the equation is linear. Thus it represents a hyperplane unless the coefficients of *all* the x_j's are zero. In this event, the n given points all lie in a k-flat, $k < n - 1$. Consequently, given n points, we write the above determinant and expand. If a proper linear equation results, it is an equation of the hyperplane on those points. If no proper linear equation results, we know the points do not determine a hyperplane but rather lie in a space of dimension smaller than $n - 1$.

For example, the equation of the line in $\mathscr{E}^2$ on $(2, 4)$ and $(4, 2)$ is

$$\begin{vmatrix} x_1 & x_2 & 1 \\ 2 & 4 & 1 \\ 4 & 2 & 1 \end{vmatrix} = 0 \quad \text{or} \quad x_1 + x_2 = 6.$$

Again, suppose we want the x_3-intercept of the plane in $\mathscr{E}^3$ which lies on $(a, 0, 0)$, $(0, b, 0)$, and $(1, 1, 1)$. In the equation of the plane, namely

$$\begin{vmatrix} x_1 & x_2 & x_3 & 1 \\ a & 0 & 0 & 1 \\ 0 & b & 0 & 1 \\ 1 & 1 & 1 & 1 \end{vmatrix} = 0,$$

we put $x_1 = x_2 = 0$ and solve for x_3. We can do this neatly by expanding the determinant in the equation

$$\begin{vmatrix} 0 & 0 & x_3 & 1 \\ a & 0 & 0 & 1 \\ 0 & b & 0 & 1 \\ 1 & 1 & 1 & 1 \end{vmatrix} = 0$$

along the first row:

$$x_3 \begin{vmatrix} a & 0 & 1 \\ 0 & b & 1 \\ 1 & 1 & 1 \end{vmatrix} - \begin{vmatrix} a & 0 & 0 \\ 0 & b & 0 \\ 1 & 1 & 1 \end{vmatrix} = 0.$$

From this,

$$x_3 = \frac{ab}{ab - a - b},$$

provided, of course, that $ab \neq a + b$.

7.11   Exercises

1. What is the determinant of this product of matrices?

$$
\begin{bmatrix} 1 & 0 & 0 \\ 0 & 2 & 0 \\ 0 & 0 & 3 \end{bmatrix}
\begin{bmatrix} 2 & 0 & 1 \\ 0 & 1 & 0 \\ 1 & 0 & 1 \end{bmatrix}
\begin{bmatrix} 1 & 1 & \frac{1}{3} \\ 1 & \frac{1}{2} & 0 \\ 1 & 0 & 0 \end{bmatrix}.
$$

2. Given that

$$
\begin{bmatrix} 1 & -2 & 0 \\ 2 & 1 & 0 \\ 0 & 0 & 1 \end{bmatrix} A
\begin{bmatrix} 1 & 2 & 0 \\ -2 & 1 & 0 \\ 0 & 0 & 1 \end{bmatrix} = 5I,
$$

what is det A?

*3. If $X = [x_1, x_2, \ldots, x_n]^\mathsf{T}$ and $Y = [y_1, y_2, \ldots, y_n]^\mathsf{T}$, show by means of Theorem 7.6.4 that

$$
(X^\mathsf{T}X)(Y^\mathsf{T}Y) - (X^\mathsf{T}Y)^2 = \sum_{1 \le i < j \le n} \begin{vmatrix} x_i & x_j \\ y_i & y_j \end{vmatrix}^2,
$$

from which

$$
(X^\mathsf{T}X)(Y^\mathsf{T}Y) \ge (X^\mathsf{T}Y)^2,
$$

in case the components of X and Y are real. This last is one form of the famous Cauchy–Schwarz inequality. Prove that equality holds if and only if there exist scalars α and β such that $\alpha X = \beta Y$.

4. Use Theorem 7.6.4 to evaluate

$$
\det
\begin{bmatrix} 1 & 0 & 0 & a \\ 0 & 1 & 0 & b \\ 0 & 0 & 1 & c \end{bmatrix}
\begin{bmatrix} 1 & 0 & 0 \\ 0 & 1 & 0 \\ 0 & 0 & 1 \\ a & b & c \end{bmatrix}.
$$

5. Reduce this matrix to triangular form by elementary transformations that leave the determinant invariant:

$$
\begin{bmatrix} 2 & -1 & 4 \\ 3 & \frac{1}{2} & 7 \\ 4 & 2 & 8 \end{bmatrix}.
$$

6. Let

$$
A = \begin{bmatrix} a_1 & a_2 & \cdots & a_n \\ 1 & 1 & \cdots & 1 \end{bmatrix}.
$$

Use the product AA^T and Theorem 7.6.4 to prove that

$$\sum_{j=1}^{n} a_j^2 \geq \frac{1}{n}\left(\sum_{j=1}^{n} a_j\right)^2.$$

Give an example to show that equality can hold here.

7. Compute the determinants of these matrices:

(a)
$$\begin{bmatrix} 0 & 1 & 0 & 0 & 0 \\ 1 & 0 & 0 & 0 & 0 \\ a & b & 0 & 1 & 2 \\ c & d & 2 & 0 & 1 \\ e & f & 1 & 2 & 0 \end{bmatrix},$$

(b)
$$\begin{bmatrix} 0 & 0 & 0 & 3 & 2 & 1 \\ 0 & 0 & 0 & 2 & 1 & 0 \\ 0 & 0 & 0 & 1 & 0 & -1 \\ 0 & 2 & 1 & 0 & 0 & 0 \\ 0 & 1 & 2 & 0 & 0 & 0 \\ 4 & 0 & 0 & 0 & 0 & 0 \end{bmatrix},$$

(c)
$$\begin{bmatrix} 0 & 0 & 0 & 0 & 1 & 2 \\ 0 & 0 & 0 & 0 & 2 & 1 \\ 0 & 0 & 1 & 2 & 0 & 0 \\ 0 & 0 & 2 & 1 & 0 & 0 \\ a & b & c & d & e & f \\ \alpha & \beta & \gamma & \delta & g & h \end{bmatrix}.$$

8. Given that $a^3 + b^3 = 1$, find by determinants the inverse of

$$\begin{bmatrix} a & b & 0 \\ 0 & a & b \\ b & 0 & a \end{bmatrix}.$$

9. Use the determinantal formula to compute A^{-1}, where

$$A = \begin{bmatrix} 0 & 0 & 1 & 1 \\ 0 & 0 & 1 & 0 \\ 1 & 1 & 0 & 0 \\ 1 & 0 & 0 & 0 \end{bmatrix}.$$

10. For what values of t, a, b will the following matrices fail to have inverses? For all other values of t, a, b, what are their inverses?

(a)
$$\begin{bmatrix} 1 & t & 0 \\ 0 & 1 & -1 \\ t & 0 & 1 \end{bmatrix},$$

(b)
$$\begin{bmatrix} a & b & 0 \\ 0 & b & b \\ a & 0 & a \end{bmatrix},$$

(c)
$$\begin{bmatrix} a & 0 & b & 0 \\ 0 & a & 0 & b \\ b & 0 & a & 0 \\ 0 & b & 0 & a \end{bmatrix}.$$

11. (a) Given that $ABA = I$, where A, B, and I are of order n, show that A^{-1} and B^{-1} exist. What does $(A^2B)^{-1}$ reduce to in this case?

    (b) Prove that if $RAC = I$, where all four matrices have order n, then A^{-1} exists and $A^{-1} = CR$.

12. Compute the inverse of

$$\alpha \begin{bmatrix} 1 & 0 \\ 0 & 1 \end{bmatrix} + \beta \begin{bmatrix} 0 & -1 \\ 1 & 0 \end{bmatrix},$$

where α and β are real numbers and write the inverse in this same form. Interpret by using the concept of isomorphism. (See Exercise 31, Section 1.18.)

13. Find A^{-1} if a, b, c, d are real numbers such that

$$a^2 + b^2 + c^2 + d^2 = 1$$

and

$$A = \begin{bmatrix} a + ib & c + id \\ -c + id & a - ib \end{bmatrix}.$$

14. Given that $A^* = A^{-1}$ and that

$$A = \begin{bmatrix} 0 & 0 & \alpha & i\alpha \\ 0 & 0 & i\alpha & \alpha \\ \alpha & i\alpha & 0 & 0 \\ i\alpha & \alpha & 0 & 0 \end{bmatrix}, \qquad i^2 = -1,$$

what are the possible values for α?

15. Show that the square matrix

$$k \left\{ \begin{matrix} \begin{bmatrix} 1 & 0 & \cdots & 0 & \vdots & 1 & 1 & \cdots & 1 \\ 0 & 1 & \cdots & 0 & \vdots & 1 & 1 & \cdots & 1 \\ \vdots & & & & \vdots & & & & \\ 0 & 0 & \cdots & 1 & \vdots & 1 & 1 & \cdots & 1 \\ \end{bmatrix} \end{matrix} \right. }$$

is nonsingular if and only if $\det B$ is *unequal* to k times the sum of its cofactors.

*16. Prove that the inverse of a nonsingular, upper triangular matrix T is also upper triangular, and that the diagonal entries of the inverse are the reciprocals of the corresponding entries of T.

17. Obtain inverses for the tridiagonal matrices

$$
\begin{bmatrix} 1 & -1 \\ -1 & 2 \end{bmatrix}, \quad
\begin{bmatrix} 1 & -1 & 0 \\ -1 & 2 & -1 \\ 0 & -1 & 2 \end{bmatrix}, \quad
\begin{bmatrix} 1 & -1 & 0 & 0 \\ -1 & 2 & -1 & 0 \\ 0 & -1 & 2 & -1 \\ 0 & 0 & -1 & 2 \end{bmatrix}.
$$

Then guess at the form of the inverse of an $n \times n$ matrix of the same structure and prove that your formula is correct.

18. Given that $a \neq 0$, $b \neq 0$, find the rank of the matrix

$$
\begin{bmatrix}
1 & 0 & 2 & a & 0 \\
0 & b & 3 & b & 1 \\
0 & b & 3 & b & 1 \\
1 & a & 3 & a & 0 \\
0 & b & 4 & 0 & 1
\end{bmatrix}.
$$

*19. Prove that the vectors

$$
X = \begin{bmatrix} x_1 \\ x_2 \\ \vdots \\ x_n \end{bmatrix} \quad \text{and} \quad Y = \begin{bmatrix} y_1 \\ y_2 \\ \vdots \\ y_n \end{bmatrix}
$$

over a field $\mathcal{F}$ are linearly dependent if and only if

$$
\det \begin{bmatrix} x_i & y_i \\ x_j & y_j \end{bmatrix} = 0
$$

whenever $1 \leq i < j \leq n$. Generalize.

*20. (a) Show that any minor of order r in a matrix product AB is a sum of products of minors of order r of A and minors of order r of B.

(b) Use this result to prove again that the rank of a product of two matrices cannot exceed the rank of either factor.

(c) Prove by induction that the determinant of a minor of order r of a product $M_1 M_2 \cdots M_p$ is a sum of products of minors of order r of $M_1, M_2, \ldots, M_p$, that is, that it has the form

$$
\sum D_{1i_1} D_{2i_2} \cdots D_{pi_p},
$$

where D_{ki_k} is a minor of M_k.

21. Show that, if A is a real $m \times n$ matrix $(m \leq n)$ which has rank m, then the rank of AA^{T} is also m, so that AA^{T} is, in fact, a nonsingular, symmetric matrix of order m. Show also that the elements on the main diagonal of AA^{T} are nonnegative and, finally, that, if A has rank $< m$, AA^{T} is singular. (This result is useful in statistical applications.)

22. Show that A^*, A^*A, and AA^* all have the same rank as A.

23. Prove that not all determinants of order r in any set of r rows of a nonsingular matrix may be zero.

24. Prove that, if $k > n/2$, this matrix is singular:

$$\begin{bmatrix} 0 & \cdots & 0 & a_{1,k+1} & \cdots & a_{1n} \\ \vdots & & & \vdots & & \\ 0 & \cdots & 0 & a_{k,k+1} & \cdots & a_{kn} \\ a_{k+1,1} & \cdots & a_{k+1,k} & a_{k+1,k+1} & \cdots & a_{k+1,n} \\ \vdots & & & \vdots & & \\ a_{n1} & \cdots & a_{nk} & a_{k+1,n} & \cdots & a_{nn} \end{bmatrix}.$$

25. Show that the rank of $\mathscr{A}$, the adjoint matrix of a square matrix A of order n, is (a) n when A is nonsingular, (b) 1 when A is of rank $n-1$, (c) 0 when A is of rank $< n-1$.

26. Solve by Cramer's rule:

(a)
$$\begin{aligned} x_1 + 2x_2 + 3x_3 &= 6 \\ 2x_1 - 2x_2 + 5x_3 &= 5 \\ 4x_1 - x_2 - 3x_3 &= 0, \end{aligned}$$

(b)
$$\begin{aligned} x_1 + x_2 &= 3 \\ x_2 + 2x_3 &= 2 \\ x_3 + 3x_4 &= 1 \\ 4x_1 + x_4 &= 0. \end{aligned}$$

27. For what values of t will the system

$$\begin{bmatrix} 1 & t & 0 \\ 0 & 1 & -1 \\ t & 0 & 1 \end{bmatrix} \begin{bmatrix} x_1 \\ x_2 \\ x_3 \end{bmatrix} = \begin{bmatrix} 1 \\ 1 \\ 1 \end{bmatrix}$$

have a unique solution?

28. Given that

$$a_1 x_1 + a_2 x_2 + \cdots + a_n x_n = 0$$

$$b_1 x_1 + b_2 x_2 + \cdots + b_n x_n = 0,$$

prove that for all values of α and β,

$$\begin{vmatrix} a_1 & \alpha \\ b_1 & \beta \end{vmatrix} x_1 + \begin{vmatrix} a_2 & \alpha \\ b_2 & \beta \end{vmatrix} x_2 + \cdots + \begin{vmatrix} a_n & \alpha \\ b_n & \beta \end{vmatrix} x_n = 0.$$

What is the interpretation of the latter equation?

29. Show that no nonzero vector is orthogonal to all these vectors:

$$\begin{bmatrix} 1 \\ 1 \\ 0 \end{bmatrix}, \quad \begin{bmatrix} 1 \\ 0 \\ 1 \end{bmatrix}, \quad \begin{bmatrix} 0 \\ 1 \\ 1 \end{bmatrix}.$$

30. Find the components of a vector in $\mathscr{E}^4$ orthogonal to each of

$$\begin{bmatrix} a \\ b \\ 1 \\ 0 \end{bmatrix}, \quad \begin{bmatrix} 0 \\ a \\ b \\ 1 \end{bmatrix}, \quad \begin{bmatrix} 1 \\ 0 \\ a \\ b \end{bmatrix}.$$

Are there any values of a and b for which there is no solution?

31. Show that the set of all vectors in $\mathscr{E}^4$ which are orthogonal to each of the vectors

$$\begin{bmatrix} 1 \\ 1 \\ 1 \\ 1 \end{bmatrix}, \quad \begin{bmatrix} 1 \\ -1 \\ 1 \\ -1 \end{bmatrix}, \quad \begin{bmatrix} 1 \\ 0 \\ -1 \\ 0 \end{bmatrix},$$

constitutes a vector space. (There are several ways to do this.) Then find a basis for this space.

32. Solve by Cramer's rule and determine for what values of t the solution is valid. For what real values of t is $x_3 = 0$?

$$tx_1 + x_2 - x_3 = 1$$
$$-x_1 + tx_2 + x_3 = 1$$
$$x_1 - x_2 + tx_3 = 1.$$

33. Show that, if det $A = 0$, the system $AX = 0$ has the (possibly trivial) solution $[A_{i1}, A_{i2}, \ldots, A_{in}]^{\mathsf{T}}$. Then, if $X_1, X_2, \ldots, X_{n+1}$ are any $n + 1$ n-vectors, show that

$$\det [X_2, X_3, \ldots, X_{n+1}]X_1 - \det [X_1, X_3, \ldots, X_{n+1}]X_2 + \cdots$$
$$+ (-1)^n \det [X_1, X_2, \ldots, X_n]X_{n+1} = 0$$

by noting that

$$\det \begin{bmatrix} X_1 & X_2 & \cdots & X_{n+1} \\ \hline 0 & 0 & \cdots & 0 \end{bmatrix} = 0.$$

34. Under what conditions can the system

$$A_{(n,\,n)} X_{(n,\,p)} + B_{(n,\,m)} Y_{(m,\,p)} = C_{(n,\,p)}$$
$$D_{(m,\,n)} X_{(n,\,p)} + E_{(m,\,m)} Y_{(m,\,p)} = G_{(m,\,p)}$$

be solved uniquely for the matrices X and Y? What is the solution?

35. Find, by Cramer's rule, the point common to the three planes in $\mathscr{E}^3$ whose equations are

$$x_1 + x_2 + x_3 = a$$
$$x_1 - x_2 \qquad\;\; = 0$$
$$x_1 + x_2 - x_3 = 0.$$

What is the locus of this point as a varies? Illustrate with a figure.

36. Find the equation of the plane in $\mathscr{E}^3$ on the points $(0, 0, 0)$, $(2, 1, 1)$, and $(-2, 1, 0)$.

37. Interpret geometrically:

(a)
$$\begin{vmatrix} x_1 - a_{n1} & x_2 - a_{n2} & \cdots & x_n - a_{nn} \\ a_{11} - a_{n1} & a_{12} - a_{n2} & \cdots & a_{1n} - a_{nn} \\ \vdots & & & \\ a_{n-1,1} - a_{n1} & a_{n-1,2} - a_{n2} & \cdots & a_{n-1,n} - a_{nn} \end{vmatrix} = 0,$$

(b)
$$\begin{vmatrix} x_1 & x_2 & \cdots & x_n \\ a_{11} & a_{12} & \cdots & a_{1n} \\ \vdots & & & \\ a_{n-1,1} & a_{n-1,2} & \cdots & a_{n-1,n} \end{vmatrix} = 0.$$

38. Interpret geometrically:

$$\begin{vmatrix} x_1^2 + x_2^2 & x_1 & x_2 & 1 \\ a_1^2 + a_2^2 & a_1 & a_2 & 1 \\ b_1^2 + b_2^2 & b_1 & b_2 & 1 \\ c_1^2 + c_2^2 & c_1 & c_2 & 1 \end{vmatrix} = 0.$$

39. Show that the area K of a triangle with vertices (x_1, x_2), (y_1, y_2), (z_1, z_2), listed in counterclockwise order, is given by

$$K = \tfrac{1}{2}\begin{vmatrix} x_1 & x_2 & 1 \\ y_1 & y_2 & 1 \\ z_1 & z_2 & 1 \end{vmatrix}.$$

Start with Figure 7-2. Thereafter, show that the triangle with vertices

$$(x_1 + h, x_2 + k), (y_1 + h, y_2 + k), (z_1 + h, z_2 + k)$$

is congruent to the given one, so that having the triangle in the first quadrant is no essential limitation and so that the formula holds in any case.

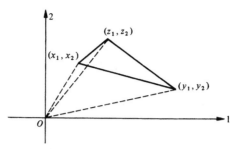

FIGURE 7-2. Area of a Triangle.

40. Show that if the point (c_1, c_2, c_3) is on the straight line determined by distinct points (a_1, a_2, a_3) and (b_1, b_2, b_3) in $\mathscr{E}^3$, then the equation

$$\begin{vmatrix} x_1 & x_2 & x_3 & 1 \\ a_1 & a_2 & a_3 & 1 \\ b_1 & b_2 & b_3 & 1 \\ c_1 & c_2 & c_3 & 1 \end{vmatrix} = 0$$

reduces to $0 = 0$, which makes sense because there isn't a unique plane on the three points in this case.

41. Use Exercise 37(a) to solve Exercise 21, Section 5.7, for the case $k = n - 1$.

42. Use Exericse 37(b) to solve Exercise 22, Section 5.7, for the case $k = n - 1$.

43. The volume V of a tetrahedron whose six sides are known is found from the formula

$$288V^2 = \begin{vmatrix} 0 & c^2 & b^2 & \alpha^2 & 1 \\ c^2 & 0 & a^2 & \beta^2 & 1 \\ b^2 & a^2 & 0 & \gamma^2 & 1 \\ \alpha^2 & \beta^2 & \gamma^2 & 0 & 1 \\ 1 & 1 & 1 & 1 & 0 \end{vmatrix},$$

where the sides a, b, c emanate from one vertex and α, β, γ are the opposite sides respectively. Find by means of this formula the volume of a regular tetrahedron with sides all equal to s.

44. Show that

$$\begin{vmatrix} y & 1 & x & x^2 \\ y_0 & 1 & x_0 & x_0^2 \\ y_0' & 0 & 1 & 2x_0 \\ y_0'' & 0 & 0 & 2 \end{vmatrix} = 0,$$

is the equation of a parabola which passes through the point (x_0, y_0) with slope y_0' and with second derivative y_0'' at that point.

45. Prove that, if one chooses a_1, a_2, a_3 so that the point (a_1, a_2, a_3) lies in the plane $y_1 + y_2 = y_3$ in 3-space, then the three points

$$(0, a_1), \qquad \left(\frac{1}{2}, \frac{a_3}{2}\right), \qquad \text{and} \qquad (1, a_2),$$

if plotted in the $x_1 x_2$-plane, are collinear.

46. On what surface will the point (a_1, a_2, a_3) lie if the points

$$(0, a_1), \qquad (1. a_2), \qquad \text{and} \qquad \left(\frac{a_3}{a_3 + 1}, \frac{a_3^2}{a_3 + 1}\right)$$

plotted in the $x_1 x_2$-plane are collinear? (This problem and the preceding one are adapted from A. S. Levens: *Nomography*, John Wiley & Sons, New York, 1959, in which determinants are used extensively to construct nomograms.)

7.12 More About the Rank of a Matrix

There are many additional interesting and useful theorems relating to the rank of a matrix. A few of these are given in this section. We begin with

Theorem 7.12.1: *If an $m \times n$ matrix A has rank r, then the major determinants of a submatrix consisting of any r rows of A are proportional to the corresponding major determinants of any other such submatrix.*

We assume $r < m$ and $r < n$. (If $r = m$ or $r = n$, the theorem is trivially true.)

We illustrate the theorem by means of an example before proceeding to the proof. In the matrix of rank 2,

$$\begin{bmatrix} 2 & 1 & -5 \\ 3 & 1 & 1 \\ 0 & -1 & 17 \end{bmatrix},$$

the submatrix of the first two rows has the major determinants

$$\begin{vmatrix} 2 & 1 \\ 3 & 1 \end{vmatrix} = -1, \qquad \begin{vmatrix} 2 & -5 \\ 3 & 1 \end{vmatrix} = 17, \qquad \begin{vmatrix} 1 & -5 \\ 1 & 1 \end{vmatrix} = 6.$$

The submatrix consisting of the last two rows has the corresponding majors,

$$\begin{vmatrix} 3 & 1 \\ 0 & -1 \end{vmatrix} = -3, \qquad \begin{vmatrix} 3 & 1 \\ 0 & 17 \end{vmatrix} = 51, \qquad \begin{vmatrix} 1 & 1 \\ -1 & 17 \end{vmatrix} = 18,$$

and from the first and last rows we obtain in the same way,

$$\begin{vmatrix} 2 & 1 \\ 0 & -1 \end{vmatrix} = -2, \qquad \begin{vmatrix} 2 & -5 \\ 0 & 17 \end{vmatrix} = 34, \qquad \begin{vmatrix} 1 & -5 \\ -1 & 17 \end{vmatrix} = 12.$$

The meaning of the theorem should now be clear. Sometimes, in applying the theorem, we have to remember that the zero n-vector and any other n-vector are "proportional." Now, the proof.

Let a set of r linearly independent columns of A be denoted by L_1, $L_2, \ldots, L_r$. Then every column of A is a linear combination of these r columns, and hence we may write

$$
\begin{aligned}
A &= \left[\sum_{i=1}^{r} k_{i1}L_i, \sum_{i=1}^{r} k_{i2}L_i, \ldots, \sum_{i=1}^{r} k_{in}L_i \right] \\
&= [L_1, L_2, \ldots, L_r] \cdot \begin{bmatrix} k_{11} & k_{12} & \cdots & k_{1n} \\ k_{21} & k_{22} & \cdots & k_{2n} \\ \vdots & & & \\ k_{r1} & k_{r2} & \cdots & k_{rn} \end{bmatrix}.
\end{aligned}
$$

The factors of this product have rank *at most* r, but since A has rank r, they also have rank *at least* r (Theorem 6.2.2). Hence each factor has rank exactly r.

Let us now select a set of r rows from A. We can effect this by picking *the same rows* from $[L_1, L_2, \ldots, L_r]$. If we denote the result of this last step by $[L'_1, L'_2, \ldots, L'_r]$, then the matrix of the r rows selected from A is given by

$$[L'_1, L'_2, \ldots, L'_r] \cdot \begin{bmatrix} k_{11} & \cdots & k_{1n} \\ \vdots & & \\ k_{r1} & \cdots & k_{rn} \end{bmatrix}.$$

Similarly, let

$$[L''_1, L''_2, \ldots, L''_r] \cdot \begin{bmatrix} k_{11} & \cdots & k_{1n} \\ \vdots & & \\ k_{r1} & \cdots & k_{rn} \end{bmatrix}$$

represent any other set of r rows of A.

Corresponding major determinants of these two sets of r rows may now be formed by picking corresponding sets of r columns from $[k_{ij}]_{(r, n)}$. By the theorem on the determinant of the product of two square matrices, these majors are

$$|L'_1, L'_2, \ldots, L'_r| \cdot \begin{vmatrix} k_{1j_1} & \cdots & k_{1j_r} \\ \vdots & & \\ k_{rj_1} & \cdots & k_{rj_r} \end{vmatrix}, \quad |L'_1, L'_2, \ldots, L'_r| \cdot \begin{vmatrix} k_{1p_1} & \cdots & k_{1p_r} \\ \vdots & & \\ k_{rp_1} & \cdots & k_{rp_r} \end{vmatrix}, \ldots,$$

and

$$|L''_1, L''_2, \ldots, L''_r| \cdot \begin{vmatrix} k_{1j_1} & \cdots & k_{1j_r} \\ \vdots & & \\ k_{rj_1} & \cdots & k_{rj_r} \end{vmatrix}, \quad |L''_1, L''_2, \ldots, L''_r| \cdot \begin{vmatrix} k_{1p_1} & \cdots & k_{1p_r} \\ \vdots & & \\ k_{rp_1} & \cdots & k_{rp_r} \end{vmatrix}, \ldots,$$

the proportionality of which is evident.

A similar argument may be developed to prove the result for the majors of sets of r columns.

The previously given example of this theorem illustrates also the following special case:

Corollary 7.12.2: *In a square matrix A of order n and rank $n - 1$, the cofactors of the elements of any two parallel lines are proportional.*

This result also follows at once from the fact that if $\det A = 0$, the cofactors of the elements of any row provide a solution of the system of equations $AX = 0$, whereas if A has rank $n - 1$, all solutions are multiples of any one nontrivial solution. Similarly, the cofactors of the elements of any column provide a solution of $A^{\mathsf{T}}X = 0$.

An equivalent way to state the corollary is to say that when A has rank $n - 1$, the matrices of cofactors:

$$\begin{bmatrix} A_{i1} & A_{i2} & \cdots & A_{in} \\ A_{k1} & A_{k2} & \cdots & A_{kn} \end{bmatrix} \quad \text{and} \quad \begin{bmatrix} A_{1j} & A_{1p} \\ A_{2j} & A_{2p} \\ \vdots & \vdots \\ A_{nj} & A_{np} \end{bmatrix}$$

have rank < 2 for all choices of i, k, j, p. Hence all their second-order major determinants vanish. This gives us

Corollary 7.12.3: *If A is a square matrix of order n and of rank $n - 1$, we have the identities among the cofactors,*

$$A_{ij} A_{kp} = A_{ip} A_{kj},$$

where the subscripts all range from 1 to n.

In particular, when $j = i$ and $p = k$, we have

$$A_{ii} A_{kk} = A_{ik} A_{ki}.$$

(The result still holds but is uninteresting if the rank of A is $< n - 1$.)

The next theorem is important in our study of quadratic forms, which follows in a later chapter.

We first define a **principal minor of order** r of a square matrix A of order n to be the determinant of the $r \times r$ submatrix that remains when $n - r$ rows and $n - r$ columns *with the same indices* are deleted from A. For example, if A is of order 3, the principal minor of order 2 resulting from the deletion of row 2 and column 2 is

$$\begin{vmatrix} a_{11} & a_{13} \\ a_{31} & a_{33} \end{vmatrix}.$$

Theorem 7.12.4: *If A is a Hermitian matrix of rank $r > 0$, then at least one principal minor of order r of A is not zero.*

Since A has rank r, suppose B is a submatrix of order r such that

$$\det B = \begin{vmatrix} a_{i_1 j_1} & \cdots & a_{i_1 j_r} \\ \vdots & & \\ a_{i_r j_1} & \cdots & a_{i_r j_r} \end{vmatrix} \neq 0.$$

If this is a principal minor, the theorem is proved for A. If it is not a principal minor, the fact that A is Hermitian and the fact that $\det B^* = \overline{\det B}$ imply that also

$$\det B^* = \begin{vmatrix} a_{j_1 i_1} & \cdots & a_{j_1 i_r} \\ \vdots & & \\ a_{j_r i_1} & \cdots & a_{j_r i_r} \end{vmatrix} \neq 0.$$

Now the determinants

$$\begin{vmatrix} a_{i_1 i_1} & \cdots & a_{i_1 i_r} \\ \vdots & & \\ a_{i_r i_1} & \cdots & a_{i_r i_r} \end{vmatrix}, \qquad \begin{vmatrix} a_{i_1 j_1} & \cdots & a_{i_1 j_r} \\ \vdots & & \\ a_{i_r j_1} & \cdots & a_{i_r j_r} \end{vmatrix},$$

and

$$\begin{vmatrix} a_{j_1 i_1} & \cdots & a_{j_1 i_r} \\ \vdots & & \\ a_{j_r i_1} & \cdots & a_{j_r i_r} \end{vmatrix}, \qquad \begin{vmatrix} a_{j_1 j_1} & \cdots & a_{j_1 j_r} \\ \vdots & & \\ a_{j_r j_1} & \cdots & a_{j_r j_r} \end{vmatrix},$$

are corresponding pairs of "majors" as in Theorem 7.12.1, and are, therefore, proportional. Hence, since $\det B^* = \overline{\det B}$,

$$(7.12.1) \quad \begin{vmatrix} a_{i_1 i_1} & \cdots & a_{i_1 i_r} \\ \vdots & & \\ a_{i_r i_1} & \cdots & a_{i_r i_r} \end{vmatrix} \cdot \begin{vmatrix} a_{j_1 j_1} & \cdots & a_{j_1 j_r} \\ \vdots & & \\ a_{j_r j_1} & \cdots & a_{j_r j_r} \end{vmatrix} = \begin{vmatrix} a_{i_1 j_1} & \cdots & a_{i_1 j_r} \\ \vdots & & \\ a_{i_r j_1} & \cdots & a_{i_r j_r} \end{vmatrix}^2 .$$

The outer bars on the right denote absolute value. The principal minors on the left are necessarily real. (Why?) The right member is unequal to zero by hypothesis, so that neither of the two principal minors on the left can be zero and the proof of the theorem is complete.

For example, the symmetric and, therefore, Hermitian matrix

$$\begin{bmatrix} 0 & 0 & 1 & 2 \\ 0 & 0 & 2 & 4 \\ 1 & 2 & 0 & 0 \\ 2 & 4 & 0 & 0 \end{bmatrix}$$

of order 4 and rank 2 has four nonzero principal minors of order 2.

The reader should think through this theorem for the special case $r = 1$ and see that the reasoning still applies.

Corollary 7.12.5: *The nonzero principal minors of order r of a Hermitian matrix A of rank r all have the same sign and their sum is unequal to zero.*

By Theorem 7.12.4, there exists at least one nonzero principal minor of order r. Suppose that the corresponding submatrix contains rows and columns with indices $i_1, i_2, \ldots, i_r$. If the principal minor involving the indices $j_1, j_2, \ldots, j_r$ also is different from zero, then, by (7.12.1), we have

$$\begin{vmatrix} a_{i_1 i_1} & \cdots & a_{i_1 i_r} \\ \vdots & & \\ a_{i_r i_1} & \cdots & a_{i_r i_r} \end{vmatrix} \cdot \begin{vmatrix} a_{j_1 j_1} & \cdots & a_{j_1 j_r} \\ \vdots & & \\ a_{j_r j_1} & \cdots & a_{j_r j_r} \end{vmatrix} > 0,$$

so that the second principal minor has the same sign as the first. Thus all nonzero principal minors of order r have the same sign and hence their sum is not zero.

7.13 Definitions

The expansions (7.3.1) and (7.3.2) of det A in terms of the elements of some row or of some column of A are special cases of an expansion in terms of the elements of a number of rows or of a number of columns of A. This more general result, known as **Laplace's expansion,** is what we shall now establish. We begin by reviewing and extending some definitions used earlier.

If we strike out $n - r$ rows and $n - r$ columns from a square matrix A of order n, the remaining elements form a **square submatrix** of A. The determinant of such a square submatrix is called a **minor determinant** of order r of A or an $(n - r)$th minor determinant of A. The word "determinant" is commonly omitted here. For example,

$$\begin{vmatrix} a_{11} & a_{13} \\ a_{21} & a_{23} \end{vmatrix},$$

is a minor of order 2, or a first minor of the matrix

$$\begin{bmatrix} a_{11} & a_{12} & a_{13} \\ a_{21} & a_{22} & a_{23} \\ a_{31} & a_{32} & a_{33} \end{bmatrix},$$

since it is the determinant of the submatrix obtained when the third row and the second column of the given matrix are deleted. It is useful to define the **minor of order zero** of an arbitrary square matrix A to be 1.

The $n - r$ rows one deletes in order to obtain a minor of order r may be chosen in $C_{n-r}^n = n!/((n - r)!r!)$ ways. The same is true for the $n - r$ columns to be deleted so that there are $(C_{n-r}^n)^2$ such minors. (The symbol C_{n-r}^n denotes "the number of combinations of n things $n - r$ at a time.") For example, a third-order matrix has $(C_{3-2}^3)^2 = 9$ minors of order 2; a matrix of order 4 has $(C_{4-2}^4)^2 = 36$ minors of order 2. Since $C_r^n = C_{n-r}^n$, it follows that there are equally many minors of orders r and $n - r$.

When the rows and columns struck out of a square matrix A have the same indices, the remaining submatrix is located symmetrically with respect to the main diagonal of A, and we call the corresponding minor a **principal minor** of A. The minor

$$\begin{vmatrix} a_{11} & \cdots & a_{1r} \\ \vdots & & \\ a_{r1} & \cdots & a_{rr} \end{vmatrix}$$

is called the **leading principal minor of order** r.

The **zero*th* minor** of A is just det A. A **first minor** M_{ij} of A is the determinant of the submatrix of order $n - 1$ of A obtained by striking out the ith row and the jth column of A. From Theorem 7.4.1 we see then that the cofactor of a_{ij} in det A is given by

$$A_{ij} = (-1)^{i+j}M_{ij}.$$

An $(n - 1)$th minor of A is just an element a_{ij} of A.

If the integers $i_1 i_2 \ldots i_r$ identify the rows and $j_1 j_2 \ldots j_r$ the columns struck out of A in constructing the minor, then we shall denote the minor by the symbol

$$M_{i_1 i_2 \cdots i_r, j_1 j_2 \cdots j_r}$$

or, more briefly, by $M_{(i)(j)}$.

A slight deviation from this notation is used for a first minor, which is commonly denoted by the symbol M_{ij}, as pointed out above. Here the comma has been omitted from between the two subscripts. Note that the initial set of subscripts identifies the rows and the second set identifies the columns *not represented* in the minor.

If $i_1 i_2 \ldots i_r$, $n > r \geq 1$, is any set of r numbers chosen from 1, 2, ..., n and arranged in natural order, and if $k_1 k_2 \ldots k_{n-r}$ is the set of $n - r$ remaining integers, also arranged in natural order, then $i_1 i_2 \ldots i_r$ and $k_1 k_2 \ldots k_{n-r}$ are called **complementary set of indices.** For example, if $n = 8$, then 147 and 23568 are such complementary sets of indices.

If $i_1 i_2 \ldots i_r$ and $k_1 k_2 \ldots k_{n-r}$ are complementary sets of row indices, and if $j_1 j_2 \ldots j_r$ and $p_1 p_2 \ldots p_{n-r}$ are complementary sets of column indices, then the two minor determinants $M_{i_1 i_2 \cdots i_r, j_1 j_2 \cdots j_r}$ and $M_{k_1 k_2 \cdots k_{n-r}, p_1 p_2 \cdots p_{n-r}}$, or, as we shall write them, $M_{(i)(j)}$ and $M_{(k)(p)}$, are called **complementary minors** of A. For example, if $n = 5$, we have as complementary minors of A,

$$M_{145,135} = \begin{vmatrix} a_{22} & a_{24} \\ a_{32} & a_{34} \end{vmatrix} \quad \text{and} \quad M_{23,24} = \begin{vmatrix} a_{11} & a_{13} & a_{15} \\ a_{41} & a_{43} & a_{45} \\ a_{51} & a_{53} & a_{55} \end{vmatrix}.$$

Notice that if we strike out of A the complete rows and columns of A represented in one minor, there remain just the elements of the complementary minor.

We make a final definition. If $M_{(i)(j)}$ and $M_{(k)(p)}$ are complementary minors, then the quantity $A_{(k)(p)}$ defined by

$$A_{(k)(p)} = (-1)^{\Sigma k + \Sigma p} M_{(k)(p)}$$

is called the **algebraic complement** or **cofactor** of $M_{(i)(j)}$. Evidently the cofactor of $M_{(k)(p)}$ is

$$A_{(i)(j)} = (-1)^{\Sigma i + \Sigma j} M_{(i)(j)}.$$

If in particular we let $M_{(k)(p)}$ be the element a_{ij}, then we see that the cofactor of a_{ij} is

$$A_{ij} = (-1)^{i+j} M_{ij},$$

which agrees with the result of Section 7.4.

The reader may show that $(-1)^{\Sigma i + \Sigma j} = (-1)^{\Sigma k + \Sigma p}$, where complementary sets of subscripts are involved as above, so that the cofactors of $M_{(i)(j)}$ and $M_{(k)(p)}$ are respectively $M_{(k)(p)}$ and $M_{(i)(j)}$ or $-M_{(k)(p)}$ and $-M_{(i)(j)}$ depending upon whether $\sum i + \sum j$ is even or odd.

Examination of the definition of $A_{(k)(p)}$ shows that the exponent on (-1) is simply the sum of the indices of the rows and columns *present* in $M_{(i)(j)}$. This observation makes writing down the cofactor of $M_{(i)(j)}$ a simple matter. For example, consider the matrix

$$\begin{bmatrix} 1 & 0 & 2 & 0 \\ 2 & 0 & 1 & 0 \\ 0 & 4 & 0 & 3 \\ 0 & 3 & 0 & 4 \end{bmatrix}.$$

The cofactor of $\begin{vmatrix} 1 & 0 \\ 2 & 0 \end{vmatrix}$, the upper left principal minor of order 2, is

$$(-1)^{(1+2)+(1+2)} \begin{vmatrix} 0 & 3 \\ 0 & 4 \end{vmatrix}$$

since $\begin{vmatrix} 0 & 3 \\ 0 & 4 \end{vmatrix}$ is the complementary minor of $\begin{vmatrix} 1 & 0 \\ 2 & 0 \end{vmatrix}$ in this case.

7.14   The Laplace Expansion

The sequence of theorems to follow will provide a derivation of the Laplace expansion of a determinant. To motivate these theorems and also to provide further illustrations of the foregoing definitions, we shall give an example of how Laplace's expansion operates before launching into the proof.

Laplace's theorem says that, if we select any r rows of A, form all possible r-rowed minors from these r rows, multiply each of these minors by its cofactor, and then add the results, we obtain det A.

In the matrix given in Section 7.13, let the first two rows be the r rows from which to form minors. We denote this fact by a dashed line:

$$\begin{bmatrix} 1 & 0 & 2 & 0 \\ 2 & 0 & 1 & 0 \\ \hdashline 0 & 4 & 0 & 3 \\ 0 & 3 & 0 & 4 \end{bmatrix}.$$

Since the formation of a minor from the first two rows requires the selection of two columns from the four, we will have $C_2^4 = 6$ minors possible. We shall select the columns systematically thus: 1 and 2, 1 and 3, 1 and 4, 2 and 3, 2 and 4, 3 and 4. Then applying Laplace's theorem as stated above, and the verbal rule for finding the sign of the cofactor given in Section 7.13, we have

$$
\begin{vmatrix} 1 & 0 & 2 & 0 \\ 2 & 0 & 1 & 0 \\ 0 & 4 & 0 & 3 \\ 0 & 3 & 0 & 4 \end{vmatrix} = \begin{vmatrix} 1 & 0 \\ 2 & 0 \end{vmatrix} \cdot (-1)^{1+2+1+2} \begin{vmatrix} 0 & 3 \\ 0 & 4 \end{vmatrix}
$$

$$
+ \begin{vmatrix} 1 & 2 \\ 2 & 1 \end{vmatrix} \cdot (-1)^{1+2+1+3} \begin{vmatrix} 4 & 3 \\ 3 & 4 \end{vmatrix} + \begin{vmatrix} 1 & 0 \\ 2 & 0 \end{vmatrix} \cdot (-1)^{1+2+1+4} \begin{vmatrix} 4 & 0 \\ 3 & 0 \end{vmatrix}
$$

$$
+ \begin{vmatrix} 0 & 2 \\ 0 & 1 \end{vmatrix} \cdot (-1)^{1+2+2+3} \begin{vmatrix} 0 & 3 \\ 0 & 4 \end{vmatrix} + \begin{vmatrix} 0 & 0 \\ 0 & 0 \end{vmatrix} \cdot (-1)^{1+2+2+4} \begin{vmatrix} 0 & 0 \\ 0 & 0 \end{vmatrix}
$$

$$
+ \begin{vmatrix} 2 & 0 \\ 1 & 0 \end{vmatrix} \cdot (-1)^{1+2+3+4} \begin{vmatrix} 0 & 4 \\ 0 & 3 \end{vmatrix} = 21.
$$

The full expansion was written here for illustrative purposes but, clearly, to obtain the end result, we would only have needed to write the single nonzero term of the sum. Someone with only a little experience in these matters would have noticed this at once and would have been able to write the result, " 21," by inspection. The example shows that the Laplace expansion is useful for evaluating certain special numerical determinants, but it is even more useful in deriving other theorems.

We shall now establish several preliminary theorems by way of preparing for the main result. In the proofs of these theorems we have frequent occasion to refer to the interchange of adjacent integers in a permutation. Such an interchange will be called briefly an **adjacent transposition.**

Theorem 7.14.1: *If $i_1 i_2 \ldots i_r$ and $k_1 k_2 \ldots k_{n-r}$ are complementary sets of indices, then the permutation $i_1 i_2 \cdots i_r k_1 k_2 \cdots k_{n-r}$ may be obtained from the natural order $1, 2, \ldots, n$ by $(\sum i) - r(r+1)/2$ adjacent transpositions.*

(The given permutation may therefore be reduced to the natural order by the same number of adjacent transpositions.)

For if we start with the natural order $1, 2, \ldots, n$, it takes $i_1 - 1$ adjacent transpositions to put i_1 into the first position and then, because $i_2 > i_1$, it takes $i_2 - 2$ more to put i_2 in the second position, etc. Finally, it takes $i_r - r$ adjacent transpositions to put i_r in the rth position. The natural order of the remaining integers has not been disturbed, so that no further transpositions are needed to construct the desired permutation. Thus we use altogether $(i_1 - 1) + (i_2 - 2) + \cdots + (i_r - r) = (\sum i) - r(r+1)/2$ adjacent transpositions since $1 + 2 + \cdots + r = r(r+1)/2$.

Theorem 7.14.2: *If* $i_1 i_2 \ldots i_r$ *and* $k_1 k_2 \ldots k_{n-r}$ *are complementary sets of row indices, and if* $j_1 j_2 \ldots j_r$ *and* $p_1 p_2 \ldots p_{n-r}$ *are complementary sets of column indices, then*

$$(-1)^{\Sigma i + \Sigma j} a_{i_1 j_1} \cdots a_{i_r j_r} a_{k_1 p_1} \cdots a_{k_{n-r} p_{n-r}}$$

is a term in the expansion of det A.

Because (i), (k) and (j), (p) here are complementary sets of row and column subscripts, every row and every column of A is represented once and only once in the above product. Hence we need only show that the sign is correct. To decide what sign would be attached to the product

$$a_{i_1 j_1} \cdots a_{i_r j_r} a_{k_1 p_1} \cdots a_{k_{n-r} p_{n-r}}$$

in the expansion of det A, we note that by the preceding theorem, it will take $(\sum i) - r(r+1)/2$ adjacent transpositions of these factors to restore their row subscripts to the natural order $1, 2, \ldots, n$. Then we can determine the sign to be attached to this product by investigating the permutation represented by the column subscripts. There are $(\sum j) - r(r+1)/2$ transpositions needed to obtain the *initially given* set of column subscripts from the natural order. Since the unscrambling of the row subscripts then imposes $(\sum i) - r(r+1)/2$ *additional* transpositions on the column subscripts, the sign of the term should be determined by

$$(-1)^{\Sigma i - r(r+1)/2 + \Sigma j - r(r+1)/2} = (-1)^{\Sigma i + \Sigma j},$$

which is just what was used above, so that the theorem is proved.

Theorem 7.14.3: *With complementary sets of row and column indices as in Theorem 7.14.2, every term in the product*

$$M_{(k)(p)} A_{(i)(j)} \equiv M_{(k)(p)} (-1)^{\Sigma i + \Sigma j} M_{(i)(j)}$$

of a minor of A and its cofactor is a term in the expansion of det A.

Since every row and column subscript will appear once and only once in each such term, it is again only the sign of such a term that needs to be investigated. A general term of this product may be written thus:

$$(-1)^{\Sigma i + \Sigma j} [(-1)^{\mu_j'} a_{i_1 j_1'} a_{i_2 j_2'} \cdots a_{i_r j_r'}] \cdot [(-1)^{\mu_{p'}} a_{k_1 p_1'} a_{k_2 p_2'} \cdots a_{k_{n-r} p_{n-r}'}],$$

where the bracketed expressions are terms from the row expansions of the two minors and where μ_j' is the number of adjacent transpositions required to produce the permutation $j_1' j_2' \ldots j_r'$ from $j_1 j_2 \ldots j_r$, with a similar definition for $\mu_{p'}'$.

Now to determine the sign that the product

$$a_{i_1 j_1'} a_{i_2 j_2'} \cdots a_{i_r j_r'} a_{k_1 p_1'} a_{k_2 p_2'} \cdots a_{k_{n-r} p_{n-r}'}$$

would have in the expansion of det A, we put the i's and the k's back into the natural order $1, 2, \ldots, n$ by transposing factors. This may be accomplished by $(\sum i) - r(r + 1)/2$ adjacent transpositions, and it induces further scrambling of the column scripts. Checking through the matter chronologically, we see that the final arrangement of the column scripts may be obtained by a total of

$$\left((\textstyle\sum j) - \frac{r(r + 1)}{2} \right) + \mu_j' + \mu_p' + \left((\textstyle\sum i) - \frac{r(r + 1)}{2} \right)$$

transpositions from the natural order, and hence the sign attached to the product in question should be determined by

$$(-1)^{\Sigma i + \Sigma j + \mu_j' + \mu_p'},$$

as is indeed the case in the signed product above. Therefore each term in the product is actually a term from det A.

Theorem 7.14.4: *If $i_1 i_2 \ldots i_r$ is any fixed set of r row indices in the natural order, and if $j_1 j_2 \ldots j_r$ runs over every set of r column indices, each set also in the natural order, then we have*

$$\det A = \sum_{(j)} M_{(k)(p)} A_{(i)(j)},$$

where the i's and k's and j's and p's are complementary sets of row and column indices respectively, and where the summation extends over all the sets of r column indices described above.

This is called the **Laplace expansion** of det A in terms of minors of order r formed from the given r rows. A similar result holds, of course, for columns.

To prove the result, we note first that the expansion of every term in this summation will contain only terms of det A by the preceding theorem. Since no two sets of subscripts $j_1 j_2 \ldots j_r$ are the same, it follows next that these terms of det A are all different. Finally, the two minors appearing in each summand have respectively $r!$ and $(n - r)!$ terms, and there are $C_r^n = n!/ (r!(n - r)!)$ ways to select $j_1 j_2 \ldots j_r$ from $1, 2, \ldots, n$. Hence there are altogether

$$r!\,(n-r)!\,\frac{n!}{r!\,(n-r)!} = n!$$

terms of det A in the complete expansion of the above sum, just as there are $n!$ terms in det A. These three facts taken together prove the theorem.

As a second illustration of Laplace's expansion we give the following example involving literal elements rather than integers. The student should check the signs of the various terms.

We make the definition

$$p_{ij} = \begin{vmatrix} x_i & x_j \\ y_i & y_j \end{vmatrix}, \qquad i, j = 1, 2, 3, 4.$$

Then we have $p_{ii} \equiv 0$ and $p_{ij} \equiv -p_{ji}$ in any case. Among the nonidentically vanishing p_{ij}'s, there exists a simple identity. It is found by applying Laplace's expansion to the following vanishing determinant:

$$0 = \begin{vmatrix} x_1 & x_2 & x_3 & x_4 \\ y_1 & y_2 & y_3 & y_4 \\ \hdashline x_1 & x_2 & x_3 & x_4 \\ y_1 & y_2 & y_3 & y_4 \end{vmatrix} = \begin{vmatrix} x_1 & x_2 \\ y_1 & y_2 \end{vmatrix} \cdot \begin{vmatrix} x_3 & x_4 \\ y_3 & y_4 \end{vmatrix} - \begin{vmatrix} x_1 & x_3 \\ y_1 & y_3 \end{vmatrix} \cdot \begin{vmatrix} x_2 & x_4 \\ y_2 & y_4 \end{vmatrix}$$

$$+ \begin{vmatrix} x_1 & x_4 \\ y_1 & y_4 \end{vmatrix} \cdot \begin{vmatrix} x_2 & x_3 \\ y_2 & y_3 \end{vmatrix} + \begin{vmatrix} x_2 & x_3 \\ y_2 & y_3 \end{vmatrix} \cdot \begin{vmatrix} x_1 & x_4 \\ y_1 & y_4 \end{vmatrix}$$

$$- \begin{vmatrix} x_2 & x_4 \\ y_2 & y_4 \end{vmatrix} \cdot \begin{vmatrix} x_1 & x_3 \\ y_1 & y_3 \end{vmatrix} + \begin{vmatrix} x_3 & x_4 \\ y_3 & y_4 \end{vmatrix} \cdot \begin{vmatrix} x_1 & x_2 \\ y_1 & y_2 \end{vmatrix},$$

or

$$p_{12}p_{34} - p_{13}p_{24} + p_{14}p_{23} + p_{23}p_{14} - p_{24}p_{13} + p_{34}p_{12} \equiv 0,$$

which reduces without difficulty to

$$(7.14.1) \qquad p_{12}p_{34} + p_{13}p_{42} + p_{14}p_{23} \equiv 0,$$

when we use the fact that $p_{ij} \equiv -p_{ji}$. This identity is important in the study of line geometry, where these p's are used as coordinates (the Plücker coordinates of a line).

7.15 The Determinant of a Product of Two Square Matrices

As an application of the Laplace expansion, we prove this basic result, which was proved in another way in Section 7.6.

Theorem 7.15.1: *The determinant of the product of two square matrices of order n is equal to the product of their determinants.*

Let the matrices be A and B and observe that

$$(7.15.1) \qquad \begin{bmatrix} A & 0 \\ -I & B \end{bmatrix} \begin{bmatrix} I & B \\ 0 & I \end{bmatrix} = \begin{bmatrix} A & AB \\ -I & 0 \end{bmatrix}.$$

Now since, for all C and D of order n,

$$\begin{bmatrix} I & C \\ 0 & I \end{bmatrix} \begin{bmatrix} I & D \\ 0 & I \end{bmatrix} = \begin{bmatrix} I & D + C \\ 0 & I \end{bmatrix},$$

we have

$$\begin{bmatrix} I & B \\ 0 & I \end{bmatrix} = \prod_{i,j=1}^{n} \begin{bmatrix} I & \tilde{B}_{ij} \\ 0 & I \end{bmatrix},$$

where $\tilde{B}_{ij}$ denotes a matrix of order n whose ij-entry is b_{ij} and whose other entries are all zeros. Each factor in this product represents an elementary transformation that does not alter the determinant. Hence the determinant of the product on the left in (7.15.1) is just the determinant of the first factor. Applying the Laplace expansion to (7.15.1), we have therefore

$$(-1)^{n(n+1)} \det A \det B = (-1)^{n(2n+1)} \det AB \det(-I),$$

so

$$\det A \det B = \det AB.$$

7.16   The Adjoint Matrix

If A is a matrix of order n and A_{ij} is the cofactor of a_{ij} in det A, then the matrix

$$(7.16.1) \qquad \mathscr{A} = [A_{ij}]^{\mathsf{T}} = \begin{bmatrix} A_{11} & A_{21} & \cdots & A_{n1} \\ A_{12} & A_{22} & \cdots & A_{n2} \\ \vdots & & & \\ A_{1n} & A_{2n} & \cdots & A_{nn} \end{bmatrix} = \text{adj } A$$

is what we have called the **adjoint matrix of** A and det $\mathscr{A}$ is called the **adjoint determinant of** A. We have already proved that

(7.16.2) $$\det \mathscr{A} = (\det A)^{n-1}.$$

This result is generalized in

Theorem 7.16.1: *If $\mathscr{M}_{(p)(k)}$ is a minor of order r of $\mathscr{A}$ and if $M_{(i)(j)}$ is a minor of order $n - r$ of A such that the i's and k's as well as the j's and p's are complementary sets of subscripts, then*

(7.16.3) $$\mathscr{M}_{(p)(k)} = (-1)^{\Sigma i + \Sigma j}(\det A)^{r-1}M_{(i)(j)} = (\det A)^{r-1}A_{(i)(j)}.$$

(Note that in $\mathscr{A}$, and hence in $\mathscr{M}$ also, the p's identify rows and the k's identify columns.)

The theorem is proved by considering the following product:

$$
\begin{bmatrix}
a_{i_1 j_1} & \cdots & a_{i_1 j_r} & a_{i_1 p_1} & \cdots & a_{i_1 p_{n-r}} \\
\vdots & & & \vdots & & \\
a_{i_r j_1} & \cdots & a_{i_r j_r} & a_{i_r p_1} & \cdots & a_{i_r p_{n-r}} \\
\hline
a_{k_1 j_1} & \cdots & a_{k_1 j_r} & a_{k_1 p_1} & \cdots & a_{k_1 p_{n-r}} \\
\vdots & & & \vdots & & \\
a_{k_{n-r} j_1} & \cdots & a_{k_{n-r} j_r} & a_{k_{n-r} p_1} & \cdots & a_{k_{n-r} p_{n-r}}
\end{bmatrix}
$$

$$
\cdot
\begin{bmatrix}
A_{i_1 j_1} & \cdots & A_{i_r j_1} & 0 & 0 & \cdots & 0 \\
\vdots & & & \vdots & & & \\
A_{i_1 j_r} & \cdots & A_{i_r j_r} & 0 & 0 & \cdots & 0 \\
\hline
A_{i_1 p_1} & \cdots & A_{i_r p_1} & 1 & 0 & \cdots & 0 \\
\vdots & & & \vdots & & & \\
A_{i_1 p_{n-r}} & \cdots & A_{i_r p_{n-r}} & 0 & 0 & \cdots & 1
\end{bmatrix}
$$

$$
=
\begin{bmatrix}
\det A & 0 & \cdots & 0 & a_{i_1 p_1} & \cdots & a_{i_1 p_{n-r}} \\
0 & \det A & \cdots & 0 & \vdots & & \\
0 & 0 & \cdots & \det A & a_{i_r p_1} & \cdots & a_{i_r p_{n-r}} \\
\hline
0 & 0 & \cdots & 0 & a_{k_1 p_1} & \cdots & a_{k_1 p_{n-r}} \\
\vdots & & & & \vdots & & \\
0 & 0 & \cdots & 0 & a_{k_{n-r} p_1} & \cdots & a_{k_{n-r} p_{n-r}}
\end{bmatrix}.
$$

Now, taking determinants of both sides and employing Exercise 56 of Section 7.5, Theorem 7.14.1, and Theorems 7.15.1 and 7.14.4, we obtain the identity

$$(-1)^{\Sigma i + \Sigma j} \det A \cdot \mathscr{M}_{(p)(k)} = (\det A)^r \cdot M_{(i)(j)}.$$

Canceling det A from both sides and multiplying both sides by $(-1)^{\Sigma i + \Sigma j}$, we have, finally,

$$\mathcal{M}_{(p)(k)} = (-1)^{\Sigma i + \Sigma j}(\det A)^{r-1} M_{(i)(j)} = (\det A)^{r-1} A_{(i)(j)},$$

as stated.

Certain special cases are frequently useful. If we put $r = n - 1$ and let $\mathcal{A}_{pk}$ denote the *cofactor* of the element in the pth row and kth column of $\mathcal{A}$, then

$$\mathcal{A}_{pk} = (\det A)^{n-2} a_{kp}.$$

Putting $r = 2$, we find

$$\begin{vmatrix} A_{i_1 j_1} & A_{i_2 j_1} \\ A_{i_1 j_2} & A_{i_2 j_2} \end{vmatrix} = \mathcal{M}_{p_1 \cdots p_{n-2}, \, k_1 \cdots k_{n-2}} = A_{i_1 i_2, \, j_1 j_2} \det A.$$

The reader should think through the special cases $r = n$ and $r = 1$ with some care.

7.17   The Row-and-Column Expansion

The theorem of this section has, among others, applications to statistics and to the study of quadric surfaces.

In the statement of the theorem, $A_{1i, \, 1j}$ represents the cofactor of the second-order minor

$$\begin{vmatrix} a_{11} & a_{1j} \\ a_{i1} & a_{ij} \end{vmatrix}$$

of the matrix $A = [a_{ij}]_n$, the notation being that explained in Section 7.13.

Theorem 7.17.1: $\det A = a_{11} A_{11} - \sum_{i, \, j=2}^{n} a_{i1} a_{1j} A_{1i, \, 1j}.$

This is called the *expansion of* det A *in terms of the elements of the first row and the elements of the first column.*

By (7.3.1), the expansion in terms of the elements of the first row is

(7.17.1) $$\det A = a_{11} A_{11} + \sum_{j=2}^{n} a_{1j} A_{1j}.$$

Now

$$A_{1j} = (-1)^{1+j} M_{1j},$$

and expanding M_{1j} in terms of elements of its first column, we see that

$$M_{1j} = \sum_{i=2}^{n} a_{i1}(-1)^{(i-1)+1}M_{1i,\,1j},$$

so that

$$A_{1j} = \sum_{i=2}^{n} a_{i1}(-1)^{i+j+1}M_{1i,\,1j}$$

$$= -\sum_{i=2}^{n} a_{i1}((-1)^{1+i+1+j}M_{1i,\,1j}),$$

or

$$A_{1j} = -\sum_{i=2}^{n} a_{i1}A_{1i,\,1j}.$$

Substitution for A_{1j} in (7.17.1) now gives the desired result.

7.18 The Diagonal Expansion of the Determinant of a Matrix

It is easy to verify by direct computation that

$$\begin{vmatrix} d_1 & b_{12} \\ b_{21} & d_2 \end{vmatrix} = \begin{vmatrix} 0 & b_{12} \\ b_{21} & 0 \end{vmatrix} + d_1 d_2,$$

and that

$$\begin{vmatrix} d_1 & b_{12} & b_{13} \\ b_{21} & d_2 & b_{23} \\ b_{31} & b_{32} & d_3 \end{vmatrix} = \begin{vmatrix} 0 & b_{12} & b_{13} \\ b_{21} & 0 & b_{23} \\ b_{31} & b_{32} & 0 \end{vmatrix} + d_1 \begin{vmatrix} 0 & b_{23} \\ b_{32} & 0 \end{vmatrix}$$

$$+ d_2 \begin{vmatrix} 0 & b_{13} \\ b_{31} & 0 \end{vmatrix} + d_3 \begin{vmatrix} 0 & b_{12} \\ b_{21} & 0 \end{vmatrix} + d_1 d_2 d_3.$$

These two examples suggest

Theorem 7.18.1: Let

$$A = B + D$$

denote an arbitrary $n \times n$ matrix, where

$$B = \begin{bmatrix} 0 & b_{12} & b_{13} & \cdots & b_{1,n-1} & b_{1n} \\ b_{21} & 0 & b_{23} & \cdots & b_{2,n-1} & b_{2n} \\ \vdots & & & & & \\ b_{n1} & b_{n2} & b_{n3} & \cdots & b_{n,n-1} & 0 \end{bmatrix}$$

and

$$D = \begin{bmatrix} d_1 & 0 & \cdots & 0 \\ 0 & d_2 & \cdots & 0 \\ \vdots & & & \\ 0 & 0 & \cdots & d_n \end{bmatrix}.$$

Then

$$\det A = \det B + \sum_{i=1}^{n} d_i B_{ii} + \sum_{1 \le i < j \le n} d_i d_j B_{ij,\,ij}$$
$$+ \sum_{1 \le i < j < k \le n} d_i d_j d_k B_{ijk,\,ijk} + \cdots + d_1 d_2 \cdots d_n.$$

This is called the **diagonal expansion** of det A.

To prove this, let us first write

$$A = [B_1 + D_1, B_2 + D_2, \ldots, B_n + D_n],$$

where B_j denotes the jth column of B and D_j denotes the jth column of D. By applying Theorem 7.3.3 to the first column of A, we obtain

$$\det A = \det [B_1, B_2 + D_2, \ldots, B_n + D_n]$$
$$+ \det [D_1, B_2 + D_2, \ldots, B_n + D_n].$$

Then applying Theorem 7.3.3 to the second columns of these matrices, and to the third columns of the resulting matrices, etc., we obtain finally,

$$\det A = \sum \det [C_1, C_2, \ldots, C_n],$$

where each C_j is either B_j or D_j, so that there are altogether 2^n determinants in the sum.

Now by putting each $C_j = B_j$, we obtain det B as one term of the sum. By putting $C_k = D_k$, $C_j = B_j$ for $j \ne k$, we obtain all terms of the form $d_k B_{kk}$, etc. Finally, by putting each $C_j = D_j$, we obtain the term $d_1 d_2 \cdots d_n$. Note that no product of $n-1$ d_j's will appear because every principal minor of order 1 of B is 0.

7.19 Exercises

1. Expand in minors of the first two rows

$$\begin{vmatrix} 2 & 0 & 1 & 0 & 1 \\ 1 & 0 & 1 & 0 & 2 \\ \hline 1 & 2 & 0 & 1 & 1 \\ 1 & 1 & 0 & 2 & 1 \\ 1 & 1 & 0 & 1 & 1 \end{vmatrix}.$$

2. Expand in minors of the last two columns

$$\begin{vmatrix} 0 & -2 & 4 & 1 & 1 \\ 1 & 2 & 1 & 1 & 2 \\ 1 & 2 & 1 & 2 & 2 \\ 0 & 1 & -2 & 3 & 3 \\ 1 & 0 & 1 & 4 & 4 \end{vmatrix}.$$

3. Use the Laplace expansion to prove that if $A_1, A_2, \ldots, A_n$ are square matrices which are used to build up a square matrix A,

$$A = \begin{bmatrix} A_1 & 0 & \cdots & 0 \\ 0 & A_2 & \cdots & 0 \\ \vdots & & & \\ 0 & 0 & \cdots & A_n \end{bmatrix},$$

where the 0's stand for blocks of elements all of which are zero, then

$$\det A = \det A_1 \det A_2 \cdots \det A_n.$$

Matrices of this type are useful in theoretical physics, numerical analysis, and in other applications.

4. Show that if A_{ij} is the cofactor of a_{ij} in det A, and if $n > 1$, then

$$\begin{vmatrix} 0 & u_1 & u_2 & \cdots & u_n \\ u_1 & a_{11} & a_{12} & \cdots & a_{1n} \\ u_2 & a_{21} & a_{22} & \cdots & a_{2n} \\ \vdots & & & & \\ u_n & a_{n1} & a_{n2} & \cdots & a_{nn} \end{vmatrix} = -\sum_{i,j=1}^{n} A_{ij} u_i u_j.$$

5. Put

$$q_{ij} = \begin{vmatrix} u_i & v_i \\ u_j & v_j \end{vmatrix} = \begin{vmatrix} u_i & u_j \\ v_i & v_j \end{vmatrix}$$

and show that, if $A = [a_{ij}]_n$, then

$$\begin{vmatrix} a_{11} & a_{12} & \cdots & a_{1n} & u_1 & v_1 \\ a_{21} & a_{22} & \cdots & a_{2n} & u_2 & v_2 \\ \vdots & & & & & \\ a_{n1} & a_{n2} & \cdots & a_{nn} & u_n & v_n \\ u_1 & u_2 & \cdots & u_n & 0 & 0 \\ v_1 & v_2 & \cdots & v_n & 0 & 0 \end{vmatrix} = \sum_{\substack{1 \le i < j \le n \\ 1 \le k < m \le n}} q_{ij} q_{km} A_{ij,\, km}.$$

The matrices whose determinants are found in Exercises 4 and 5 are called **bordered matrices**. Such matrices are useful in a variety of applications.

6. Show that

$$\begin{vmatrix} a_1 & b_1 & c_1 & d_1 & 0 & 0 \\ a_2 & b_2 & c_2 & d_2 & 0 & 0 \\ a_3 & b_3 & c_3 & d_3 & 0 & 0 \\ 0 & 0 & 0 & d_1 & e_1 & f_1 \\ 0 & 0 & 0 & d_2 & e_2 & f_2 \\ 0 & 0 & 0 & d_3 & e_3 & f_3 \end{vmatrix} = \begin{vmatrix} a_1 & b_1 & c_1 & d_1 & 0 & 0 \\ a_2 & b_2 & c_2 & d_2 & 0 & 0 \\ a_3 & b_3 & c_3 & d_3 & 0 & 0 \\ -a_1 & -b_1 & -c_1 & 0 & e_1 & f_1 \\ -a_2 & -b_2 & -c_2 & 0 & e_2 & f_2 \\ -a_3 & -b_3 & -c_3 & 0 & e_3 & f_3 \end{vmatrix}.$$

Now define

$$(abc) = \det[abc] = \det \begin{bmatrix} a_1 & b_1 & c_1 \\ a_2 & b_2 & c_2 \\ a_3 & b_3 & c_3 \end{bmatrix}$$

and similarly in other cases. Then show that

$$(abc)(def) = (dbc)(aef) + (adc)(bef) + (abd)(cef).$$

To what does this identity reduce if $[def]$ is an identity matrix? If all corresponding entries d_i and f_i are identical? [Compare (7.14.1) with this last case.]

7. Prove Theorem 7.6.4 by applying the Laplace expansion to

$$\begin{vmatrix} A_{(m,\, n)} & 0_{(m,\, m)} \\ -I_n & B_{(n,\, m)} \end{vmatrix}.$$

8. Restate Theorem 7.17.1 in such a way that it will apply to any row and column with the same index.

9. Show that, if $A = [a_{ij}]_n$ and if A_{ij} is the cofactor of a_{ij} in det A, then

$$\det [a_{ij} + x]_n = \det A + x \sum_{i,\,j=1}^{n} A_{ij}.$$

10. How many identities of the form given in Theorem 7.16.1 are there when $n = 2$? $n = 3$? $n = 4$?

11. Evaluate

$$\det \begin{bmatrix} \lambda & -1 & 2 \\ 1 & \lambda & 4 \\ -2 & -4 & \lambda \end{bmatrix}$$

by the method of Theorem 7.18.1.

12. Let

$$A = \begin{bmatrix} a & -b \\ b & a \end{bmatrix}, \qquad B = \begin{bmatrix} \alpha & -\beta \\ \beta & \alpha \end{bmatrix}.$$

Use Theorem 7.6.1 to prove that

$$(a^2 + b^2)(\alpha^2 + \beta^2) = (a\alpha - b\beta)^2 + (a\beta + b\alpha)^2.$$

What theorem about absolute values of complex numbers now follows?

13. Let A be an $m \times n$ matrix and let the ith row of $\tilde{A}$ be defined by

$$\tilde{A}^{(i)} = \lambda_{i1}A^{(1)} + \lambda_{i2}A^{(2)} + \cdots + \lambda_{im}A^{(m)}, \qquad i = 1, 2, \ldots, m.$$

Give a formula for a minor of order r of $\tilde{A}$ in terms of minors of order r of A.

14. Let A be an $n \times n$ matrix and let

$$B = \begin{bmatrix} a_{i_1 j_1} & \cdots & a_{i_1 j_n} \\ \vdots & & \\ a_{i_n j_1} & \cdots & a_{i_n j_n} \end{bmatrix},$$

where $i_1, i_2, \ldots, i_n$ and $j_1, j_2, \ldots, j_n$ are permutations of $1, 2, \ldots, n$. Express the elements of B^{-1} in terms of minors of A and det A.

15. Let $\tilde{A} = [M_{ij}]_{n \times n}$, where the M_{ij} are the first minors of $A_{n \times n}$. Prove that det $\tilde{A} = \det \mathscr{A}$. Prove also that if $\tilde{M}_{(i)(j)}$ is a minor of $\tilde{A}$, then

$$\tilde{M}_{(i)(j)} = (-1)^{\Sigma i + \Sigma j} \mathscr{A}_{(i)(j)}.$$

16. Show that, if det $A = 1$, then A is the adjoint matrix of $\mathscr{A}$.

17. Show that, if A, B, D are of order n and if $AB = D$, then $\mathscr{D} = \mathscr{B}\mathscr{A}$; that is, the adjoint of the product of two matrices is the product of their adjoints in reverse order.

18. Prove that, if $\mathscr{A}_{ji}$ denotes a cofactor of the adjoint matrix $\mathscr{A}$ of A, then

$$\mathscr{A}_{ji} = a_{ij} \cdot (\det A)^{n-2}.$$

19. Prove that if A is nonsingular,

$$(A^{-1})_{(p)(k)} = \frac{1}{\det A} M_{(i)(j)}.$$

Illustrate with examples for $n = 3$, $r = 1, 2$.

20. Given that D is square, of the same order as A, and nonsingular, use the fact that

$$\det \begin{bmatrix} A & B \\ C & D \end{bmatrix} = \det \begin{bmatrix} A & B \\ C & D \end{bmatrix} \begin{bmatrix} I & 0 \\ -D^{-1}C & I \end{bmatrix}$$

to prove that

$$\det \begin{bmatrix} A & B \\ C & D \end{bmatrix} = \det (AD - BD^{-1}CD),$$

so that, if D^{-1} exists and C and D commute,

$$\det \begin{bmatrix} A & B \\ C & D \end{bmatrix} = \det (AD - BC).$$

21. By deriving a similar result for the matrix

$$\begin{bmatrix} A & X \\ X^{\mathsf{T}} & a \end{bmatrix},$$

where a is a nonzero scalar, invent a numerical procedure for computing the determinant of a symmetric matrix.

22. Apply Exercise 21 to

$$\det \begin{bmatrix} A & X \\ X^{\mathsf{T}} & 1 \end{bmatrix} \quad \text{and} \quad \det \begin{bmatrix} A & X \\ X^{\mathsf{T}} & -1 \end{bmatrix};$$

then evaluate the difference of these two determinants in two ways and thus derive the identity

$$\det A = \tfrac{1}{2}(\det (A + XX^{\mathsf{T}}) + \det (A - XX^{\mathsf{T}})).$$

23. Obtain a formula for

$$\det \begin{bmatrix} 0 & x & \cdots & x & x \\ x & 0 & \cdots & x & x \\ \vdots & & & & \\ x & x & \cdots & 0 & x \\ x & x & \cdots & x & 0 \end{bmatrix}_{n \times n}$$

by induction. Then use the diagonal expansion to obtain a formula for

$$\det \begin{bmatrix} a & x & \cdots & x & x \\ x & a & \cdots & x & x \\ \vdots & & & & \\ x & x & \cdots & a & x \\ x & x & \cdots & x & a \end{bmatrix}_{n \times n}.$$

Show that the inverse of this last matrix, when it exists, is of the same form. Compare with Exercise 15, Section 7.5.

Linear Transformations

8.1 Mappings

One of the most important ideas in mathematics is that of a **mapping** or **function**, namely, a pairing of the members a of a set $\mathscr{A}$ with members b of a set $\mathscr{B}$ such that when the first member a of a pair (a, b) is given, the second member b is uniquely determined. Such a set of pairs is called a **mapping from $\mathscr{A}$ to $\mathscr{B}$** (Figure 8-1) or a **function on $\mathscr{A}$ with values in $\mathscr{B}$.**

For example, the equation

$$(8.1.1) \qquad \begin{bmatrix} y_1 \\ y_2 \end{bmatrix} = \begin{bmatrix} 1 & 0 & -1 \\ -1 & 1 & 0 \end{bmatrix} \begin{bmatrix} x_1 \\ x_2 \\ x_3 \end{bmatrix}$$

assigns to each vector X in $\mathscr{E}^3$ a unique corresponding vector Y in $\mathscr{E}^2$.

The member b of the pair (a, b) is called the **image** of a under the mapping. The member a of the pair is called the **counterimage** of b. If we denote the mapping by the symbol f, we write

$$a \xrightarrow{f} b$$

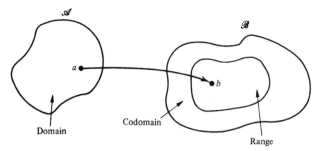

FIGURE 8-1. Mapping from $\mathscr{A}$ to $\mathscr{B}$.

to mean, "a is mapped by f onto b," or, more familiarly, we write

$$b = f(a).$$

If $b = f(a)$, the pair (a, b) is said to **belong to the mapping** f.

The set $\mathscr{A}$ is called the **domain** of the mapping and the set $\mathscr{B}$ is called its **codomain**. The subset of $\mathscr{B}$ which contains all the images of the elements of $\mathscr{A}$ and no other elements is called the **range** $\mathscr{R}$ of the mapping. For example, if $\mathscr{M}$ is the set of all square matrices with elements in a number field $\mathscr{F}$, the determinant function assigns to each matrix M of $\mathscr{M}$ a unique scalar b of $\mathscr{F}$:

$$b = \det M.$$

Here we have a mapping with domain $\mathscr{M}$ and range $\mathscr{F}$.

The formula

(8.1.2)
$$\begin{bmatrix} y_1 \\ y_2 \end{bmatrix} = \begin{bmatrix} 2 & 1 \\ 1 & 2 \end{bmatrix} \begin{bmatrix} x_1 \\ x_2 \end{bmatrix}$$

defines a mapping from the set of vectors in $\mathscr{E}^2$ to the set of vectors in $\mathscr{E}^2$, which illustrates the fact that the sets $\mathscr{A}$ and $\mathscr{B}$ of the definition need not be distinct. Since the coefficient matrix is nonsingular, we can rewrite this equation in the form

(8.1.3)
$$\begin{bmatrix} x_1 \\ x_2 \end{bmatrix} = \frac{1}{3} \begin{bmatrix} 2 & -1 \\ -1 & 2 \end{bmatrix} \begin{bmatrix} y_1 \\ y_2 \end{bmatrix}.$$

In this case, given an image vector $\begin{bmatrix} y_1 \\ y_2 \end{bmatrix}$, we can compute its unique counterimage $\begin{bmatrix} x_1 \\ x_2 \end{bmatrix}$.

Whenever, as in the preceding example, each element b of the range of a mapping is the image of precisely one member of $\mathscr{A}$, the mapping is called a **one-to-one mapping**: Only one element a of $\mathscr{A}$ maps onto an element b of the range. On the other hand, the determinant function is a **many-to-one mapping**, that is, one in which more than one element a of the domain can have the same image b.

All three of the mappings used as examples so far have the property that every element of the codomain is the image of some element of the domain. In such a case, we say that $\mathscr{A}$ is mapped **onto** $\mathscr{B}$ and we call the mapping an **onto mapping**. (Thus "onto" is used as an adjective by mathematicians, not just as a preposition.) If the range of the mapping is not all of $\mathscr{B}$ but is rather a proper subset of $\mathscr{B}$, we say that $\mathscr{A}$ is mapped **strictly into** $\mathscr{B}$ and that the mapping is a **strictly-into mapping**. If we do not know or do not wish to specify whether a mapping from $\mathscr{A}$ to $\mathscr{B}$ is onto or strictly-into, we call it an **into mapping**. Thus *into* can denote either *strictly-into* or *onto*. If we replace $\mathscr{B}$ by the range $\mathscr{R}$, the resulting mapping from $\mathscr{A}$ to $\mathscr{R}$ is, of course, onto.

An example of a strictly-into mapping is given by the determinant function applied to the set $\mathscr{A}$ of $n \times n$ matrices A with entries such that $|a_{ij}| \leq 1$, the set $\mathscr{B}$ being the set of all real numbers. Since the absolute value of every term in the expansion of $\det A$ is ≤ 1 and since $|\det A|$ is equal to or less than the sum of the absolute values of the terms of the expansion, we have at once that

$$|\det A| \leq 1 + 1 + \cdots + 1 = n!$$

Hence the mapping is a strictly-into mapping.

We have noted that the mapping from $\mathscr{E}^2$ to $\mathscr{E}^2$ defined by (8.1.2) is a one-to-one, onto mapping and that, via the rule (8.1.3), we can find for each Y in $\mathscr{E}^2$ the unique counterimage X in $\mathscr{E}^2$ with which Y is paired in the mapping. By contrast, in the case of the mapping from the set of square real matrices with $|a_{ij}| \leq 1$ to the field of real numbers via the determinant function, each member of the range arises from many members of the domain so that the mapping is not one-to-one and there is no unique counterimage. Also, since not all real numbers are the determinants of matrices in the set, some elements of the codomain have no counterimage at all. These examples illustrate the next observation.

When we have a *one-to-one* mapping f of a set $\mathscr{A}$ *onto* a set $\mathscr{B}$, and only when both these conditions are satisfied, each member of $\mathscr{B}$ has one and only one counterimage in $\mathscr{A}$. The mapping from $\mathscr{B}$ to $\mathscr{A}$ which pairs each element b of $\mathscr{B}$ with its unique counterimage in $\mathscr{A}$ is called the **inverse** of f and is denoted by f^{-1}. The pair (b, a) belongs to the mapping f^{-1} if and only if the pair (a, b) belongs to f. We write in this case

$$b \xrightarrow{f^{-1}} a \qquad \text{or} \qquad a = f^{-1}(b).$$

Note that if f^{-1} exists, f^{-1} is also one-to-one and onto and the inverse of f^{-1} is f. When f^{-1} exists, the function f is said to be **invertible**.

An example is the mapping from the vectors of $\mathscr{E}^2$ to the vectors of $\mathscr{E}^2$ defined by

$$\begin{bmatrix} y_1 \\ y_2 \end{bmatrix} = \begin{bmatrix} 3 & 2 \\ 1 & 1 \end{bmatrix} \begin{bmatrix} x_1 \\ x_2 \end{bmatrix} + \begin{bmatrix} 1 \\ -1 \end{bmatrix},$$

which has the inverse defined by

$$\begin{bmatrix} x_1 \\ x_2 \end{bmatrix} = \begin{bmatrix} 1 & -2 \\ -1 & 3 \end{bmatrix} \begin{bmatrix} y_1 \\ y_2 \end{bmatrix} + \begin{bmatrix} -3 \\ 4 \end{bmatrix}.$$

The reader should show how this last equation was obtained and should check its correctness by substitution.

8.2 Linear Mappings

Consider now a mapping f from a vector space $\mathscr{V}$ to a vector space $\mathscr{W}$, where the scalars in both cases belong to the same field, $\mathscr{F}$. Such a mapping f is said to be a **linear mapping** if and only if, for all vectors V_1 and V_2 in $\mathscr{V}$ and for all scalars α in $\mathscr{F}$,

(8.2.1) $$f(V_1 + V_2) = f(V_1) + f(V_2)$$

and

(8.2.2) $$f(\alpha V_1) = \alpha f(V_1).$$

Thus the mapping from the set $\mathscr{M}$ of all real $n \times n$ matrices to the set $\mathscr{R}$ of all real numbers defined by

$$\det A = b,$$

where A belongs to $\mathscr{M}$ and b belongs to $\mathscr{R}$ is not linear, because ordinarily

$$\det(A + B) \neq \det A + \det B.$$

The reader should provide an illustrative example. Note also that $\det(\alpha A) = \alpha^n \det A$, not $\alpha \det A$, so that, in this case, both conditions are violated.

On the other hand, the mapping from $\mathscr{M}$ to $\mathscr{M}$ defined by

$$f(A) = A^{\mathsf{T}}$$

is linear because

$$(A + B)^{\mathsf{T}} = A^{\mathsf{T}} + B^{\mathsf{T}}$$

and

$$(\alpha A)^{\mathsf{T}} = \alpha A^{\mathsf{T}}.$$

The mapping from $\mathscr{F}^n$ to $\mathscr{F}^m$ defined by

$$Y_{m \times 1} = A_{m \times n} X_{n \times 1},$$

where A is a matrix over $\mathscr{F}$, is also linear because, for all X_1 and X_2 in $\mathscr{F}^n$ and for all scalars α,

$$A(X_1 + X_2) = AX_1 + AX_2$$

and

$$A(\alpha X_1) = \alpha(AX_1).$$

This example is really a little theorem. Its converse is more impressive:

Theorem 8.2.1: *Every linear mapping from $\mathscr{F}^n$ to $\mathscr{F}^m$ may be defined by an equation*

(8.2.3) $$Y_{m \times 1} = A_{m \times n} X_{n \times 1}.$$

To prove the theorem, denote the mapping by f and let

(8.2.4) $$A_j = f(E_j),$$

where the E_j's are the elementary vectors of $\mathscr{F}^n$. Then, since for every X in $\mathscr{F}^n$,

$$X = x_1 E_1 + x_2 E_2 + \cdots + x_n E_n,$$

by repeated use of (8.2.1) we have

$$f(X) = f(x_1 E_1) + f(x_2 E_2) + \cdots + f(x_n E_n),$$

so that, by applying (8.2.2) to each term, we obtain

$$f(X) = x_1 f(E_1) + x_2 f(E_2) + \cdots + x_n f(E_n).$$

Next, by (8.2.4), we have

$$f(X) = x_1 A_1 + x_2 A_2 + \cdots + x_n A_n.$$

If we now put $f(X) = Y$, we can rewrite this last equation as

(8.2.5) $$Y = [A_1, A_2, \ldots, A_n]X,$$

which has the form stated in the theorem.

The proof not only establishes the required result but also proves

Theorem 8.2.2: *A linear mapping from $\mathscr{F}^n$ to $\mathscr{F}^m$ is determined uniquely by the images of the elementary vectors $E_1, E_2, \ldots, E_n$ of $\mathscr{F}^n$, the images of these vectors constituting respectively the columns $A_1, A_2, \ldots, A_n$ of the matrix A of the mapping. Moreover, if $Y = AX$ and $Y = BX$ effect the same mapping, then $A = B$.*

For example, the linear mapping from $\mathscr{E}^3$ to $\mathscr{E}^3$ such that

$$f(E_1) = \begin{bmatrix} 1 \\ -1 \\ 2 \end{bmatrix}, \qquad f(E_2) = \begin{bmatrix} 2 \\ 1 \\ 3 \end{bmatrix}, \qquad f(E_3) = \begin{bmatrix} -1 \\ 0 \\ 1 \end{bmatrix}$$

may be represented by the matrix equation

$$Y = \begin{bmatrix} 1 & 2 & -1 \\ -1 & 1 & 0 \\ 2 & 3 & 1 \end{bmatrix} X.$$

Linear mappings of vector spaces are commonly called **linear operators** or **linear transformations** or **linear homogeneous transformations**. A linear operator mapping $\mathscr{F}^n$ into $\mathscr{F}^n$ is called a **linear operator on $\mathscr{F}^n$**.

The preceding theorem leads at once to the following result:

Theorem 8.2.3: *Given any n linearly independent vectors $A_1, A_2, \ldots, A_n$ of $\mathscr{F}^n$, there exists a unique invertible linear operator on $\mathscr{F}^n$ by which their images are the elementary vectors.*

The operator $Y = A^{-1}X$, where $A = [A_1, A_2, \ldots, A_n]$, is the unique operator that effects the required mapping.

Theorems 8.2.2 and 8.2.3 combine to yield

Theorem 8.2.4: *Given any two sets of n linearly independent vectors of $\mathscr{F}^n$, there exists a unique invertible linear operator on $\mathscr{F}^n$ which maps the vectors of one set respectively onto those of the other.*

Let $A_1, A_2, \ldots, A_n$ and $B_1, B_2, \ldots, B_n$ be the two sets of independent vectors. Then the operator $Y = A^{-1}X$ maps the A_j's onto the E_j's respectively, and the operator $Z = BY$ maps the E_j's onto the B_j's respectively. Hence the operator $Z = BA^{-1}X$ maps the A_j's directly onto the B_j's respectively. Since A and B are nonsingular, so is BA^{-1}, and the operator $Z = BA^{-1}X$ is invertible. Uniqueness follows from Theorem 8.2.2.

To illustrate the last theorem in $\mathscr{E}^2$, we find the linear operator which maps the vectors $[1, 1]^{\mathsf{T}}$, $[1, -1]^{\mathsf{T}}$ respectively onto the vectors $[1, 2]^{\mathsf{T}}$, $[2, -1]^{\mathsf{T}}$.

We first let $X = AY$ denote the operator which maps $[1, 0]^{\mathsf{T}}$ and $[0, 1]^{\mathsf{T}}$ respectively onto $[1, 1]^{\mathsf{T}}$, $[1, -1]^{\mathsf{T}}$. Then, by Theorem 8.2.2,

$$
X = \begin{bmatrix} 1 & 1 \\ 1 & -1 \end{bmatrix} Y \quad \text{or} \quad Y = \begin{bmatrix} \frac{1}{2} & \frac{1}{2} \\ \frac{1}{2} & -\frac{1}{2} \end{bmatrix} X.
$$

The operator $Z = BY$ which maps $[1, 0]^{\mathsf{T}}$ and $[0, 1]^{\mathsf{T}}$ respectively onto $[1, 2]^{\mathsf{T}}$, $[2, -1]^{\mathsf{T}}$ is, again by Theorem 8.2.2,

$$
Z = \begin{bmatrix} 1 & 2 \\ 2 & -1 \end{bmatrix} Y.
$$

Combining the two operators, we obtain

$$
Z = \begin{bmatrix} 1 & 2 \\ 2 & -1 \end{bmatrix} \cdot \begin{bmatrix} \frac{1}{2} & \frac{1}{2} \\ \frac{1}{2} & -\frac{1}{2} \end{bmatrix} X = \begin{bmatrix} \frac{3}{2} & -\frac{1}{2} \\ \frac{1}{2} & \frac{3}{2} \end{bmatrix} X,
$$

which maps $[1, 1]^{\mathsf{T}}$ and $[1, -1]^{\mathsf{T}}$ respectively onto $[1, 2]^{\mathsf{T}}$ and $[2, -1]^{\mathsf{T}}$.

Alternatively, we can write

$$
\begin{bmatrix} z_1 \\ z_2 \end{bmatrix} = \begin{bmatrix} a & b \\ c & d \end{bmatrix} \cdot \begin{bmatrix} x_1 \\ x_2 \end{bmatrix},
$$

from which, by substituting pairs of corresponding vectors, we obtain

$$
\begin{bmatrix} 1 \\ 2 \end{bmatrix} = \begin{bmatrix} a & b \\ c & d \end{bmatrix} \cdot \begin{bmatrix} 1 \\ 1 \end{bmatrix}, \qquad \begin{bmatrix} 2 \\ -1 \end{bmatrix} = \begin{bmatrix} a & b \\ c & d \end{bmatrix} \cdot \begin{bmatrix} 1 \\ -1 \end{bmatrix}.
$$

These are equivalent to the two pairs of scalar equations

$$
a + b = 1, \qquad c + d = 2,
$$
$$
a - b = 2, \qquad c - d = -1.
$$

Solving these equations for a, b, c, d, we have, again,

$$
Z = \begin{bmatrix} \frac{3}{2} & -\frac{1}{2} \\ \frac{1}{2} & \frac{3}{2} \end{bmatrix} X.
$$

8.3 Some Properties of Linear Operators on Vector Spaces

The **image of a subset** $\mathscr{S}$ of $\mathscr{F}^n$ by a linear operator is defined to be the set of all images of vectors of $\mathscr{S}$. We have then

Theorem 8.3.1: *Given a linear operator $Y = AX$ mapping $\mathscr{F}^n$ into $\mathscr{F}^m$, the image of a subspace $\mathscr{V}$ of $\mathscr{F}^n$ by this operator is a subspace of $\mathscr{F}^m$. This image space is spanned by the images of the vectors of any basis for $\mathscr{V}$.*

Let the subspace $\mathscr{V}$ have the basis $V_1, V_2, \ldots, V_k$. Then the vectors X of $\mathscr{V}$ are the set of all linear combinations of the form

$$X = \alpha_1 V_1 + \alpha_2 V_2 + \cdots + \alpha_k V_k.$$

The image of each such vector is given by

$$AX = \alpha_1 AV_1 + \alpha_2 AV_2 + \cdots + \alpha_k AV_k,$$

which is a linear combination of the vectors $AV_1, AV_2, \ldots, AV_k$. That is, the vectors of $\mathscr{V}$ have as images the vectors of the subspace spanned by AV_1, $AV_2, \ldots, AV_k$, and the theorem is proved.

Since, in the preceding proof, the image space is the set of linear combinations of k vectors, which may or may not be linearly independent, we have

Corollary 8.3.2: *The dimension of the image of a subspace of dimension k of $\mathscr{F}^n$ by a linear operator is not greater than k.*

Since the equation

$$Y = A_{m \times n} X$$

can be rewritten in the form

$$(8.3.1) \qquad Y = x_1 A_1 + x_2 A_2 + \cdots + x_n A_n,$$

where the A_j's are the columns of A, we have

Theorem 8.3.3: *The range of a linear operator $Y = AX$ which maps $\mathscr{F}^n$ into $\mathscr{F}^m$ is the subspace of $\mathscr{F}^m$ spanned by the columns of A. The dimension of this subspace is $r(A)$.*

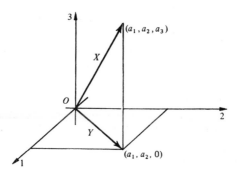

FIGURE 8-2. A Projection.

As an illustration of the three preceding results, note that the linear operator

$$Y = \begin{bmatrix} 1 & 0 & 0 \\ 0 & 1 & 0 \\ 0 & 0 & 0 \end{bmatrix} X$$

maps the vectors of $\mathscr{E}^3$ onto the two-dimensional subspace spanned by

$$\begin{bmatrix} 1 \\ 0 \\ 0 \end{bmatrix}, \quad \begin{bmatrix} 0 \\ 1 \\ 0 \end{bmatrix}, \quad \begin{bmatrix} 0 \\ 0 \\ 0 \end{bmatrix},$$

that is, onto the vectors of the 1,2-plane (Figure 8-2). The image of an arbitrary vector $[a_1, a_2, a_3]^{\mathsf{T}}$ by this operator is just its projection $[a_1, a_2, 0]^{\mathsf{T}}$ on the 1,2-plane (Figure 8-2). This is a many-to-one mapping. The reader should describe geometrically the set of all counterimages of a given vector $[a_1, a_2, 0]^{\mathsf{T}}$.

As the figure of this example suggests, *we may interpret the operator as a mapping of points (the terminal points of vectors) as well as one of vectors.*

Theorem 8.3.4: If $Y = AX$ represents a linear operator mapping $\mathscr{F}^n$ into $\mathscr{F}^m$ and if $r(A) = n$, then the dimension of the image of any subspace $\mathscr{V}$ of $\mathscr{F}^n$ is the same as that of $\mathscr{V}$.

Let $\mathscr{V}$ be of dimension $k \le n$ and let $V_1, V_2, \ldots, V_k$ be a basis for $\mathscr{V}$ so that $AV_1, AV_2, \ldots, AV_k$ span the image space. Any equation

$$\alpha_1 AV_1 + \alpha_2 AV_2 + \cdots + \alpha_k AV_k = 0,$$

where the α's are scalars, implies

$$A(\alpha_1 V_1 + \alpha_2 V_2 + \cdots + \alpha_k V_k) = 0.$$

Since A has rank n, the only solution of $AX = 0$ is the trivial solution. That is,

$$\alpha_1 V_1 + \alpha_2 V_2 + \cdots + \alpha_k V_k = 0.$$

Since the V_j's are independent, this implies that all the α_j's are 0, which in turn proves that the vectors $AV_1, AV_2, \ldots, AV_k$ are linearly independent, so the image space also has dimension k.

A linear operator mapping $\mathscr{F}^n$ into $\mathscr{F}^n$ is called simply a **linear operator on $\mathscr{F}^n$**. Such an operator is said to be **nonsingular** if and only if the dimension of the image of *every* subspace is the same as that of the subspace itself. Otherwise it is said to be **singular**. Note that the preceding example illustrates a singular operator since the image of $\mathscr{E}^3$ itself is only two dimensional. Are there any subspaces of $\mathscr{E}^3$ whose dimension is preserved by the projection in this example?

Theorem 8.3.5: *A linear operator $Y = AX$ on $\mathscr{F}^n$ is nonsingular if and only if A is nonsingular.*

If $Y = AX$ is nonsingular, $\mathscr{F}^n$ must have $\mathscr{F}^n$, not some subspace of $\mathscr{F}^n$, as its image. Now the columns of A span the image of $\mathscr{F}^n$, which is $\mathscr{F}^n$ itself. The columns of A must therefore be linearly independent, so A must be nonsingular.

Conversely, let A be nonsingular. Then A has rank n, so by Theorem 8.3.4, the dimension of the image of every k-dimensional subspace is k, and the theorem follows.

When a linear operator on $\mathscr{F}^n$ represented by $Y = AX$ is nonsingular, so that A^{-1} exists, then the inverse mapping is given by

$$X = A^{-1}Y.$$

The set of vectors which are mapped onto the zero vector by the operator $Y = AX$, that is, the set of solutions of $AX = 0$, is called the **null space** or **kernel of the operator**. The dimension of the null space of an operator is called the **nullity** of the operator and is governed by the inequalities developed in Section 6.5. From preceding definitions and (7.8.1), we have at once

Theorem 8.3.6: *A linear operator on $\mathscr{F}^n$ is nonsingular if and only if its null space consists of the zero vector alone.*

8.4   Exercises

1. Show that the range of the mapping from $\mathscr{E}^3$ to $\mathscr{E}^3$ defined by

$$\begin{bmatrix} y_1 \\ y_2 \\ y_3 \end{bmatrix} = \begin{bmatrix} 2 & -1 & 0 \\ -1 & 2 & 1 \\ -1 & -1 & -1 \end{bmatrix} \begin{bmatrix} x_1 \\ x_2 \\ x_3 \end{bmatrix}$$

is the set of vectors for which $y_1 + y_2 + y_3 = 0$. Illustrate with a figure.

2. What is the range of the mapping from $\mathscr{E}^3$ to $\mathscr{E}^3$ defined by

$$\begin{bmatrix} y_1 \\ y_2 \\ y_3 \end{bmatrix} = \begin{bmatrix} x_1 \\ x_2 \\ x_3 \end{bmatrix} + \begin{bmatrix} h_1 \\ h_2 \\ h_3 \end{bmatrix},$$

where h_1, h_2, and h_3 are fixed? Describe the mapping geometrically. Is it one-to-one? Is it onto? Is it linear? Illustrate some specific cases with figures.

3. A linear operator mapping $\mathscr{E}^3$ into $\mathscr{E}^2$ is to map

$$\begin{bmatrix} 1 \\ 0 \\ 0 \end{bmatrix} \text{ onto } \begin{bmatrix} 1 \\ 0 \end{bmatrix}, \quad \begin{bmatrix} 0 \\ 1 \\ 0 \end{bmatrix} \text{ onto } \begin{bmatrix} 0 \\ 1 \end{bmatrix}, \quad \text{and} \quad \begin{bmatrix} 0 \\ 0 \\ 1 \end{bmatrix} \text{ onto } \begin{bmatrix} 2 \\ 2 \end{bmatrix}.$$

Represent the operator in matrix form.

4. Find the range of the mapping from $\mathscr{E}^3$ to $\mathscr{E}^4$ defined by

$$\begin{bmatrix} y_1 \\ y_2 \\ y_3 \\ y_4 \end{bmatrix} = \begin{bmatrix} 1 & -1 & 2 \\ 2 & 0 & 1 \\ 3 & -1 & 3 \\ 1 & 1 & -1 \end{bmatrix} \begin{bmatrix} x_1 \\ x_2 \\ x_3 \end{bmatrix}.$$

5. Find the inverse of the linear operator on $\mathscr{E}^3$ defined by

$$\begin{bmatrix} y_1 \\ y_2 \\ y_3 \end{bmatrix} = \begin{bmatrix} \dfrac{1}{\sqrt{2}} & 0 & \dfrac{1}{\sqrt{2}} \\ -\dfrac{1}{\sqrt{2}} & 0 & \dfrac{1}{\sqrt{2}} \\ 0 & 1 & 0 \end{bmatrix} \begin{bmatrix} x_1 \\ x_2 \\ x_3 \end{bmatrix}.$$

6. Show that the image Y of a vector X by the operator of Exercise 5 has the same length as X.

*7. Show that a linear operator on $\mathscr{F}^n$ is nonsingular if and only if the images of $E_1, E_2, \ldots, E_n$ are linearly independent vectors.

*8. Show that a linear operator mapping $\mathscr{F}^n$ into $\mathscr{F}^m$ maps every linear combination of vectors of $\mathscr{F}^n$ onto the same linear combination of the images of these vectors.

*9. Show that the images, by a linear operator mapping $\mathscr{F}^n$ into $\mathscr{F}^m$, of dependent vectors are also dependent vectors. In particular, the image of 0 is 0.

10. A linear operator on $\mathscr{E}^3$ maps

$$\begin{bmatrix} 1 \\ 0 \\ 0 \end{bmatrix} \text{ onto } \begin{bmatrix} 0 \\ 2 \\ 1 \end{bmatrix}, \quad \begin{bmatrix} 1 \\ 1 \\ 0 \end{bmatrix} \text{ onto } \begin{bmatrix} 1 \\ 0 \\ 2 \end{bmatrix}, \quad \text{and} \quad \begin{bmatrix} 1 \\ 1 \\ 1 \end{bmatrix} \text{ onto } \begin{bmatrix} 2 \\ 1 \\ 0 \end{bmatrix}.$$

Onto what vectors does it map E_2 and E_3? (You don't need to find the matrix of the operator to answer the question.)

11. (a) A linear operator on $\mathscr{E}^2$ maps

$$\begin{bmatrix} 1 \\ 1 \end{bmatrix} \text{ onto } \begin{bmatrix} -2 \\ 0 \end{bmatrix} \quad \text{and} \quad \begin{bmatrix} 1 \\ -1 \end{bmatrix} \text{ onto } \begin{bmatrix} 0 \\ 2 \end{bmatrix}.$$

Find the operator and its inverse.

(b) Do the same for the operator that maps

$$\begin{bmatrix} 2 \\ 1 \end{bmatrix} \text{ onto } \begin{bmatrix} 1 \\ 2 \end{bmatrix} \quad \text{and} \quad \begin{bmatrix} 1 \\ 2 \end{bmatrix} \text{ onto } \begin{bmatrix} 2 \\ 1 \end{bmatrix}.$$

12. Find the linear operator on $\mathscr{E}^3$ which maps the vectors

$$\begin{bmatrix} -1 \\ 1 \\ 1 \end{bmatrix}, \quad \begin{bmatrix} 1 \\ -1 \\ 1 \end{bmatrix}, \quad \begin{bmatrix} 1 \\ 1 \\ -1 \end{bmatrix}$$

onto E_1, E_2, E_3 respectively.

13. Determine the linear operator $Y = AX$ which maps the vectors E_1, E_2, E_3 of $\mathscr{E}^3$ onto the vectors $[1, 0]^T, [0, 1]^T, [1, 1]^T$ of $\mathscr{E}^2$ respectively. Show that every vector of $\mathscr{E}^2$ is the image by this operator of infinitely many vectors of $\mathscr{E}^3$. If each of the vectors of $\mathscr{E}^3$ which have the same image $[a, b]^T$ is represented as a geometrical vector $\overrightarrow{OP}$, what is the locus of the endpoints P?

14. Determine the linear operator $Y = AX$ which maps the vectors $[1, 2]^T$ and $[2, 1]^T$ of $\mathscr{E}^2$ respectively onto the vectors $[7, 0, -8]^T$ and $[5, 3, 2]^T$ of $\mathscr{E}^3$. Onto what vectors does it map the elementary vectors of $\mathscr{E}^2$? Does any vector in $\mathscr{E}^2$ map onto $[1, 9, 22]^T$? Onto $[1, 0, 0]^T$? Which vectors in $\mathscr{E}^3$ are images by this operator of vectors in $\mathscr{E}^2$?

15. Show that the complete solution of the system of equations

$$y_1 + 2y_2 - 5y_3 + y_4 = 0$$
$$2y_1 - y_2 - 5y_3 - 3y_4 = 0$$

is, in fact, the range of this linear operator on $\mathscr{E}^4$:

$$Y = \begin{bmatrix} 3 & 1 & 2 & 1 \\ 5 & -1 & 2 & 3 \\ 2 & 0 & 1 & 1 \\ -3 & 1 & -1 & -2 \end{bmatrix} X.$$

Explain.

16. Show that any k-dimensional subspace $(k < n)$ of $\mathscr{F}^n$ may be mapped onto the k-dimensional subspace with equations $y_1 = y_2 = \cdots = y_{n-k} = 0$ by a nonsingular linear operator.

17. Show that any k-dimensional subspace of $\mathscr{F}^n$ can be mapped onto any other such subspace by a nonsingular operator.

18. The basis vectors of a k-dimensional subspace of $\mathscr{F}^n$ may be used as the columns of an $n \times k$ matrix of rank k. To what results concerning matrices are Exercises 16 and 17 therefore equivalent?

19. Given two sets of vectors $A_1, A_2, \ldots, A_n$ and $B_1, B_2, \ldots, B_n$ of $\mathscr{F}^n$, under what conditions does there exist a linear operator which maps the A's onto the B's respectively? When is this operator nonsingular?

20. Given an r-dimensional subspace $\mathscr{V}$ and a k-dimensional subspace $\mathscr{W}$ of $\mathscr{F}^n$, show that if $k \le r$ there exists a linear operator which maps $\mathscr{V}$ onto $\mathscr{W}$. Show that if $k = r < n$, the operator may be either singular or nonsingular but if $k < r$, it is necessarily singular.

21. Interpret each of the following linear operators geometrically in $\mathscr{E}^3$:
    (a) $Y = 0 \cdot X$, where 0 denotes the zero matrix.
    (b) $Y = IX$.
    (c) $Y = \lambda X, \lambda \ne 0$.
    (d) $Y = D[\lambda_1, \lambda_2, \lambda_3]X$.

22. Show that each of the following vector functions is a linear operator, and represent it in the form $Y = AX$. In each case, the domain of the function is $\mathscr{F}^n$.
    (a) $Y(X) = (V^{\mathsf{T}}X)V$, where V is a fixed vector of $\mathscr{F}^n$.
    (b) $Y(X) = [x_1, \frac{1}{2}(x_1 + x_2), \frac{1}{3}(x_1 + x_2 + x_3), \ldots, \frac{1}{n}(x_1 + x_2 + \cdots + x_n)]^{\mathsf{T}}$.
    (c) $Y(X) = [x_{j_1}, x_{j_2}, \ldots, x_{j_n}]^{\mathsf{T}}$, where $j_1, j_2, \ldots, j_n$ is a permutation of $1, 2, \ldots, n$.
    (d) $Y(X) = [x_1, x_2, \ldots, x_k, 0, 0, \ldots, 0]^{\mathsf{T}}$, where $k < n$.
    (e) $Y(X) = BX + CX$, where B and C are $m \times n$ matrices over $\mathscr{F}$.
    (f) $Y(X) = B(CX)$, where B is an $m \times r$ matrix and C is an $r \times n$ matrix over $\mathscr{F}$.
    (g) $Y(X) = \alpha X$, where α is a scalar in $\mathscr{F}$.
    (h) $Y(X) = \alpha(BX)$, where α is a scalar in $\mathscr{F}$ and B is an $m \times n$ matrix over $\mathscr{F}$.

23. Show that each of the following vector functions is *not* a linear operator:
    (a) $f(X) = AX + B$, where A is an $n \times n$ matrix over $\mathscr{F}$ and B is a fixed nonzero n-vector over $\mathscr{F}$.
    (b) $g(X) = [\sqrt{x_1^2}, \sqrt{x_2^2}, \ldots, \sqrt{x_n^2}]^{\mathsf{T}}$, where $\mathscr{F}$ is the real field.
    (c) $h(X) = (A^{\mathsf{T}}X)X$, where A is a fixed n-vector.
    Interpret (a), (b), (c) geometrically in $\mathscr{E}^2$.

*24. Show that, if $Y = AX$ represents a nonsingular linear operator, then the y's are linearly independent functions of the x's.

25. The **cross product** Z of two vectors X and Y in $\mathscr{E}^3$ is defined by

$$Z = X \times Y = [(x_2 y_3 - x_3 y_2), (x_3 y_1 - x_1 y_3), (x_1 y_2 - x_2 y_1)]^\mathsf{T}.$$

Show that the operator $Y = A \times X$, where A is a fixed 3-vector, is a singular linear operator. Then show that the **scalar triple product** $(A \times B)^\mathsf{T} C$ is given by $\det[A, B, C]$.

26. Let $AX = B$ be a consistent system, and let $X = CY$ be a nonsingular linear operator. Show that the system $(AC)Y = B$ is also consistent, and that the solutions of the two systems are in one-to-one correspondence.

27. Let $AX = B$ be a consistent system, and let P and Q denote nonsingular matrices such that PAQ is in the normal form as defined in Chapter 5. Show that, if P and Q are known, the given system of equations may be solved readily with the aid of the substitution $X = QY$. What are the arbitrary parameters in the solution?

28. Let $AX = B$ be a consistent system of equations, and put $X = CY$. Here A is $m \times n$, C is $n \times p$, and Y is $p \times 1$. If the system $(AC)Y = B$ is consistent, then each solution Y leads to a solution $X = CY$ of the first system. Show that, vice versa, each solution X of the first system leads to a solution of the second if and only if the columns of C span the solution space of the first system. Comment on the significance of the nonsingularity of C in Exercise 26.

8.5 Linear Transformations of Coordinates

Up to now, we have interpreted an equation $Y = A_{n \times n} X$, A over $\mathscr{F}$, as a linear operator on $\mathscr{F}^n$, that is, as a mapping that operates on vectors of $\mathscr{F}^n$ to produce vectors of $\mathscr{F}^n$ or, equivalently, as a mapping that operates on points of $\mathscr{F}^n$ to produce points of $\mathscr{F}^n$. In this interpretation, the basis or reference system remains fixed.

If $V_1, V_2, \ldots, V_k$ constitute a basis for a vector space $\mathscr{V}$ and if

$$X = \alpha_1 V_1 + \alpha_2 V_2 + \cdots + \alpha_k V_k,$$

then $\alpha_1, \alpha_2, \ldots, \alpha_k$ are called the **coordinates** of X with respect to the given basis. Moreover, once the basis is chosen, these coordinates are uniquely determined for every X in $\mathscr{V}$. In particular, $E_1, E_2, \ldots, E_n$ are called the **standard basis** or the **natural basis** for $\mathscr{F}^n$ and since, for each X in $\mathscr{F}^n$,

$$\begin{bmatrix} x_1 \\ x_2 \\ \vdots \\ x_n \end{bmatrix} = x_1 E_1 + x_2 E_2 + \cdots + x_n E_n,$$

the components of X are its coordinates with respect to the standard basis.

Note that, given the origin, a given ordered basis for $\mathscr{E}^n$ may be used to determine a coordinate system for $\mathscr{E}^n$ in the following way: The j-axis has the direction of the jth basis vector, which identifies the positive sense and the unit of length on this axis. For example, the vectors E_1, E_2, E_3 generate the familiar, orthogonal coordinate system in $\mathscr{E}^3$ (Figure 8-3).

Equally well, we can choose any three linearly independent vectors A_1, A_2, A_3 of $\mathscr{E}^3$ as basis vectors and represent X as a linear combination,

$$(8.5.1) \qquad X = y_1 A_1 + y_2 A_2 + y_3 A_3,$$

which may also be written

$$(8.5.2) \qquad X = AY,$$

where $A = [A_1, A_2, A_3]$. In this equation, the columns of A are the new basis vectors and, given any vector X in $\mathscr{E}^3$, the equivalent equation

$$(8.5.3) \qquad Y = A^{-1}X$$

yields explicitly the coordinates y_1, y_2, y_3, of X with respect to the basis A_1, A_2, A_3. We have computed these coordinates in earlier chapters by solving the system of equations represented by (8.5.1) by sweepout.

This interpretation generalizes at once to $\mathscr{F}^n$. Let X be any vector of $\mathscr{F}^n$ referred to the standard basis. Let $A_1, A_2, \ldots, A_n$ be an arbitrary basis for

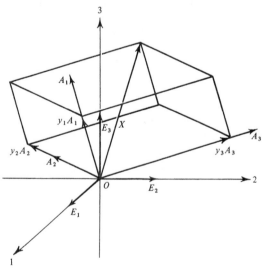

FIGURE 8-3. $X = x_1 E_1 + x_2 E_2 + x_3 E_3 = y_1 A_1 + y_2 A_2 + y_3 A_3$.

$\mathscr{F}^n$, referred to the standard reference system, so that there exist $y_1, y_2, \ldots, y_n$ such that

(8.5.4)
$$X = y_1 A_1 + y_2 A_2 + \cdots + y_n A_n$$

or

(8.5.5)
$$X = AY,$$

where

$$A = [A_1, A_2, \ldots, A_n],$$

that is, where *the columns of A are the old coordinates of the vectors constituting the new basis.* Then, for each X, the equation

(8.5.6)
$$Y = A^{-1}X$$

yields the coordinates $y_1, y_2, \ldots, y_n$ of X with respect to the new basis.

We thus have an alternative geometrical interpretation of a linear transformation $Y = AX$: We regard vectors and points of $\mathscr{F}^n$ as remaining fixed while the reference system is being changed. The transformation equations now define the relationship between the coordinates of any vector X with respect to the standard basis and its coordinates with respect to the new basis. A linear transformation interpreted this way is called a **linear transformation of coordinates** or a **change of basis** in $\mathscr{F}^n$.

To illustrate, consider the linear transformation of coordinates in $\mathscr{E}^2$ whose equations are

$$X = \begin{bmatrix} 5 & -1 \\ -3 & 6 \end{bmatrix} Y, \qquad Y = \frac{1}{27} \begin{bmatrix} 6 & 1 \\ 3 & 5 \end{bmatrix} X.$$

The new basis vectors are

$$A_1 = \begin{bmatrix} 5 \\ -3 \end{bmatrix} \qquad \text{and} \qquad A_2 = \begin{bmatrix} -1 \\ 6 \end{bmatrix}$$

in the standard reference system. The corresponding Y's are, of course, $\begin{bmatrix} 1 \\ 0 \end{bmatrix}$ and $\begin{bmatrix} 0 \\ 1 \end{bmatrix}$, respectively. Now the lengths of A_1 and A_2 determine the units of length on the y_1-axis and the y_2-axis respectively, since these are the unit vectors of the new system. These units are respectively $\sqrt{34}$ and $\sqrt{37}$ times

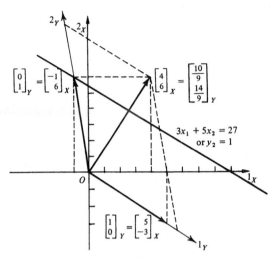

FIGURE 8-4. Transformation of Coordinates.

the original unit of length (Figure 8-4). Consider the vector $\begin{bmatrix} 4 \\ 6 \end{bmatrix}$ in the standard basis. We have, for the coordinates with respect to the new basis,

$$Y = \frac{1}{27}\begin{bmatrix} 6 & 1 \\ 3 & 5 \end{bmatrix}\begin{bmatrix} 4 \\ 6 \end{bmatrix} = \begin{bmatrix} \frac{10}{9} \\ \frac{14}{9} \end{bmatrix}.$$

Note that these smaller coordinate values are consistent with the larger units of length. In Figure 8-4, the subscripts X and Y on vectors identify the reference systems with respect to which their coordinates have been computed.

In view of the one-to-one correspondence between points and vectors, we can also answer questions such as this one: If a line has the equation

$$3x_1 + 5x_2 = 27$$

or

$$[3, 5]X = 27,$$

in the standard reference system, what is its equation in the new reference system? Since

$$X = \begin{bmatrix} 5 & -1 \\ -3 & 6 \end{bmatrix}Y,$$

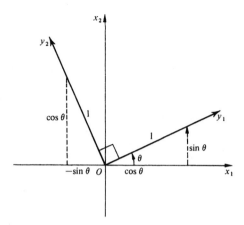

FIGURE 8-5. Rotation of Axes.

we have

$$[3, 5]\begin{bmatrix} 5 & -1 \\ -3 & 6 \end{bmatrix} Y = 27$$

or

$$y_2 = 1,$$

so the line is parallel to the y_1-axis and passes through the point $(0, 1)_Y$.

The preceding transformation does not preserve formulas for length or angle but it does preserve the test for parallelism of lines. (Can you prove these statements?)

By way of contrast, consider the linear transformation of coordinates (rotation of axes through the angle θ) defined by

$$(8.5.7) \quad Y = \begin{bmatrix} \cos \theta & \sin \theta \\ -\sin \theta & \cos \theta \end{bmatrix} X \quad \text{or} \quad X = \begin{bmatrix} \cos \theta & -\sin \theta \\ \sin \theta & \cos \theta \end{bmatrix} Y,$$

where X is referred to the standard basis. The coordinates of the new unit vectors with respect to the standard basis may be obtained by substituting $\begin{bmatrix} 1 \\ 0 \end{bmatrix}$ and $\begin{bmatrix} 0 \\ 1 \end{bmatrix}$ for Y in the second of these equations and are the columns $\begin{bmatrix} \cos \theta \\ \sin \theta \end{bmatrix}$ and $\begin{bmatrix} -\sin \theta \\ \cos \theta \end{bmatrix}$ respectively. These two vectors are orthogonal unit vectors, so the new reference system is again orthonormal (Figure 8-5). Since

$$\begin{bmatrix} \cos\theta & -\sin\theta \\ \sin\theta & \cos\theta \end{bmatrix}^{\mathsf{T}} \begin{bmatrix} \cos\theta & -\sin\theta \\ \sin\theta & \cos\theta \end{bmatrix} = I_2,$$

we have $Y^{\mathsf{T}}Y = X^{\mathsf{T}}X$, so the formula for length is preserved. Also, if Y_1 corresponds to X_1 and Y_2 corresponds to X_2, we have $Y_1^{\mathsf{T}}Y_2 = X_1^{\mathsf{T}}X_2$, so the formula for angular measure and the test for orthogonality are also preserved. The formulas for length and angle are said to be **invariant** under rotations of axes.

A rotation of the axes through the angle θ is an example of an orthogonal transformation of coordinates. (See Section 8.14.)

The procedures used to accomplish a change of basis in $\mathscr{F}^n$ apply equally well in a subspace $\mathscr{V}$ of $\mathscr{F}^n$. Let $A_1, A_2, \ldots, A_k$ and $B_1, B_2, \ldots, B_k$ be two bases for $\mathscr{V}$. Then we have a set of relations of the form

$$(8.5.8) \qquad B_j = \sum_{i=1}^{k} s_{ij} A_i, \qquad j = 1, 2, \ldots, k,$$

defining the B's in terms of the A's. We can combine these k equations into the single matrix equation

$$(8.5.9) \qquad\qquad\qquad B = AS,$$

where

$$B = [B_1, B_2, \ldots, B_k], \qquad A = [A_1, A_2, \ldots, A_k], \qquad S = [s_{ij}]_{k \times k}.$$

Since B has rank k, it follows that S has rank k also, so S is nonsingular.

Now let X be any vector of $\mathscr{V}$. Then there exist unique sets of scalars $y_1, y_2, \ldots, y_k$ and $z_1, z_2, \ldots, z_k$ such that

$$(8.5.10) \qquad X = \sum_{i=1}^{k} y_i A_i \qquad \text{and} \qquad X = \sum_{j=1}^{k} z_j B_j,$$

or, if we put $Y = [y_1, y_2, \ldots, y_k]^{\mathsf{T}}$ and $Z = [z_1, z_2, \ldots, z_k]^{\mathsf{T}}$,

$$(8.5.11) \qquad\qquad X = AY \qquad \text{and} \qquad X = BZ.$$

Substituting from (8.5.8) into the second equation of (8.5.10) and rearranging, we get

$$(8.5.12) \qquad\qquad X = \sum_{i=1}^{k} \left(\sum_{j=1}^{k} s_{ij} z_j \right) A_i.$$

Since the y_i in (8.5.10) are unique, we therefore conclude that

$$(8.5.13) \qquad y_i = \sum_{j=1}^{k} s_{ij} z_j, \qquad i = 1, 2, \ldots, k,$$

so, in matrix notation,

$$Y = SZ \qquad \text{or} \qquad Z = S^{-1} Y.$$

Thus the change of basis in $\mathscr{V}$ induces a **linear transformation of coordinates** relating Y and Z. In summary, we have

Theorem 8.5.1: *Let $\mathscr{V}$ be a subspace of dimension k, $k \leq n$, of $\mathscr{F}^n$ and let the bases of $\mathscr{V}$ defined by the columns of the matrices $A = [A_1, A_2, \ldots, A_k]$ and $B = [B_1, B_2, \ldots, B_k]$ be related by the equation*

$$(8.5.14) \qquad B = AS.$$

Let Z and Y be the k-vectors over $\mathscr{F}$ which give the coordinates of an arbitrary vector X of $\mathscr{V}$ with respect to the B and A bases respectively. Then Z and Y are related by the linear transformation of coordinates

$$(8.5.15) \qquad Z = S^{-1} Y.$$

For example, in $\mathscr{E}^3$ consider the subspace $\mathscr{V}$ with basis $[1, 1, 0]^\mathsf{T}$ and $[0, 1, 1]^\mathsf{T}$. The vectors $[1, 2, 1]^\mathsf{T}$ and $[1, 0, -1]^\mathsf{T}$ are also in this space and, being independent, form a basis for it (Figure 8-6). Moreover,

$$\begin{bmatrix} 1 \\ 2 \\ 1 \end{bmatrix} = \begin{bmatrix} 1 \\ 1 \\ 0 \end{bmatrix} + \begin{bmatrix} 0 \\ 1 \\ 1 \end{bmatrix} \qquad \text{and} \qquad \begin{bmatrix} 1 \\ 0 \\ -1 \end{bmatrix} = \begin{bmatrix} 1 \\ 1 \\ 0 \end{bmatrix} - \begin{bmatrix} 0 \\ 1 \\ 1 \end{bmatrix}.$$

Now if

$$X = y_1 \begin{bmatrix} 1 \\ 1 \\ 0 \end{bmatrix} + y_2 \begin{bmatrix} 0 \\ 1 \\ 1 \end{bmatrix} = z_1 \begin{bmatrix} 1 \\ 2 \\ 1 \end{bmatrix} + z_2 \begin{bmatrix} 1 \\ 0 \\ -1 \end{bmatrix},$$

we obtain, by substitution on the right,

$$y_1 \begin{bmatrix} 1 \\ 1 \\ 0 \end{bmatrix} + y_2 \begin{bmatrix} 0 \\ 1 \\ 1 \end{bmatrix} = (z_1 + z_2) \begin{bmatrix} 1 \\ 1 \\ 0 \end{bmatrix} + (z_1 - z_2) \begin{bmatrix} 0 \\ 1 \\ 1 \end{bmatrix},$$

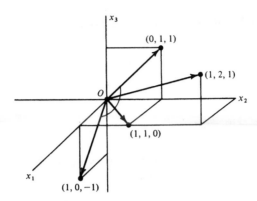

FIGURE 8-6. Change of Basis.

from which, since the coefficients are unique by Theorem 5.5.8,

$$y_1 = z_1 + z_2,$$
$$y_2 = z_1 - z_2,$$

or

$$\begin{bmatrix} y_1 \\ y_2 \end{bmatrix} = \begin{bmatrix} 1 & 1 \\ 1 & -1 \end{bmatrix} \begin{bmatrix} z_1 \\ z_2 \end{bmatrix}, \qquad \begin{bmatrix} z_1 \\ z_2 \end{bmatrix} = \frac{1}{2} \begin{bmatrix} 1 & 1 \\ 1 & -1 \end{bmatrix} \begin{bmatrix} y_1 \\ y_2 \end{bmatrix},$$

which is in agreement with (8.5.15). (Write A and B; then check this.)

The most important special case of Theorem 8.5.1 occurs when $k = n$. In this case A and B are nonsingular n-square matrices so (8.5.9) yields

(8.5.16) $$S = A^{-1}B.$$

Hence we have

Theorem 8.5.2: *If $A_1, A_2, \ldots, A_n$ and $B_1, B_2, \ldots, B_n$ are bases of $\mathscr{F}^n$ and if Y and Z give the coordinates of an arbitrary vector X of $\mathscr{F}^n$ with respect to these two bases respectively, then*

(8.5.17) $$Y = (A^{-1}B)Z.$$

In the special case when $B_1, B_2, \ldots, B_n$ is the natural basis so that $B = I_n$ and $Z = X$, equation (8.5.17) reduces to (8.5.3).

We illustrate again with an example from $\mathscr{E}^3$. Suppose we wish to express the coordinates z_1, z_2, z_3 of a vector X with respect to the coordinate system $[1, 1, 0]^T$, $[1, 1, 1]^T$, $[1, 0, 0]^T$ in terms of its coordinates y_1, y_2, y_3 with respect to the coordinate system $[1, 0, 1]^T$, $[0, 1, 0]^T$, $[0, 0, 1]^T$. We have

$$X = \begin{bmatrix} 1 & 1 & 1 \\ 1 & 1 & 0 \\ 0 & 1 & 0 \end{bmatrix} Z \quad \text{or} \quad Z = \begin{bmatrix} 0 & 1 & -1 \\ 0 & 0 & 1 \\ 1 & -1 & 0 \end{bmatrix} X,$$

and

$$X = \begin{bmatrix} 1 & 0 & 0 \\ 0 & 1 & 0 \\ 1 & 0 & 1 \end{bmatrix} Y.$$

Hence

$$Z = \begin{bmatrix} 0 & 1 & -1 \\ 0 & 0 & 1 \\ 1 & -1 & 0 \end{bmatrix} \cdot \begin{bmatrix} 1 & 0 & 0 \\ 0 & 1 & 0 \\ 1 & 0 & 1 \end{bmatrix} Y = \begin{bmatrix} -1 & 1 & -1 \\ 1 & 0 & 1 \\ 1 & -1 & 0 \end{bmatrix} Y.$$

Note how this result is derived by using only basic principles. No complicated set of formulas needs to be memorized.

8.6 Transformation of a Linear Operator

Let the equation

(8.6.1) $$X_2 = AX_1$$

represent a linear operator on $\mathscr{F}^n$, the reference system being the standard basis (our first interpretation of a linear transformation). Let

(8.6.2) $$X = BY, \quad Y = B^{-1}X$$

be a transformation of coordinates which defines the coordinates y_1, y_2, $\ldots$, y_n of the vector X with respect to the new basis defined by the columns of B.

What is the representation of the operator (8.6.1) in this new reference system? Since X_1 and X_2 will ordinarily have altered coordinates in the new reference system, one should expect to replace A by a different matrix in order to effect the same geometrical mapping. If the new names of X_1 and X_2 are Y_1 and Y_2 respectively, so that

$$X_1 = BY_1, \quad X_2 = BY_2,$$

we have, from (8.6.1),

$$BY_2 = ABY_1,$$

so

$$(8.6.3) \qquad\qquad Y_2 = (B^{-1}AB)Y_1.$$

This proves

Theorem 8.6.1: *The operator on $\mathcal{F}^n$ represented by the matrix A in the standard reference system is represented by the matrix $B^{-1}AB$ in the reference system whose basis vectors are defined, in the standard reference system, by the columns of B.*

Thus, not only does every vector of $\mathcal{F}^n$ have a particular set of coordinates in every reference system for $\mathcal{F}^n$, but also every linear operator on $\mathcal{F}^n$ has a particular representation in every reference system for $\mathcal{F}^n$.

The mapping of A onto the matrix $B^{-1}AB$ is called a **similarity transformation** of A, and A and $B^{-1}AB$ are said to be **similar.** An important problem of matrix algebra is the determination of the canonical forms to which a matrix can be reduced by similarity transformations. The particularly important case where B is orthogonal will be treated in Chapter 9.

To illustrate the theorem, let an operator on $\mathscr{E}^2$ be

$$X_2 = \begin{bmatrix} 1 & 2 \\ 2 & 1 \end{bmatrix} X_1$$

and let the transformation of coordinates be the special case of (8.5.7) defined by

$$X = \begin{bmatrix} \dfrac{1}{\sqrt{2}} & -\dfrac{1}{\sqrt{2}} \\[3mm] \dfrac{1}{\sqrt{2}} & \dfrac{1}{\sqrt{2}} \end{bmatrix} Y;$$

that is, let the axes be rotated through $45°$.

In the new reference system, the operator is

$$Y_2 = \begin{bmatrix} \dfrac{1}{\sqrt{2}} & \dfrac{1}{\sqrt{2}} \\[3mm] -\dfrac{1}{\sqrt{2}} & \dfrac{1}{\sqrt{2}} \end{bmatrix} \begin{bmatrix} 1 & 2 \\ 2 & 1 \end{bmatrix} \begin{bmatrix} \dfrac{1}{\sqrt{2}} & -\dfrac{1}{\sqrt{2}} \\[3mm] \dfrac{1}{\sqrt{2}} & \dfrac{1}{\sqrt{2}} \end{bmatrix} Y_1$$

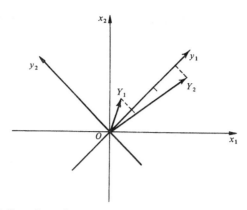

FIGURE 8-7. Transformation of a Linear Operator.

or

$$Y_2 = \begin{bmatrix} 3 & 0 \\ 0 & -1 \end{bmatrix} Y_1.$$

(See Figure 8-7).

In the new reference system, the operator is easily described geometrically. To get the image Y_2 of Y_1, one multiplies the first component of Y_1 by 3 and the second component by -1. This is a small illustration of how a properly chosen transformation can simplify and clarify a problem.

A similar procedure may be applied to an operator which maps vectors of $\mathscr{F}^n$ onto vectors of $\mathscr{F}^m$. Let the operator be

(8.6.4) $$Y = A_{m \times n} X,$$

where each of X and Y is referred to the natural reference system in its space. Suppose now that we introduce new reference systems in $\mathscr{F}^n$ and in $\mathscr{F}^m$ respectively by means of the linear transformations of coordinates

$$X = B\tilde{X}, \qquad Y = C\,\tilde{Y}.$$

Then the operator $Y = AX$ is represented by

$$C\tilde{Y} = A(B\tilde{X})$$

or

(8.6.5) $$\tilde{Y} = (C^{-1}AB)\tilde{X}$$

in the new reference systems. Note how this generalizes (8.6.3).

Since C^{-1} and B are nonsingular, $C^{-1}AB$ has the same rank as A. This is consistent with the fact that (8.6.4) and (8.6.5) describe the same geometrical transformation, the only difference being the reference systems.

8.7   Exercises

1. Show that the operator on $\mathscr{E}^2$ defined by

$$Y = \begin{bmatrix} k & 0 \\ 0 & 1 \end{bmatrix} X, \qquad k > 0,$$

multiplies the area of the rectangle with vertices (a_1, b_2), (b_1, b_2), (b_1, a_2), (a_1, a_2) by k and hence multiplies the area of any simple closed curve by k. This operator is called an **elongation** if $k > 1$ and a **contraction** if $k < 1$. Illustrate with a figure.

2. Use Exercise 1 to transform a circle into an ellipse and hence show that the area of an ellipse with semiaxes a and b is πab. Determine by a similar argument the formula for the volume of an ellipsoid.

3. What is the effect of a real operator

$$Y = \begin{bmatrix} k & 0 \\ 0 & k \end{bmatrix} X, \qquad k \neq 0.$$

on an arbitrary plane figure? What aspects of the figure remain unaltered, that is, are invariant with respect to, this operator? This operator is called a **similarity transformation** and the origin is the **center of similitude**.

4. What is the effect of a real transformation

$$Y = \begin{bmatrix} \alpha & 0 \\ 0 & \beta \end{bmatrix} X, \qquad \alpha = \pm 1, \quad \beta = \mp 1,$$

on an arbitrary plane figure? What aspects of the figure are invariant with respect to the operator? Such an operator is called a **reflection**—in the 1-axis if $\beta = -1$ and in the 2-axis if $\alpha = -1$.

5. Show that the distance between two points in $\mathscr{E}^n$ remains invariant under every **translation**, that is, under every transformation of the form

$$Y = X + H,$$

where H is a fixed vector. Illustrate with figures in $\mathscr{E}^2$ and $\mathscr{E}^3$.

6. Show that a rotation about the origin in $\mathscr{E}^2$ is represented by the equation

$$X = \begin{bmatrix} \cos \theta & -\sin \theta \\ \sin \theta & \cos \theta \end{bmatrix} Y,$$

where θ is the angle of rotation. Then show that every such rotation leaves the distance between any two points P and Q invariant.

7. What figures are invariant under a given rotation about the origin? Under *every* rotation about the origin?

8. In $\mathscr{E}^2$, a **shear transformation** parallel to the 1-axis is any operator defined by an equation

$$Y = \begin{bmatrix} 1 & k \\ 0 & 1 \end{bmatrix} X.$$

(A shear parallel to the 2-axis is defined analogously.) Put $k = 2$ and determine the effect of the shear on
(a) a line $x_1 = a_1$,
(b) the line $x_1 + 2x_2 = 6$,
(c) the circle $x_1^2 + x_2^2 = 4$.
Shear transformations are important in the study of deformations in mechanics.

9. Find the image of the triangle with vertices $(0, 0)$, (a, b), (c, d), by the operator

$$X_2 = \begin{bmatrix} 1 & -1 \\ 1 & 1 \end{bmatrix} X_1.$$

Illustrate with a figure. What effect does this operator have on the length of an arbitrary vector? How is the area of the image triangle related to that of the original triangle? (Recall the determinant formula for the area of a triangle here.) What effect does the transformation have on areas in general? Why?

10. What is the effect of the operator on $\mathscr{E}^3$ defined by

$$Y = \begin{bmatrix} 1 & 0 & k_1 \\ 0 & 1 & k_2 \\ 0 & 0 & 1 \end{bmatrix} X, \qquad k_1 > 0, k_2 > 0$$

on a line parallel to the x_3-axis? On the cylinder $x_1^2 + x_2^2 = 1$? Is the image of the cylinder still a circular cylinder?

11. If a triangle $T: (x_{11}, x_{12}), (x_{21}, x_{22}), (x_{31}, x_{32})$ in $\mathscr{E}^2$ is transformed by the non-singular linear transformation

$$Y = \begin{bmatrix} a_{11} & a_{12} \\ a_{21} & a_{22} \end{bmatrix} X,$$

how is the area of the image triangle related to that of T? How is the area of any plane figure for which area is defined affected by this transformation?

12. If $Y = AX$ is an operator on $\mathscr{F}^n$, the set of all vectors X such that $AX = X$ is called the **fixed space** of the operator. Show that this space is a vector space and determine its dimension.

*13. Let $Y = AX$ be an operator from $\mathscr{F}^n$ to $\mathscr{F}^m$. The set of all vectors X of $\mathscr{F}^n$ whose image is the zero vector of $\mathscr{F}^m$ is the null space or kernel of the operator and the dimension $N(A)$ of the null space is the nullity of the operator

(Section 8.3). Restate the results of Section 6.5 in terms of subspaces and their dimensions.

14. Determine the null space of the transformation

$$Y = \begin{bmatrix} 1 & 2 & 1 \\ 2 & 2 & 0 \\ 1 & 0 & -1 \end{bmatrix} X.$$

15. Let $Y = AX$ represent an operator from $\mathscr{F}^n$ to $\mathscr{F}^m$. Show that the image of every p-dimensional subspace $\mathscr{V}$ of $\mathscr{F}^n$ has dimension at least $p - N(A)$, where $N(A)$ is the nullity of A.

16. Interpret geometrically in $\mathscr{E}^2$:

   (a) the operator $X' = \begin{bmatrix} 0 & -1 \\ 1 & 0 \end{bmatrix} X$;

   (b) the linear transformation of coordinates

$$Y = \begin{bmatrix} 0 & -1 \\ 1 & 0 \end{bmatrix} X.$$

17. If it is desired to use

$$A_1 = \begin{bmatrix} 1 \\ 1 \\ 1 \end{bmatrix}, \qquad A_2 = \begin{bmatrix} 1 \\ 1 \\ -2 \end{bmatrix}, \qquad A_3 = \begin{bmatrix} -1 \\ 1 \\ 0 \end{bmatrix},$$

   expressed in the standard basis, as a new basis for $\mathscr{E}^3$, what equation gives the new coordinates of a given vector X in terms of its coordinates with respect to the standard basis?

18. If a linear transformation of coordinates maps the standard basis representations

$$X_1 = \begin{bmatrix} 1 \\ 2 \\ -1 \end{bmatrix}, \qquad X_2 = \begin{bmatrix} 3 \\ 1 \\ 0 \end{bmatrix}, \qquad X_3 = \begin{bmatrix} -1 \\ 0 \\ 0 \end{bmatrix}$$

   respectively onto

$$Y_1 = \begin{bmatrix} 1 \\ 0 \\ 0 \end{bmatrix}, \qquad Y_2 = \begin{bmatrix} -1 \\ 1 \\ 0 \end{bmatrix}, \qquad Y_3 = \begin{bmatrix} -2 \\ 0 \\ 1 \end{bmatrix},$$

   find a matrix equation that defines the transformation and identify the new basis vectors.

19. If a transformation of coordinates in $\mathscr{E}^4$ is such that the standard basis vectors have new coordinates given by

$$\begin{bmatrix} 1 \\ 1 \\ 1 \\ 0 \end{bmatrix}, \quad \begin{bmatrix} 1 \\ 1 \\ 0 \\ 1 \end{bmatrix}, \quad \begin{bmatrix} 1 \\ 0 \\ 1 \\ 1 \end{bmatrix}, \quad \begin{bmatrix} 0 \\ 1 \\ 1 \\ 1 \end{bmatrix}$$

respectively, write a matrix equation for the transformation.

20. Let the equation

$$\begin{bmatrix} y_1 \\ y_2 \end{bmatrix} = \begin{bmatrix} 1 & 2 \\ 0 & 1 \end{bmatrix} \begin{bmatrix} x_1 \\ x_2 \end{bmatrix}$$

define a transformation of coordinates in $\mathscr{E}^2$ in the usual way. Draw a figure showing both the standard and the new reference systems. Then determine what points have the same coordinates in both systems.

21. Given a linear operator defined by the equation

$$Y = \begin{bmatrix} 1 & 3 \\ 3 & 2 \end{bmatrix} X$$

with respect to the natural reference system of $\mathscr{E}^2$, find an equation representing the same operator with respect to the basis $[2, 1]^{\mathsf{T}}$, $[1, 2]^{\mathsf{T}}$.

22. Suppose that in $\mathscr{E}^2$

$$X = y_1 \begin{bmatrix} 1 \\ 1 \end{bmatrix} + y_2 \begin{bmatrix} -1 \\ 1 \end{bmatrix},$$

and also

$$X = z_1 \begin{bmatrix} 2 \\ 1 \end{bmatrix} + z_2 \begin{bmatrix} -1 \\ 2 \end{bmatrix}.$$

What is the relation between

$$Y = \begin{bmatrix} y_1 \\ y_2 \end{bmatrix} \quad \text{and} \quad Z = \begin{bmatrix} z_1 \\ z_2 \end{bmatrix}?$$

23. Given the linear transformation of coordinates

$$Y = \begin{bmatrix} 1 & 2 & 3 \\ 2 & 0 & 2 \\ 3 & 2 & 1 \end{bmatrix} X,$$

find the new coordinates (y's) of the old basis vectors (E's) and the old coordinates (x's) of the new elementary vectors.

24. Interpret the linear transformation of coordinates in $\mathscr{E}^n$,

$$X = D[d_1, d_2, \ldots, d_n]Y, \qquad d_1 d_2 \cdots d_n \neq 0,$$

geometrically if $n = 3$. What vector X is represented by $Y = [1, 1, \ldots, 1]^\mathsf{T}$?

25. What is the form of the operator in Exercise 9 in the reference system defined by

$$\begin{bmatrix} \cos\theta \\ \sin\theta \end{bmatrix}, \qquad \begin{bmatrix} -\sin\theta \\ \cos\theta \end{bmatrix} ?$$

How do you explain the result geometrically?

26. Prove that the identity operator $Y = IX$ has the same representation in every reference system.

*27. Show that the determinant and the trace of a linear operator on $\mathscr{F}^n$ are invariant with respect to nonsingular linear transformations of coordinates.

*28. If $Y_2 = AY_1$ represents a linear operator on $\mathscr{F}^n$ in the reference system $B_1, B_2, \ldots, B_n$, what is its representation in the natural reference system? In the reference system $C_1, C_2, \ldots, C_n$?

*29. Using the notation of Theorem 8.5.1, show that, if also

$$[C_1, C_2, \ldots, C_k] = [B_1, B_2, \ldots, B_k]Q,$$

then the coordinate vector $W = [w_1, w_2, \ldots, w_k]^\mathsf{T}$ of a vector X with respect to the C's is related to the vectors Y and Z of the theorem by the transformations $Z = QW$ and $Y = SQW$.

30. Given a nonsingular operator on $\mathscr{F}^n$ with matrix A, what will be the matrix of the operator in the reference system defined by the columns of A? By the columns of A^{-1}? In what reference systems will the matrix of the operator still be A?

31. Under what conditions will an operator on $\mathscr{F}^n$ be represented by the same matrix A in two coordinate systems, $B_1, B_2, \ldots, B_n$ and $C_1, C_2, \ldots, C_n$?

32. Obtain the matrix of the coordinate transformation in $\mathscr{F}^n$ which replaces the natural basis by the basis

$$B_k = \sum_{j=1}^{k} E_j, \qquad k = 1, 2, \ldots, n.$$

Then find the matrix of the inverse transformation.

33. Obtain the matrix of the transformation of coordinates which replaces the basis $1, x, \ldots, x^{n-1}$ of the n-dimensional space of real polynomials of degree $n - 1$ in x by the basis $1, (x - a), (x - a)^2, \ldots, (x - a)^{n-1}$.

34. Let $A_1, A_2, \ldots, A_n$ be a basis for $\mathscr{F}^n$ and let

$$A = [A_1, A_2, \ldots, A_n], \qquad A^{-1} = \begin{bmatrix} B^{(1)} \\ B^{(2)} \\ \vdots \\ B^{(n)} \end{bmatrix}.$$

Prove that if X is any vector of $\mathscr{F}^n$, then

$$X = \sum (B^{(j)}X)A_j,$$

so $B^{(1)}X, B^{(2)}X, \ldots, B^{(n)}X$ are the coordinates of X with respect to $A_1, A_2, \ldots, A_n$ respectively.

35. Let $f(A_1, A_2, \ldots, A_m)$ denote a polynomial function of matrices $A_1, A_2, \ldots, A_m$ of order n, where f and the A_j's are all over $\mathscr{F}$. If $\tilde{A}_j = B^{-1}A_j B, j = 1, 2, \ldots, m$, and $\tilde{f} = B^{-1}fB$, prove that

$$\tilde{f}(A_1, A_2, \ldots, A_m) = f(\tilde{A}_1, \tilde{A}_2, \ldots, \tilde{A}_m).$$

36. Given that F is a differentiable function of $x_1, x_2, \ldots, x_n$ and that $Y = AX$, where $\det A \neq 0$, show that

$$\left[\frac{\partial F}{\partial y_1}, \frac{\partial F}{\partial y_2}, \ldots, \frac{\partial F}{\partial y_n}\right]^{\mathsf{T}} = (A^{-1})^{\mathsf{T}}\left[\frac{\partial F}{\partial x_1}, \frac{\partial F}{\partial x_2}, \ldots, \frac{\partial F}{\partial x_n}\right]^{\mathsf{T}}.$$

37. In vector analysis the cross product $X \times Y$ of two vectors X and Y in $\mathscr{E}^3$ is often defined to be the vector obtained by expanding a symbolic determinant

$$X \times Y = \begin{vmatrix} i & j & k \\ x_1 & x_2 & x_3 \\ y_1 & y_2 & y_3 \end{vmatrix} = \begin{vmatrix} x_2 & x_3 \\ y_2 & y_3 \end{vmatrix} i - \begin{vmatrix} x_1 & x_3 \\ y_1 & y_3 \end{vmatrix} j + \begin{vmatrix} x_1 & x_2 \\ y_1 & y_2 \end{vmatrix} k,$$

where the "unit vectors" i, j, k are the same as the elementary vectors E_1, E_2, E_3. Using a similar determinantal notation, show how the cross product generalizes to a vector product of $n - 1$ n-vectors $X_1, X_2, \ldots, X_{n-1}$. How is this latter product affected by permutations of the X's?

38. Show that, as X sweeps out $\mathscr{E}^3$ but Y remains fixed, the cross product $Z = X \times Y$ sweeps out a plane. Identify this plane, and show that the relationship between Z and X is a singular linear operator. Generalize to $\mathscr{E}^n$ with the aid of the preceding exercise.

39. The volume of the parallelepiped in $\mathscr{E}^3$ determined by three vectors $X_j, j = 1, 2, 3$, is given by $|X_1^{\mathsf{T}}(X_2 \times X_3)|$, where the cross product has been defined above. If $Y_j = AX_j$, where A is a nonsingular operator, how is the volume $|Y_1^{\mathsf{T}}(Y_2 \times Y_3)|$ related to that determined by the X's? Generalize to n dimensions if you can.

40. Let $X_1, X_2, \ldots, X_n$ constitute a basis for $\mathscr{U}^n$ and let $X_i^* X_j = g_{ij}$. If $X = \sum \xi_i X_i$ and $Y = \sum \eta_j X_j$, show that

$$X^*Y = \sum_{i, j} g_{ij} \bar{\xi}_i \eta_j.$$

This formula gives the form of the inner product in the reference system $X_1, X_2, \ldots, X_n$.

*41. Let $X_2 = AX_1$ define a linear operator on $\mathscr{F}^n$. Let $\mathscr{V}$ be a subspace of $\mathscr{F}^n$ such that the image by this operator of every vector in $\mathscr{V}$ is still in $\mathscr{V}$. Then $\mathscr{V}$ is called an **invariant subspace** with respect to this operator. (Note that this definition does not require individual vectors to be invariant.)

(a) Invent concrete examples in $\mathscr{E}^2$ and $\mathscr{E}^3$ and illustrate with figures.

(b) Show that if $\mathscr{V}$ is invariant with respect to the operator defined by $X_2 = AX_1$, then it is also invariant with respect to every operator of the form $X_2 = f(A)X_1$, where $f(A)$ is a polynomial function of A.

8.8   The Algebra of Linear Operators

Given the operators $Y = AX$ and $Z = BY$ on $\mathscr{F}^n$, they permit us to map X onto Z by first mapping X onto Y, then mapping Y onto Z. Substituting from the first equation into the second, we get $Z = (BA)X$, which represents an operator that maps X directly onto Z and which is called the **product** of the two first-given operators. Such a product of operators is always defined when the matrices B and A which represent them are conformable, so that the product BA is defined. Note that the matrix of the first operator appears on the *right* in this product.

If now $W = CZ$, and if we substitute for Z from the previous equation, we get $W = C(BA)X$. If, on the other hand, we first form the product of $W = CZ$ and $Z = BY$ to obtain $W = (CB)Y$, and thereafter form the product with $Y = AX$, we get $W = (CB)AX$. Since $C(BA) = (CB)A$, the two final products are the same. This fact and the fact that ordinarily $AB \neq BA$ allow us to conclude

Theorem 8.8.1: *The multiplication of linear operators on $\mathscr{F}^n$ is associative but is not in general commutative.*

It is important to recognize that the most essential aspect of the representation $Y = AX$ of a linear operator is the matrix A. The symbols Y and X are just convenient symbols for the vectors which correspond under the mapping. If the domain is the same for both Z and X, the equations $W = AZ$ and $Y = AX$ represent the same operator. The only occasion when the symbols for the vectors are of special significance is when we wish to form the product of the operators. In this case, these symbols serve to indicate the order in which the multiplication is to be carried out, as in the preceding discussion.

The **identity operator** on $\mathscr{F}^n$ is the mapping such that for all X in $\mathscr{F}^n$, the image of X is X. (The reader should show that this mapping is linear.) If $AX = X$ for all X, by putting X successively equal to $E_1, E_2, \ldots, E_n$, we find that $A_j = E_j, j = 1, 2, \ldots, n$, so that $A = I_n$. Conversely, if $A = I_n$, the operator with matrix A is the identity operator. This proves

Theorem 8.8.2: *The identity operator on $\mathscr{F}^n$ is represented by the matrix I_n and by no other.*

The product of the operator represented by the matrix A and the identity operator has matrix AI or IA and hence is simply the operator represented by the matrix A.

We have already seen that if A is nonsingular, the operator represented by the matrix A^{-1} is the inverse of the operator represented by the matrix A. The product of these two operators is represented by the matrix AA^{-1} or $A^{-1}A$, namely I_n, and thus is the identity operator.

Suppose now that $Y = AX$ has an inverse. Then, since it is invertible, this transformation must be one to one. If A were to have rank less than n, then the equation $AX = 0$ would have more than one solution, so the vector zero would have more than one counterimage and the mapping would not be one to one. Hence A has rank n, A^{-1} exists, and the inverse mapping is $X = A^{-1}Y$. Thus we have

Theorem 8.8.3: *A linear operator on $\mathscr{F}^n$ represented by a matrix A has an inverse if and only if A is nonsingular, and the inverse operator is represented by A^{-1}.*

In the algebra of linear operators, the identity mapping and inverses of mappings are of particular significance.

The fact that a matrix represents an operator is often useful in interpreting and proving theorems about matrices. For example, the image $\mathscr{U}$ of $\mathscr{F}^p$ by the operator $Y = A_{m \times n} B_{n \times p} X = A(BX)$ is the image by $Y = AZ$ of the image $\mathscr{V}$ of $\mathscr{F}^p$ by $Z = BX$.

Since $\mathscr{V}$ is a subspace of $\mathscr{F}^n$, the image space $\mathscr{U}$ is a subspace of the image $\mathscr{W}$ of $\mathscr{F}^n$ by $Y = AZ$ (Figure 8-8). The subspace $\mathscr{W}$ of $\mathscr{F}^m$ has dimension $r(A)$ since it is spanned by the columns of A and the subspace $\mathscr{U}$ of $\mathscr{W}$ has dimension $r(AB)$ since it is spanned by the columns of AB. Hence $r(AB) \le r(A)$.

Also the dimension of $\mathscr{V}$ is not greater than $r(B)$ since $\mathscr{V}$ is spanned by the columns of B. Thus the dimension of $\mathscr{U}$, which by Corollary 8.3.2 is

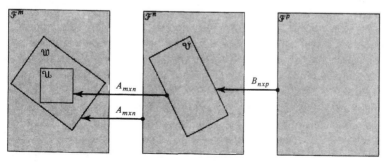

FIGURE 8-8. Geometrical Interpretation of a Product of Operators.

not greater than that of $\mathscr{V}$, is also $\leq r(B)$. That is, $r(AB) \leq r(B)$. Thus we have a geometric proof of the fact that the rank of the product of two matrices cannot exceed that of either factor.

8.9   Groups of Operators

A fundamental aspect of mathematics is the classification of sets of mathematical objects according to important properties which they possess. The concepts "field" and "vector space" which we have already introduced are examples of this. We now introduce another important concept of this type. A set G of mathematical objects, for which equality means identity, and each ordered pair (a, b) of which can be combined by an operation (denoted here by $\circ$) to give a unique "product" $a \circ b$, is called a **group** if and only if the following properties hold:

(a) G is *closed* with respect to the operation $\circ$; that is, if a and b are arbitrary elements of G, $a \circ b$ is also an element of G.
(b) The *associative law* holds, that is, for all a, b, c of G, $a \circ (b \circ c) = (a \circ b) \circ c$.
(c) G contains an *identity element*, that is, an element e such that for all a of G, $a \circ e = e \circ a = a$.
(d) For each element a of G, G contains an element a^{-1}, called the *inverse* of a, such that $a \circ a^{-1} = a^{-1} \circ a = e$.

If the commutative law also holds, that is, if $a \circ b = b \circ a$ for all a and b of G, then G is called a **commutative** or **Abelian group**.

If two groups G_1 and G_2 have the same operation $\circ$, and if every element of G_1 is also an element of G_2, G_1 is called a **subgroup** of G_2. If a group G contains only a finite number n of elements, it is called a **finite group** and is said to be of **order** n. Our principal concern here is with certain infinite groups of linear operators or linear transformations of coordinates.

There are many familiar examples of groups. The set of all complex numbers, the operation $\circ$ being ordinary addition, is an example. Indeed, the sum of two complex numbers is again a complex number, addition is associative, $a + 0 = 0 + a = a$, so that 0 is the identity element, and for each complex number a, $-a$ is its inverse. Thus all four requirements are satisfied.

Another example is provided by the set of four complex numbers 1, -1, i, $-i$, the operation now being multiplication. It is easy to check the closure property, multiplication is associative, 1 is the identity element, and the inverses of these numbers are respectively their reciprocals 1, -1, $-i$, i, all of which are in the set.

The vectors of a vector space $\mathscr{V}$ form a group with respect to the operation of addition. The closure property follows from the definition of a vector space, the addition of matrices is associative, the identity element is the zero vector, which is in every vector space, and the additive inverse of a vector X of the space is the vector $-X$ which, again by the definition of a vector space, is also in $\mathscr{V}$.

In the case of linear operators on $\mathscr{F}^n$, the operation $\circ$ is taken to be multiplication, as defined in a previous section. This operation is associative because matrix multiplication is associative. The identity operator $Y = I_n X$ is the identity element, since the product of this operator with another whose matrix is A has the matrix $I_n A$ or AI_n, that is, A. Finally, if A is nonsingular, operators with matrices A and A^{-1} are inverses in the group sense, since the matrix of their product is either AA^{-1} or $A^{-1}A$, that is, I_n, the matrix of the identity operator. In view of these observations, a set of linear operators on $\mathscr{F}^n$ is a group of linear operators if and only if

(a) the product of any two operators of the set is an operator of the set,
(b) the identity operator belongs to the set, and
(c) the inverse of every operator of the set is also in the set.

This last requirement implies that *a group of linear operators may contain no singular operators.*

We now turn to some examples, noting first that the set of all non-singular linear operators on $\mathscr{F}^n$ constitutes a group with respect to the operation of multiplication:

(a) If $Y = AX$ and $Z = BY$ are two such operators, their product $Z = (BA)X$ is also a nonsingular operator, for BA is nonsingular whenever A and B are.
(b) The identity operator $Y = I_n X$ is nonsingular and hence belongs to the set.
(c) If $Y = AX$ is in the set, A^{-1} exists and is nonsingular, so the inverse operator $Z = A^{-1}Y$ is also in the set.

This group of operators is called the **full linear group** over $\mathscr{F}$. The basic property of this group of operators follows from Theorem 8.3.4: *Any operator of the full linear group maps each subspace of $\mathscr{F}^n$ onto a subspace of the same dimension*; that is, the property of being a vector space and the dimension of a vector space are *invariant* under operators of this group.

Among the various special operators of the full linear group are the finitely many operators which simply permute the n variables. These permutations constitute what is known as the **symmetric group**. Since there are just $n!$ such permutations, the symmetric group has order $n!$ To prove that we actually have a group, we proceed as follows: Any such operator may be written in the form

$$x'_1 = x_{i_1}$$
$$x'_2 = x_{i_2}$$
$$\vdots$$
$$x'_n = x_{i_n},$$

where the x_{i_j}'s are $x_1, x_2, \ldots, x_n$ in some order. This may be written in matrix form thus:

$$X' = PX, \qquad P = [E_{i_1}, E_{i_2}, \ldots, E_{i_n}].$$

The matrix P, a **permutation matrix**, has a single entry 1 in each row and in each column, all other entries being zero. Among these operators is the identity operator $X' = IX$ corresponding to the identical permutation which leaves the variables in the natural order. If we permute n variables and then permute them again, we still have a permutation of these variables. Hence the product of two permutations $X'' = P_2 X'$, $X' = P_1 X$, is a permutation $X'' = (P_2 P_1)X$. (This may also be shown by examining the elements of $P_2 P_1$.) Finally we observe that, if P is a permutation matrix, so is P^T, for it must have a single entry 1 in each *column* and in each *row*. Moreover, we have

$$P^\mathsf{T} P = I_n$$

for, if the ith row of P^T has its 1 in the kth column, then the ith column of P has its 1 in the kth row, and the product of the ith row of P^T and the ith column of P will be 1. On the other hand, no other column of P has a 1 in the kth row, and hence the product of the ith row of P^T and any *other* column of P must be 0. Thus $P^\mathsf{T} P$ is indeed the identity matrix, and hence the inverse of the operator $X' = PX$ is the operator $X'' = P^\mathsf{T} X'$, which is also a permutation. Thus the set of all $n!$ permutations is indeed a group of operators. This group is of great usefulness in a variety of applications.

The above discussion concerning groups of operators applies equally well to groups of linear transformations of coordinates since in this case, too, a product of transformations corresponds to a product of matrices. In either case, therefore, we are fundamentally concerned with groups of matrices, but different geometrical interpretations are given to matrix multiplication.

It would be difficult to overemphasize the importance of the group concept in modern mathematics, for it enters in some form into nearly every branch of the subject. However, we shall use it here in only the most elementary fashion.

8.10 Exercises

1. Compute the product of the operators

$$Z = \begin{bmatrix} 4 & 1 & 1 \\ 5 & 2 & 0 \end{bmatrix} Y, \qquad Y = \begin{bmatrix} 2 & -1 \\ -5 & 3 \\ -2 & 1 \end{bmatrix} X.$$

Explain geometrically what has occurred.

2. Note that

$$\begin{bmatrix} \sqrt{2} & 0 \\ 0 & \sqrt{2} \end{bmatrix} \begin{bmatrix} \dfrac{1}{\sqrt{2}} & -\dfrac{1}{\sqrt{2}} \\ \dfrac{1}{\sqrt{2}} & \dfrac{1}{\sqrt{2}} \end{bmatrix} = \begin{bmatrix} 1 & -1 \\ 1 & 1 \end{bmatrix}.$$

How, therefore, may the effect of the operator defined by the matrix on the right be described geometrically, inasmuch as it is the product of two operators whose geometrical properties are known?

3. Show that every operator on $\mathscr{E}^2$ can be decomposed into a product of shears, reflections, and elongations or contractions. (The result is not unique, nor is it actually necessary to use all three types of transformations.)

4. Show that, in a given group, there exists only one identity element and only one inverse for a given element.

5. Show that each of the following sets of linear operators on a vector space is a group:
   (a) The set of all real nonsingular operators $X' = DX$, where D is a diagonal matrix.
   (b) The set of all real operators $X' = AX$ with $\det A = \pm 1$.
   (c) The set of all real operators $X' = AX$ with $\det A = 1$.
   (d) The set of all complex operators $X' = AX$ with $|\det A| = 1$.

6. Show that, with respect to the operation of matrix multiplication, the set of all real matrices

$$\begin{bmatrix} a & -b \\ b & a \end{bmatrix}, \qquad a^2 + b^2 \neq 0,$$

constitutes a group.

7. Show that the set of matrices

$$\begin{bmatrix} 1 & 0 \\ 0 & 1 \end{bmatrix}, \quad \begin{bmatrix} -1 & 0 \\ 0 & -1 \end{bmatrix}, \quad \begin{bmatrix} 0 & 1 \\ -1 & 0 \end{bmatrix}, \quad \begin{bmatrix} 0 & -1 \\ 1 & 0 \end{bmatrix},$$

$$\begin{bmatrix} 0 & 1 \\ 1 & 0 \end{bmatrix}, \quad \begin{bmatrix} 0 & -1 \\ -1 & 0 \end{bmatrix}, \quad \begin{bmatrix} 1 & 0 \\ 0 & -1 \end{bmatrix}, \quad \begin{bmatrix} -1 & 0 \\ 0 & 1 \end{bmatrix},$$

forms a group of which the first four form a subgroup. Can you find the other two subgroups of order 4 and the five subgroups of order 2?

8. Two groups G and H are said to be **isomorphic** if and only if there exists a one-to-one correspondence f between their elements such that if $f(g_1) = h_1$ and $f(g_2) = h_2$, then $f(g_1 g_2) = h_1 h_2$. (In accord with standard usage, the group operation is denoted by simple juxtaposition here.) Is the group whose elements are the matrices

$$\begin{bmatrix} 1 & 0 \\ 0 & 1 \end{bmatrix}, \quad \begin{bmatrix} -1 & 0 \\ 0 & -1 \end{bmatrix}, \quad \begin{bmatrix} 0 & 1 \\ -1 & 0 \end{bmatrix}, \quad \begin{bmatrix} 0 & -1 \\ 1 & 0 \end{bmatrix},$$

$$\begin{bmatrix} i & 0 \\ 0 & -i \end{bmatrix}, \quad \begin{bmatrix} -i & 0 \\ 0 & i \end{bmatrix}, \quad \begin{bmatrix} 0 & i \\ -i & 0 \end{bmatrix}, \quad \begin{bmatrix} 0 & -i \\ -i & 0 \end{bmatrix},$$

isomorphic to the group of Exercise 7?

9. A group G is said to be **generated** by the set of elements $g_1, g_2, \ldots, g_k$ if and only if
   (a) every element g of G is representable in the form $g = g_{i_1} g_{i_2} \cdots g_{i_n}$, where the i_j belong to $\{1, 2, \ldots, k\}$ and need not be distinct and where n depends on g, and
   (b) for every integer n and every choice of the i_j from $\{1, 2, \ldots, k\}$, the product $g_{i_1} g_{i_2} \cdots g_{i_n}$ belongs to G. The elements $g_1, g_2, \ldots, g_k$ are called a set of **generators** of G. Every group has at least one set of generators, namely the set of all its elements.

   Show that the group of Exercise 8 is generated by the two matrices

$$\begin{bmatrix} 0 & -1 \\ 1 & 0 \end{bmatrix} \quad \text{and} \quad \begin{bmatrix} 0 & i \\ i & 0 \end{bmatrix}.$$

10. Find a minimal set of generators of the group given in Exercise 7.

11. Show that the set of all matrices $\begin{bmatrix} 1 & 0 \\ c & 1 \end{bmatrix}$, where c is an arbitrary complex number, is a group with respect to multiplication. What are some infinite subgroups of this group?

12. Show that the set of all nonsingular upper triangular matrices of order n over a field $\mathscr{F}$ constitutes a group with respect to multiplication.

13. Show that, over a field $\mathscr{F}$, the nonsingular matrices of order n which commute with a fixed matrix of order n form a group.

14. Do the elementary row transformations of a given matrix form a group?

15. By actual consideration of their elements, show that the product of two permutation matrices of order n is again a permutation matrix.

16. Show that every permutation matrix can be factored into a product of certain of the elementary matrices $P_{12}, P_{23}, \ldots, P_{n-1,n}$, where P_{ij} is a permutation matrix which corresponds to the interchange of the ith and jth variables (columns). (Some of these matrices may have to be used more than once in the product.)

17. Prove that, for every permutation matrix P of order n, there exists a positive integer q such that $P^q = I_n$, and hence that $P^{q-1} = P^{-1}$. How can you find the smallest positive integer that will serve? (*Hint:* Every power of P is a permutation matrix (of which there are only $n!$). Hence there must exist integers $r > s > 0$ such that $P^r = P^s$, etc.)

18. Show that the determinant of a permutation matrix is $+1$ or -1 depending on whether the corresponding permutation is even or odd.

19. An arbitrary permutation matrix P may be written in the form $P = [E_{i_1}, E_{i_2}, \ldots, E_{i_n}]$, where the E's are elementary n-vectors and $i_1, i_2, \ldots, i_n$ is a permutation of $1, 2, \ldots, n$. Establish a general rule for writing P^{-1} in terms of the E's. For example, if $P = [E_3, E_1, E_2]_3$, then $P^{-1} = [E_2, E_3, E_1]_3$, etc.

20. Prove that an $n \times n$ matrix commutes with every $n \times n$ permutation matrix if and only if it has the form $aI + bE$, where E is a matrix all of whose entries are 1's and a and b are scalars.

21. Determine under what conditions the matrix $aI + bE$ defined in Exercise 20 is nonsingular. Then compute the inverse. (*Hint:* Use the preceding exercise to show that the inverse has the same form. Then find scalars α and β such that $\alpha I + \beta E$ is the desired inverse.) Do the nonsingular matrices of this form constitute a group?

22. Let us call a matrix of the form $[E_i, E_{i+1}, \ldots, E_n, E_1, \ldots, E_{i-1}]$ a **cyclic permutation matrix**. Show that the set of cyclic permutation matrices of order n constitutes a subgroup of the set of all $n \times n$ permutation matrices. Show also that all the matrices of this group may be represented as powers of a single matrix.

23. The cyclic permutation matrices

$$I = \begin{bmatrix} 1 & 0 & 0 \\ 0 & 1 & 0 \\ 0 & 0 & 1 \end{bmatrix}, \qquad U = \begin{bmatrix} 0 & 1 & 0 \\ 0 & 0 & 1 \\ 1 & 0 & 0 \end{bmatrix}, \qquad U^2 = \begin{bmatrix} 0 & 0 & 1 \\ 1 & 0 & 0 \\ 0 & 1 & 0 \end{bmatrix}$$

are used in the study of three-phase electric current. Show that the set of matrix polynomials $\alpha I + \beta U + \gamma U^2$, where α, β, γ are arbitrary complex numbers, constitutes a commutative ring with a unit element. Show that there exist divisors of zero in this ring.

24. Let us call a matrix of the form $[\alpha_1 E_{j_1}, \alpha_2 E_{j_2}, \ldots, \alpha_n E_{j_n}]$, where the α's are nonzero scalars from a field $\mathscr{F}$, a **weighted permutation matrix** over $\mathscr{F}$. Show that the set of all weighted permutation matrices over an arbitrary field $\mathscr{F}$ constitutes a group with respect to multiplication.

8.11 Unitary and Orthogonal Matrices

In Chapter 1 we made a number of definitions concerning matrices with complex elements. The transposed conjugate $(\overline{A})^{\mathsf{T}}$ of a matrix A was called the **tranjugate** and was denoted by A^*. A square matrix A was defined to be **Hermitian** if and only if $A = A^*$. It was pointed out that if A is real, $A^* = A^{\mathsf{T}}$, so that a real Hermitian matrix is symmetric. A few simple properties of the tranjugate and of Hermitian matrices were given in the exercises in Section 1.18. More are given below. A careful study of all these will shortly repay itself.

For use in coming sections, we define an $n \times n$ matrix U to be **unitary** if and only if it has the property that $U^*U = I_n$; that is, U is unitary if and only if its tranjugate U^* is its inverse. Since $U^* = U^{-1}$, we have also $UU^* = I_n$.

A *real* unitary matrix has the property that $U^\mathsf{T}U = UU^\mathsf{T} = I_n$. A real unitary matrix is said to be **orthogonal.**

From the condition

$$U^*U = \begin{bmatrix} U_1{}^* \\ U_2{}^* \\ \vdots \\ U_n{}^* \end{bmatrix} \cdot [U_1, U_2, \ldots, U_n] = [U_i{}^* U_j]_n = [\delta_{ij}]_n = I_n,$$

where $U_1, U_2, \ldots, U_n$ are the columns of U, and from a similar examination of UU^*, we obtain

Theorem 8.11.1: *A square matrix U is unitary if and only if its columns* (rows) *are mutually orthogonal unit vectors.*

Since $\det U^* = \det (\overline{U})^\mathsf{T} = \det \overline{U} = \overline{\det U}$, we see that, if U is unitary, then $\overline{\det U} \cdot \det U = |\det U|^2 = 1$. This yields

Theorem 8.11.2: *If U is unitary, the absolute value of $\det U$ is 1.*

If U is orthogonal, $\det U$ must therefore be ± 1. If $\det U$ is 1, the matrix U is said to be **unimodular.** The unimodular unitary matrices of order n are analogous to the matrices of rotations of axes in $\mathscr{E}^3$. An orthogonal matrix with $\det U = 1$ is called **proper.** If $\det U = -1$, it is called **improper.**

Any set of n linearly independent, mutually orthogonal, normalized (that is, unit) vectors of $\mathscr{U}^n$ is called an **orthonormal reference system** or an **orthonormal basis** for $\mathscr{U}^n$. We have then, from Theorem 8.11.1,

Theorem 8.11.3: *The columns of a unitary matrix U constitute an orthonormal basis for $\mathscr{U}^n$. The same applies to the columns of U^*.*

8.12 Exercises

1. Under what conditions will the matrix

$$\begin{bmatrix} a+b & b-a \\ a-b & b+a \end{bmatrix},$$

where a and b represent real numbers, be orthogonal?

2. Given that

$$A = \begin{bmatrix} a & -3b & c_1 \\ 2a & 2b & c_2 \\ -a & b & c_3 \end{bmatrix},$$

choose a, b, c_1, c_2, and c_3 so that A is orthogonal. Is the solution unique?

*3. Prove that if the vectors $X_1, X_2, \ldots, X_k$ of $\mathscr{U}^n$ are mutually orthogonal, then they are linearly independent.

4. Show that if both the rows and the columns of a complex matrix A are sets of mutually orthogonal vectors, then A must be square.

5. Show that these are orthogonal matrices:

(a) $\begin{bmatrix} \dfrac{1}{\sqrt{2}} & 0 & \dfrac{1}{\sqrt{2}} \\[2ex] \dfrac{1}{\sqrt{2}} & 0 & -\dfrac{1}{\sqrt{2}} \\[2ex] 0 & 1 & 0 \end{bmatrix}$,

(b) $\begin{bmatrix} \dfrac{1}{\sqrt{6}} & \dfrac{2}{\sqrt{5}} & \dfrac{1}{\sqrt{30}} \\[2ex] -\dfrac{1}{\sqrt{6}} & 0 & \dfrac{5}{\sqrt{30}} \\[2ex] \dfrac{2}{\sqrt{6}} & -\dfrac{1}{\sqrt{5}} & \dfrac{2}{\sqrt{30}} \end{bmatrix}$,

(c) $\begin{bmatrix} \dfrac{1}{\sqrt{14}} & \dfrac{3}{\sqrt{14}} & \dfrac{2}{\sqrt{14}} \\[2ex] \dfrac{3}{\sqrt{10}} & -\dfrac{1}{\sqrt{10}} & 0 \\[2ex] -\dfrac{1}{\sqrt{35}} & -\dfrac{3}{\sqrt{35}} & \dfrac{5}{\sqrt{35}} \end{bmatrix}$,

(d) $\begin{bmatrix} \dfrac{1}{\sqrt{2}} & -\dfrac{1}{\sqrt{2}} & 0 \\[2ex] \dfrac{1}{\sqrt{6}} & \dfrac{1}{\sqrt{6}} & -\dfrac{2}{\sqrt{6}} \\[2ex] \dfrac{1}{\sqrt{3}} & \dfrac{1}{\sqrt{3}} & \dfrac{1}{\sqrt{3}} \end{bmatrix}$.

*6. Determine all unit n-vectors X whose components are exclusively 0's and 1's. Then determine all $n \times n$ orthogonal matrices whose entries are exclusively 0's and 1's.

7. Show that, if U is orthogonal and det $U = +1$, then each element of U is equal to its cofactor in det U, and if det $U = -1$, then each element of U is equal to the negative of its cofactor in det U.

8. Prove that, if $X \neq 0$ is a real n-vector, then the matrix

$$I_n - \frac{2}{X^\mathsf{T} X} X X^\mathsf{T}$$

is orthogonal and symmetric.

9. Under what conditions are the matrices

$$\begin{bmatrix} 0 & \alpha & 0 & i\beta \\ \alpha & 0 & i\beta & 0 \\ 0 & i\beta & 0 & \alpha \\ i\beta & 0 & \alpha & 0 \end{bmatrix}, \quad \begin{bmatrix} 0 & 0 & \gamma & i\delta \\ 0 & 0 & i\delta & \gamma \\ \gamma & i\delta & 0 & 0 \\ i\delta & \gamma & 0 & 0 \end{bmatrix}, \quad \begin{bmatrix} 0 & \eta & 0 & \zeta \\ \eta & 0 & \zeta & 0 \\ 0 & \zeta & 0 & \eta \\ \zeta & 0 & \eta & 0 \end{bmatrix}$$

unitary if $\alpha, \beta, \gamma, \delta, \eta, \zeta$, are real numbers? (The resulting unitary matrices are called "scattering matrices" in microwave circuit theory.)

10. Let the real equation

$$[A, B]_{m \times n} X_{n \times 1} = C_{m \times 1}, \qquad n \geq m,$$

where A is an $m \times m$ orthogonal matrix, be multiplied by A^T. What is the effect on the corresponding system of equations?

11. Solve this system by inspection, using the idea of Exercise 10:

$$\frac{1}{\sqrt{2}} x_1 - \frac{1}{\sqrt{2}} x_2 + x_3 = 1$$

$$\frac{1}{\sqrt{2}} x_1 + \frac{1}{\sqrt{2}} x_2 - x_3 = 1.$$

*12. Prove that the system of linear equations represented by $AX = B$ is consistent if and only if B is orthogonal to every solution of the corresponding transposed homogeneous system $A^\mathsf{T} Y = 0$.

13. Given that ω is a complex cube root of unity, for what values of α, β, γ will the matrix

$$\begin{bmatrix} \dfrac{\alpha}{\sqrt{3}} & \dfrac{\beta}{\sqrt{3}} & \dfrac{\gamma}{\sqrt{3}} \\[2mm] \dfrac{\alpha}{\sqrt{3}} & \dfrac{\beta\omega^2}{\sqrt{3}} & \dfrac{\gamma\omega}{\sqrt{3}} \\[2mm] \dfrac{\alpha}{\sqrt{3}} & \dfrac{\beta\omega}{\sqrt{3}} & \dfrac{\gamma\omega^2}{\sqrt{3}} \end{bmatrix}, \qquad \left(\omega = \frac{-1 + i\sqrt{3}}{2} \right)$$

be both unitary and unimodular?

*14. Show that, if U is unitary, then so are $\overline{U}$, U^T, and U^k for every integer k.

15. Given that V and W are $n \times n$ unitary matrices which commute, show that

$$\frac{1}{\sqrt{2}} \begin{bmatrix} V & -W \\ W^* & V^* \end{bmatrix}$$

is also unitary.

16. Show that, if U is both unitary and Hermitian, then $U^2 = I_n$, that is, U is involutory. In fact, any two of these properties imply the third.

*17. Show that any unitary matrix can be transformed into a unimodular, unitary matrix by multiplying any line thereof by a suitable scalar $\alpha + i\beta$.

*18. Prove in detail that $\det U^* = \overline{\det U}$.

*19. Show that a sum of Hermitian matrices is Hermitian.

20. Show that $A + \overline{A}$ has only real elements, and that $A - \overline{A}$ has only pure imaginary elements.

21. Show that k n-vectors $A_1, A_2, \ldots, A_k$ are linearly independent if and only if their conjugates $\overline{A}_1, \overline{A}_2, \ldots, \overline{A}_k$ are linearly independent.

*22. Show that a matrix A and its tranjugate A^* have the same rank.
*23. Show that, if H is Hermitian, det H is real.
 24. Show that every permutation matrix is orthogonal.
 25. Show that, if a matrix U of order 2 is both unitary and unimodular, it can be written in the form

$$\begin{bmatrix} \alpha & -\beta \\ \overline{\beta} & \overline{\alpha} \end{bmatrix},$$

where $\overline{\alpha}\alpha + \overline{\beta}\beta = 1$.

 26. Show that, if X and Y are n-vectors, $XY^\mathsf{T} + YX^\mathsf{T}$ is symmetric and that $XY^* + YX^*$ is Hermitian.
 27. Show that the Pauli spin matrices of Exercise 22, Section 1.9, are unitary, involutory, and Hermitian.
 28. Let $Z = X + iY$, $H = A + iB$, where X and Y are real n-vectors and A and B are real matrices of order n. Show that, if H is Hermitian, the scalar Z^*HZ is real for all Z. Show similarly that, if H is skew-Hermitian, the scalar Z^*HZ is a pure imaginary.
*29. Prove that, if U is unitary and U^*AU and U^*BU are both diagonal matrices, then A and B commute.

8.13 Unitary Transformations

If U is a unitary matrix, then the linear vector function defined by $Y = UX$, whose domain and range are $\mathscr{U}^n$, is called a **unitary operator** on $\mathscr{U}^n$. On the other hand, the same equation can be used to represent a **unitary transformation of coordinates**, that is, one in which the new reference vectors, just as $E_1, E_2, \ldots, E_n$, constitute an orthonormal basis for $\mathscr{U}^n$.

It is convenient at times to refer to the equation $Y = UX$ as representing a **linear transformation** which may then be interpreted either as an operator or as a transformation of coordinates. In this way, we can use one algebraic argument to prove two geometric theorems.

The unitary transformations play a role in $\mathscr{U}^n$ analogous to that of the distance-preserving rotations and reflections in $\mathscr{E}^3$. We have, in fact,

Theorem 8.13.1: In $\mathscr{U}^n$ *a linear transformation* $Y = AX$ *leaves the length of all vectors invariant if and only if it is a unitary transformation.*

Geometrically, this means that, if the transformation represents an operator, the image of a given vector always has the same length as the given vector. If the transformation is a linear transformation of coordinates, then the formula for the length of a given vector is the same in the new reference system as it is in the old reference system.

It will suffice to show that X^*X is invariant for all X if and only if A is unitary. The condition $Y^*Y = X^*X$, where $Y = AX$, is satisfied for every vector X if and only if $X^*A^*AX = X^*X$ for all X.

Now if $A^*A = I$, then $X^*A^*AX = X^*X$ for all X. Conversely, suppose that $X^*A^*AX = X^*X$ for all X. Then we may put $X = E_j$, which yields

$$(8.13.1) \qquad E_j^T A^*A E_j = (A^*A)_{jj} = E_j^T E_j = 1,$$

where $(A^*A)_{jj}$ denotes the jj-entry of A^*A. Thus the diagonal entries of A^*A are all 1.

Next we put $X = E_j + E_k$, $j \neq k$, and obtain, since the length of X is invariant,

$$(E_j^T + E_k^T)A^*A(E_j + E_k) = (E_j^T + E_k^T)(E_j + E_k) = 2.$$

Recalling that $E_j^T A^*A E_k = (A^*A)_{jk}$, we find with the aid of (8.13.1) that

$$(8.13.2) \qquad (A^*A)_{jk} + (A^*A)_{kj} = 0.$$

Finally, we put $X = E_j + iE_k$, where $i^2 = -1$, and obtain

$$(E_j^T - iE_k^T)A^*A(E_j + iE_k) = (E_j^T - iE_k^T)(E_j + iE_k) = 2,$$

so that now

$$(8.13.3) \qquad (A^*A)_{jk} - (A^*A)_{kj} = 0.$$

The results (8.13.2) and (8.13.3) imply that, for all $j \neq k$,

$$(A^*A)_{jk} = 0.$$

The off-diagonal entries of A^*A are thus all zero, so the proof that A is unitary is complete.

The proof clearly does not depend on whether the transformation represents a linear operator or a linear transformation of coordinates.

The scalar $X^*X = \sum \bar{x}_j x_j$ is also called the **Hermitian unit form**, so the previous theorem says that *a transformation is unitary if and only if it leaves the Hermitian unit form invariant.*

Theorem 8.13.2: *The unitary operators on $\mathcal{U}^n$ constitute a group; so do the unitary transformations of coordinates in $\mathcal{U}^n$.*

To prove this, we need only show that the unitary *matrices* of order n form a group with respect to multiplication:

(a) If A and B are unitary, then $A^*A = I_n$, $B^*B = I_n$, and hence $(AB)^*(AB) = B^*(A^*A)B = B^*B = I_n$, so AB is also unitary.

(b) If A is unitary, then $A^*A = I_n$, so $A^* = A^{-1}$, and hence $(A^{-1})^*A^{-1} = (A^*)^*A^{-1} = AA^{-1} = I_n$, so A^{-1} is also unitary.
(c) The identity matrix is unitary since $I_n^* I_n = I_n$.
These three properties establish the desired result.

The study of unitary n-space may be regarded as the study of those properties of vectors which are invariant under unitary transformations. The next two theorems illustrate this point.

Theorem 8.13.3: *In $\mathcal{U}^n$, the inner product X^*Y is invariant under a unitary transformation of coordinates.*

If U is unitary and $X = UW$, $Y = UZ$, then $X^*Y = W^*U^*UZ = W^*Z$, which proves the theorem.
An important consequence of this theorem is that the test for orthogonality of two vectors is independent of the choice of the orthonormal reference system for $\mathcal{U}^n$.
The same computations allow us also to conclude

Theorem 8.13.4: *In $\mathcal{U}^n$, a unitary operator maps orthogonal vectors onto orthogonal vectors and nonorthogonal vectors onto nonorthogonal vectors.*

8.14 Orthogonal Transformations

We have pointed out that, if U is unitary and real, then $U^*U = I_n$ reduces to $U^TU = I_n$ and U is called an orthogonal matrix. A transformation $Y = UX$ is then called an **orthogonal transformation**. The previous theorems have as special cases:

Theorem 8.14.1: *In $\mathcal{E}^n$, a linear transformation leaves the length of each vector invariant if and only if it is an orthogonal transformation.*

The proof of this theorem is similar to the proof of Theorem 8.13.1. However, instead of using the imaginary unit i, one uses the fact that A^TA is symmetric.

Theorem 8.14.2: *The orthogonal operators on $\mathcal{E}^n$ constitute a group; so do the orthogonal transformations of coordinates in $\mathcal{E}^n$.*

If we allow all the operators in question to have the vectors of $\mathcal{U}^n$ as their common domain, it now follows that the real orthogonal operators are a subgroup of the group of unitary operators and the latter, in turn, are a subgroup of the full linear group over the complex field (Section 8.9).

We have finally the special cases:

Theorem 8.14.3: *In $\mathscr{E}^n$, the inner product $X^{\mathsf{T}}Y$ is invariant under an orthogonal transformation of coordinates.*

Theorem 8.14.4: *In $\mathscr{E}^n$, an orthogonal operator maps orthogonal vectors onto orthogonal vectors and nonorthogonal vectors onto nonorthogonal vectors.*

The reader should write out proofs of each of these.

8.15   The Eulerian Angles

If $X = AY$ represents an orthogonal transformation of coordinates in $\mathscr{E}^3$, then since the columns of A are *unit* vectors, and also represent the new reference vectors in the original reference system, *the columns of A are given by the direction cosines of the new reference axes with respect to the old reference system.* Similarly, in the equivalent equation $Y = A^{\mathsf{T}}X$, *the columns of A^{T} are given by the direction cosines of the old reference axes with respect to the new reference system.* These observations ordinarily make it easy to write down the matrix of the transformation or its inverse. For example, an orthogonal transformation of coordinates in which the new reference axes pass through the points $P_1: (1, 1, 1)$, $P_2: (0, 2, -2)$, $P_3: (-2, 1, 1)$ respectively may be written in the form

$$
X = \begin{bmatrix}
\dfrac{1}{\sqrt{3}} & 0 & \dfrac{-2}{\sqrt{6}} \\[2ex]
\dfrac{1}{\sqrt{3}} & \dfrac{1}{\sqrt{2}} & \dfrac{1}{\sqrt{6}} \\[2ex]
\dfrac{1}{\sqrt{3}} & \dfrac{-1}{\sqrt{2}} & \dfrac{1}{\sqrt{6}}
\end{bmatrix} Y,
$$

where the columns are the normalized forms of $\overrightarrow{OP_1}$, $\overrightarrow{OP_2}$, $\overrightarrow{OP_3}$, that is, the direction cosines of these vectors in the x-reference system.

To illustrate how these observations are useful in a practical application, we discuss the Eulerian angles used in classical mechanics.

In many problems in mechanics, it is necessary to specify the instantaneous position of a rigid body which is free to rotate about a fixed point O, that is, free to move in such a way that every point in it remains always at the same distance from O.

We consider a line L passing through O and allow L to move with the body, but always require it to remain in the same position relative to the rigid body. If now the position of L relative to the reference system at O is fixed, the only further motion possible for the body is rotation about L.

The motion of the body may be described mathematically by using L as one axis of a reference system, centered also at O, the other two axes of which are, like L, fixed relative to the body and, therefore, move with it. When we specify the position of this reference system relative to the original reference system centered at O, we have then specified the position of the body. In this type of motion, every point of the body remains at a constant distance from O.

Let OX_1, OX_2, OX_3 denote the axes of the fixed reference system and let OW_1, OW_2, OW_3 denote the axes of the movable reference system. We wish to define the orthogonal transformation $W = AX$, which expresses the w-coordinates of an arbitrary point in terms of its x-coordinates.

First let new axes OY_1, OY_2, $OY_3 \equiv OX_3$ be obtained by a rotation about OX_3 through the angle ϕ in the x_1x_2-plane, from OX_2 to the plane determined by OX_3 and OW_3 (Figure 8-9). Next we obtain new axes $OZ_1 \equiv OY_1$, OZ_2, OZ_3 by rotation about OY_1 through the angle θ from OX_3 to OW_3. Finally, we rotate about OZ_3 through the angle ψ between OZ_1 and OW_1. From the figure it is clear that this step brings us to the w-reference system. The angles ϕ, θ, ψ are to be chosen in the interval from 0 to 2π.

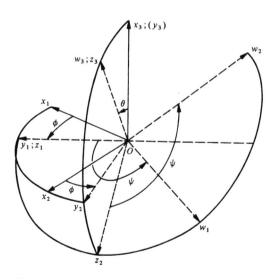

FIGURE 8-9. Eulerian Angles

Writing the matrix representations of the above rotations according to the rules stated above, we obtain respectively,

$$(8.15.1) \qquad Y = \begin{bmatrix} \cos\phi & \sin\phi & 0 \\ -\sin\phi & \cos\phi & 0 \\ 0 & 0 & 1 \end{bmatrix} X,$$

$$(8.15.2) \qquad Z = \begin{bmatrix} 1 & 0 & 0 \\ 0 & \cos\theta & \sin\theta \\ 0 & -\sin\theta & \cos\theta \end{bmatrix} Y,$$

$$(8.15.3) \qquad W = \begin{bmatrix} \cos\psi & \sin\psi & 0 \\ -\sin\psi & \cos\psi & 0 \\ 0 & 0 & 1 \end{bmatrix} Z.$$

The reader should verify in detail the correctness of the entries in these matrices, and then should compute the product transformation $W = AX$.

The angles ψ, θ, ϕ used here are known as the *Eulerian angles*. There are other definitions of the Eulerian angles in the literature. Some authors use clockwise angles, some use left-handed reference systems, and some interchange the meaning of ϕ and ψ. (The usage here is the same as in H. Goldstein: *Classical Mechanics*, Addison-Wesley, Reading, Mass., 1950, and in H. C. Corben and P. Stehle: *Classical Mechanics*, 2nd ed., John Wiley & Sons, New York, 1960.) Whatever notation the reader may encounter, he should be able to write correctly the matrices of the transformations simply by following the pattern established here and taking due account at each step of the differences in notation.

It should be pointed out that the Eulerian angles exhibit a certain defect. When the axes OX_3 and OW_3 coincide, the angle ϕ is undefined. If we assign it a value ϕ_0, we find at the next step that $\theta = 0$ and, at the last step, that $\psi = \alpha - \phi_0$, where α is the angle from OX_1 to OW_1. The product A in this case reduces to

$$\begin{bmatrix} \cos\alpha & \sin\alpha & 0 \\ -\sin\alpha & \cos\alpha & 0 \\ 0 & 0 & 1 \end{bmatrix},$$

as one would expect.

8.16   The Triangularization of a Real Matrix

In the previous section we decomposed an orthogonal transformation of coordinates in $\mathscr{E}^3$ into the product of three plane rotations. The particular rotations chosen there were those most used in mechanics. There is a similar procedure in $\mathscr{E}^n$ which follows as a corollary of the next theorem. To make the process systematic, we define $P_{ik}(\theta)$, $i \neq k$, to be an $n \times n$ matrix such that in the

ith row and ith column appears $\cos \theta$;
ith row and kth column appears $-\sin \theta$;
kth row and ith column appears $\sin \theta$;
kth row and kth column appears $\cos \theta$.

The remaining elements are 0, except for the other elements on the principal diagonal, which are all 1.

It is readily seen that $P_{ik}(\theta)$ is orthogonal, that det $P_{ik}(\theta) = 1$, and that

$$P_{ki}(\theta) = P_{ik}(-\theta) = P_{ik}^{\mathsf{T}}(\theta) = P_{ik}^{-1}(\theta).$$

In the previous section, we employed $P_{21}(\phi)$, $P_{32}(\theta)$, and $P_{21}(\psi)$.

We now prove a theorem from which a number of important results are derived.

Theorem 8.16.1: *Every square real matrix A may be written in the form*

(8.16.1) $$A = P_{21}P_{31} \cdots P_{n1}P_{32} \cdots P_{n2} \cdots P_{n, n-1}B,$$

where

$$P_{ik} = P_{ik}(\theta_{ik}),$$

the θ's depending on A, and where B is upper triangular with $b_{ii} \geq 0$ for $i < n$.

To prove this, let us write first of all

$$P_{21}^{-1}(\theta_{21})A = S,$$

and in the product require that $s_{21} = 0$; that is,

$$a_{11} \sin \theta_{21} + a_{21} \cos \theta_{21} = 0.$$

Hence, if $a_{11} \neq 0$, θ_{21} is defined by

$$\tan \theta_{21} = -\frac{a_{21}}{a_{11}}.$$

If $a_{11} = 0$, but $a_{21} \neq 0$, we take $\theta_{21} = \pi/2$ so that $-a_{21} \neq 0$ will appear in the 11-position in S. If $a_{11} = a_{21} = 0$, we take $\theta_{21} = 0$, and P_{21} is the identity matrix.

Next we put

$$P_{31}^{-1}(\theta_{31})S = T,$$

and require the vanishing of t_{31}:

$$t_{31} = s_{11} \sin \theta_{31} + s_{31} \cos \theta_{31} = 0.$$

Again, if $s_{11} \neq 0$, θ_{31} is defined by

$$\tan \theta_{31} = -\frac{s_{31}}{s_{11}}.$$

If $s_{11} = 0$ but $s_{31} \neq 0$, we take $\theta_{31} = \pi/2$. If $s_{11} = s_{31} = 0$, we take $\theta_{31} = 0$. In the matrix T we have

$$t_{21} = 0 \cdot s_{11} + 1 \cdot s_{21} = 0,$$

so that the work of the previous step has not been undone.

We can continue this and obtain all zeros in the first column except possibly for the 11-entry, which is to be ≥ 0. If the last step in the first column is given by

$$P_{n1}^{-1}(\theta_{n1})R = Q,$$

then we require

$$q_{n1} = r_{11} \sin \theta_{n1} + r_{n1} \cos \theta_{n1} = 0,$$

$$q_{11} = r_{11} \cos \theta_{n1} - r_{n1} \sin \theta_{n1} \geq 0.$$

Here θ_{n1} is determined just as before, except that it may be necessary to add or subtract π in order to satisfy the inequality.

Now we operate on the second column in the same way, using

$$P_{32}^{-1}(\theta_{32}), \ldots, P_{n2}^{-1}(\theta_{n2})$$

to reduce to zero the elements of the second column below the main diagonal. The zeros in the first column are not destroyed by these operations.

Continuing thus with the first $n - 1$ columns, we have, finally,

(8.16.2) $$P_{n, n-1}^{-1}(\theta_{n, n-1}) \cdots P_{21}^{-1}(\theta_{21})A = B,$$

where B is triangular, and from which (8.16.1) follows at once, so the proof is complete. Since we do not work with the last column, there is no way to control the sign of b_{nn} in this process.

Since the P_{ij}'s are proper orthogonal and the product of proper orthogonal matrices is proper orthogonal, we have

Corollary 8.16.2: *Every real matrix A can be expressed in the form*

$$(8.16.3) \qquad\qquad A = PB,$$

where P is a proper orthogonal matrix and B is upper triangular, with $b_{ii} \geq 0$, $i = 1, 2, \ldots, n-1$.

If A itself is orthogonal, then more can be said about B. We have

Corollary 8.16.3: *If A is orthogonal and if* det $A = 1$, *then in* (8.16.1), $B = I_n$. *If* det $A = -1$, *then in B, $b_{ij} = \delta_{ij}$, except that $b_{nn} = -1$.*

Indeed, from (8.16.3) it follows in either case that, since A is orthogonal, B is orthogonal. Hence, from the first column of B, we conclude $b_{11}^2 = 1$, so that, since $b_{11} \geq 0$, $b_{11} = 1$. Also, from the first row of B, we have

$$b_{11}^2 + b_{12}^2 + \cdots + b_{1n}^2 = 1,$$

from which now follows $b_{12} = b_{13} = \cdots = b_{1n} = 0$. Then from column 2 and row 2 we have $b_{22} = 1$, $b_{23} = b_{24} = \cdots = b_{2n} = 0$. Similarly for the first $n-1$ rows and columns of B. Finally, $b_{nn}^2 = 1$. Since det $A =$ det B because det $P_{ik} = 1$ in every case, we therefore have $b_{nn} = \pm 1$ according as det $A = \pm 1$, and the proof of the corollary is complete.

An orthogonal transformation with matrix B in which $b_{ij} = \delta_{ij}$, except that $b_{ii} = -1$ for exactly one value of i, is called a **reflection** (in the ith coordinate hyperplane). An orthogonal transformation $P_{ij}(\theta)$ is called a **plane rotation**. Corollary 8.16.3 may therefore be restated thus:

Corollary 8.16.4: *Every orthogonal transformation of coordinates may be factored into a product of plane rotations and at most one reflection.*

Finally we prove

Corollary 8.16.5: *If A is nonsingular, then the representation $A = PB$ of Corollary 8.16.2 is unique.*

If $A = PB = P'B'$, with P, P' both proper orthogonal and B, B' upper triangular, then $I = P^{\mathsf{T}}P'B'B^{-1}$. Thus the orthogonal matrix I is represented as the product of the proper orthogonal matrix PP' and the upper triangular matrix $B'B^{-1}$. Hence, by Corollary 8.16.3, $B'B^{-1} = I$, and, therefore, $P^{\mathsf{T}}P' = I$. Hence $B' = B$ and $P' = P$, and the corollary is proved.

8.17 Exercises

1. Show algebraically that the product of two rotations of axes in $\mathscr{E}^2$ is a rotation of axes in $\mathscr{E}^2$.

2. Prove that in $\mathscr{E}^2$, every proper orthogonal transformation of coordinates is a rotation of axes about the origin.

3. Show that in $\mathscr{E}^2$ the product of a rotation of axes and a reflection is an improper orthogonal transformation and that every improper orthogonal transformation in $\mathscr{E}^2$ is representable as such a product.

4. Prove that the operator on $\mathscr{E}^2$ with matrix

$$\begin{bmatrix} \cos\theta & \sin\theta \\ \sin\theta & -\cos\theta \end{bmatrix}$$

is involutorial (is its own inverse) and explain geometrically, using the concepts of rotation and reflection.

5. Compute the product of the three transformations (8.15.1) through (8.15.3).

6. Write the equation of the transformation of coordinates in $\mathscr{E}^3$ which

    (a) rotates the axes about the 1-axis through an angle α_1,

    (b) then rotates the new axes about the new 2-axis through an angle α_2,

    (c) and finally rotates the resulting axes, about the 3-axis obtained by (b), through an angle α_3. Show that every proper orthogonal transformation of coordinates in $\mathscr{E}^3$ can be obtained in this way. Illustrate with a figure that shows how the angles $\alpha_1, \alpha_2, \alpha_3$ must be chosen to accomplish the result.

7. How is the cross product in $\mathscr{E}^3$, $X \times Y$, affected by an orthogonal transformation of coordinates?

8. Prove that if a linear operator on $\mathscr{E}^n$ maps an orthonormal system $U_1, U_2, \ldots, U_n$ onto an orthonormal system $V_1, V_2, \ldots, V_n$, then the matrix of the operator is orthogonal.

9. Determine all $n \times n$ orthogonal matrices of 0's and 1's. Show that they form a group with respect to multiplication. What group is it?

10. Determine all $n \times n$ orthogonal matrices of 0's and 1's which are involutory, that is, whose squares are the identity.

11. Show that the unimodular, unitary transformations form a subgroup of the unitary group of transformations of n variables.

*12. Show in the manner of the proof of Theorem 8.13.1 that the inner product X^*Y is invariant under linear transformation if and only if the transformation is unitary and that the invariance of distance then follows as a corollary.

13. Write a proof of Theorem 8.14.1 which does not use the imaginary unit i but exploits the fact that $A^\mathsf{T}A$ is symmetric.

14. Triangularize the matrix

$$A = \begin{bmatrix} 1 & 2 & -1 \\ 1 & 3 & 0 \\ -\sqrt{2} & 2\sqrt{2} & \dfrac{\sqrt{2}}{2} \end{bmatrix}$$

by the method of Theorem 8.16.1.

15. Prove that the length of a line segment in $\mathscr{E}^n$ is unaltered by a combined translation and orthogonal operator, $Y = AX + Y_0$ (A orthogonal).

16. Find the equation in $\mathscr{E}^n$ of the line $X = X_0 + t(X_1 - X_0)$ in the reference system determined by the transformation of coordinates

$$Y = A(X - X_0),$$

where A is orthogonal. Illustrate with an example and figure in $\mathscr{E}^3$.

17. An operator $\tilde{X} = AX$ on $\mathscr{E}^n$ is called a **similarity transformation** if and only if $\tilde{X}^{\mathsf{T}}\tilde{Y} = cX^{\mathsf{T}}Y$, where $c > 0$. For what matrices A is $\tilde{X} = AX$ a similarity transformation? What is the effect of such an operator on the distance between two points? What is its effect on the angle between two vectors? Why is it reasonable to use the name "similarity transformation" in this case? (The term is also used in other ways in other contexts.)

18. List as many properties of operators, vectors, and figures as you can that are invariant under unitary transformations of coordinates.

8.18   Orthogonal Vector Spaces

We now extend to $\mathscr{U}^n$ some more of the geometry of $\mathscr{E}^3$.

We define a vector to be **orthogonal** to the subspace $\mathscr{V}$ of $\mathscr{U}^n$ if and only if it is orthogonal to every vector of $\mathscr{V}$. We have

Theorem 8.18.1: *A vector V is orthogonal to a subspace $\mathscr{V}$ of $\mathscr{U}^n$ if and only if it is orthogonal to every vector of a basis for $\mathscr{V}$.*

Let $A_1, A_2, \ldots, A_k$ be a basis for $\mathscr{V}$. Then if V is orthogonal to $\mathscr{V}$, $V^*A_i = 0$, $i = 1, 2, \ldots, k$. Conversely, if $V^*A_i = 0$, $i = 1, 2, \ldots, k$, then

$$V^*\left(\sum_{i=1}^{k} \alpha_i A_i\right) = \sum_{i=1}^{k} \alpha_i(V^*A_i) = 0,$$

and the theorem follows.

We next define two vector spaces in $\mathscr{U}^n$ to be **mutually orthogonal spaces** if and only if every vector in either space is orthogonal to every vector in the other. We prove next

Theorem 8.18.2: *Two subspaces of $\mathscr{U}^n$ are orthogonal if and only if every vector in a basis of either space is orthogonal to every vector in a basis of the other.*

Let the spaces be $\mathscr{V}$, with a basis $A_1, A_2, \ldots, A_m$, and $\mathscr{W}$, with a basis $B_1, B_2, \ldots, B_p$. First assume that $\mathscr{V}$ and $\mathscr{W}$ are orthogonal. Then every A_i is orthogonal to every B_j, by the definition of orthogonal spaces.

Conversely, let every A_i be orthogonal to every B_j, so that $A_i^* B_j = 0$ for $i = 1, 2, \ldots, m$; $j = 1, 2, \ldots, p$. Then

$$(\alpha_1 A_1 + \alpha_2 A_2 + \cdots + \alpha_m A_m)^*(\beta_1 B_1 + \beta_2 B_2 + \cdots + \beta_p B_p)$$
$$= \sum_{i=1}^{m} \sum_{j=1}^{p} \alpha_i \beta_j A_i^* B_j = 0.$$

Thus every vector of $\mathscr{V}$ is orthogonal to every vector of $\mathscr{W}$ and the theorem is completely proved.

A simple example is the pair of vector spaces $\mathscr{V}$ and $\mathscr{W}$, where $\mathscr{V}$ is the space of all vectors in a fixed plane on O and $\mathscr{W}$ is the space of all vectors on a fixed line on O, the line and the plane being orthogonal (Figure 8-10).

Theorem 8.18.3: *The set $\mathscr{V}^\perp$ (read $\mathscr{V}$-orthocomplement) of all vectors orthogonal to every vector of a given subspace $\mathscr{V}$ of $\mathscr{U}^n$ is a vector space orthogonal to $\mathscr{V}$. The sum of the dimensions of $\mathscr{V}$ and $\mathscr{V}^\perp$ is n.*

Let $\mathscr{V}$ have dimension k and let $A_1, A_2, \ldots, A_k$ constitute a basis for $\mathscr{V}$. Then, by Theorem 8.18.1, the vectors X belonging to the set $\mathscr{V}^\perp$ constitute the set of all solutions of the system of equations

$$A_1^* X = 0, A_2^* X = 0, \ldots, A_k^* X = 0.$$

Since the A's are independent, the coefficient matrix $[A_1, A_2, \ldots, A_k]^*$ of this system has rank k and hence there are $n - k$ linearly independent solutions. Thus $\mathscr{V}^\perp$ is a vector space of dimension $n - k$ and the theorem now follows with the aid of Theorem 8.18.2.

The vector spaces represented in Figure 8-10 illustrate this theorem in $\mathscr{E}^3$.

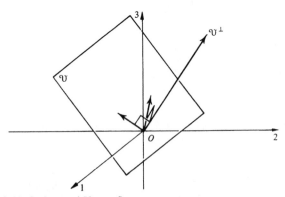

FIGURE 8-10 Orthogonal Vector Spaces.

Two mutually orthogonal subspaces of $\mathscr{U}^n$ are called **complementary orthogonal vector spaces**, briefly, **orthocomplements**, if and only if the sum of their dimensions is n. In particular, $\mathscr{V}$ and $\mathscr{V}^\perp$ are orthocomplements.

Theorem 8.18.4: *In $\mathscr{U}^n$, the sum of complementary orthogonal subspaces $\mathscr{V}$ and $\mathscr{V}^\perp$ of $\mathscr{U}^n$ is $\mathscr{U}^n$.*

Let $A_1, A_2, \ldots, A_k$ and $B_1, B_2, \ldots, B_{n-k}$ be orthonormal bases for $\mathscr{V}$ and $\mathscr{V}^\perp$. If

$$X = \alpha_1 A_1 + \alpha_2 A_2 + \cdots + \alpha_k A_k + \beta_1 B_1 + \beta_2 B_2 + \cdots + \beta_{n-k} B_{n-k} = 0,$$

then, for each i and j, $A_i^* X = \alpha_i = 0$ and $B_j^* X = \beta_j = 0$, so the set of n-vectors $A_1, A_2, \ldots, A_k, B_1, B_2, \ldots, B_{n-k}$ is a linearly independent set and hence constitutes a basis (actually an orthonormal basis) for $\mathscr{U}^n$. This proves the theorem.

Theorem 8.18.5: *Let A be a matrix over the complex field and let $\mathscr{V}$ be the vector space consisting of all solutions of the equation $AX = 0$. Then the complementary orthogonal space $\mathscr{V}^\perp$ is spanned by the columns of A^*.*

Let $A^{(i)}$ denote the ith row of A. If X is any solution of $AX = 0$, then $A^{(i)} X = 0$, that is, $(A^{(i)*})^* X = 0$. Since $A^{(i)*}$ is the ith column of A^*, this last equation implies that every vector X of $\mathscr{V}$ is orthogonal to every vector of the column space of A^* (Theorem 8.18.1). Thus the two spaces are orthogonal. Moreover, their dimensions are respectively $n - r(A)$ and $r(A)$, so they are complementary orthogonal vector spaces.

When A is real, so that $A^* = A^\mathsf{T}$, the orthocomplement of the subspace of $\mathscr{E}^n$ defined by $AX = 0$ is spanned by the columns of A^T.

For example, consider in $\mathscr{E}^4$ the subspace defined by the equation

$$AX = \begin{bmatrix} 1 & 1 & 0 & 0 \\ 0 & 1 & 1 & 0 \\ 0 & 0 & 1 & 1 \end{bmatrix} X = 0.$$

The solution space $\mathscr{V}$ has dimension $4 - 3 = 1$ and is the set of all scalar multiples of

$$\begin{bmatrix} 1 \\ -1 \\ 1 \\ -1 \end{bmatrix}.$$

The complementary orthogonal space $\mathscr{V}^{\perp}$ is spanned by the columns of A^{T}:

$$\begin{bmatrix} 1 \\ 1 \\ 0 \\ 0 \end{bmatrix}, \quad \begin{bmatrix} 0 \\ 1 \\ 1 \\ 0 \end{bmatrix}, \quad \begin{bmatrix} 0 \\ 0 \\ 1 \\ 1 \end{bmatrix}$$

and has dimension 3. The last four vectors written are linearly independent and hence span $\mathscr{E}^4$.

Theorem 8.18.6: *If $\mathscr{V}$ and $\mathscr{V}^{\perp}$ are orthocomplements in $\mathscr{U}^n$, then every vector X in $\mathscr{U}^n$ has a unique decomposition of the form $X = X_1 + X_2$, where X_1 belongs to $\mathscr{V}$ and X_2 belongs to $\mathscr{V}^{\perp}$.*

Let $A_1, A_2, \ldots, A_k$ and $B_1, B_2, \ldots, B_{n-k}$ be orthonormal bases for $\mathscr{V}$ and $\mathscr{V}^{\perp}$, respectively. Then, as in the proof of Theorem 8.18.4, these n vectors are an independent set and hence are a basis for $\mathscr{U}^n$. X therefore has a unique representation of the form

$$X = \sum_{i=1}^{k} \alpha_i A_i + \sum_{j=1}^{n-k} \beta_j B_j.$$

In this expression, the first sum represents a vector in $\mathscr{V}$ and the second represents a vector in $\mathscr{W}$; that is,

$$X_1 = \sum_{i=1}^{k} \alpha_i A_i \quad \text{and} \quad X_2 = \sum_{j=1}^{n-k} \beta_j B_j.$$

This completes the proof of the theorem.

The decomposition given in this theorem leads to a useful and interesting geometrical result. We define X_1 to be the **projection** of X on $\mathscr{V}$ and X_2 to be the projection of X on $\mathscr{V}^{\perp}$. Using this definition, we can state

Theorem 8.18.7: *Let X denote a fixed vector of $\mathscr{U}^n$ and let Y denote an arbitrary vector of a fixed subspace $\mathscr{V}$ of $\mathscr{U}^n$. Then the length of $X - Y$ has a minimum value which is attained when Y is the projection of X on $\mathscr{V}$.*

The geometric significance of the theorem in $\mathscr{E}^3$ is indicated in Figure 8-11.

Let $X = X_1 + X_2$ represent the unique decomposition of X into the sum of a vector X_1 of $\mathscr{V}$ and X_2 of $\mathscr{V}^{\perp}$. Then X_1 denotes the projection of X on $\mathscr{V}$. We write

$$X - Y = (X - X_1) + (X_1 - Y),$$

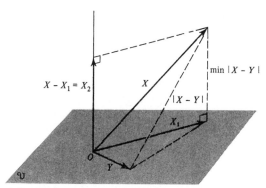

FIGURE 8-11. Minimal Property of Projections.

and observe that, because X_1 and Y belong to $\mathscr{V}$, so does $X_1 - Y$. Also, $X - X_1 = X_2$, which is in $\mathscr{V}^\perp$. Hence $X - X_1$ and $X_1 - Y$ are orthogonal; that is,

$$(X - X_1)^*(X_1 - Y) = 0.$$

Therefore,

$$|X - Y| = \sqrt{(X - Y)^*(X - Y)} = \sqrt{|X - X_1|^2 + |X_1 - Y|^2},$$

and, since X, X_1 are fixed, this will be a minimum when $|X_1 - Y| = 0$, that is, when $Y = X_1$. This completes the proof.

Suppose now that we have given two subspaces $\mathscr{V}$ and $\mathscr{W}$ of $\mathscr{U}^n$ whose dimensions are k_1 and k_2 respectively, whose intersection has dimension r, and whose sum has dimension s. Let orthonormal bases for the intersection, for $\mathscr{V}$, and for $\mathscr{W}$ be respectively

$$A_1, A_2, \ldots, A_r,$$
$$A_1, A_2, \ldots, A_r, B_1, \ldots, B_{k_1-r},$$
$$A_1, A_2, \ldots, A_r, C_1, \ldots, C_{k_2-r}.$$

Then a basis for the sum of the two spaces is, by Section 5.10,

$$A_1, A_2, \ldots, A_r, B_1, \ldots, B_{k_1-r}, C_1, \ldots, C_{k_2-r}.$$

Let these vectors constitute the columns of a matrix S. Then, by Theorem 8.18.5, the orthocomplement of the sum space is the set of all solutions X of the equation $S^*X = 0$. These vectors are precisely the ones which satisfy *both* equations

$$[A_1, \ldots, A_r, B_1, \ldots, B_{k_1-r}]^*X = 0$$
$$[A_1, \ldots, A_r, C_1, \ldots, C_{k_2-r}]^*X = 0,$$

that is, which constitute the intersection of the orthocomplements of $\mathscr{V}$ and $\mathscr{W}$. Hence we have

Theorem 8.18.8: *The orthocomplement of the sum of two vector spaces is the intersection of their orthocomplements.*

The concept of orthogonality also appears in a useful theorem concerning the consistency of a system of linear equations, $AX = B$.

Theorem 8.18.9: *In the complex field, a system $AX = B$ of m equations in n unknowns is consistent if and only if $B^*Y_0 = 0$ for every solution Y_0 of the homogeneous system $A^*Y = 0$.*

Suppose, in fact, that $AX = B$ is consistent and let X_0 be any solution. Then if $A^*Y_0 = 0$,

$$B^*Y_0 = (AX_0)^*Y_0 = X_0^*(A^*Y_0) = 0.$$

Conversely, suppose that $B^*Y = 0$ for all Y such that $A^*Y = 0$. This means that B belongs to the complementary orthogonal space of the vector space defined by $A^*Y = 0$. Hence, by Theorem 8.18.5, B belongs to the vector space spanned by the columns of $(A^*)^* = A$, so the last column of $[A, B]$ is a linear combination of the earlier columns. Thus A and $[A, B]$ must have the same rank and the system $AX = B$ is consistent.

In the real field, $AX = B$ is consistent if and only if B is orthogonal to every solution of $A^TY = 0$. This last system of equations is called the **transposed homogeneous system** *corresponding to the system $AX = B$.*

8.19   Exercises

1. Let $\mathscr{V}_1$ of dimension k_1 and $\mathscr{V}_2$ of dimension k_2 be subspaces of $\mathscr{U}^n$ and let $k_1 + k_2$ be $< n$. Prove that there exist vectors X orthogonal to both $\mathscr{V}_1$ and $\mathscr{V}_2$.

2. If the subspaces $\mathscr{V}_1$ and $\mathscr{V}_2$ of $\mathscr{U}^n$ have dimensions k_1 and k_2 respectively, if $k_1 < k_2$, and if $\mathscr{V}_1 \cap \mathscr{V}_2 = \mathscr{W}$, prove that there exist vectors X in $\mathscr{V}_2$ which are orthogonal to $\mathscr{W}$.

3. Prove that if $A_1, A_2, \ldots, A_{k_1}$ and $B_1, B_2, \ldots, B_{k_2}$ are bases for any two orthogonal subspaces $\mathscr{V}_1$ and $\mathscr{V}_2$ of $\mathscr{U}^n$, then all $k_1 + k_2$ A's and B's are linearly independent, so $k_1 + k_2 \leq n$.

4. Prove that if $A^*(I_n - A) = 0$, the sum of the ranks of A and $I_n - A$ is n.

5. Find an orthonormal basis for the orthocomplement of the subspace $\mathscr{V}$ of $\mathscr{E}^4$ spanned by

$$\begin{bmatrix} 1 \\ -1 \\ 0 \\ 1 \end{bmatrix}, \quad \begin{bmatrix} 1 \\ 0 \\ 1 \\ 0 \end{bmatrix}.$$

6. Given that $X = [1, 0, 1, 1]^T$, find X_1 and X_2 such that $X = X_1 + X_2$, where X_1 and X_2 belong to the orthocomplementary spaces of Exercise 5.

7. If Y is an arbitrary vector of the subspace $\mathscr{V}$ of Exercise 5 and if X is as given in Exercise 6, find the minimum value of $|X - Y|$.

8. Prove that, if $P^*Q = 0$, then the image spaces of the operators with matrices P and Q are orthogonal. Here P and Q are $n \times n$ matrices over $\mathscr{C}$.

9. Given that $A_1, A_2, \ldots, A_k$ constitute an orthonormal basis for a subspace $\mathscr{V}$ of $\mathscr{U}^n$, and that U is a unitary $k \times k$ matrix, what can be said about the vectors B_1, $B_2, \ldots, B_k$ determined by the relation

$$[B_1, B_2, \ldots, B_k] = [A_1, A_2, \ldots, A_k]U?$$

10. Show that if a vector X of $\mathscr{U}^n$ is orthogonal to each of the linearly independent vectors $A_1, A_2, \ldots, A_k$ of $\mathscr{U}^n$, then $X, A_1, A_2, \ldots, A_k$ are a linearly independent set, so X does not belong to the vector space spanned by $A_1, A_2, \ldots, A_k$.

*11. Prove that $A_1, A_2, \ldots, A_k$ are mutually orthogonal unit n-vectors if and only if

$$[A_1, A_2, \ldots, A_k]^* [A_1, A_2, \ldots, A_k] = I_k.$$

*12. Given that $U_1, U_2, \ldots, U_k$ are mutually orthogonal unit n-vectors, prove that one can find n-vectors $U_{k+1}, \ldots, U_n$ such that $U_1, U_2, \ldots, U_n$ form an orthonormal basis for $\mathscr{U}^n$. Give examples in $\mathscr{E}^3$ for $k = 1$ and $k = 2$.

13. Given that the rows of $A_{m \times n}$ are mutually orthogonal unit vectors, show that $AX_{n \times 1} = B_{m \times 1}$ has the particular solution $X = \sum_{i=1}^{n} b_i A^{(i)*}$. What can you say about the relative sizes of m and n?

14. Reduce the system of equations

$$
\begin{aligned}
x_1 - x_2 + x_3 \qquad &= 1 \\
x_1 - x_2 \qquad + 2x_4 &= 2 \\
-2x_1 \qquad + 2x_3 + x_4 &= 1
\end{aligned}
$$

to an equivalent system in which the rows of coefficients of the x's are mutually orthogonal unit vectors and then obtain the particular solution mentioned in Exercise 13.

8.20 Projections

Given an operator $Y = A_{m \times n} X$ from $\mathscr{F}^n$ to $\mathscr{F}^m$, the set of all vectors X such that $AX = 0$ is the null space of the operator. If $r(A) = n$, then the null space consists of the zero vector alone. If $r(A) < n$, the null space is nontrivial. Projections, which are important in more advanced aspects of linear algebra and in applications such as quantum mechanics, statistics, and so on, provide a good illustration of the case $r(A) < n$. We now examine some of the basic properties of these transformations.

An operator $Y = AX$ on $\mathscr{F}^n$ is a **projection** if and only if $A^2 = A$. As an example, consider again the operator defined by

$$Y = \begin{bmatrix} 1 & 0 & 0 \\ 0 & 1 & 0 \\ 0 & 0 & 0 \end{bmatrix} X$$

which projects every vector $[a_1, a_2, a_3]^{\mathsf{T}}$ onto the vector $[a_1, a_2, 0]^{\mathsf{T}}$ in the 1,2-plane (see Figure 8-2). The geometric interpretation of this operator is exactly what is meant by "projection" in the familiar sense. The set of all vectors which project onto the zero vector is the set of all vectors $k[0, 0, 1]^{\mathsf{T}}$; that is, the null space of this operator is the x_3-axis. The range of the operator is the set of all vectors in the 1,2-plane.

Since $A^2 = A$ for every projection, we have $A(I - A) = 0$. This equation is satisfied if $A = 1$ or 0. Otherwise, since only a singular matrix can be a divisor of zero, both A and $I - A$ must be singular. We therefore have

Theorem 8.20.1: *Every projection except the identity is singular and hence has a nontrivial null space.*

Suppose next that a vector V belongs to the range of the projection. Then there exists a vector X such that $AX = V$. Since

$$A(AX) = A^2 X = AX,$$

it follows that for every vector V of the range,

$$AV = V.$$

Moreover, if X is any vector such that $AX = X$, then X is its own image and hence belongs to the range of the projection. Since the range of the projection is simply the column space of A, we have proved

Theorem 8.20.2: *The vectors of $\mathscr{F}^n$ which are invariant under a projection are precisely the vectors of the column space of the matrix of the projection, that is, the vectors of its range.*

Now if $V \neq 0$ and $AV = V$, then $AV \neq 0$. Hence no nonzero vector of the range belongs to the null space. Since the intersection of the null space and the range is thus the zero vector alone, the dimension of the sum of these two spaces is the sum of their dimensions; that is, it is $(n - r(A)) + r(A) = n$. Hence, by Theorem 5.10.4, we have

Theorem 8.20.3: *The null space and the range of a projection of $\mathscr{F}^n$ are complementary vector spaces; that is, their intersection is the zero vector and their sum is $\mathscr{F}^n$. Every vector of $\mathscr{F}^n$ has a unique representation as the sum of a vector of the range and a vector of the null space of the projection.*

We shall define, for an arbitrary operator on a vector space, the **fixed space** to be the set of all vectors each of which is left fixed by the operator. The fixed space is a vector space which is a subset of the range of the operator. As we have seen in Theorem 8.20.2, the fixed space and the range of a projection are identical. We can now obtain a converse to the preceding theorem.

Suppose that $Y = AX$ represents any operator on $\mathscr{F}^n$ such that the sum of its null space and its fixed space is $\mathscr{F}^n$ so that every vector X of $\mathscr{F}^n$ can be written in the form

$$X = P + Q,$$

where P belongs to the fixed space and Q belongs to the null space. Since $AP = P$ and $AQ = 0$, we have

$$AX = A(P + Q) = AP = P.$$

Hence, for all X in $\mathscr{F}^n$,

$$A^2X = AP = A(P + Q) = AX$$

so

$$A^2 = A.$$

The operator $Y = AX$ is therefore a projection. Thus we have proved

Theorem 8.20.4: *Every operator on $\mathscr{F}^n$ such that the sum of its null space and its fixed space is $\mathscr{F}^n$, is a projection.*

The two preceding theorems together characterize the projections of $\mathscr{F}^n$ as being precisely those operators on $\mathscr{F}^n$ whose null and fixed spaces sum to $\mathscr{F}^n$.

We have already pointed out that the fixed space of a projection of $\mathscr{F}^n$ with matrix A is the same as the range, which is the set of all vectors AX. Consider now the set of all vectors $X - AX$. Since $A^2 = A$, we have $A(X - AX) = 0$, so every vector $X - AX$ belongs to the null space of the projection. Conversely, if X belongs to the null space, $X = X - AX$. Thus the null space is the set of all vectors $X - AX$. Since $\mathscr{F}^n$ is the sum of the range and the null space of the projection, the decomposition $X = AX + (X - AX)$ of X into the sum of a vector of the range and a vector of the null space is unique. We summarize in

Theorem 8.20.5: *The fixed space (range) of a projection $Y = AX$ of $\mathscr{F}^n$ is the set of all vectors AX, that is, the column space of A. Its null space is the set of all vectors $X - AX$, that is, the column space of $I - A$. The formula*

$$X = AX + (X - AX)$$

gives the unique decomposition of X into the sum of a vector of the range and a vector of the null space of the projection.

We prove finally

Theorem 8.20.6: *If $\mathscr{F}^n$ is the sum of two of its subspaces, $\mathscr{V}$ of dimension k and $\mathscr{W}$ of dimension $n - k$, then there is a unique projection having $\mathscr{V}$ as its fixed space (range) and $\mathscr{W}$ as its null space.*

First note that the intersection of $\mathscr{V}$ and $\mathscr{W}$ has dimension zero since their sum has dimension n. Hence, if $B_1, B_2, \ldots, B_k$ is a basis for $\mathscr{V}$ and $B_{k+1}, B_{k+2}, \ldots, B_n$ is a basis for $\mathscr{W}$, these vectors are independent, so $B = [B_1, B_2, \ldots, B_n]$ is nonsingular. We now seek all operators $X' = AX$ for which $\mathscr{V}$ is the fixed space and $\mathscr{W}$ is the null space, that is, such that

$$A[B_1, B_2, \ldots, B_k, B_{k+1}, \ldots, B_n] = [B_1, B_2, \ldots, B_k, 0, 0, \ldots, 0].$$

Since B is nonsingular, this implies that

$$(8.20.1) \qquad A = [B_1, B_2, \ldots, B_k, 0, 0, \ldots, 0]B^{-1}.$$

If another pair of bases for $\mathscr{V}$ and $\mathscr{W}$ leads in the same way to the matrix $\tilde{A}$, then for every vector X of $\mathscr{V}$, $AX = X$ and $\tilde{A}X = X$, while for every vector X of $\mathscr{W}$, $AX = 0$ and $\tilde{A}X = 0$. This implies that for every X in $\mathscr{F}^n$, $AX = \tilde{A}X$, so $A = \tilde{A}$. It follows that the matrix A of (8.20.1) is independent of the choice of bases for $\mathscr{V}$ and $\mathscr{W}$, so the operator is unique. The theorem now follows from Theorem 8.20.4.

An important aspect of the proof just given is that it shows how to write the matrix of a projection when its fixed and null spaces are given. For example, if the fixed and null spaces of a projection of $\mathscr{E}^3$ have bases

$$\begin{bmatrix} 1 \\ 1 \\ 1 \end{bmatrix}, \qquad \begin{bmatrix} 1 \\ -1 \\ 1 \end{bmatrix}, \qquad \text{and} \qquad \begin{bmatrix} 0 \\ 0 \\ 1 \end{bmatrix}$$

respectively, then

$$A = \begin{bmatrix} 1 & 1 & 0 \\ 1 & -1 & 0 \\ 1 & 1 & 0 \end{bmatrix} \begin{bmatrix} 1 & 1 & 0 \\ 1 & -1 & 0 \\ 1 & 1 & 1 \end{bmatrix}^{-1} = \begin{bmatrix} 1 & 0 & 0 \\ 0 & 1 & 0 \\ 1 & 0 & 0 \end{bmatrix}.$$

Other properties of projections appear in following exercises.

8.21 Orthogonal Projections in $\mathcal{U}^n$

Of special interest in applications are the **orthogonal projections**, that is, those for which the range and the null space are orthogonal. (Here the name derives from the geometric property. The matrix of a projection other than the identity, being singular, cannot be orthogonal or unitary.)

Suppose that A is the matrix of an orthogonal projection of $\mathcal{U}^n$. Then by Theorem 8.20.5, for all X and Y in $\mathcal{U}^n$, we have

$$(AX)^*(Y - AY) = 0 \quad \text{and} \quad (AY)^*(X - AX) = 0.$$

These equations imply that

$$X^*A^*Y = (AX)^*(AY) = ((AY)^*(AX))^* = (Y^*A^*X)^* = X^*AY.$$

Since $X^*A^*Y = X^*AY$ for *all* X and Y in $\mathcal{U}^n$, it follows that $A^* = A$, so A is Hermitian.

Conversely, let $A^2 = A$ and $A^* = A$. Then, for all X and Y in $\mathcal{U}^n$,

$$(AX)^*(Y - AY) = X^*A^*Y - X^*A^*AY = 0$$

so a Hermitian projection is orthogonal. In summary, we have

Theorem 8.21.1: *A projection of $\mathcal{U}^n$ is orthogonal if and only if its matrix is Hermitian. In particular, a projection of $\mathscr{E}^n$ is orthogonal if and only if its matrix is symmetric.*

When the range of an orthogonal projection is given, its null space is the complementary orthogonal space of the range. Thus, by Theorem 8.20.6, the range of an orthogonal projection determines the projection completely. To see how the range determines the matrix A of the projection, let $U_1, U_2, \ldots, U_k$ be an orthonormal basis for the range. Then for all X in $\mathcal{U}^n$, AX belongs to the range and hence there exist scalars α_i such that

$$AX = \sum_1^k \alpha_i U_i.$$

The vector

$$X - AX = X - \sum_1^k \alpha_i U_i$$

belongs to the null space of the projection and hence is orthogonal to every vector of the range; that is, for all X,

$$U_j^*\left(X - \sum_1^k \alpha_i U_i\right) = 0, \qquad j = 1, 2, \ldots, k.$$

From this,

$$\alpha_j = U_j^* X, \qquad j = 1, 2, \ldots, k,$$

so

$$AX = \sum_1^k (U_i^* X)U_i = \sum_1^k U_i(U_i^* X) = \sum_1^k (U_i U_i^*)X$$

for *all* X, which implies that

(8.21.1) $$A = \sum_1^k U_i U_i^*.$$

The same argument used in the proof of Theorem 8.20.6 shows that A is unique; that is, it is independent of the choice of orthonormal basis for the range. We have, therefore,

Theorem 8.21.2: *If the range of an orthogonal projection of $\mathscr{U}^n$ has an orthonormal basis $U_1, U_2, \ldots, U_k$, then the matrix of the projection is $A = \sum_1^k U_i U_i^*$.*

A geometrical illustration of these matters in $\mathscr{E}^3$ is given in Figure 8-12. As an arithmetic example, observe that the projection of $\mathscr{E}^3$ onto the 1,2-plane (Figure 8-2) has, by (8.21.1), the matrix

$$A = \begin{bmatrix} 1 \\ 0 \\ 0 \end{bmatrix}[1 \quad 0 \quad 0] + \begin{bmatrix} 0 \\ 1 \\ 0 \end{bmatrix}[0 \quad 1 \quad 0] = \begin{bmatrix} 1 & 0 & 0 \\ 0 & 1 & 0 \\ 0 & 0 & 0 \end{bmatrix}.$$

The concept of an orthogonal projection permits the generalization of certain familiar geometric concepts from $\mathscr{E}^2$ and $\mathscr{E}^3$ to $\mathscr{U}^n$. Thus, in $\mathscr{E}^3$, the angle θ between a line and a plane is defined to be the angle between the line and its orthogonal projection on the plane (Figure 8-12). Let the subspace $\mathscr{V}$ of $\mathscr{U}^n$ be the range of an orthogonal projection. Then we define

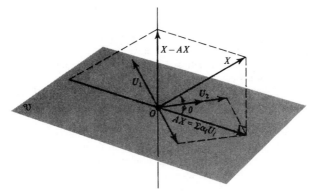

FIGURE 8-12. Orthogonal Projection in $\mathscr{E}^3$.

the **angle** $\theta(X, \mathscr{V})$ between a vector $X \neq 0$ and $\mathscr{V}$ in $\mathscr{U}^n$ to be the angle between X and the orthogonal projection of X on $\mathscr{V}$, so if A is the matrix of the projection, determined from a basis for $\mathscr{V}$ as in Theorem 8.21.2, and if $AX \neq 0$,

$$\cos \theta = \frac{X^*AX}{|X| \cdot |AX|}.$$

If $AX = 0$, X is orthogonal to $\mathscr{V}$ (Why?) and we define $\theta = \pi/2$.

Since $A = A^*$ and $A = A^2$, we have

$$X^*AX = (AX)^*(AX) = |AX|^2.$$

Thus when $X \neq 0$ and $AX \neq 0$, the angle θ between X and $\mathscr{V}$ is determined by

(8.21.2) $$\cos \theta = \frac{|AX|}{|X|}, \qquad 0 \leq \theta \leq \frac{\pi}{2},$$

exactly as one would expect (Figure 8-12).

8.22 Exercises

1. Determine bases for the null space and the fixed space of each of these projections:

(a) $\begin{bmatrix} \frac{1}{2} & 0 & \frac{1}{2} \\ 0 & 1 & 0 \\ \frac{1}{2} & 0 & \frac{1}{2} \end{bmatrix}$,

(b) $\begin{bmatrix} \frac{1}{9} & -\frac{2}{9} & \frac{2}{9} \\ -\frac{2}{9} & \frac{4}{9} & -\frac{4}{9} \\ \frac{2}{9} & -\frac{4}{9} & \frac{4}{9} \end{bmatrix}$.

2. Show that, if A is the matrix of a projection, then $I - A$ is also the matrix of a projection.
3. Show that the range of the projection with matrix A is the null space of the projection with matrix $I - A$ and vice versa.
4. Show that, if A and B are matrices of projections and commute, then AB also represents a projection. Generalize.
5. Show that, if A and B are matrices of projections and if A and B anticommute, then $A + B$ is the matrix of a projection. Generalize.
6. Let A and B be the matrices of projections of $\mathcal{U}^n$. Prove that their ranges are mutually orthogonal if and only if $A^*B = B^*A = 0$.
7. Let A and B denote the matrices of orthogonal projections of $\mathcal{U}^n$. Prove that AB is also the matrix of an orthogonal projection if and only if $AB = BA$.
8. Let $A_1, A_2, \ldots, A_k$ denote matrices of orthogonal projections of $\mathcal{U}^n$. Prove that $\sum A_j$ is also the matrix of an orthogonal projection if $A_i A_j = 0$ when $i \neq j$. Illustrate with an example and a figure in $\mathcal{E}^3$.
9. Find the matrix of the orthogonal projection of $\mathcal{E}^4$ whose range is the plane of intersection of the hyperplanes

$$x_1 + x_2 - x_3 - x_4 = 0$$

$$x_1 - x_2 + x_3 - x_4 = 0.$$

10. Let P and Q be matrices of projections of $\mathcal{F}^n$. We define $P \subseteq Q$ (P included in Q) if and only if $PQ = P$. Prove that in $\mathcal{E}^3$

$$\begin{bmatrix} 1 & 0 & 0 \\ 0 & 0 & 0 \\ 0 & 0 & 0 \end{bmatrix} \subseteq \begin{bmatrix} 1 & 0 & 0 \\ 0 & 1 & 0 \\ 0 & 0 & 0 \end{bmatrix},$$

and discuss geometrically.

11. Prove that the inclusion relation for matrices of projections has these properties:
    (a) $P \subseteq P$ (the reflexive property);
    (b) $P \subseteq Q, Q \subseteq P$ imply $P = Q$ (the antisymmetric property);
    (c) $P \subseteq Q, Q \subseteq R$ imply $P \subseteq R$ (the transitive property).
12. Show that $Y = AX$ is a projection if and only if there is a reference system $B_1, B_2, \ldots, B_n$ such that for some $r \leq n$,

$$B^{-1}AB = \begin{bmatrix} I_r & 0 \\ \hline 0 & 0 \end{bmatrix}.$$

13. Show that an operator $\tilde{X} = AX$ leaves both the null space and the fixed space of the projection $Y = BX$ invariant if and only if $AB = BA$.
14. Show that, if A is the matrix of an operator with respect to the natural reference system and A is idempotent, then in every reference system the matrix of the operator is idempotent; that is, the property of being a projection does not depend on the reference system.
15. Determine the set of all operators on $\mathcal{F}^n$ that leave fixed every vector of the subspace of $\mathcal{F}^n$ with basis $B_1, B_2, \ldots, B_k$. Which of these are projections?

CHAPTER
9

The Characteristic
Value Problem

9.1 Definition of the Characteristic Value Problem

Given a real matrix A of order 3, consider the linear operator $X_2 = AX_1$, which maps $\mathscr{E}^3$ into $\mathscr{E}^3$. Ordinarily the vectors X_2 and X_1, where $X_2 = AX_1$, are not collinear. An important fact in many applications is that, for certain vectors $X \neq 0$, AX and X *are* collinear; that is, there exists a scalar λ such that $AX = \lambda X$.

The **characteristic value problem** or the **eigenvalue problem** is the following: Given a real or complex $n \times n$ matrix A, for what vectors $X \neq 0$ and for what scalars λ is it true that

$$(9.1.1) \qquad AX = \lambda X;$$

that is, for what vectors X are AX and X proportional?

A nonzero vector X which satisfies (9.1.1) is called a **characteristic vector** (Figure 9-1) or **eigenvector** or **latent vector** of A and the associated value λ is called a **characteristic root** or **eigenvalue** or **latent root** of A.

For example, since

$$\begin{bmatrix} 1 & 2 \\ 2 & 1 \end{bmatrix} \begin{bmatrix} 1 \\ 1 \end{bmatrix} = 3 \begin{bmatrix} 1 \\ 1 \end{bmatrix},$$

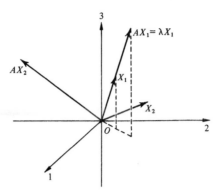

FIGURE 9-1 Characteristic and Noncharacteristic Vectors.

$\begin{bmatrix} 1 \\ 1 \end{bmatrix}$ is a characteristic vector and 3 is the associated characteristic root of the matrix $\begin{bmatrix} 1 & 2 \\ 2 & 1 \end{bmatrix}$.

Now (9.1.1) holds if and only if

$$(9.1.2) \qquad\qquad (A - \lambda I)X = 0,$$

where the scalar λ has been replaced by the scalar matrix λI in order to make it possible to extract the factor X. This equation represents a homogeneous system of n equations in n unknowns which has a nontrivial solution for X if and only if the coefficient matrix has rank $r < n$, that is, if and only if

$$(9.1.3) \qquad\qquad \det(A - \lambda I) = 0,$$

or, equivalently,

$$(9.1.4) \qquad \begin{vmatrix} a_{11} - \lambda & a_{12} & \cdots & a_{1n} \\ a_{21} & a_{22} - \lambda & \cdots & a_{2n} \\ \vdots & & & \\ a_{n1} & a_{n2} & \cdots & a_{nn} - \lambda \end{vmatrix} = 0.$$

This equation is called the **characteristic equation** or **secular equation** of A. The left member of this equation is called the **characteristic polynomial**, $\phi(\lambda)$, of A. It will be studied extensively in Section 9.5.

As inspection of the main diagonal shows, this determinant will expand into a polynomial of degree n in the variable λ. If we solve the equation, we get n roots, $\lambda_1, \lambda_2, \ldots, \lambda_n$ (not necessarily all different). These are the **characteristic roots** of A. If λ_j is one of these roots, we solve the equation

$$(A - \lambda_j I)X = 0$$

for X. The nontrivial solutions are the characteristic vectors X of A that are associated with the characteristic root λ_j; that is, they are the nonzero vectors X such that

$$AX = \lambda_j X.$$

The vector space consisting of all solutions of this equation includes the zero vector in addition to the characteristic vectors associated with λ_j and is called the **characteristic subspace** associated with λ_j.

Since it is possible for equation (9.1.4) to have only one distinct root and for the corresponding equation (9.1.2) to have only one independent solution, it is possible that there is only one characteristic subspace and that it is one-dimensional. An example of this is given in the next section.

9.2 Four Examples

Consider again the matrix

$$A = \begin{bmatrix} 2 & 1 \\ 1 & 2 \end{bmatrix}.$$

We want to determine all λ's and all X's such that

$$\begin{bmatrix} 2 & 1 \\ 1 & 2 \end{bmatrix} X = \lambda X.$$

Here we need

$$\det(A - \lambda I) = \begin{vmatrix} 2 - \lambda & 1 \\ 1 & 2 - \lambda \end{vmatrix} = (2 - \lambda)^2 - 1 = (1 - \lambda)(3 - \lambda) = 0,$$

so the characteristic roots are $\lambda = 1$ and $\lambda = 3$.

Putting $\lambda = 1$ in $(A - \lambda I)X = 0$, we get

$$\begin{bmatrix} 1 & 1 \\ 1 & 1 \end{bmatrix} \begin{bmatrix} x_1 \\ x_2 \end{bmatrix} = 0$$

or, in scalar form, a system that reduces to the single equation

$$x_1 + x_2 = 0,$$

which has the complete solution

$$\begin{bmatrix} x_1 \\ x_2 \end{bmatrix} = t \begin{bmatrix} -1 \\ 1 \end{bmatrix}.$$

For any $t \neq 0$, we have a characteristic vector associated with the root $\lambda = 1$. In fact, for all t,

$$\begin{bmatrix} 2 & 1 \\ 1 & 2 \end{bmatrix} \begin{bmatrix} -t \\ t \end{bmatrix} = 1 \begin{bmatrix} -t \\ t \end{bmatrix},$$

as required. That is, associated with the root $\lambda = 1$ there is a one-dimensional characteristic subspace. In this particular instance, because $\lambda = 1$, every individual vector of this subspace is left invariant by the operator.

Now we put $\lambda = 3$ in $(A - \lambda I)X = 0$ and get

$$\begin{bmatrix} -1 & 1 \\ 1 & -1 \end{bmatrix} \begin{bmatrix} x_1 \\ x_2 \end{bmatrix} = 0,$$

which is equivalent to the single scalar equation

$$x_1 - x_2 = 0$$

with the complete solution

$$\begin{bmatrix} x_1 \\ x_2 \end{bmatrix} = t \begin{bmatrix} 1 \\ 1 \end{bmatrix}.$$

For any $t \neq 0$, we get a characteristic vector associated with the root $\lambda = 3$. In fact, for all t,

$$\begin{bmatrix} 2 & 1 \\ 1 & 2 \end{bmatrix} \begin{bmatrix} t \\ t \end{bmatrix} = 3 \begin{bmatrix} t \\ t \end{bmatrix}.$$

Associated with the characteristic root $\lambda = 3$, we thus have a one-dimensional characteristic subspace such that A maps every vector of this subspace onto a vector of the subspace (Figure 9-2). However, in this instance, except for the vector 0, individual vectors of the subspace are not left invariant by A.

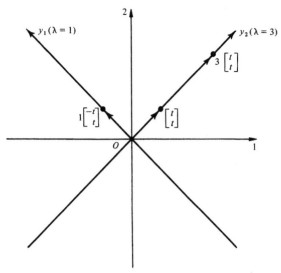

FIGURE 9-2. Characteristic Subspaces of $\begin{bmatrix} 2 & 1 \\ 1 & 2 \end{bmatrix}$.

Note that the subspaces determined by the roots $\lambda = 1$ and $\lambda = 3$ respectively contain the unit characteristic vectors

$$\begin{bmatrix} -\dfrac{1}{\sqrt{2}} \\[2mm] \dfrac{1}{\sqrt{2}} \end{bmatrix} \quad \text{and} \quad \begin{bmatrix} \dfrac{1}{\sqrt{2}} \\[2mm] \dfrac{1}{\sqrt{2}} \end{bmatrix}.$$

These unit vectors determine an orthogonal transformation of coordinates,

$$X = \begin{bmatrix} -\dfrac{1}{\sqrt{2}} & \dfrac{1}{\sqrt{2}} \\[2mm] \dfrac{1}{\sqrt{2}} & \dfrac{1}{\sqrt{2}} \end{bmatrix} Y.$$

Applying this to the operator

$$X_2 = \begin{bmatrix} 2 & 1 \\ 1 & 2 \end{bmatrix} X_1,$$

we get, in the usual way,

$$
Y_2 = \begin{bmatrix} -\dfrac{1}{\sqrt{2}} & \dfrac{1}{\sqrt{2}} \\[2ex] \dfrac{1}{\sqrt{2}} & \dfrac{1}{\sqrt{2}} \end{bmatrix}^{-1} \begin{bmatrix} 2 & 1 \\ 1 & 2 \end{bmatrix} \begin{bmatrix} -\dfrac{1}{\sqrt{2}} & \dfrac{1}{\sqrt{2}} \\[2ex] \dfrac{1}{\sqrt{2}} & \dfrac{1}{\sqrt{2}} \end{bmatrix} Y_1 = \begin{bmatrix} 1 & 0 \\ 0 & 3 \end{bmatrix} Y_1 .
$$

Thus the operator has a particularly simple representation in the coordinate system determined by its unit characteristic vectors. We shall see presently when and why this happens.

Consider next the identity matrix of order 2. Its characteristic equation, $(1 - \lambda)^2 = 0$, has the double root $\lambda = 1$. Equation (9.1.2) now becomes simply $0 \cdot X = 0$, of which every 2-vector X is a solution. In this case, therefore, with the double root of the characteristic equation there is associated the two-dimensional vector space consisting of all 2-vectors.

In contrast to the preceding example, let

$$
A = \begin{bmatrix} 1 & -2 \\ 0 & 1 \end{bmatrix}.
$$

The characteristic equation is again $(1 - \lambda)^2 = 0$. The characteristic vectors are now determined by the equation

$$
\begin{bmatrix} 0 & -2 \\ 0 & 0 \end{bmatrix} \cdot \begin{bmatrix} x_1 \\ x_2 \end{bmatrix} = \begin{bmatrix} 0 \\ 0 \end{bmatrix},
$$

that is, by the single equation $x_2 = 0$, the complete solution of which is

$$
\begin{bmatrix} x_1 \\ x_2 \end{bmatrix} = t \begin{bmatrix} 1 \\ 0 \end{bmatrix}.
$$

In this case there is only a one-dimensional vector space of characteristic vectors associated with the double root.

As a final example, consider

$$
A = \begin{bmatrix} 1 & -1 \\ 1 & 1 \end{bmatrix}, \qquad \begin{vmatrix} 1 - \lambda & -1 \\ 1 & 1 - \lambda \end{vmatrix} = (1 - \lambda)^2 + 1, \qquad \lambda = 1 \pm i.
$$

Thus a real matrix may well have complex characteristic roots, and hence also complex characteristic vectors, for we cannot have A and $X \neq 0$ real in the equation $AX = \lambda X$ if λ is not also real. In this case, the operator $X_2 = AX_1$ has no nonzero, real, characteristic subspaces. This example illustrates the fact that the whole characteristic value problem is normally a problem in the complex number field. This is assumed throughout this chapter except when we treat real, symmetric matrices, in which case all the characteristic roots are real, as we shall show.

9.3 Two Basic Theorems

A fundamental and often useful result is given in

Theorem 9.3.1: *Let* $\lambda_1, \lambda_2, \ldots, \lambda_k$ *be distinct characteristic roots of a matrix* A, *and let* $X_1, X_2, \ldots, X_k$ *be characteristic vectors associated with these roots respectively. Then* $X_1, X_2, \ldots, X_k$ *are linearly independent.*

If the equation

$$(9.3.1) \qquad \alpha_1 X_1 + \alpha_2 X_2 + \cdots + \alpha_k X_k = 0$$

holds, where the α's are scalars, then repeated multiplication by A, combined with the fact that $AX_i = \lambda_i X_i$, yields the additional equations

$$(9.3.2)\qquad\begin{aligned} &\alpha_1\lambda_1 X_1 &&+ \alpha_2\lambda_2 X_2 &&+ \cdots + \alpha_k\lambda_k X_k &&= 0 \\ &\alpha_1\lambda_1^2 X_1 &&+ \alpha_2\lambda_2^2 X_2 &&+ \cdots + \alpha_k\lambda_k^2 X_k &&= 0 \\ &\quad\vdots \\ &\alpha_1\lambda_1^{k-1} X_1 &&+ \alpha_2\lambda_2^{k-1} X_2 &&+ \cdots + \alpha_k\lambda_k^{k-1} X_k &&= 0. \end{aligned}$$

The equations (9.3.1) and (9.3.2) can be combined in partitioned matrix form thus:

$$[\alpha_1 X_1, \alpha_2 X_2, \ldots, \alpha_k X_k] \begin{bmatrix} 1 & \lambda_1 & \lambda_1^2 & \cdots & \lambda_1^{k-1} \\ 1 & \lambda_2 & \lambda_2^2 & \cdots & \lambda_2^{k-1} \\ \vdots & & & & \\ 1 & \lambda_k & \lambda_k^2 & \cdots & \lambda_k^{k-1} \end{bmatrix} = 0.$$

Since the λ's are all unequal, the second factor here is a nonsingular Vandermonde matrix (see Exercise 23, Section 7.5). If we multiply on the right by the inverse of this Vandermonde matrix, we conclude that

$$[\alpha_1 X_1, \alpha_2 X_2, \ldots, \alpha_k X_k] = 0,$$

which, since no X_j is 0, implies that every α_j is 0. Hence the X_j's are independent and the proof is complete.

If all n of the characteristic roots are distinct, and if

$$AV_i = \lambda_i V_i, \qquad V_i \neq 0, \qquad i = 1, 2, \ldots, n,$$

then

$$[AV_1, AV_2, \ldots, AV_n] = [\lambda_1 V_1, \lambda_2 V_2, \ldots, \lambda_n V_n],$$

so that, if

$$V = [V_1, V_2, \ldots, V_n],$$

we have

$$AV = V \operatorname{diag}[\lambda_1, \lambda_2, \ldots, \lambda_n].$$

Since the V_i are linearly independent by the preceding theorem, V^{-1} exists and hence

$$V^{-1} AV = \operatorname{diag}[\lambda_1, \lambda_2, \ldots, \lambda_n].$$

This amounts to applying the nonsingular transformation of coordinates $X = VY$ to the operator $X_2 = AX_1$ and allows us to state

Theorem 9.3.2: *When the characteristic roots of A are distinct, there exists at least one coordinate system in which the operator $X_2 = AX_1$ is represented by a diagonal matrix whose diagonal entries are the characteristic roots of A.*

In such a coordinate system, the operator has the representation

$$Y_2 = \operatorname{diag}[\lambda_1, \lambda_2, \ldots, \lambda_n] Y_1$$

or, in scalar form,

$$y_{i2} = \lambda_i y_{i1}, \qquad i = 1, 2, \ldots, n$$

so that, if the λ_i are all real, it may be interpreted as a product of changes of scale along the several axes. This particularly simple representation is often desirable as it permits correspondingly simple analyses of a broad variety of applications in both the physical and the social sciences. In most applications, the characteristic roots are all distinct, so this kind of reduction of the operator with matrix A to diagonal form is possible. When A is a real symmetric matrix, the λ_i's are all real and the reduction is always possible, regardless of whether or not the λ_i are distinct. Moreover, the V_i may be chosen so as to constitute an orthonormal set, which makes V orthogonal. All this we prove in later sections.

9.4 Exercises

1. Determine the characteristic roots and the associated characteristic subspaces for these matrices:

(a) $\begin{bmatrix} 1 & 2 & 0 \\ 0 & 1 & 0 \\ 0 & 0 & 1 \end{bmatrix}$,

(b) $\begin{bmatrix} 1 & 0 & 0 \\ 0 & 2 & 0 \\ 0 & 0 & 1 \end{bmatrix}$,

(c) $\begin{bmatrix} 0 & 1 & 0 \\ 1 & 0 & 0 \\ 0 & 0 & 1 \end{bmatrix}$,

(d) $\begin{bmatrix} 0 & i & i \\ -i & 0 & i \\ -i & -i & 0 \end{bmatrix}$,

(e) $\begin{bmatrix} 0 & 1 & 2 \\ 1 & 0 & -1 \\ 2 & -1 & 0 \end{bmatrix}$,

(f) $\begin{bmatrix} \dfrac{1}{\sqrt{2}} & \dfrac{1}{\sqrt{2}} & 0 \\ \dfrac{1}{\sqrt{2}} & -\dfrac{1}{\sqrt{2}} & 0 \\ 0 & 0 & 1 \end{bmatrix}$,

(g) $\begin{bmatrix} 1 & 1 & 1 & 1 \\ 0 & 1 & 1 & 1 \\ 0 & 0 & -1 & -1 \\ 0 & 0 & 0 & -1 \end{bmatrix}$,

(h) $\begin{bmatrix} 0 & 1 & 0 & 0 \\ 0 & 0 & 1 & 0 \\ 0 & 0 & 0 & 1 \\ -1 & 4 & -6 & 4 \end{bmatrix}$.

*2. Show that $\lambda = 0$ is a characteristic root of A if and only if $\det A = 0$.

*3. Show that the characteristic roots of a triangular matrix are just the diagonal entries of that matrix.

4. Show that if $AX = \lambda X$, $X \neq 0$, then λ^2 is a characteristic root of A^2 (with associated characteristic vector X). Then show that λ^p is a characteristic root of A^p for every positive integer p.

5. Show that if $AX = \lambda X$, $|X| = 1$, then

$$\lambda = X^*AX.$$

What is the corresponding result when X is not a unit vector?

6. Give an example to show that, if $AX = \lambda_1 X$ and $AY = \lambda_2 Y$ with $\lambda_1 \neq \lambda_2$, then $\alpha X + \beta Y$ is not, in general, a characteristic vector of A.

7. Show that if X is a characteristic vector of A associated with the characteristic root λ, then so is $A^p X$ for every positive integer p, unless $A^p X = 0$.

8. By using the matrix

$$
\begin{bmatrix}
0 & 1 & 0 & \cdots & 0 \\
0 & 0 & 1 & \cdots & 0 \\
\vdots & & & & \\
0 & 0 & 0 & \cdots & 1 \\
-a_0 & -a_1 & -a_2 & \cdots & -a_{n-1}
\end{bmatrix}
$$

show that any given polynomial $(-1)^n(a_0 + a_1\lambda + \cdots + a_{n-1}\lambda^{n-1} + \lambda^n)$ of degree n in λ may be regarded as the characteristic polynomial of a matrix of order n. This matrix is called the **companion matrix** of the given polynomial.

9. Show that the characteristic polynomial of the $n \times n$ matrix

$$
\begin{bmatrix}
0 & 1 & 1 & \cdots & 1 & 1 \\
1 & 0 & 1 & \cdots & 1 & 1 \\
\vdots & & & & & \\
1 & 1 & 1 & \cdots & 1 & 0
\end{bmatrix}
$$

is $(-1)^{n-1}(n - 1 - \lambda)(1 + \lambda)^{n-1}$.

10. For what values of h will the matrix

$$
\begin{bmatrix}
h & 1 & 0 \\
1 & h & 0 \\
0 & 0 & 1
\end{bmatrix}
$$

have multiple characteristic roots?

11. For what values of k do the matrices

$$
\begin{vmatrix}
1 & 4k \\
4 & 3
\end{vmatrix}, \qquad
\begin{vmatrix}
1 & 2 \\
2 & k
\end{vmatrix}
$$

have (a) real and distinct, (b) real and equal, (c) complex characteristic roots?

12. Under what conditions on the elements will the matrix

$$
\begin{vmatrix}
a & b \\
c & d
\end{vmatrix}
$$

have equal characteristic roots?

*13. Prove that, if A and B are $n \times n$ matrices and if there exist n linearly independent vectors $X_1, X_2, \ldots, X_n$ and scalars $\lambda_1, \lambda_2, \ldots, \lambda_n$ such that $AX_j = \lambda_j X_j$ and $BX_j = \lambda_j X_j$, $j = 1, 2, \ldots, n$, then $A = B$.

14. Find a transformation that will reduce this operator to diagonal form:

$$
X_2 = \begin{bmatrix}
4 & -1 & -2 \\
2 & 1 & -2 \\
1 & -1 & 1
\end{bmatrix} X_1.
$$

15. The three characteristic vectors

$$\begin{bmatrix} 1 \\ -1 \\ 1 \end{bmatrix}, \quad \begin{bmatrix} 1 \\ 1 \\ 0 \end{bmatrix}, \quad \begin{bmatrix} 1 \\ -1 \\ 0 \end{bmatrix}$$

of a 3×3 matrix A are associated respectively with the characteristic roots 1, -1, and 0. Find A.

16. Find the characteristic roots and characteristic vectors of the real matrix

$$A = \begin{bmatrix} a & -b \\ b & a \end{bmatrix}.$$

Normalize the characteristic vectors and in each case compute X^*AX.

17. Show that if $A^*X = \lambda X$, $AY = \mu Y$, and $\bar{\lambda} \neq \mu$, then $X^*Y = 0$.

*18. Show that the characteristic vectors of A^{-1} are the same as those of A. Show also that $\det(A^{-1} - \mu I) = (-\mu)^n \det A^{-1} \det(A - (1/\mu)I)$ and hence that, if the characteristic roots of A are $\lambda_1, \lambda_2, \ldots, \lambda_n$, distinct or not, those of A^{-1} are $(1/\lambda_1), (1/\lambda_2), \ldots, (1/\lambda_n)$.

19. Show that the characteristic roots of kA are $k\lambda_1, k\lambda_2, \ldots, k\lambda_n$, where $\lambda_1, \lambda_2, \ldots, \lambda_n$ are those of A.

*20. If U is a unitary matrix and $U^*AU = D[d_1, d_2, \ldots, d_n]$, where D is a diagonal matrix whose diagonal elements are the d_j's, show that the d_j's are the characteristic roots of A and that the columns of U are characteristic vectors corresponding to the d's respectively. (*Hint:* Subtract λI from each member of the equation.)

21. Calculate the characteristic polynomial of the matrix $[1 + \delta_{ij}]_{n \times n}$.

22. Given that P is the matrix of a projection, show that its characteristic roots are all 0 or 1. Then show that the characteristic vectors associated with the root 1 span the range of P. What vectors correspond to the root 0?

23. Determine the characteristic roots of the matrix

$$\begin{bmatrix} a_1 & a_2 & \cdots & a_n \\ a_1 & a_2 & \cdots & a_n \\ \vdots & & & \\ a_1 & a_2 & \cdots & a_n \end{bmatrix}.$$

24. A matrix of the type

$$\begin{bmatrix} 0 & 1 & 0 & 0 \\ 0 & 0 & \mu & 0 \\ 0 & 0 & 0 & \mu \\ -b & -c & -a\mu & 0 \end{bmatrix}$$

appears in the study of hydrodynamic stability. Determine the real characteristic subspaces under the assumption that $\mu = 4$, $a = -1$, $b = 0$, $c = 0$. (Values chosen to make the computations reasonable.)

25. Let

$$A = \begin{bmatrix} A_1 & 0 & 0 & 0 \\ 0 & A_2 & 0 & 0 \\ 0 & 0 & A_3 & 0 \\ 0 & 0 & 0 & A_4 \end{bmatrix}$$

be a real, symmetric, pseudodiagonal matrix. Under what conditions will the characteristic vectors of A have many zeros in them?

*26. Show that if all the characteristic roots of a symmetric matrix A are zero, then A is also zero. Show by means of a counterexample that if A is not assumed symmetric, the conclusion does not follow.

27. Let A be $m \times n$ and B be $n \times m$. Prove that AB and BA have the same nonzero characteristic roots.

28. Prove that if λ is a characteristic root of A and if $\lambda^p = 1$, then $A^p - I$ is singular.

29. Show that if $AX = \lambda_1 X$ and $AX = \lambda_2 X$ where $X \neq 0$, then $\lambda_1 = \lambda_2$; that is, the characteristic subspaces of a matrix A have only the zero vector in common.

30. Let $X + iY$, X and Y real, be a characteristic vector associated with the root $\lambda = \alpha + i\beta$ of a real, orthogonal matrix A. Prove that $X^{\mathsf{T}}Y = 0$. (Note that $\alpha^2 + \beta^2 = 1$.) Give an illustrative example.

31. Prove that if $X + iY$, where X and Y are real, is a unit characteristic vector associated with the characteristic root $\lambda = i\alpha$, α real, of a real, skew-symmetric matrix, then $X^{\mathsf{T}}Y = 0$, $X^{\mathsf{T}}X = Y^{\mathsf{T}}Y = \frac{1}{2}$, and $X^{\mathsf{T}}AX = Y^{\mathsf{T}}AY = 0$.

32. Let

$$I = \begin{bmatrix} 1 & 0 & 0 & 0 \\ 0 & 1 & 0 & 0 \\ 0 & 0 & 1 & 0 \\ 0 & 0 & 0 & 1 \end{bmatrix}, \quad \mathscr{I} = \begin{bmatrix} 0 & 1 & 0 & 0 \\ -1 & 0 & 0 & 0 \\ 0 & 0 & 0 & -1 \\ 0 & 0 & 1 & 0 \end{bmatrix},$$

$$\mathscr{J} = \begin{bmatrix} 0 & 0 & 1 & 0 \\ 0 & 0 & 0 & 1 \\ -1 & 0 & 0 & 0 \\ 0 & -1 & 0 & 0 \end{bmatrix}, \quad \mathscr{K} = \begin{bmatrix} 0 & 0 & 0 & 1 \\ 0 & 0 & -1 & 0 \\ 0 & 1 & 0 & 0 \\ -1 & 0 & 0 & 0 \end{bmatrix}.$$

Complete the multiplication table

×	I	$\mathscr{I}$	$\mathscr{J}$	$\mathscr{K}$
I				
$\mathscr{I}$				
$\mathscr{J}$				
$\mathscr{K}$				

and then show that:
(a) The set of all matrices of the form

$$Q = aI + b\mathscr{I} + c\mathscr{J} + d\mathscr{K},$$

where a, b, c, d are arbitrary real numbers is a ring (see Section 2.14).
(b) $QQ^{\mathsf{T}} = (a^2 + b^2 + c^2 + d^2)I$ and $Q + Q^{\mathsf{T}} = 2aI$.
(c) $(Q - \lambda I)(Q - \lambda I)^{\mathsf{T}} = ((a^2 + b^2 + c^2 + d^2) - 2a\lambda + \lambda^2)I$ and, hence, the characteristic roots of Q are $a \pm i\sqrt{b^2 + c^2 + d^2}$, each counted twice.
(d) $Q^{-1} = (a^2 + b^2 + c^2 + d^2)^{-1}(aI - b\mathscr{I} - c\mathscr{J} - d\mathscr{K})$ if $Q \neq 0$.
33. Show that, if

$$I = \begin{bmatrix} 1 & 0 \\ 0 & 1 \end{bmatrix}, \quad R = \begin{bmatrix} i & 0 \\ 0 & -i \end{bmatrix}, \quad S = \begin{bmatrix} 0 & 1 \\ -1 & 0 \end{bmatrix}, \quad T = \begin{bmatrix} 0 & i \\ i & 0 \end{bmatrix},$$

then the set of all matrices P defined by

$$P = aI + bR + cS + dT,$$

where a, b, c, d are arbitrary real numbers, is a ring isomorphic to the ring of matrices defined in the preceding exercise.

The rings of matrices in the two preceding exercises are called **quaternion rings of matrices** since they are isomorphic to the ring of quaternions (see G. Birkhoff and S. MacLane: *A Survey of Modern Algebra*, 3rd ed., Macmillan, New York, 1965, pp. 222 ff.).

9.5 The Characteristic Polynomial and Its Roots

The polynomial

$$(9.5.1) \qquad \phi(\lambda) = \det(A - \lambda I) = \begin{vmatrix} a_{11} - \lambda & a_{12} & \cdots & a_{1n} \\ a_{21} & a_{22} - \lambda & \cdots & a_{2n} \\ \vdots & & & \\ a_{n1} & a_{n2} & \cdots & a_{nn} - \lambda \end{vmatrix}$$

is called the **characteristic polynomial** of A. The product of the diagonal entries, namely

$$(9.5.2) \qquad (a_{11} - \lambda)(a_{22} - \lambda) \cdots (a_{nn} - \lambda),$$

is a term in the expansion of the determinant. From this product, we see that the highest-degree term in λ is $(-\lambda)^n$. It is convenient to write $\phi(\lambda)$ in descending powers of $-\lambda$:

$$(9.5.3) \qquad \phi(\lambda) = (-\lambda)^n + b_{n-1}(-\lambda)^{n-1} + \cdots + b_1(-\lambda) + b_0.$$

If, in the expansion of (9.5.2), we form the product of the term a_{ii} from one factor and the terms $-\lambda$ from each of the other $n-1$ factors, then sum the results, we get

$$(a_{11} + a_{22} + \cdots + a_{nn})(-\lambda)^{n-1}.$$

There is no other way to get terms involving $(-\lambda)^{n-1}$ since every term other than (9.5.2) in the expansion of the determinant contains at most $n-2$ of the diagonal entries. Hence

$$(9.5.4) \qquad b_{n-1} = a_{11} + a_{22} + \cdots + a_{nn} = \text{tr } A.$$

By putting $\lambda = 0$ in (9.5.1) and (9.5.3), we get $\phi(0) = \det A$ and $\phi(0) = b_0$ respectively, so

$$(9.5.5) \qquad\qquad\qquad b_0 = \det A.$$

These formulas for b_{n-1} and b_0 are special cases of

Theorem 9.5.1: *The coefficient of* $(-\lambda)^r$ *in* $\phi(\lambda)$ *is the sum of the principal minor determinants of order* $n-r$ *of* A. *In particular, the coefficients of* $(-\lambda)^n$, $(-\lambda)^{n-1}$, *and* $(-\lambda)^0$ *are respectively* 1, $b_{n-1} = \text{tr } A$, *and* $b_0 = \det A$.

To prove this, note that

$$A - \lambda I = [A_1 - \lambda E_1, A_2 - \lambda E_2, \ldots, A_n - \lambda E_n],$$

where A_j is the jth column of A and E_j is the jth elementary n-vector. By repeated use of Theorem 7.3.3, we see that $\det(A - \lambda I)$ is the sum of 2^n determinants, the jth column in each of which is either A_j or $-\lambda E_j$, so that

$$\begin{aligned}
\det(A - \lambda I) = {} & \det[-\lambda E_1, -\lambda E_2, \ldots, -\lambda E_n] \\
& + \det[A_1, -\lambda E_2, \ldots, -\lambda E_n] + \cdots \\
& + \det[-\lambda E_1, \ldots, -\lambda E_{n-1}, A_n] \\
& + \det[A_1, A_2, -\lambda E_3, \ldots, -\lambda E_n] + \cdots \\
& + \det[-\lambda E_1, \ldots, -\lambda E_{n-2}, A_{n-1}, A_n] \\
& + \cdots \\
& + \det[A_1, A_2, \ldots, A_{n-r}, -\lambda E_{n-r+1}, \ldots, -\lambda E_n] + \cdots \\
& + \det[-\lambda E_1, \ldots, -\lambda E_r, A_{r+1}, \ldots, A_n] \\
& + \cdots \\
& + \det[A_1, A_2, \ldots, A_n].
\end{aligned}$$

If we use Theorem 7.2.8 to factor out the coefficients $-\lambda$, what remains in each case is a principal minor determinant of A and the theorem follows. (To see this better, write out the preceding formula in full for $n = 3$.)

This theorem may be used, if we wish, to expand $\phi(\lambda)$. For example,

$$
\begin{vmatrix} 2 - \lambda & 0 & 1 \\ 0 & 4 - \lambda & -1 \\ 1 & -1 & -2 - \lambda \end{vmatrix} = (-\lambda)^3 + (2 + 4 - 2)(-\lambda)^2
$$

$$
+ \left(\begin{vmatrix} 2 & 0 \\ 0 & 4 \end{vmatrix} + \begin{vmatrix} 2 & 1 \\ 1 & -2 \end{vmatrix} + \begin{vmatrix} 4 & -1 \\ -1 & -2 \end{vmatrix} \right)(-\lambda)
$$

$$
+ \begin{vmatrix} 2 & 0 & 1 \\ 0 & 4 & -1 \\ 1 & -1 & -2 \end{vmatrix}
$$

$$
= (-\lambda)^3 + 4\lambda^2 + 6\lambda - 22.
$$

Now let the n roots of the characteristic equation $\phi(\lambda) = 0$ be $\lambda_1, \lambda_2, \ldots, \lambda_n$. Then we can write, by the factor theorem,

$$(9.5.6) \qquad \phi(\lambda) = (\lambda_1 - \lambda)(\lambda_2 - \lambda) \cdots (\lambda_n - \lambda).$$

The coefficient of $(-\lambda)^{n-1}$ in this product will be, by the same reasoning we used to obtain (9.5.4),

$$b_{n-1} = \lambda_1 + \lambda_2 + \cdots + \lambda_n,$$

so

$$(9.5.7) \qquad \operatorname{tr} A = \lambda_1 + \lambda_2 + \cdots + \lambda_n.$$

Similarly, from (9.5.6), we find

$$\phi(0) = b_0 = \lambda_1 \lambda_2 \cdots \lambda_n,$$

so

$$(9.5.8) \qquad \det A = \lambda_1 \lambda_2 \cdots \lambda_n.$$

Making comparisons of the remaining coefficients of the expansions of the products in (9.5.4) and (9.5.6), we find that

$$b_r = \sum \lambda_{j_1} \lambda_{j_2} \cdots \lambda_{j_{n-r}}, \qquad r = 0, 1, \ldots, n - 1.$$

Here the sum is extended over the $C(n, n - r)$ combinations $j_1, j_2, \ldots, j_{n-r}$ of $1, 2, \ldots, n$, taken $n - r$ at a time. For $r = 0, 1, \ldots, n - 2$, these expressions are called the **higher traces** of the matrix. Thus we have proved

Theorem 9.5.2: *The trace of A is the sum of the characteristic roots of A and the determinant of A is the product of the characteristic roots of A. In general, the coefficient b_r of $\phi(\lambda)$ is the sum of the products of the roots of $\phi(\lambda)$, $n - r$ at a time.*

For example, if

$$A = \begin{bmatrix} 2 & -1 & 5 \\ -1 & -4 & 3 \\ 5 & 3 & 2 \end{bmatrix},$$

we have

$$\operatorname{tr} A = \lambda_1 + \lambda_2 + \lambda_3 = 0$$

and

$$\det A = \lambda_1 \lambda_2 \lambda_3 = 34.$$

The characteristic roots $\lambda_1, \lambda_2, \ldots, \lambda_n$ may or may not be distinct. Let the distinct values be labeled $\lambda_1, \lambda_2, \ldots, \lambda_p$, with multiplicities $r_1, r_2, \ldots, r_p$ respectively, so that we may write

$$\phi(\lambda) = (\lambda_1 - \lambda)^{r_1}(\lambda_2 - \lambda)^{r_2} \cdots (\lambda_p - \lambda)^{r_p},$$

where

$$r_1 + r_2 + \cdots + r_p = n.$$

With each of the distinct values λ_i, there is associated a characteristic subspace of dimension $n - \operatorname{rank}(A - \lambda_i I)$. Since the set of distinct characteristic roots always contains at least one member λ_1 and since $\operatorname{rank}(A - \lambda_i I) < n$, every square matrix A has at least one characteristic subspace, whose dimension is > 0. We can say more about the dimensions of the characteristic subspaces. We begin with

Lemma 9.5.3: *If λ_1 is an r_1-fold characteristic root of A, then 0 is an r_1-fold characteristic root of $A - \lambda_1 I$.*

Since $\lambda = \lambda_1$ is an r_1-fold characteristic root of A, precisely r_1 of the factors $\lambda_i - \lambda$ must be identical to $\lambda_1 - \lambda$, so we can write

(9.5.9) $\phi(\lambda) = (\lambda_1 - \lambda)^{r_1} \psi(\lambda),$

where $\psi(\lambda)$ represents the product of the remaining factors $\lambda_j - \lambda$, so $\psi(\lambda_1) \neq 0$. Hence the characteristic polynomial of $A - \lambda_1 I$ is

$$(9.5.10) \quad \det((A - \lambda_1 I) - \lambda I) = \det(A - (\lambda_1 + \lambda)I)$$
$$= \phi(\lambda_1 + \lambda)$$
$$= (\lambda_1 - (\lambda_1 + \lambda))^{r_1} \psi(\lambda_1 + \lambda) \quad \text{by (9.5.9)}$$
$$= (0 - \lambda)^{r_1} \psi(\lambda_1 + \lambda).$$

Now when $\lambda = 0$, $\psi(\lambda_1 + \lambda)$ becomes $\psi(\lambda_1)$, which is not zero. Thus no additional factors $-\lambda$ appear in $\psi(\lambda_1 + \lambda)$ and 0 is precisely an r_1-fold characteristic root of $A - \lambda_1 I$, as was to be shown.

Now recall that in the expansion of $\det((A - \lambda_1 I) - \lambda I)$, the coefficient of $(-\lambda)^{r_1}$ is the sum of the principal minor determinants of order $n - r_1$ of $A - \lambda_1 I$. Since this coefficient is also the constant term in the expansion of $\psi(\lambda_1 + \lambda)$, namely $\psi(\lambda_1) \neq 0$, there must be at least one principal minor determinant of order $n - r_1$ of $A - \lambda_1 I$ which is $\neq 0$. Hence the rank of $A - \lambda_1 I$ is *at least* $n - r_1$. It may be even higher, as this example shows:

$$A = \begin{bmatrix} 1 & 1 & 1 & 1 \\ 0 & 1 & 1 & 1 \\ 0 & 0 & 1 & 1 \\ 0 & 0 & 0 & 1 \end{bmatrix}, \quad \begin{array}{l} \det(A - \lambda I) = (1 - \lambda)^4, \lambda_1 = 1, r_1 = 4, \\ n - r_1 = 0, \operatorname{rank}(A - 1 \cdot I) = 3 > n - r_1. \end{array}$$

In summary, we have

Theorem 9.5.4: *If λ_1 is an r_1-fold characteristic root of a matrix A, then the rank of $A - \lambda_1 I$ is not less than $n - r_1$ and the dimension of the associated characteristic subspace is not greater than $n - (n - r_1) = r_1$.*

If ρ_j is the dimension of the characteristic subspace associated with λ_j, then

$$(9.5.11) \quad 1 \leq \rho_j \leq r_j, \quad j = 1, 2, \ldots, p,$$

so that, summing these inequalities, we have

$$(9.5.12) \quad p \leq \sum_1^p \rho_j \leq \sum_1^p r_j = n.$$

A useful special case follows directly from (9.5.11):

Theorem 9.5.5: *If λ_1 is a simple root of $\phi(\lambda) = 0$ so that $r_1 = 1$, the rank of $A - \lambda_1 I$ is $n - 1$ and the dimension of the associated characteristic subspace is 1.*

9.6   Similar Matrices

Given the matrices A and C, we say that C is **similar** to A if and only if there exists a matrix B such that $C = B^{-1}AB$. We saw in Chapter 8 how a change of basis gives rise to similar matrices. Since $A = I^{-1}AI$, A is similar to itself. That is, similarity is a *reflexive* relation. Since $C = B^{-1}AB$ implies that $A = (B^{-1})^{-1}C(B^{-1})$, C similar to A implies A similar to C. That is, similarity is a *symmetric* relation so that we may properly use the symmetrical expression, "A and C are similar." Finally, let A be similar to C and C to F so that for some B and D, $A = B^{-1}CB$ and $C = D^{-1}FD$. Then $A = (DB)^{-1}F(DB)$ and A is similar to F, so similarity is a *transitive* relation. In summary, we have

Theorem 9.6.1: *Similarity of matrices is an equivalence relation.*

In the relation $C = B^{-1}AB$, all three matrices may be over the same number field, but this is not necessarily the case. We may have A real, B and C complex, for example, or C real, A and B complex. Such cases will arise in what follows.

Similar matrices have important properties in common, as is indicated by

Theorem 9.6.2: *Let A and C be similar matrices and let $C = B^{-1}AB$. Then A and C have identical characteristic polynomials and hence have identical characteristic roots. Moreover, if X is a characteristic vector of A associated with the root λ, then $B^{-1}X$ is a characteristic vector of C, again associated with the root λ.*

The characteristic polynomial of C is given by

$$
\begin{aligned}
\det(C - \lambda I) &= \det(B^{-1}AB - \lambda I) \\
&= \det(B^{-1}AB - B^{-1}\lambda IB) \\
&= \det(B^{-1}(A - \lambda I)B) \\
&= \det B^{-1}\det(A - \lambda I)\det B \\
&= \det B^{-1}\det B\det(A - \lambda I) \quad \text{since determinants are scalars} \\
&= \det(B^{-1}B(A - \lambda I)) \\
&= \det(A - \lambda I).
\end{aligned}
$$

Thus C and A have the same characteristic polynomial and therefore have precisely the same characteristic roots $\lambda_1, \lambda_2, \ldots, \lambda_n$, so that

$$\operatorname{tr} A = \lambda_1 + \lambda_2 + \cdots + \lambda_n = \operatorname{tr} C$$

and

$$\det A = \lambda_1 \lambda_2 \cdots \lambda_n = \det C.$$

Now suppose that X is a characteristic vector of A associated with the root λ so that $AX = \lambda X$. Then, since $CB^{-1} = B^{-1}A$, we have

$$C(B^{-1}X) = (CB^{-1})X = (B^{-1}A)X = B^{-1}(AX) = B^{-1}(\lambda X) = \lambda(B^{-1}X),$$

so that $B^{-1}X$ is a characteristic vector of C associated with the characteristic root λ.

All this has a geometrical interpretation. Let A be the matrix of a linear operator $X_2 = AX_1$ in the standard reference system. Recall that if we introduce the transformation of coordinates $X = BY$ or $Y = B^{-1}X$, we get the representation of the operator in the new reference system: $Y_2 = (B^{-1}AB)Y_1$. The previous theorem thus implies

Theorem 9.6.3: In every reference system, the operator A has the same characteristic polynomial, the same characteristic roots, the same trace, the same determinant, and the same characteristic subspaces.

That the characteristic subspaces are the same follows from the fact that $B^{-1}X$ and X are representations of the same vector in the new and in the old reference systems, respectively. Since they are independent of the choice of reference system, the characteristic polynomial, the characteristic roots, the trace, the determinant, the higher traces, and the characteristic subspaces are called **invariants** of the operator; that is, they are properties of the operator rather than of the particular matrix used to represent it.

9.7 Exercises

1. If A is of order 3, if the sum of its characteristic roots is 9, if their product is 24, and if $\lambda_3 = 3$, what are the possible values of λ_1 and λ_2 ?
*2. Show that $\lambda = 0$ is a characteristic root of A if and only if $\det A = 0$.
3. Prove, *without finding the characteristic roots*, that if

$$A = \begin{bmatrix} -1 & 1 & -1 \\ 1 & -1 & 1 \\ -1 & 1 & 2 \end{bmatrix},$$

then two of the characteristic roots of A differ only in sign.
4. Use Theorem 9.5.1 to find the characteristic polynomial of

$$\begin{bmatrix} 1 & 1 & 0 & 0 \\ 1 & -1 & 0 & 1 \\ 0 & 0 & 2 & 1 \\ 0 & 1 & 1 & 2 \end{bmatrix}.$$

*5. Show that A and A^{T} have the same characteristic polynomial.

6. Show that if A is of order n, then

$$\det(A - I) = \sum_{k=0}^{n} (-1)^{n-k} \sigma_k,$$

where σ_k is the sum of the principal minors of order k of A. This includes the principal minor of order 0, which is by definition 1.

7. A linear operator on $\mathscr{E}^3$ maps

$$\begin{bmatrix} 1 \\ 2 \\ 1 \end{bmatrix} \text{ onto } \begin{bmatrix} 1 \\ 2 \\ 1 \end{bmatrix}, \quad \begin{bmatrix} 2 \\ 1 \\ 0 \end{bmatrix} \text{ onto } \begin{bmatrix} -4 \\ -2 \\ 0 \end{bmatrix}, \quad \text{and} \quad \begin{bmatrix} 1 \\ 1 \\ 1 \end{bmatrix} \text{ onto } \begin{bmatrix} 0 \\ 0 \\ 0 \end{bmatrix}.$$

Without finding A itself, find the characteristic polynomial of A and also its trace and determinant.

*8. Show that if the rank of A is r, then at least $n - r$ characteristic roots of A are zero. Give an example to show that the words "at least" are justified.

*9. If $AX_j = \lambda_j X_j, j = 1, 2, \ldots, n$, and if the X_j's are linearly independent, show that $\lambda_1, \lambda_2, \ldots, \lambda_n$ are the characteristic roots of A.

10. Show that, for each r, the sums of the corresponding principal minors of order r of AB and BA are equal, where A and B are of order n, so that AB and BA have exactly the same characteristic roots but not necessarily the same characteristic vectors.

11. Show that, if the characteristic roots of A are distinct and those of B are distinct, then $AB = BA$ if and only if A and B have the same characteristic vectors.

12. Show that, if $AB = BA$, $AX_0 = \lambda X_0$, and $BX_0 \neq 0$, then BX_0 is a characteristic vector of A associated with the characteristic root λ. Then show that, if the vector space of characteristic vectors of A associated with λ is *one-dimensional*, X_0 must also be a characteristic vector of B associated with an appropriate characteristic root μ of B.

13. Given that A is $m \times n$, show that $A^{\mathsf{T}}A$ and AA^{T} have the same characteristic roots, except possibly for zero roots. Does the same result hold for A^*A and AA^*?

14. By differentiating $\det(A - \lambda I)$ repeatedly, prove that the rth derivative of $\phi(\lambda)$ is given by $(-1)^r r!$ times the sum of the principal minors of order $n - r$ of the characteristic matrix if $r \leq n$ and by 0 if $r \geq n$. (The principal minor of order 0 is by definition 1.) Then use this result and Taylor's theorem to prove Theorem 9.5.1.

15. Show that, if $a_{ij} \geq 0$ for all i and j and $\sum_{i=1}^{n} a_{ij} \leq 1$ for each j, then

$$\det(A - I) \geq 0, \quad \text{if } n \text{ is even,}$$

$$\leq 0, \quad \text{if } n \text{ is odd.}$$

16. How is the characteristic polynomial of A^* related to the characteristic polynomial $\phi(\lambda)$ of A? What is the nature of the coefficients of $\phi(\lambda)$ in the event that A is Hermitian? Skew-Hermitian?

17. Prove that, if X is any nonzero vector of $\mathscr{U}^n$, the nonzero characteristic root of XX^* is $X^*X = \sum |x_i|^2$.

18. Prove that if there exists B such that $B^{-1}AB$ is diagonal, then every matrix similar to A is also similar to the same diagonal matrix.

19. Prove that the relation of similarity separates the set of all $n \times n$ matrices over a field into mutually exclusive classes of similar matrices.

20. Prove that for all scalars α, αI_n is the only $n \times n$ matrix that is similar to αI_n.

21. Show that if $A \neq 0$ is nilpotent, then A is not similar to any diagonal matrix.

22. Show by considering the determinant and the trace that there is no real matrix A such that

$$A^{-1} \begin{bmatrix} -1 & 2 \\ -2 & 0 \end{bmatrix} A$$

is diagonal.

23. Show that, if there exists a nonsingular matrix C such that

$$C^{-1}AC = \text{diag}[\lambda_1, \ldots, \lambda_n] \quad \text{and} \quad C^{-1}BC = \text{diag}[\mu_1, \ldots, \mu_n],$$

then $AB = BA$ and the characteristic roots of AB are $\lambda_1\mu_1, \lambda_2\mu_2, \ldots, \lambda_n\mu_n$.

9.8 The Characteristic Roots of a Hermitian Matrix

A Hermitian matrix has been defined as one which is equal to its tranjugate: $A = A^*$. Concerning such matrices, which include real symmetric matrices as a special case, we prove first

Theorem 9.8.1: *The characteristic roots of a Hermitian matrix are all real.*

Let A be Hermitian, let λ be any root of the characteristic equation of A, and let Y be an associated *unit* characteristic vector. Then from the equation

$$(9.8.1) \qquad AY = \lambda Y,$$

we deduce that

$$(9.8.2) \qquad Y^*AY = \lambda Y^*Y = \lambda.$$

But we also have, since $A = A^*$ and Y^*AY is a scalar,

$$Y^*AY = Y^*A^*Y = (Y^*AY)^* = \overline{Y^*AY}.$$

Thus the scalar Y^*AY is equal to its own conjugate and hence is real. That is, by (9.8.2), λ is real.

As a special case, we have

Theorem 9.8.2: *The characteristic roots of a real symmetric matrix are all real.*

Because in this case all the coefficients in the system of equations $AX = \lambda X$ are real, we have also

Theorem 9.8.3: *The characteristic subspaces of a real symmetric matrix may all be spanned by real vectors.*

We prove next

Theorem 9.8.4: *If X and Y are characteristic vectors associated with distinct characteristic roots λ and μ of a Hermitian matrix A, then X and Y are orthogonal.*

Suppose, in fact, that

$$AX = \lambda X \qquad \text{and} \qquad AY = \mu Y, \qquad \lambda \neq \mu.$$

Then from these equations, we have

$$Y^*AX = \lambda Y^*X \qquad \text{and} \qquad X^*AY = \mu X^*Y.$$

Forming the transposed conjugate of the first of these equations, we have

$$X^*AY = \lambda X^*Y,$$

since $A = A^*$ and λ is real. It then follows that

$$\mu X^*Y = \lambda X^*Y,$$

so that, since $\lambda \neq \mu$, we must have

$$X^*Y = 0;$$

that is, X and Y are orthogonal.

Again, an important special case is:

Theorem 9.8.5: *Characteristic vectors associated with distinct characteristic roots of a real symmetric matrix are orthogonal.*

9.9 The Diagonal Form of a Hermitian Matrix

In the following, the notation $D[\lambda_1, \lambda_2, \ldots, \lambda_n]$ for a diagonal matrix with diagonal elements $\lambda_1, \lambda_2, \ldots, \lambda_n$ will be used.

We prove first an existence theorem:

Theorem 9.9.1: *If A is Hermitian, there exists a unitary matrix U such that U*AU is a diagonal matrix whose diagonal elements are the characteristic roots of A: $U*AU = D[\lambda_1, \lambda_2, \ldots, \lambda_n]$.*

We shall prove the theorem by induction on the order of A, and then later show how the matrix U may be computed.

First we observe that, if A is Hermitian and if V is any matrix of order n, then $V*AV$ is also Hermitian since $(V*AV)* = V*A*(V*)* = V*AV$ because $A* = A$. From this we have the result that if $V*AV = B$ and $b_{21} = b_{31} = \cdots = b_{n1} = 0$, then, if A is Hermitian, $b_{12} = b_{13} = \cdots = b_{1n} = 0$ also.

A second observation useful in our proof is that, if V is unitary and if $V*AV = B$, where B is diagonal, then the diagonal elements of B are necessarily the characteristic roots of A (Exercise 20, Section 9.4).

In case A is of order 2, we must show there exists a unitary matrix U of order 2 such that

$$U*AU = D[\lambda_1, \lambda_2],$$

where λ_1 and λ_2 are the characteristic roots of A. That is, since $U* = U^{-1}$, we must show there exists a unitary matrix U such that

$$\begin{bmatrix} a_{11} & a_{12} \\ a_{21} & a_{22} \end{bmatrix} \cdot \begin{bmatrix} u_{11} & u_{12} \\ u_{21} & u_{22} \end{bmatrix} = \begin{bmatrix} u_{11} & u_{12} \\ u_{21} & u_{22} \end{bmatrix} \cdot \begin{bmatrix} \lambda_1 & 0 \\ 0 & \lambda_2 \end{bmatrix}.$$

Forming both products and equating their first columns, we obtain the equations

$$(a_{11} - \lambda_1)u_{11} + a_{12}u_{21} = 0$$
$$a_{21}u_{11} + (a_{22} - \lambda_1)u_{21} = 0,$$

which have nontrivial solutions for u_{11}, u_{21} since λ_1 is a characteristic root of A. We select one of these solutions for the first column of U. Since the equations are homogeneous, this solution may be assumed normalized, that is, we may assume $[u_{11}, u_{21}]^\mathsf{T}$ is a unit 2-vector, so that

$$\bar{u}_{11}u_{11} + \bar{u}_{21}u_{21} = 1.$$

The Characteristic Value Problem /Ch. 9

Now we use for the second column of U any unit vector orthogonal to the first, so that

$$\bar{u}_{12}u_{12} + \bar{u}_{22}u_{22} = 1, \qquad \bar{u}_{11}u_{12} + \bar{u}_{21}u_{22} = 0.$$

Because it then has mutually orthogonal unit vectors as columns, U is a unitary matrix. Also, the product U^*AU has as its first column the vector $[\lambda_1, 0]^\mathsf{T}$. By the first observation above, it then follows that the 1,2-entry of this product is also zero, and, therefore, from the second observation we conclude that the lower right element is necessarily λ_2. This proves the theorem for $n = 2$.

Now let $n - 1$ be any integer ≥ 2 such that the theorem holds true for all Hermitian matrices of order $n - 1$, and let A be any Hermitian matrix of order n. We shall determine first a unitary matrix V such that

$$V^*AV = \begin{bmatrix} \lambda_1 & 0 \\ \hline 0 & B \end{bmatrix},$$

that is, such that

$$AV = V \begin{bmatrix} \lambda_1 & 0 \\ \hline 0 & B \end{bmatrix},$$

where λ_1 is any characteristic root of A and B is a matrix of order $n - 1$.

Once again, we equate the elements of the first columns of the left and right members of this last equation and obtain the equations

$$(a_{11} - \lambda_1)v_{11} + a_{12}v_{21} + \cdots + a_{1n}v_{n1} = 0$$
$$a_{21}v_{11} + (a_{22} - \lambda_1)v_{21} + \cdots + a_{2n}v_{n1} = 0$$
$$\vdots$$
$$a_{n1}v_{11} + a_{n2}v_{21} + \cdots + (a_{nn} - \lambda_1)v_{n1} = 0,$$

for the determination of the first column $V_1 = [v_{11}, v_{21}, \ldots, v_{n1}]^\mathsf{T}$ of V. Since λ_1 is a characteristic root of A, this homogeneous system has nontrivial solutions, one of which we normalize and use for V_1.

The remaining columns of V we simply choose in such a way that $V_1, V_2, \ldots, V_n$ form a set of mutually orthogonal unit vectors so that V is indeed unitary (Theorem 5.8.1).

It now follows from the first observation at the beginning of this proof that V^*AV actually has the required form

$$\begin{bmatrix} \lambda_1 & 0 \\ \hline 0 & B \end{bmatrix}.$$

Since V^*AV is Hermitian, B must be Hermitian also. Furthermore, by an argument like that used in the second observation above, it follows that the characteristic roots of B are just the remaining characteristic roots $\lambda_2, \lambda_3, \ldots, \lambda_n$ of A. Hence, by the induction hypothesis, there exists a unitary matrix W of order $n-1$ such that

$$W^*BW = D[\lambda_2, \lambda_3, \ldots, \lambda_n].$$

It then follows that

$$\begin{bmatrix} 1 & \vdots & 0 \\ \cdots & \vdots & \cdots \\ 0 & \vdots & W^* \end{bmatrix} V^*AV \begin{bmatrix} 1 & \vdots & 0 \\ \cdots & \vdots & \cdots \\ 0 & \vdots & W \end{bmatrix} = \begin{bmatrix} 1 & \vdots & 0 \\ \cdots & \vdots & \cdots \\ 0 & \vdots & W^* \end{bmatrix} \cdot \begin{bmatrix} \lambda_1 & \vdots & 0 \\ \cdots & \vdots & \cdots \\ 0 & \vdots & B \end{bmatrix} \cdot \begin{bmatrix} 1 & \vdots & 0 \\ \cdots & \vdots & \cdots \\ 0 & \vdots & W \end{bmatrix}$$

$$= \begin{bmatrix} \lambda_1 & \vdots & 0 \\ \cdots & \vdots & \cdots \\ 0 & \vdots & W^*BW \end{bmatrix} = D[\lambda_1, \lambda_2, \ldots, \lambda_n].$$

Thus the unitary matrix

$$U = V \begin{bmatrix} 1 & \vdots & 0 \\ \cdots & \vdots & \cdots \\ 0 & \vdots & W \end{bmatrix}$$

has the property stated in the theorem, which is therefore true for all positive integers $n \geq 2$.

Specializing once again to the real case, we have

Theorem 9.9.2: *If A is a real symmetric matrix, there exists an orthogonal matrix U such that U^TAU is a diagonal matrix whose diagonal elements are the characteristic roots of A.*

The geometric interpretation of this result is simple. Let A be the matrix of a real symmetric linear operator in the natural reference system. If U is the matrix of an orthogonal transformation of coordinates, then the matrix representing the operator in the new coordinate system is U^TAU, since $U^{-1} = U^T$ here. Because all characteristic roots of a real symmetric matrix are real, the theorem says that, given a symmetric linear operator over the real field, we can always find a reference system in $\mathscr{E}^n$ in which the matrix of that operator is diagonal. In this reference system, the effect of the operator is simply to multiply each component of a vector by a factor which is one of the characteristic roots of the operator. (See Figure 8-7, page 333.)

Theorem 9.9.3: *The unitary matrix U of Theorem 9.9.1 may be chosen so that the characteristic roots $\lambda_1, \lambda_2, \ldots, \lambda_n$ appear in any desired order in the diagonal matrix U^*AU.*

This follows from the mode of proof of Theorem 9.9.1, and it also follows readily from the fact that a permutation matrix, which one may use to permute the rows or columns of a given matrix, is unitary. The reader may supply the details.

9.10   The Diagonalization of a Hermitian Matrix

Theorem 9.10.1 permits us now to deduce a chain of theorems which show how the matrix U of that theorem may be computed. First we prove

Theorem 9.10.1: *If λ_1 is a k_1-fold characteristic root of a Hermitian matrix A, then the rank of $A - \lambda_1 I_n$ is $n - k_1$.*

By Theorems 9.9.1 and 9.9.3, there exists a unitary matrix U such that

$$U^*AU = D[\lambda_1, \lambda_1, \dots, \lambda_1; \lambda_{k_1+1}, \dots, \lambda_n],$$

the λ_1 occurring k_1 times and $\lambda_{k_1+1}, \dots, \lambda_n$ all being distinct from λ_1. Since U is unitary, subtracting $\lambda_1 I_n$ from both sides of this equation gives

$$U^*(A - \lambda_1 I_n)U = D[0, 0, \dots, 0; (\lambda_{k_1+1} - \lambda_1), \dots, (\lambda_n - \lambda_1)].$$

From this, since U is nonsingular, it follows that the rank of $A - \lambda_1 I_n$ is the same as that of the matrix on the right, which is precisely $n - k_1$ since $(\lambda_{k_1+1} - \lambda_1), \dots, (\lambda_n - \lambda_1)$ are all unequal to zero.

An immediate consequence of Theorem 9.10.1 is

Theorem 9.10.2: *If λ_1 is a k_1-fold characteristic root of a Hermitian matrix A, there exist k_1 linearly independent characteristic vectors of A associated with λ_1, that is, with λ_1 there is associated a k_1-dimensional characteristic subspace.*

We have next

Theorem 9.10.3: *With every Hermitian matrix A we can associate an orthonormal set of n characteristic vectors.*

The characteristic vectors associated with a given characteristic root of A form a vector space for which we can construct an orthonormal basis, by the Gram–Schmidt process if necessary. For each A, there are altogether n vectors in the bases so constructed, by the preceding theorem. Since characteristic vectors associated with distinct characteristic roots of a Hermitian matrix are orthogonal, it follows then that these n basis vectors will serve as the orthonormal set mentioned in the theorem.

This result indicates how the diagonalization process may be effected. For let the n vectors of such an orthonormal system be $U_1, U_2, \ldots, U_n$. Then

(9.10.1) $$AU_j = \lambda_j U_j, \qquad j = 1, 2, \ldots, n,$$

where, in each case, λ_j is the characteristic root associated with U_j. Hence, if

$$U = [U_1, U_2, \ldots, U_n],$$

equations (9.10.1) may be combined in the single equation

$$[AU_1, AU_2, \ldots, AU_n] = [\lambda_1 U_1, \lambda_2 U_2, \ldots, \lambda_n U_n],$$

or

$$AU = UD[\lambda_1, \lambda_2, \ldots, \lambda_n],$$

or, since U is unitary by construction,

$$U^*AU = D[\lambda_1, \lambda_2, \ldots, \lambda_n].$$

We summarize in

Theorem 9.10.4: *If $U_1, U_2, \ldots, U_n$ is an orthonormal system of characteristic vectors associated respectively with the characteristic roots $\lambda_1, \lambda_2, \ldots, \lambda_n$ of a Hermitian matrix A, and if U is the unitary matrix $[U_1, U_2, \ldots, U_n]$, then*

$$U^*AU = D[\lambda_1, \lambda_2, \ldots, \lambda_n].$$

The vectors $U_1, U_2, \ldots, U_n$ are often called a set of **principal axes** of A and the transformation with matrix U used to diagonalize A is called a **principal-axis transformation**.

9.11 Examples

We illustrate the preceding theorem with some examples of the real symmetric case.

(a) Let

$$A = \begin{bmatrix} 1 & 2 \\ 2 & 1 \end{bmatrix},$$

so that $\phi(\lambda) = (\lambda - 3)(\lambda + 1)$. Then normalized characteristic vectors associated with the characteristic roots 3 and -1 are respectively

$$\begin{bmatrix} \dfrac{1}{\sqrt{2}} \\ \dfrac{1}{\sqrt{2}} \end{bmatrix} \quad \text{and} \quad \begin{bmatrix} \dfrac{-1}{\sqrt{2}} \\ \dfrac{1}{\sqrt{2}} \end{bmatrix},$$

so that we may put

$$U = \begin{bmatrix} \dfrac{1}{\sqrt{2}} & \dfrac{-1}{\sqrt{2}} \\ \dfrac{1}{\sqrt{2}} & \dfrac{1}{\sqrt{2}} \end{bmatrix}.$$

It is then easy to verify that $U^*AU = D[3, -1]$ by computing the product.
  (b) Let

$$A = \begin{bmatrix} 5 & 2 & 0 & 0 \\ 2 & 2 & 0 & 0 \\ 0 & 0 & 5 & -2 \\ 0 & 0 & -2 & 2 \end{bmatrix}.$$

Then the characteristic roots are 1, 1, 6, 6. Putting $\lambda = 1$ in $AX = \lambda X$, we obtain the system of equations

$$\begin{aligned} 4x_1 + 2x_2 \qquad\qquad &= 0 \\ 2x_1 + x_2 \qquad\qquad &= 0 \\ 4x_3 - 2x_4 &= 0 \\ -2x_3 + x_4 &= 0, \end{aligned}$$

with the complete solution

$$X = k_1 \begin{bmatrix} 1 \\ -2 \\ 0 \\ 0 \end{bmatrix} + k_2 \begin{bmatrix} 0 \\ 0 \\ 1 \\ 2 \end{bmatrix}.$$

The basis vectors here are already orthogonal so that we need only normalize them to obtain two columns of U:

$$U_1 = \left[\frac{1}{\sqrt{5}}, \frac{-2}{\sqrt{5}}, 0, 0\right]^{\mathsf{T}}, \qquad U_2 = \left[0, 0, \frac{1}{\sqrt{5}}, \frac{2}{\sqrt{5}}\right]^{\mathsf{T}}.$$

Proceeding in the same fashion with the root $\lambda = 6$, we obtain for the other two columns of U,

$$U_3 = \left[\frac{2}{\sqrt{5}}, \frac{1}{\sqrt{5}}, 0, 0\right]^{\mathsf{T}}, \qquad U_4 = \left[0, 0, \frac{-2}{\sqrt{5}}, \frac{1}{\sqrt{5}}\right]^{\mathsf{T}}.$$

Then if $U = [U_1, U_2, U_3, U_4]$, it is easy to check that

$$U^*AU = D[1, 1, 6, 6].$$

9.12 Triangularization of an Arbitrary Matrix

Not every matrix can be reduced to diagonal form by a unitary transformation. On the other hand, it is always possible to attain a triangular form:

Theorem 9.12.1: *Every square matrix A over the complex field can be reduced by a unitary transformation to upper triangular form with the characteristic roots on the diagonal.*

The theorem is proved by induction. We begin with the case $n = 2$. Let A have characteristic roots λ_1, λ_2. With λ_1 we can associate a unit characteristic vector U_1. (It may be that A has only one independent characteristic vector, but it has at least one.) Let U_2 be any unit vector orthogonal to U_1. Then $U = [U_1, U_2]$ is unitary. Moreover,

$$U^*AU = \begin{bmatrix} U_1^* \\ U_2^* \end{bmatrix} \cdot [AU_1, AU_2] = \begin{bmatrix} U_1^*AU_1 & U_1^*AU_2 \\ U_2^*AU_1 & U_2^*AU_2 \end{bmatrix}.$$

Since $AU_1 = \lambda_1 U_1$, $U_1^*U_1 = 1$, and $U_2^*U_1 = 0$, this reduces to

$$U^*AU = \begin{bmatrix} \lambda_1 & U_1^*AU_2 \\ 0 & U_2^*AU_2 \end{bmatrix}.$$

Since A and U^*AU have the same characteristic roots, it now follows that $U_2^*AU_2 = \lambda_2$, and the theorem is proved for $n = 2$.

Now suppose that the result holds for all matrices of order $n - 1$, and let A be of order n. Let λ_1 be any characteristic root of A, and let U_1 denote a corresponding unit characteristic vector. Let $U_2, \ldots, U_n$ be so chosen that $U_1, U_2, \ldots, U_n$ is an orthonormal set, that is, so that $U = [U_1, U_2, \ldots, U_n]$ is unitary. Then, just as in the case where $n = 2$, we obtain

$$U^*AU = \begin{bmatrix} \lambda_1 & U_1^*AU_2 & \cdots & U_1^*AU_n \\ 0 & U_2^*AU_2 & \cdots & U_2^*AU_n \\ \vdots & & & \\ 0 & U_n^*AU_2 & \cdots & U_n^*AU_n \end{bmatrix} = \begin{bmatrix} \lambda_1 & B \\ \hline 0 & C \end{bmatrix}.$$

By the induction hypothesis, there exists a unitary matrix W of order $n - 1$ which will triangularize the lower right submatrix of order $n - 1$. Let

$$V = \begin{bmatrix} 1 & 0 \\ \hline 0 & W \end{bmatrix}.$$

Then V is unitary and

$$V^*(U^*AU)V = \begin{bmatrix} 1 & 0 \\ \hline 0 & W^* \end{bmatrix} \cdot \begin{bmatrix} \lambda_1 & B \\ \hline 0 & C \end{bmatrix} \cdot \begin{bmatrix} 1 & 0 \\ \hline 0 & W \end{bmatrix} = \begin{bmatrix} \lambda_1 & B \\ \hline 0 & W^*CW \end{bmatrix},$$

where W^*CW is upper triangular. Introducing symbols b_{ij} to simplify the notation, we can rewrite this in the form

$$(9.12.1) \qquad (UV)^*A(UV) = \begin{bmatrix} \lambda_1 & b_{12} & b_{13} & \cdots & b_{1n} \\ 0 & \lambda_2 & b_{23} & \cdots & b_{2n} \\ \vdots & & & & \\ 0 & 0 & 0 & \cdots & \lambda_n \end{bmatrix}.$$

Since UV is unitary, the characteristic roots of the triangular matrix are the same as those of A; that is, the diagonal elements are indeed the λ's, and the theorem is proved.

This theorem is often used in deriving other important results, as is illustrated in the next section.

9.13   Normal Matrices

A matrix A is said to be **normal** if and only if

$$A^*A = AA^*.$$

Simple examples are unitary, Hermitian, and skew-Hermitian matrices. So are diagonal matrices with arbitrary diagonal elements, which are not necessarily unitary, Hermitian, or skew-Hermitian. We prove first

Theorem 9.13.1: *If U is unitary, then A is normal if and only if U^*AU is normal.*

We have

$$(U^*AU)^* (U^*AU) = U^*A^*AU,$$

and

$$(U^*AU)(U^*AU)^* = U^*AA^*U.$$

Now $A^*A = AA^*$ if and only if $U^*A^*AU = U^*AA^*U$, which proves the theorem.

The importance of the concept of normality is indicated by

Theorem 9.13.2: *A matrix A over the complex field can be diagonalized by a unitary transformation if and only if A is normal.*

First, suppose that U is unitary and $U^*AU = D[\lambda_1, \lambda_2, \ldots, \lambda_n]$. Then $A = UDU^*$ and $A^*A = UD^*DU$, while $AA^* = UDD^*U$. But D and D^*, being diagonal, commute. Hence $A^*A = AA^*$ and A is normal.

Conversely, suppose that A is normal. Then, by Theorem 9.12.1 there exists a unitary matrix U such that $U^*AU = B$, where B is upper triangular and, as is readily checked, B is also normal. Let

$$B = \begin{bmatrix} \lambda_1 & b_{12} & b_{13} & \cdots & b_{1n} \\ 0 & \lambda_2 & b_{23} & \cdots & b_{2n} \\ \vdots & & & & \\ 0 & 0 & 0 & \cdots & \lambda_n \end{bmatrix}.$$

Since $B^*B = BB^*$, we have from the 1, 1-entries that

$$\overline{\lambda_1}\lambda_1 = \lambda_1\overline{\lambda_1} + b_{12}\overline{b}_{12} + \cdots + b_{1n}\overline{b}_{1n}$$

or

$$0 = |b_{12}|^2 + |b_{13}|^2 + \cdots + |b_{1n}|^2,$$

which implies that $b_{12} = b_{13} = \cdots = b_{1n} = 0$. In the same way, comparison of the 2, 2-entries now implies that $b_{23} = b_{24} = \cdots = b_{2n} = 0$, and so on, for all entries above the diagonal. That is, the assumption that A is normal implies that B is diagonal and the theorem is proved.

We have thus characterized fully the class of matrices which can be diagonalized by unitary transformations.

9.14 Exercises

1. Prove that, if $A^* = -A$, that is, if A is skew-Hermitian, the nonzero characteristic roots of A are pure imaginaries which occur in conjugate pairs if A is real.
2. Prove that, if $X_1 + iX_2$, where X_1 and X_2 are real, is a complex characteristic vector associated with a complex characteristic root $\lambda_1 + i\lambda_2$ of a *real* matrix A, then AX_1 and AX_2 are both linear combinations of X_1 and X_2.
3. Determine the characteristic roots and vectors of the Hermitian matrix

$$\begin{bmatrix} 1 & 0 & 0 \\ 0 & 0 & \omega^2 \\ 0 & \omega & 0 \end{bmatrix},$$

where ω is a complex cube root of unity: $\omega = e^{2\pi i/3}$.
*4. By Descartes' rule of signs, show that a Hermitian or real symmetric matrix has all positive characteristic roots if and only if the coefficients of $\varphi(\lambda)$ alternate in sign. Show also that Descartes' rule gives the exact numbers of positive and negative roots.
5. Prove that the characteristic roots of A^*A are all non-negative.
6. Show that a real, skew-symmetric matrix has an even or odd number of zero characteristic roots according as its order is even or odd and, hence, that its rank is even. (Use Exercise 1.)
7. Prove that the characteristic roots of a real symmetric matrix A are all equal if and only if A is scalar.
8. Prove that a real symmetric matrix is the matrix of a projection if and only if each characteristic root of A is 0 or 1.
9. Given that A is a real orthogonal matrix and that $X + iY$ is a characteristic vector associated with the root $\alpha + i\beta$, $\beta \neq 0$, prove that $X^TY = 0$ and that $X^TX = Y^TY$.
10. Find the orthogonal matrices which diagonalize the matrices

(a) $\begin{bmatrix} 1 & 3 \\ 3 & 1 \end{bmatrix}$,   (b) $\begin{bmatrix} 1 & -1 & 0 \\ -1 & 1 & 0 \\ 0 & 0 & 1 \end{bmatrix}$, (c) $\begin{bmatrix} 4 & 1 & 0 \\ 1 & 4 & 0 \\ 0 & 0 & 4 \end{bmatrix}$,

(d) $\begin{bmatrix} 0 & \sqrt{2} & -1 \\ \sqrt{2} & 1 & -\sqrt{2} \\ -1 & -\sqrt{2} & 0 \end{bmatrix}$,   (e) $\begin{bmatrix} 1 & 3 & 4 \\ 3 & 1 & 0 \\ 4 & 0 & 1 \end{bmatrix}$,

(f) $\begin{bmatrix} 2 & 3 & 0 & 0 \\ 3 & 2 & 0 & 0 \\ 0 & 0 & 3 & 2 \\ 0 & 0 & 2 & 3 \end{bmatrix}$,

(g) $\begin{bmatrix} 0 & 0 & 1 & 0 \\ 0 & 1 & 0 & 0 \\ 1 & 0 & 0 & 0 \\ 0 & 0 & 0 & 1 \end{bmatrix}$.

11. Show that if $U_1^* A_1 U_1 = D[\lambda_1, \ldots, \lambda_k]$ and $U_2^* A_2 U_2 = D[\lambda_{k+1}, \ldots, \lambda_n]$, then

$$\begin{bmatrix} U_1 & 0 \\ \hline 0 & U_2 \end{bmatrix}^* \begin{bmatrix} A_1 & 0 \\ \hline 0 & A_2 \end{bmatrix} \begin{bmatrix} U_1 & 0 \\ \hline 0 & U_2 \end{bmatrix} = D[\lambda_1, \ldots, \lambda_k, \lambda_{k+1}, \ldots, \lambda_n].$$

12. Show that, if U is a unitary matrix such that $U^* A U = D[\lambda_1, \lambda_2, \ldots, \lambda_n]$, then $U^* A^* U = D[\bar{\lambda}_1, \bar{\lambda}_2, \ldots, \bar{\lambda}_n]$, and if A is nonsingular, $U^* A^{-1} U = D[\lambda_1^{-1}, \lambda_2^{-1}, \ldots, \lambda_n^{-1}]$.

13. Show that if

$$A = \begin{vmatrix} 1 & 1 \\ 0 & 1 \end{vmatrix},$$

there exists no nonsingular matrix B of order 2 such that $B^{-1}AB = D[d_1, d_2]$. This example shows that *not every square matrix can be diagonalized by a nonsingular transformation of coordinates.*

14. Show that if λ is a *real* characteristic root of an orthogonal matrix U, then λ must be 1 or -1. Show that for every complex root λ, $|\lambda| = 1$.

15. Show that if U is an orthogonal matrix of odd order, then U necessarily has 1 or -1 as a characteristic root according as det U is 1 or -1.

16. Prove that a real, symmetric matrix is also orthogonal if and only if its characteristic roots are all ± 1. Use the argument of the proof to help you construct an example of a real, symmetric, orthogonal matrix of order 3.

17. Show that if λ is a real characteristic root of a unitary matrix U, then $\lambda = \pm 1$. (Show first that for every root λ, $|\lambda| = 1$.) Use this result to show that U is also Hermitian if and only if its characteristic roots are all ± 1.

*18. Given that $U^* A U = D[\lambda_1, \lambda_2, \ldots, \lambda_n]$, where U is unitary, obtain a simple formula for A^p, where p is any positive integer.

19. Let A be a Hermitian matrix with characteristic roots $\lambda_1, \lambda_2, \ldots, \lambda_r, 0, 0, \ldots, 0$, where no λ_j is 0. Prove that A has rank r.

20. Let A be a Hermitian matrix and let $U_1, U_2, \ldots, U_n$ be an orthonormal set of characteristic vectors corresponding to the characteristic roots $\lambda_1, \lambda_2, \ldots, \lambda_n$ respectively. Show that the solution of

$$(A - \lambda I)X = B, \qquad \lambda \neq \lambda_j, \qquad j = 1, 2, \ldots, n,$$

is given by

$$X = \sum_{j=1}^{n} \frac{U_j^* B}{\lambda_j - \lambda} U_j.$$

The Characteristic Value Problem /Ch. 9

*21. Prove that the characteristic roots of a Hermitian matrix A are all equal if and only if A is scalar.

*22. Prove that all the characteristic roots of a Hermitian matrix A are zero if and only if A is zero. Show by means of an example that a nonzero, non-Hermitian matrix can have all its characteristic roots zero.

23. Use Theorem 9.12.1 to construct an example of the triangularization of a 3×3 non-Hermitian matrix. (That is, start with the triangular form, pick an arbitrary unitary matrix, etc.)

24. Show that, if a Hermitian matrix H has characteristic roots $\lambda_1, \lambda_2, \ldots, \lambda_n$ such that $|\lambda_j| < 1$ for all j, then as $p \to \infty$, each element of H^p approaches zero.

25. If U^*AU is diagonal, where U is unitary, and if

$$B = \begin{bmatrix} A & \vdots & 0 \\ \cdots & \vdots & \cdots\cdots\cdots\cdots\cdots \\ 0 & \vdots & D[\alpha_1, \alpha_2, \ldots, \alpha_n] \end{bmatrix},$$

what transformation will diagonalize B?

26. Find a unitary transformation U such that

$$U^* \begin{bmatrix} a & -b \\ b & a \end{bmatrix} U = \begin{bmatrix} \lambda_1 & 0 \\ 0 & \lambda_2 \end{bmatrix},$$

where a and b are real numbers.

27. Show that, if A has distinct characteristic roots, then there exists a *unique* unitary matrix U such that

$$U^*AU = D[\lambda_1, \lambda_2, \ldots, \lambda_n],$$

where the λ's appear *in a specified order* on the right.

28. Prove that, if B is nonsingular and commutes with A, and if U diagonalizes A, then BU also diagonalizes A.

29. Prove that, if A and B are Hermitian matrices with the same characteristic roots $\lambda_1, \lambda_2, \ldots, \lambda_n$, then there exists a unitary matrix U such that $U^*AU = B$, and conversely.

30. Given the real nonzero vector B, show that the system

$$BB^\mathsf{T}X = B$$

is consistent. What is its rank? Show that the complete solution of the system is

$$X = \sum_{i=1}^{n-1} \alpha_i P_i + \frac{1}{B^\mathsf{T}B} B,$$

where $P_1, P_2, \ldots, P_{n-1}$ are independent characteristic vectors of BB^T associated with the zero characteristic roots, and the α_i are parameters.

31. Construct an example of a normal matrix which is not unitary, Hermitian, skew-Hermitian, or diagonal.

*32. Show that every proper orthogonal matrix A in $\mathscr{E}^3$ represents a rotation about the characteristic vector corresponding to the root $\lambda = 1$, the angle of rotation being defined by

$$\cos \theta = \frac{(\text{tr } A) - 1}{2}.$$

(*Hint:* Introduce the characteristic vector in question as a coordinate axis.)

*33. Given that $X_1, X_2, \ldots, X_n$ is a set of mutually orthogonal, unit characteristic vectors of A, show that (1) every matrix $X_j X_j^*$ is a projection, (2) $\sum X_j X_j^* = I$, and (3) $A = \sum \lambda_j X_j X_j^*$. This last result is called the **spectral decomposition** of A.

34. Obtain the spectral decomposition of each matrix:

(a) $\begin{bmatrix} 1 & 2 \\ 2 & 1 \end{bmatrix}$, (b) $\begin{bmatrix} 0 & 1 & 0 \\ 1 & 0 & 0 \\ 0 & 0 & 1 \end{bmatrix}$.

9.15 Characteristic Roots of a Polynomial Function of a Matrix

We now prove three theorems which are useful in a variety of applications and which apply to arbitrary real or complex matrices. We begin with

Theorem 9.15.1: *If $\lambda_1, \lambda_2, \ldots, \lambda_n$ are the characteristic roots, distinct or not, of a matrix A of order n, and if $g(A)$ is any polynomial function of A, then the characteristic roots of $g(A)$ are $g(\lambda_1), g(\lambda_2), \ldots, g(\lambda_n)$.*

We know that

$$\det (A - \lambda I_n) = (\lambda_1 - \lambda)(\lambda_2 - \lambda) \cdots (\lambda_n - \lambda),$$

and we wish to prove that, for any polynomial function $g(A)$,

$$\det (g(A) - \lambda I_n) = (g(\lambda_1) - \lambda)(g(\lambda_2) - \lambda) \cdots (g(\lambda_n) - \lambda).$$

Now suppose that $g(x)$ is of degree r in x and that, for a fixed value of λ, the roots of $g(x) - \lambda = 0$ are $x_1, x_2, \ldots, x_r$. Then we have

$$g(x) - \lambda = \alpha(x - x_1)(x - x_2) \cdots (x - x_r),$$

where α is the coefficient of x^r in $g(x)$. Hence

$$g(A) - \lambda I_n = \alpha(A - x_1 I_n)(A - x_2 I_n) \cdots (A - x_r I_n),$$

so that, if $\phi(\lambda)$ is the characteristic polynomial of A,

$$\det(g(A) - \lambda I_n) = \alpha^n \det(A - x_1 I_n) \det(A - x_2 I_n) \cdots \det(A - x_r I_n)$$
$$= \alpha^n \phi(x_1) \phi(x_2) \cdots \phi(x_r)$$
$$= \alpha^n (\lambda_1 - x_1)(\lambda_2 - x_1) \cdots (\lambda_n - x_1)$$
$$\cdot (\lambda_1 - x_2)(\lambda_2 - x_2) \cdots (\lambda_n - x_2)$$
$$\vdots$$
$$\cdot (\lambda_1 - x_r)(\lambda_2 - x_r) \cdots (\lambda_n - x_r).$$

By rearranging the orders of the factors, we now obtain

$$\det(g(A) - \lambda I_n) = \alpha(\lambda_1 - x_1)(\lambda_1 - x_2) \cdots (\lambda_1 - x_r)$$
$$\cdot \alpha(\lambda_2 - x_1)(\lambda_2 - x_2) \cdots (\lambda_2 - x_r)$$
$$\vdots$$
$$\cdot \alpha(\lambda_n - x_1)(\lambda_n - x_2) \cdots (\lambda_n - x_r)$$
$$= (g(\lambda_1) - \lambda)(g(\lambda_2) - \lambda) \cdots (g(\lambda_n) - \lambda),$$

as was to be proved.

For example, since A^p, where p is a positive integer, is a polynomial function of A, the characteristic roots of A^p are $\lambda_1^p, \lambda_2^p, \ldots, \lambda_n^p$.

One might expect a similar result to hold for the roots of A^{-1}, when A^{-1} exists. However, the preceding theorem does not apply, since A^{-1} is not a polynomial function of A. On the other hand, since A^{-1} is nonsingular, so that no characteristic root μ of A^{-1} is zero, we have

$$\det(A^{-1} - \mu I) = (-\mu)^n \det A^{-1} \det(A - \mu^{-1} I).$$

For this expression to vanish, we must have

$$\det(A - \mu^{-1} I) = 0.$$

The roots of this equation in μ^{-1} are, of course, $\lambda_1, \lambda_2, \ldots, \lambda_n$. Hence the roots μ of A^{-1} must be $\lambda_1^{-1}, \lambda_2^{-1}, \ldots, \lambda_n^{-1}$, as expected.

We summarize in

Theorem 9.15.2: *If the characteristic roots of A are $\lambda_1, \lambda_2, \ldots, \lambda_n$, those of A^p, where p is any positive integer, are $\lambda_1^p, \lambda_2^p, \ldots, \lambda_n^p$. If A^{-1} exists, this conclusion holds for all negative integers as well.*

9.16 The Cayley–Hamilton Theorem

Now suppose that A is a real, symmetric matrix with characteristic roots $\lambda_1, \lambda_2, \ldots, \lambda_n$ and with characteristic equation $\phi(\lambda) = 0$. Then $\phi(A)$ is also a real, symmetric matrix and, by Theorem 9.15.1, its characteristic roots are $\phi(\lambda_1), \phi(\lambda_2), \ldots, \phi(\lambda_n)$, all of which are zero. On the other hand, if a real, symmetric matrix has all its roots zero, it must be the zero matrix. (Prove this.) Hence $\phi(A) = 0$. That is, A satisfies its own characteristic equation. This is a special case of the famous **Cayley–Hamilton theorem**:

Theorem 9.16.1: *Every matrix satisfies its own characteristic equation.*

For example, if

$$A = \begin{bmatrix} 1 & 2 & 0 \\ 2 & -1 & 0 \\ 0 & 0 & 1 \end{bmatrix},$$

then

$$\phi(\lambda) = \begin{vmatrix} 1-\lambda & 2 & 0 \\ 2 & -1-\lambda & 0 \\ 0 & 0 & 1-\lambda \end{vmatrix} = -5 + 5\lambda + \lambda^2 - \lambda^3.$$

Hence

$$\phi(A) = -5 \begin{bmatrix} 1 & 0 & 0 \\ 0 & 1 & 0 \\ 0 & 0 & 1 \end{bmatrix} + 5 \begin{bmatrix} 1 & 2 & 0 \\ 2 & -1 & 0 \\ 0 & 0 & 1 \end{bmatrix}$$

$$+ \begin{bmatrix} 1 & 2 & 0 \\ 2 & -1 & 0 \\ 0 & 0 & 1 \end{bmatrix}^2 - \begin{bmatrix} 1 & 2 & 0 \\ 2 & -1 & 0 \\ 0 & 0 & 1 \end{bmatrix}^3$$

$$= \begin{bmatrix} -5 & 0 & 0 \\ 0 & -5 & 0 \\ 0 & 0 & -5 \end{bmatrix} + \begin{bmatrix} 5 & 10 & 0 \\ 10 & -5 & 0 \\ 0 & 0 & 5 \end{bmatrix}$$

$$+ \begin{bmatrix} 5 & 0 & 0 \\ 0 & 5 & 0 \\ 0 & 0 & 1 \end{bmatrix} + \begin{bmatrix} -5 & -10 & 0 \\ -10 & 5 & 0 \\ 0 & 0 & -1 \end{bmatrix}$$

$$= 0.$$

The Characteristic Value Problem /*Ch.* 9

We now proceed with the general proof. The characteristic matrix of A is $A - \lambda I_n$. Let us use C to denote the adjoint matrix of $A - \lambda I_n$ (see Section 7.16). The cofactors of $A - \lambda I_n$ are of degree at most $n - 1$ in λ, so the same is true of the elements of C. Hence we may represent C as a matrix polynomial

$$(9.16.1) \qquad C = C_0 + C_1\lambda + C_2\lambda^2 + \cdots + C_{n-1}\lambda^{n-1},$$

where C_k is the matrix whose elements are the coefficients of λ^k in the corresponding elements of C. The following example illustrates the idea. Let

$$A = \begin{bmatrix} 1 & 2 & 0 \\ 2 & -1 & 0 \\ 0 & 0 & 1 \end{bmatrix}, \quad \text{so} \quad A - \lambda I_n = \begin{bmatrix} 1-\lambda & 2 & 0 \\ 2 & -1-\lambda & 0 \\ 0 & 0 & 1-\lambda \end{bmatrix},$$

and hence

$$
\begin{aligned}
C &= \begin{bmatrix} \lambda^2 - 1 & 2\lambda - 2 & 0 \\ 2\lambda - 2 & \lambda^2 - 2\lambda + 1 & 0 \\ 0 & 0 & \lambda^2 - 5 \end{bmatrix} \\
&= \begin{bmatrix} -1 & -2 & 0 \\ -2 & 1 & 0 \\ 0 & 0 & -5 \end{bmatrix} + \begin{bmatrix} 0 & 2 & 0 \\ 2 & -2 & 0 \\ 0 & 0 & 0 \end{bmatrix}\lambda + \begin{bmatrix} 1 & 0 & 0 \\ 0 & 1 & 0 \\ 0 & 0 & 1 \end{bmatrix}\lambda^2.
\end{aligned}
$$

From (7.7.4), we have the relation

$$(A - \lambda I_n)C = (\det(A - \lambda I_n))I_n\,;$$

that is,

$$AC - \lambda C = \phi(\lambda)I_n\,.$$

Substituting the expansion of C given in (9.16.1) and also writing $\phi(\lambda)$ in the form

$$\phi(\lambda) = \sum_{k=0}^{n} a_k\lambda^k,$$

we have

$$\sum_{k=0}^{n-1} AC_k\lambda^k - \sum_{k=0}^{n-1} C_k\lambda^{k+1} = \sum_{k=0}^{n} (a_k I_n)\lambda^k.$$

In this identity in λ, we may equate corresponding coefficients and obtain the following set of equations:

$$
\begin{aligned}
AC_0 &= a_0 I_n, \\
AC_1 \quad - C_0 &= a_1 I_n, \\
AC_2 \quad - C_1 &= a_2 I_n, \\
&\vdots \\
AC_{n-1} - C_{n-2} &= a_{n-1} I_n, \\
- C_{n-1} &= a_n I_n.
\end{aligned}
$$

In order now to eliminate the matrices C_k from these equations, we need only to multiply them on the left by

$$
I_n, A, A^2, \ldots, A^{n-1}, A^n
$$

respectively and add the results, thus obtaining

$$(9.16.2) \qquad 0 = a_0 I + a_1 A + a_2 A^2 + \cdots + a_{n-1} A^{n-1} + a_n A^n,$$

or $\phi(A) = 0$, so the theorem is proved.

9.17   The Minimum Polynomial of a Matrix

There are, of course, polynomial functions of a square matrix A, other than its characteristic function $\phi(A)$, which reduce to zero. In fact, if $f(\lambda) = \phi(\lambda) g(\lambda)$ where $g(\lambda)$ is any polynomial in λ, then $f(A) = 0$ also.

Among all not identically zero polynomials $p(\lambda)$ such that $p(A) = 0$, there must exist some of lowest degree. If this degree is μ, then $0 < \mu \le n$. Let each such polynomial of degree μ be divided by the coefficient of λ^μ. Then the result must be the same in every case, for if different polynomials were obtained, the difference of two of these, say $p_1(\lambda)$ and $p_2(\lambda)$, would be free of λ^μ, thus:

$$p_1(\lambda) - p_2(\lambda) = \alpha_{\mu-1}\lambda^{\mu-1} + \alpha_{\mu-2}\lambda^{\mu-2} + \cdots + \alpha_1\lambda + \alpha_0.$$

Substitution of A would now yield

$$p_1(A) - p_2(A) = \alpha_{\mu-1}A^{\mu-1} + \alpha_{\mu-2}A^{\mu-2} + \cdots + \alpha_1 A + \alpha_0 I_n = 0,$$

so that we would have an equation of degree less than μ satisfied by A.

The contradiction proves our claim. This unique polynomial of lowest degree μ which vanishes at A is called the **minimum polynomial** of A and is denoted by $m(\lambda)$. We summarize in

Theorem 9.17.1: *For a given square matrix A, there exists a unique polynomial $m(\lambda)$ of lowest degree μ, in which the coefficient of λ^μ is equal to unity, such that $m(A) = 0$.*

To illustrate, consider the scalar matrix

$$A = \begin{bmatrix} \alpha & 0 \\ 0 & \alpha \end{bmatrix}.$$

Here $\phi(\lambda) = (\alpha - \lambda)^2$, but the minimum polynomial is just $\lambda - \alpha$ since $A - \alpha I = 0$ here.

We prove now

Theorem 9.17.2: *Every polynomial $p(\lambda)$ such that $p(A) = 0$ is exactly divisible by $m(\lambda)$.*

To prove this, let the quotient when $p(\lambda)$ is divided by $m(\lambda)$ be $q(\lambda)$ and let the remainder, which is of degree less than μ, be $r(\lambda)$. Then we have

$$p(\lambda) \equiv m(\lambda)q(\lambda) + r(\lambda).$$

Substitution of A now yields

$$r(A) = 0.$$

Since $r(\lambda)$ is of degree less than μ, this implies $m(\lambda)$ is not a minimum polynomial unless $r(\lambda) \equiv 0$. Thus

$$p(\lambda) \equiv m(\lambda)q(\lambda),$$

and the theorem is proved.

A particular case of this result is that the minimum polynomial is a divisor of the characteristic polynomial $\phi(\lambda)$. The relation between $\phi(\lambda)$ and $m(\lambda)$ is more closely defined in the next two theorems.

Theorem 9.17.3: *Every linear factor $\lambda - \lambda_1$ of $\phi(\lambda)$ is also a factor of $m(\lambda)$.*

We know in fact that dividing $m(\lambda)$ by $\lambda - \lambda_1$ yields an identity

$$m(\lambda) \equiv (\lambda - \lambda_1)s(\lambda) + r,$$

where r is a constant. Substituting A for λ, we have

$$m(A) = (A - \lambda_1 I_n)s(A) + rI_n = 0,$$

so that, if $r \neq 0$, we have

$$(A - \lambda_1 I_n)\left(\frac{-s(A)}{r}\right) = I_n.$$

This means that $A - \lambda_1 I_n$ has an inverse, whereas, since λ_1 is a characteristic root, it is singular. The contradiction shows that r must be zero, so that

$$m(\lambda) \equiv (\lambda - \lambda_1)s(\lambda),$$

as was to be proved.

An immediate consequence is

Theorem 9.17.4: *If the characteristic roots of A are distinct,*

$$\phi(\lambda) = (-1)^n m(\lambda).$$

Indeed, in this case $\phi(\lambda) = (-1)^n \prod_{j=1}^n (\lambda - \lambda_j)$ and $m(\lambda) = \prod_{j=1}^n (\lambda - \lambda_j)$, by the preceding theorem.

One might be tempted to conclude at this point that the minimum polynomial is simply the product of the distinct factors of $\phi(\lambda)$. That this is not the case may be shown by an example. Let

$$A = \begin{bmatrix} 1 & 1 & 1 & \cdots & 1 \\ 0 & 1 & 1 & \cdots & 1 \\ 0 & 0 & 1 & \cdots & 1 \\ \vdots & & & & \\ 0 & 0 & 0 & \cdots & 1 \end{bmatrix}_n,$$

so that $\phi(\lambda) \equiv (-1)^n(\lambda - 1)^n$.

Then $A - I \neq 0$, so $\lambda - 1$ is not the minimum polynomial. However, it is easily verified that $(A - I)^n = 0$, while no lesser power of $A - I$ is zero. Thus the minimum polynomial is $(\lambda - 1)^n$ in this case.

An equation $A^p + \alpha_{p-1}A^{p-1} + \cdots + \alpha_1 A + \alpha_0 I = 0$, which is satisfied by a fixed matrix A, says that the matrices $I, A, A^2, \ldots, A^p$ are linearly dependent. Hence the minimum polynomial of A may be found by determining the smallest positive integer p for which such a dependence exists. Since $\phi(A) = 0$, we have $p \leq n$. For small values of n, in the case of matrices with integer entries, the computation is sometimes practicable by the following scheme. We begin by computing $A^2, A^3, \ldots, A^n$. We then form a row matrix B_i for each A^i (B_0 for I) such that

$$B_i = [(A^i)^{(1)}, (A^i)^{(2)}, \ldots, (A^i)^{(n)}],$$

where $(A^i)^{(j)}$ is the jth row of A^i. Thus each row B_i has n^2 elements. Now we form an $(n + 1) \times n^2$ matrix

$$B = \begin{bmatrix} B_0 \\ B_1 \\ \vdots \\ B_n \end{bmatrix}.$$

The problem is to find the smallest integer p such that the first p rows of B are dependent. To record conveniently the row operations required to determine this, we append an identity matrix of order $n + 1$ to B, thus obtaining an $(n + 1) \times (n^2 + n + 1)$ array $[B, I_{n+1}]$. We perform on the identity matrix I_{n+1} the same row operations as on B. However, the dependence of the rows of B is all that is at issue. For example, if

$$A = \begin{bmatrix} -1 & 3 & 0 & 3 \\ -1 & 3 & 1 & 3 \\ 1 & -1 & -2 & 0 \\ 1 & -1 & -1 & -1 \end{bmatrix},$$

we obtain this 5×21 matrix:

I:	1	0	0	0	0	1	0	0	0							
A:	-1	3	0	3	-1	3	1	3	1							
A^2:	1	3	0	3	2	2	-2	3	-2							
A^3:	-1	9	0	9	-3	11	3	9	3							
A^4:	1	15	0	15	4	12	-4	15	-4							

0	1	0	0	0	0	1	1	0	0	0	0
-1	-2	0	1	-1	-1	-1	0	1	0	0	0
2	3	0	-2	2	2	1	0	0	1	0	0
-3	-4	0	3	-3	-3	-1	0	0	0	1	0
4	5	0	-4	4	4	1	0	0	0	0	1

Now any column among the first n^2, which is linearly dependent on preceding columns, may be deleted without altering the dependence of the rows in any way. This permits the successive deletion of columns 3, 4, 7, 8, 9, 10, 12, 13, 14, 15, 16 in this example, so that only this matrix needs to be considered:

$$
\left[
\begin{array}{rrrrr:rrrrr}
1 & 0 & 0 & 1 & 1 & 1 & 0 & 0 & 0 & 0 \\
-1 & 3 & -1 & 3 & -2 & 0 & 1 & 0 & 0 & 0 \\
1 & 3 & 2 & 2 & 3 & 0 & 0 & 1 & 0 & 0 \\
-1 & 9 & -3 & 11 & -4 & 0 & 0 & 0 & 1 & 0 \\
1 & 15 & 4 & 12 & 5 & 0 & 0 & 0 & 0 & 1
\end{array}
\right].
$$

By sweepout, *using row operations only*, we now obtain:

$$
\left[
\begin{array}{rrrrr:rrrrr}
1 & 0 & 0 & 1 & 1 & 1 & 0 & 0 & 0 & 0 \\
0 & 3 & -1 & 4 & -1 & 1 & 1 & 0 & 0 & 0 \\
0 & 0 & 3 & -3 & 3 & -2 & -1 & 1 & 0 & 0 \\
0 & 0 & 0 & 0 & 0 & -2 & -3 & 0 & 1 & 0 \\
0 & 0 & 0 & 0 & 0 & 0 & -2 & -3 & 0 & 1
\end{array}
\right].
$$

The 5×5 matrix on the right, being in fact a product of matrices of elementary transformations, records what has been done to the abbreviated rows representing I, A, A^2, A^3, A^4 respectively. Indeed, if we interpret I, A, A^2, A^3, A^4 to mean *rows*, the product

$$
\left[
\begin{array}{rrrrr}
1 & 0 & 0 & 0 & 0 \\
1 & 1 & 0 & 0 & 0 \\
-2 & -1 & 1 & 0 & 0 \\
-2 & -3 & 0 & 1 & 0 \\
0 & -2 & -3 & 0 & 1
\end{array}
\right]
\cdot
\left[
\begin{array}{r}
I \\
A \\
A^2 \\
A^3 \\
A^4
\end{array}
\right]
$$

represents exactly what we must do to the original array in order to reduce rows four and five to rows of zeros. From the fourth row, we have, then,

$$
-2I - 3A + 0 \cdot A^2 + A^3 + 0 \cdot A^4 = 0,
$$

that is,

(9.17.1) $$ A^3 - 3A - 2I = 0, $$

so the minimal equation is

$$
\lambda^3 - 3\lambda - 2 = 0,
$$

with roots -1, -1, 2.

Notice that the fifth row yields

$$A^4 - 3A^2 - 2A = 0,$$

which is an immediate consequence of the equation (9.17.1) yielded by row four.

Although this method of finding the minimum polynomial is perfectly general, the computational difficulties increase rapidly with n, as is the case with most matrix calculations. This explains why effective procedures for use with digital computers are so essential.

9.18   Powers of Matrices

An important application of the Cayley–Hamilton theorem is in the representation of high powers of a matrix. Let us write $\phi(A) = 0$ in the form

$$(9.18.1) \qquad A^n + a_{n-1}A^{n-1} + \cdots + a_1 A + a_0 I_n = 0,$$

so that, if we have already computed $A^2, A^3, \ldots, A^{n-1}$, we can express A^n as a linear combination of these:

$$(9.18.2) \qquad A^n = -a_0 I_n - a_1 A - \cdots - a_{n-1}A^{n-1}.$$

Multiplying through by A and substituting from (9.18.2) for A^n on the right, we obtain

$$(9.18.3) \quad A^{n+1} = a_{n-1}a_0 I_n + (a_{n-1}a_1 - a_0)A + (a_{n-1}a_2 - a_1)A^2$$
$$+ \cdots + (a_{n-1}^2 - a_{n-2})A^{n-1}.$$

By continuing this process, we can express any positive integral power of A as a linear combination of $I, A, A^2, \ldots, A^{n-1}$.

If A^{-1} exists, by multiplying (9.18.1) by A^{-1} and solving for A^{-1}, we obtain

$$(9.18.4) \qquad A^{-1} = -\frac{a_1}{a_0} I - \frac{a_2}{a_0} A - \cdots - \frac{1}{a_0} A^{n-1}.$$

Multiplying (9.18.4) by A^{-1} and substituting from (9.18.4) for A^{-1}, we obtain

$$A^{-2} = \left(\frac{a_1^2}{a_0^2} - \frac{a_2}{a_0}\right)I_n + \left(\frac{a_1 a_2}{a_0^2} - \frac{a_3}{a_0}\right)A + \cdots + \left(\frac{a_1 a_{n-1}}{a_0^2} - \frac{1}{a_0}\right)A^{n-2} + \frac{a_1}{a_0^2} A^{n-1}.$$

By continuation of this process, any negative integral power of A may also be expressed as a linear combination of $I_n, A, A^2, \ldots, A^{n-1}$.

These procedures are used in deriving formulas for automatic computation.

Another possible application of the Cayley–Hamilton theorem is to the evaluation of $\phi(\lambda)$ itself. In fact, from (9.18.1) we have

$$(9.18.5) \qquad (A^n X) + a_{n-1}(A^{n-1} X) + \cdots + a_1(AX) + a_0 X = 0,$$

where X is an arbitrarily chosen fixed vector. This matrix equation is equivalent to n scalar equations in the n unknowns $a_1, a_2, \ldots, a_n$, which may be solved by any appropriate procedure provided the coefficient matrix is nonsingular. *This method of determining $\phi(\lambda)$ often involves very tedious computations.*

When the minimum polynomial is of lower degree than the characteristic polynomial, the equation $m(A) = 0$ may be used instead of $\phi(A) = 0$ in the computation of powers of A.

9.19   Exercises

1. Verify the Cayley–Hamilton theorem for the matrix

$$A = \begin{bmatrix} 0 & 1 & 0 & 0 \\ 0 & 0 & 1 & 0 \\ 0 & 0 & 0 & 1 \\ 1 & 1 & 1 & 1 \end{bmatrix}$$

and then compute A^5 and A^{-1} by the method of Section 9.18.

2. Given that

$$A = \begin{bmatrix} 1 & \sqrt{3} & 0 \\ \sqrt{3} & -1 & 0 \\ 0 & 0 & 1 \end{bmatrix}, \qquad g(A) = A^2 + A + I_3,$$

find the characteristic roots of A and of $g(A)$.

*3. Prove that if $AX = \lambda X$, then for all positive integers p, $A^p X = \lambda^p X$. Hence show that if $AX = \lambda X$ and if $g(A)$ is any polynomial function of A, then $g(A)X = g(\lambda)X$.

4. Find the powers and the characteristic roots of the powers of the cyclic permutation matrix

$$P = \begin{bmatrix} 0 & 1 & 0 & 0 & 0 \\ 0 & 0 & 1 & 0 & 0 \\ 0 & 0 & 0 & 1 & 0 \\ 0 & 0 & 0 & 0 & 1 \\ 1 & 0 & 0 & 0 & 0 \end{bmatrix}.$$

*5. A matrix A is said to be **diagonalizable** if there exists a matrix B such that $B^{-1}AB = D[d_1, d_2, \ldots, d_n]$. Let A be diagonalizable and let $\phi(\lambda) = 0$ be the characteristic equation of A. Show that $\phi(D) = 0$ and hence that $\phi(BDB^{-1}) = 0$, that is, that $\phi(A) = 0$. This proves the Cayley–Hamilton theorem for every diagonalizable matrix.

*6. Show that the set of all polynomial functions over a number field $\mathscr{F}$ of a fixed matrix A over $\mathscr{F}$ is a vector space of dimension μ over $\mathscr{F}$, where μ is the degree of the minimum polynomial of A.

7. In proving the Cayley–Hamilton theorem, we in effect assumed that, if F_0, F_1, ..., F_p and G_0, G_1, ..., G_p are fixed matrices of order n, and if

$$F_0 + F_1\lambda + \cdots + F_p\lambda^p \equiv G_0 + G_1\lambda + \cdots + G_p\lambda^p,$$

that is, if these two polynomials yield the same matrix for all values of the scalar λ, then we must have $F_0 = G_0$, $F_1 = G_1$, ..., $F_p = G_p$. This is equivalent to the fact that, if H_0, H_1, ..., H_p are matrices of order n, then $H_0 + H_1\lambda + \cdots + H_p\lambda^p \equiv 0$ if and only if $H_0 = H_1 = \cdots = H_p = 0$. Show that the last result follows from the corresponding fact for ordinary polynomials.

*8. The determination of the values of λ and the associated vectors X which satisfy the equation

$$AX = \lambda BX$$

is the **generalized characteristic value problem**. In the most important applications, B is nonsingular. Show that, if also $AB = BA$ and A and B have distinct characteristic roots, the "characteristic roots" are given by $\lambda_j = \mu_j/v_{k_j}$, where the μ_j's are the characteristic roots of A and the v_{k_j}'s are those of B, in a suitable order.

9. Show that the characteristic vectors of A are all characteristic vectors of $g(A)$, where $g(A)$ is any polynomial function of A. Show by an example (for example, a matrix A such that $A^2 = 0$) that the converse is not true for all A and $g(A)$. (One can also choose g so that $g(\lambda_1) = g(\lambda_2)$, with $\lambda_1 \neq \lambda_2$.)

10. Compute $\phi(\lambda)$ with the aid of (9.18.5) if

$$A = \begin{bmatrix} 1 & 0 & 1 \\ -1 & -1 & 0 \\ 0 & 1 & 1 \end{bmatrix}.$$

11. With the aid of Exercise 8, Section 9.4, show that every polynomial equation of degree n in an unknown matrix A of order n has at least one solution.

12. Compute the minimum polynomial of the matrix

$$\begin{bmatrix} 0 & 1 & 0 & 0 & 0 \\ 0 & 0 & 1 & 0 & 0 \\ 1 & 0 & 0 & 0 & 0 \\ 0 & 0 & 0 & 0 & 1 \\ 0 & 0 & 0 & 1 & 0 \end{bmatrix}$$

by the method given in Section 9.17.

13. Prove that $\operatorname{tr}(A^p) = \sum \lambda_j^p$, where the λ_j's are the characteristic roots of A. Then prove $\sum \lambda_j^2 = (\sum \lambda_j)^2 - 2 \sum_{i<j} \lambda_i \lambda_j$, so that

$$\sum_{i<j} \lambda_i \lambda_j = \tfrac{1}{2}((\operatorname{tr} A)^2 - \operatorname{tr}(A^2)).$$

(By Newton's formulas expressing arbitrary symmetric functions of the roots in terms of sums of powers, all symmetric functions, and hence the coefficients of the characteristic polynomial itself, can be expressed in terms of the traces of the powers of A.)

14. Find the minimum polynomial of the cyclic permutation matrix

$$\begin{bmatrix} 0 & 1 & 0 & 0 & 0 \\ 0 & 0 & 1 & 0 & 0 \\ 0 & 0 & 0 & 1 & 0 \\ 0 & 0 & 0 & 0 & 1 \\ 1 & 0 & 0 & 0 & 0 \end{bmatrix}.$$

CHAPTER
10

Quadratic, Bilinear and Hermitian Forms

10.1 Quadratic Forms

A homogeneous polynomial q of the type

$$(10.1.1) \qquad q = X^{\mathsf{T}} A X = \sum_{i,\,j=1}^{n} a_{ij} x_i x_j,$$

the coefficients of which are in a field $\mathscr{F}$, is called a **quadratic form** over $\mathscr{F}$. Such forms have many applications in the physical sciences and engineering, in the mathematics of computation, in statistics, in geometry, and so on. In most applications, the field $\mathscr{F}$ is the field of real numbers. From now on we assume that this is the case.

When the matrix product $X^{\mathsf{T}} A X$ is given, the scalar expansion given in (10.1.1) may be written by inspection, since the i,j-entry of A is the coefficient of the product $x_i x_j$. In this expansion, the similar terms $a_{ij} x_i x_j$ and $a_{ji} x_j x_i$ would naturally be combined into a single term $(a_{ij} + a_{ji}) x_i x_j$. It is clear from this fact that distinct $n \times n$ matrices A_1 and A_2 will lead to the same quadratic polynomial provided only that all corresponding sums of the type $a_{ij} + a_{ji}$ are equal. Moreover, given the quadratic form, one could not identify the corresponding matrix A by inspection. It is consistent with the various uses made of quadratic forms to eliminate this ambiguity once and for all by replacing each of the pair of coefficients a_{ij} and a_{ji} of a given form q by their mean, $(a_{ij} + a_{ji})/2$. This amounts to replacing A by $(A + A^T)/2$. Then the coefficients of a given quadratic form will define a unique symmetric matrix. In what follows, every quadratic form X^TAX with which we work will therefore be assumed to have a symmetric matrix A.

For example,

$$X^T \begin{bmatrix} 1 & 2 \\ 2 & -3 \end{bmatrix} X = x_1^2 + 4x_1x_2 - 3x_2^2$$

and

$$x_1^2 - 2x_1x_2 + 5x_1x_3 + 2x_3^2 = X^T \begin{bmatrix} 1 & -1 & \frac{5}{2} \\ -1 & 0 & 0 \\ \frac{5}{2} & 0 & 2 \end{bmatrix} X.$$

Note how missing terms lead to zero entries in the coefficient matrix.

If the x's are independent variables, the rank of A is called the **rank of the form** and det A is called the **discriminant of the form**.

A nonsingular linear transformation $X = B\tilde{X}$ maps a quadratic form X^TAX onto the form $\tilde{X}^T(B^TAB)\tilde{X}$, where B^TAB is also symmetric and has the same rank as A. We say that two quadratic forms X^TA_1X and $\tilde{X}^TA_2\tilde{X}$ are **equivalent** if and only if there is a nonsingular transformation $X = B\tilde{X}$ such that

$$X^TA_1X = \tilde{X}^TB^TA_1B\tilde{X} = \tilde{X}^TA_2\tilde{X},$$

that is, if and only if, for a suitable nonsingular matrix B,

(10.1.2) $A_2 = B^TA_1B.$

Two matrices A_1 and A_2 related as in (10.1.2), with B nonsingular, are said to be **congruent**.

As an example, consider the form

$$q = 29x_1^2 + 24x_1x_2 + 5x_2^2 = X^T \begin{bmatrix} 29 & 12 \\ 12 & 5 \end{bmatrix} X.$$

This can be rearranged in the form

$$q = (5x_1 + 2x_2)^2 + (2x_1 + x_2)^2.$$

Hence let us apply the transformation

$$\tilde{X} = \begin{bmatrix} 5 & 2 \\ 2 & 1 \end{bmatrix} X \quad \text{or} \quad X = \begin{bmatrix} 1 & -2 \\ -2 & 5 \end{bmatrix} \tilde{X}.$$

Then

$$X^{\mathsf{T}} \begin{bmatrix} 29 & 12 \\ 12 & 5 \end{bmatrix} X = \tilde{X}^{\mathsf{T}} \begin{bmatrix} 1 & -2 \\ -2 & 5 \end{bmatrix} \begin{bmatrix} 29 & 12 \\ 12 & 5 \end{bmatrix} \begin{bmatrix} 1 & -2 \\ -2 & 5 \end{bmatrix} \tilde{X}$$

$$= \tilde{X}^{\mathsf{T}} \begin{bmatrix} 1 & 0 \\ 0 & 1 \end{bmatrix} \tilde{X} = \tilde{x}_1^2 + \tilde{x}_2^2.$$

Thus the forms $29x_1^2 + 24x_1x_2 + 5x_2^2$ and $\tilde{x}_1^2 + \tilde{x}_2^2$ are equivalent and the matrices $\begin{bmatrix} 29 & 12 \\ 12 & 5 \end{bmatrix}$ and $\begin{bmatrix} 1 & 0 \\ 0 & 1 \end{bmatrix}$ are congruent.

10.2   Diagonalization of Quadratic Forms

A given quadratic form can be reduced in various ways to equivalent forms which emphasize certain of its basic properties. The most important of these reductions is diagonalization by means of an orthogonal transformation.

We proved in Section 9.9 that for every real symmetric matrix A, there exists an orthogonal matrix U such that

$$U^{\mathsf{T}}AU = D[\lambda_1, \lambda_2, \ldots, \lambda_n],$$

where $\lambda_1, \lambda_2, \ldots, \lambda_n$ are the characteristic roots of A. As a consequence, the transformation $X = U\tilde{X}$ applied to the quadratic form $X^{\mathsf{T}}AX$ gives

(10.2.1) $X^{\mathsf{T}}AX = \tilde{X}^{\mathsf{T}}(U^{\mathsf{T}}AU)\tilde{X} = \lambda_1\tilde{x}_1^2 + \lambda_2\tilde{x}_2^2 + \cdots + \lambda_n\tilde{x}_n^2.$

If the rank of A is r, then so is that of $D[\lambda_1, \lambda_2, \ldots, \lambda_n]$. Hence $n - r$ characteristic roots of A must be zero and, if $\lambda_1, \lambda_2, \ldots, \lambda_r$ denote the nonzero characteristic roots, (10.2.1) becomes simply

(10.2.2) $X^{\mathsf{T}}AX = \lambda_1\tilde{x}_1^2 + \lambda_2\tilde{x}_2^2 + \cdots + \lambda_r\tilde{x}_r^2.$

The computational aspects of this reduction are, of course, the same as those discussed in Section 9.10 in connection with the determination of the matrix U.

For example, let

$$q = 5x_1^2 + 2x_1x_2 + 5x_2^2 = X^{\mathsf{T}} \begin{bmatrix} 5 & 1 \\ 1 & 5 \end{bmatrix} X.$$

Here A has characteristic roots 6 and 4, with associated normalized characteristic vectors $[1/\sqrt{2}, 1/\sqrt{2}]^{\mathsf{T}}$ and $[1/\sqrt{2}, -1/\sqrt{2}]^{\mathsf{T}}$ respectively. Hence we put

$$X = \begin{bmatrix} \dfrac{1}{\sqrt{2}} & \dfrac{1}{\sqrt{2}} \\ \dfrac{1}{\sqrt{2}} & -\dfrac{1}{\sqrt{2}} \end{bmatrix} \tilde{X}$$

and get

$$X^{\mathsf{T}} \begin{bmatrix} 5 & 1 \\ 1 & 5 \end{bmatrix} X = \tilde{X}^{\mathsf{T}} \begin{bmatrix} \dfrac{1}{\sqrt{2}} & \dfrac{1}{\sqrt{2}} \\ \dfrac{1}{\sqrt{2}} & -\dfrac{1}{\sqrt{2}} \end{bmatrix} \begin{bmatrix} 5 & 1 \\ 1 & 5 \end{bmatrix} \begin{bmatrix} \dfrac{1}{\sqrt{2}} & \dfrac{1}{\sqrt{2}} \\ \dfrac{1}{\sqrt{2}} & -\dfrac{1}{\sqrt{2}} \end{bmatrix} \tilde{X}$$

$$= \tilde{X}^{\mathsf{T}} \begin{bmatrix} 6 & 0 \\ 0 & 4 \end{bmatrix} \tilde{X}.$$

10.3 A Geometrical Application

The diagonalization of quadratic forms has a ready geometrical interpretation. For example, in $\mathscr{E}^2$, the graph of the quadratic equation

(10.3.1) $$a_{11}x_1^2 + 2a_{12}x_1x_2 + a_{22}x_2^2 = b,$$

where b is a constant, is a proper conic section, a pair of straight lines (not necessarily distinct), a single point, or the empty set. To determine the precise nature of the graph, we first rewrite (10.3.1) in matrix form:

(10.3.2) $$X^{\mathsf{T}} \begin{bmatrix} a_{11} & a_{12} \\ a_{12} & a_{22} \end{bmatrix} X = b.$$

Then there exists an orthogonal matrix U which diagonalizes the symmetric matrix

$$A = \begin{bmatrix} a_{11} & a_{12} \\ a_{12} & a_{22} \end{bmatrix},$$

so that, if we put $X = UY$, we obtain the equation

$$Y^\mathsf{T} \begin{bmatrix} \lambda_1 & 0 \\ 0 & \lambda_2 \end{bmatrix} Y = b,$$

where λ_1 and λ_2 are the characteristic roots of A. In scalar form, we now have the familiar equation

$$(10.3.3) \qquad \lambda_1 y_1^2 + \lambda_2 y_2^2 = b.$$

If we interpret the transformation $X = UY$ as a transformation of the coordinate system, the figure constituting the graph of (10.3.1) is not changed. Only the reference system is altered. However, this alteration is such that the equation becomes so simple in the new reference system that the nature of the graph can be determined by inspection. In particular, if $\lambda_1 \neq 0$, $\lambda_2 \neq 0$, $b \neq 0$, we can rewrite (10.3.3) in the form

$$(10.3.4) \qquad \frac{y_1^2}{b/\lambda_1} + \frac{y_2^2}{b/\lambda_2} = 1.$$

If $b/\lambda_1 > 0$ and $b/\lambda_2 > 0$, equation (10.3.4) represents an **ellipse**. If also $\lambda_1 = \lambda_2$, the ellipse becomes a **circle**. If b/λ_1 and b/λ_2 are opposite in sign, equation (10.3.4) represents a **hyperbola**. If b/λ_1 and b/λ_2 are both negative, the graph of (10.3.4) is the empty set.

The nature of the graph when one or more of λ_1, λ_2, b is zero is readily determined and is left to the reader to discuss.

The graphs in the three cases cited are illustrated in Figure 10-1, where only the new axes (y_1 and y_2) are shown. These axes are called the **principal axes** of the conic and the transformation $X = UY$ is called a **principal-axis transformation**.

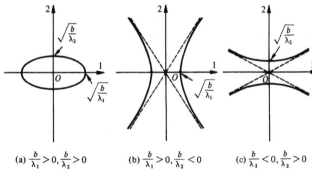

(a) $\frac{b}{\lambda_1} > 0, \frac{b}{\lambda_2} > 0$ (b) $\frac{b}{\lambda_1} > 0, \frac{b}{\lambda_2} < 0$ (c) $\frac{b}{\lambda_1} < 0, \frac{b}{\lambda_2} > 0$

FIGURE 10-1. Some Graphs of $\lambda_1 y_1^2 + \lambda_2 y_2^2 = b$.

In $\mathscr{E}^3$, the equation

$$(10.3.5) \qquad X^{\mathsf{T}}AX = X^{\mathsf{T}}\begin{bmatrix} a_{11} & a_{12} & a_{13} \\ a_{12} & a_{22} & a_{23} \\ a_{13} & a_{23} & a_{33} \end{bmatrix}X = b$$

represents what is called a **quadric surface**. Again, let U be an orthogonal matrix which diagonalizes the symmetric matrix A. Then the nonsingular mapping $X = UY$ transforms (10.3.5) into the equation

$$(10.3.6) \qquad \lambda_1 y_1^2 + \lambda_2 y_2^2 + \lambda_3 y_3^2 = b,$$

where $\lambda_1, \lambda_2,$ and λ_3 are the characteristic roots of A. If $\lambda_1, \lambda_2, \lambda_3,$ and b are all different from zero, we may write equation (10.3.6) in the form

$$(10.3.7) \qquad \frac{y_1^2}{b/\lambda_1} + \frac{y_2^2}{b/\lambda_2} + \frac{y_3^2}{b/\lambda_3} = 1.$$

If all three denominators are positive, the equation represents an **ellipsoid**. This is the most important case in practice. If one of the denominators is negative, the equation represents a **hyperboloid of one sheet**; if two are negative, it represents a **hyperboloid of two sheets**; if all three are negative, the graph is the empty set.

If $b = 0, \lambda_1\lambda_2\lambda_3 \neq 0,$ and $\lambda_1, \lambda_2,$ and λ_3 all have the same sign, the graph is just the origin, but if the signs are not all the same, the graph is a **cone**.

The nontrivial cases mentioned so far are illustrated in Figure 10-2.

The other cases in which one or more of $\lambda_1, \lambda_2, \lambda_3,$ and b are zero should be discussed by the reader.

As in $\mathscr{E}^2$, the columns of the orthogonal matrix U define the **principal axes** of the quadric.

In $\mathscr{E}^n$, the equation

$$X^{\mathsf{T}}AX = b$$

represents what is called a **hyperquadric**, an n-**quadric**, or simply a **quadric**. As in $\mathscr{E}^2$ and $\mathscr{E}^3$, we can use an appropriate orthogonal transformation of coordinates $X = UY$ to reduce the equation to the form

$$\sum_{j=1}^{n} \lambda_j y_j^2 = b.$$

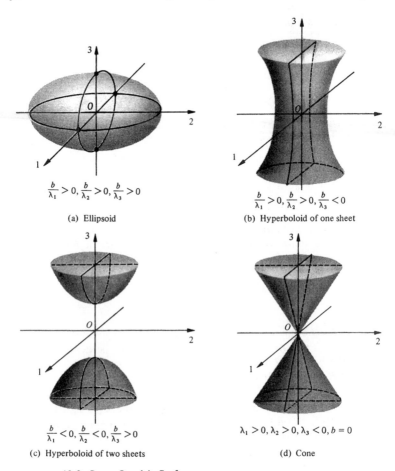

$$\frac{b}{\lambda_1} > 0, \frac{b}{\lambda_2} > 0, \frac{b}{\lambda_3} > 0$$

(a) Ellipsoid

$$\frac{b}{\lambda_1} > 0, \frac{b}{\lambda_2} > 0, \frac{b}{\lambda_3} < 0$$

(b) Hyperboloid of one sheet

$$\frac{b}{\lambda_1} < 0, \frac{b}{\lambda_2} < 0, \frac{b}{\lambda_3} > 0$$

(c) Hyperboloid of two sheets

$$\lambda_1 > 0, \lambda_2 > 0, \lambda_3 < 0, b = 0$$

(d) Cone

FIGURE 10-2. Some Quadric Surfaces.

The most important special case here is that in which all the λ's are positive. In this event, the locus of the equation is commonly called an **ellipsoid**. As in $\mathscr{E}^2$ and $\mathscr{E}^3$, the columns of U define the **principal axes** of the quadric in $\mathscr{E}^n$.

10.4 Definite Forms and Matrices

Let us take a second look at each of our last two examples:

$$q = (5x_1 + 2x_2)^2 + (2x_1 + x_2)^2$$

and

$$q = 5x_1^2 + 2x_1 x_2 + 5x_2^2 = 6\tilde{x}_1^2 + 4\tilde{x}_2^2.$$

Each of these is representable as a sum of squares of linear expressions and hence can never be negative for any real vector X. Moreover, they will take on the value zero only when $X = 0$. This sort of situation is common in applications in both the physical and the social sciences as well as in mathematics. The form typically represents some quantity, such as energy or the square of a distance, that may be zero but cannot be negative. We are thus led to make the following definition: A real quadratic form q, defined by $q(X) = X^T A X$, is **positive definite** if and only if $q(X) > 0$ for all real X except $X = 0$. If $q(X) \geq 0$ for all real X, q is said to be **positive semidefinite** or **non-negative definite**. Similarly, q is **negative definite** (**negative semidefinite**) if $q(X) < 0$ for all real X except $X = 0$ ($q(X) \leq 0$ for all real X).

The first example cited above illustrates the fact that any quadratic form that can be written as a sum of squares of linear functions is positive definite or positive semidefinite. The second example illustrates

Theorem 10.4.1: *A quadratic form $X^T A X$ is positive definite if and only if the characteristic roots of A are all positive.*

To prove this, assume first that the form is positive definite. Let λ_1, necessarily real since A is real and symmetric, be any characteristic root of A and let U_1 be a normalized, real characteristic vector such that

$$AU_1 = \lambda_1 U_1.$$

Then, since the form is positive definite,

$$0 < U_1^T(AU_1) = U_1^T(\lambda_1 U_1) = \lambda_1 U_1^T U_1 = \lambda_1.$$

Hence all characteristic roots are positive.

Now assume that all characteristic roots of A are positive and let U_1, $U_2, \ldots, U_n$ be an orthonormal set of vectors associated with the roots λ_1, $\lambda_2, \ldots, \lambda_n$ respectively. Let X be any real vector. Since the U_j are a basis for $\mathscr{E}^n$, we may write $X = \sum \alpha_i U_i$, where the α's are real. Hence

$$X^T A X = \left(\sum_i \alpha_i U_i\right)^T A\left(\sum_j \alpha_j U_j\right) = \sum_i \sum_j (\alpha_i \alpha_j U_i^T A U_j)$$

$$= \sum_i \sum_j \alpha_i \alpha_j U_i^T(\lambda_j U_j) = \sum_i \sum_j \alpha_i \alpha_j \lambda_j (U_i^T U_j)$$

$$= \sum_i \sum_j \alpha_i \alpha_j \lambda_j \delta_{ij} = \sum_i \alpha_i^2 \lambda_i.$$

Since all the λ_i are positive, and all the α_i are real, $\sum_i \alpha_i^2 \lambda_i \geq 0$. Moreover, it can be zero only if all the α_i are zero, that is, only if $X = 0$. Thus the form is seen to be positive definite.

A slight alteration of this proof yields

Theorem 10.4.2: *A quadratic form $X^{\mathsf{T}}AX$ is positive semidefinite if and only if all the characteristic roots of A are non-negative.*

Next we establish a necessary condition for positive definiteness.

Theorem 10.4.3: *If $X^{\mathsf{T}}AX$ is positive definite, then every principal minor determinant of A is positive. In particular,* det $A > 0$ *and every diagonal element $a_{ii} > 0$.*

First, assume that the form is positive definite. Then, since all characteristic roots of A are positive, we have det $A = \lambda_1 \lambda_2 \cdots \lambda_n > 0$.

Every principal minor of order r, $0 < r < n$, is the determinant of a submatrix obtainable by deleting symmetrically a suitable set of rows and columns of A. This corresponds to putting $n - r$ of the x_i equal to zero and looking at the quadratic form in the remaining variables only. Since the original form is positive definite, so is this reduced form, for the values of the reduced form will also be values of the original form. Now we apply our first observation to the reduced form and the proof is complete.

The reader may prove, in similar fashion,

Theorem 10.4.4: *If $X^{\mathsf{T}}AX$ is positive semidefinite, then every principal minor determinant of A is non-negative.*

Whereas the two preceding theorems do not make it possible to prove that a form *is* positive definite or semidefinite, they do make it possible to prove that a form *is not*. For example, the form

$$x_1^2 + 2x_1 x_2 + x_1 x_3 + 4x_2 x_3 - x_2^2 + x_3^2$$

is not positive definite or semidefinite because $a_{22} = -1$. In fact, if $X = [0, 1, 0]^{\mathsf{T}}$, the form has the value -1. It is not negative definite or semidefinite either.

We can extend Theorem 10.4.4 a bit:

Theorem 10.4.5: *If the quadratic form $X^{\mathsf{T}}AX$ is positive semidefinite and if x_i actually appears in the form, then $a_{ii} > 0$.*

Since x_i actually appears in the form, we have $a_{ij} \neq 0$ for some j. If $j = i$, then $a_{ii} \neq 0$ and hence $a_{ii} > 0$ by the preceding theorem. If $j \neq i$, then the principal minor

$$\begin{vmatrix} a_{ii} & a_{ij} \\ a_{ji} & a_{jj} \end{vmatrix} \geq 0;$$

that is, since $a_{ij} = a_{ji} \neq 0$,

$$a_{ii}a_{jj} \geq (a_{ij})^2 > 0,$$

so $a_{ii} \neq 0$ and hence again $a_{ii} > 0$ by the preceding theorem. Incidentally, $a_{jj} > 0$ also.

We show next that positive definiteness is invariant under transformation:

Theorem 10.4.6: *A quadratic form X^TAX is positive definite if and only if, for every real, nonsingular transformation $X = B\tilde{X}$, the equivalent form $\tilde{X}^T(B^TAB)\tilde{X}$ is positive definite.*

Since B is nonsingular and $X = B\tilde{X}$, $X = 0$ if and only if $\tilde{X} = 0$. Hence, since $X^TAX = \tilde{X}^T(B^TAB)\tilde{X}$, if either member is positive for all nonzero X (or $\tilde{X}$), the other must be also and the theorem is proved.

It is convenient to make the following definition at this point: a real symmetric matrix A is **positive definite** or **semidefinite** if and only if the corresponding quadratic form is positive definite or semidefinite. Hence a real, symmetric matrix A is positive definite if and only if all its characteristic roots are positive, if and only if all its leading principal minors are positive, and if and only if every matrix B^TAB, where B is nonsingular, is positive definite. Similarly for semidefinite matrices. At times it is easier to work with matrices rather than with forms. For example:

Theorem 10.4.7: *A real symmetric matrix A is positive definite if and only if A^{-1} exists and is positive definite and symmetric.*

Suppose that A is real, symmetric, and positive definite. Then every $\lambda_i > 0$. Hence $\det A = \lambda_1\lambda_2 \cdots \lambda_n > 0$. Hence A^{-1} exists. From $AA^{-1} = I$ follows $(A^{-1})^TA^T = (A^{-1})^TA = I$, so $(A^{-1})^T = A^{-1}$ and hence A^{-1} is symmetric. The characteristic roots of A^{-1} are λ_1^{-1}, $\lambda_2^{-1}, \ldots, \lambda_n^{-1}$, which are all positive. Hence A^{-1} is a positive definite, symmetric matrix.

The converse follows from the same argument, since $A = (A^{-1})^{-1}$.

An extension of the preceding theorem is

Theorem 10.4.8: *A is positive definite and symmetric if and only if A^p is positive definite and symmetric for all integers p.*

This follows principally from the fact that the characteristic roots of A^p are $\lambda_1^p, \lambda_2^p, \ldots, \lambda_n^p$ and from Theorem 10.4.7.

We prove next the often useful

Theorem 10.4.9: *A real matrix A is positive definite and symmetric if and only if there exists a nonsingular matrix B such that $A = B^TB$.*

First assume that $A = B^{\mathsf{T}}B$, where B is nonsingular. Then we have readily that A is symmetric. Moreover, $X^{\mathsf{T}}AX = X^{\mathsf{T}}B^{\mathsf{T}}BX = (BX)^{\mathsf{T}}(BX) = |BX|^2 \geq 0$. Since, therefore, $X^{\mathsf{T}}AX = 0$ if and only if $BX = 0$, and since, because B is nonsingular, $BX = 0$ if and only if $X = 0$, it follows that A is positive definite.

Conversely, assume that A is a positive definite symmetric matrix. Then there exists an orthogonal matrix C such that

$$CAC^{\mathsf{T}} = D[\lambda_1, \lambda_2, \ldots, \lambda_n].$$

Then

$$A = C^{\mathsf{T}}D[\sqrt{\lambda_1}, \sqrt{\lambda_2}, \ldots, \sqrt{\lambda_n}]^2 C$$

or

$$A = (D[\sqrt{\lambda_1}, \sqrt{\lambda_2}, \ldots, \sqrt{\lambda_n}]C)^{\mathsf{T}}(D[\sqrt{\lambda_1}, \sqrt{\lambda_2}, \ldots, \sqrt{\lambda_n}]C).$$

Now we put

$$B = D[\sqrt{\lambda_1}, \sqrt{\lambda_2}, \ldots, \sqrt{\lambda_n}]C,$$

which is nonsingular since C and D are, and thus have the stated result.

The preceding proof suggests

Theorem 10.4.10: *A real, symmetric, positive definite matrix A has a real, symmetric, positive definite pth root for each positive integer p.*

There exists an orthogonal matrix U such that

$$U^{\mathsf{T}}AU = D[\lambda_1, \lambda_2, \ldots, \lambda_n],$$

where all the λ's are positive so that the principal roots $\lambda_1^{1/p}, \lambda_2^{1/p}, \ldots, \lambda_n^{1/p}$ are all positive real numbers. Then

$$A = UD[\lambda_1^{1/p}, \ldots, \lambda_n^{1/p}]U^{\mathsf{T}} \cdot UD[\lambda_1^{1/p}, \ldots, \lambda_n^{1/p}]U^{\mathsf{T}} \cdots UD[\lambda_1^{1/p}, \ldots, \lambda_n^{1/p}]U^{\mathsf{T}}$$

$$= (UD[\lambda_1^{1/p}, \ldots, \lambda_n^{1/p}]U^{\mathsf{T}})^p.$$

Moreover, the product in parentheses is real, symmetric, and positive definite.

Up to this point we have given a number of determinantal conditions for definiteness that are necessary but not sufficient. An important necessary and sufficient condition involves examining only the leading principal minors of A. We define the **leading principal minor of order r of A** thus:

$$p_r = \begin{vmatrix} a_{11} & a_{12} & \cdots & a_{1r} \\ a_{21} & a_{22} & \cdots & a_{2r} \\ \vdots & & & \\ a_{r1} & a_{r2} & \cdots & a_{rr} \end{vmatrix}, \qquad r = 1, 2, \ldots, n.$$

It is useful to define also $p_0 = 1$ as the **leading principal minor of order zero**. We now have

Theorem 10.4.11: *A real quadratic form of rank r is positive semidefinite, or positive definite if $r = n$, and actually contains the variables $x_1, x_2, \ldots, x_r$ if and only if $p_1, p_2, \ldots, p_r$ are all positive.*

The proof of the theorem involves procedures not yet developed and is given in Section 10.9. The theorem is in principle easy to apply. For example, to establish the positive definiteness of

$$X^{\mathsf{T}} \begin{bmatrix} 2 & 1 & 1 & \cdots & 1 \\ 1 & 2 & 1 & \cdots & 1 \\ \vdots & & & & \\ 1 & 1 & 1 & \cdots & 2 \end{bmatrix}_{n \times n} X,$$

note that

$$p_1 = 2,$$

$$p_2 = \begin{vmatrix} 2 & 1 \\ 1 & 2 \end{vmatrix} = 3,$$

$$p_3 = \begin{vmatrix} 2 & 1 & 1 \\ 1 & 2 & 1 \\ 1 & 1 & 2 \end{vmatrix} = \begin{vmatrix} 1 & 0 & -1 \\ 0 & 1 & -1 \\ 1 & 1 & 2 \end{vmatrix} = \begin{vmatrix} 1 & 0 & 0 \\ 0 & 1 & 0 \\ 1 & 1 & 4 \end{vmatrix} = 4.$$

The procedure here is to subtract the bottom row from each of the others, then add the first two columns to the last. The same procedure works on all the principal minors, so that, in fact, $p_r = r + 1$. Since every p_r is greater than 0, the form is positive definite.

10.5 Exercises

*1. Show that, if A_1 and A_2 are symmetric, $X^{\mathsf{T}} A_1 X$ is identical to $X^{\mathsf{T}} A_2 X$ in the x's if and only if $A_1 = A_2$. (This proves that the symmetric matrix of a quadratic form is unique.)

2. Show that $X^{\mathsf{T}} A X \equiv X^{\mathsf{T}} A^{\mathsf{T}} X$, whether or not A is symmetric, and that $X^{\mathsf{T}} B X \equiv 0$ if B is skew-symmetric.

3. Show by means of an example that, if A_1 and A_2 are *not* both symmetric, we can have $X^T A_1 X \equiv X^T A_2 X$ even though A_1 and A_2 have different ranks. (This is another reason for agreeing to use only symmetric matrices in quadratic forms.)

4. Apply the transformation

$$X = \begin{bmatrix} 2 & -1 & -2 & 1 \\ 2 & 1 & 0 & 0 \\ 0 & 0 & 2 & -1 \\ 0 & 0 & 2 & 1 \end{bmatrix} Y$$

to the form

$$X^T \begin{bmatrix} 0 & 1 & 0 & 0 \\ 1 & 0 & 1 & 0 \\ 0 & 1 & 0 & 1 \\ 0 & 0 & 1 & 0 \end{bmatrix} X.$$

5. Diagonalize each quadratic form by means of an orthogonal transformation $(n = 3)$:

   (a) $q(X) = x_1^2 - 6x_1 x_2 + x_2^2$,

   (b) $q(X) = 2x_1 x_2 + 2x_2 x_3$,

   (c) $q(X) = 2x_1 x_2 + 2x_2 x_3 + 2x_3 x_1$.

6. Write each quadratic form of Exercise 5 as a linear combination of squares of linearly independent linear expressions.

7. Rewrite $(a_1 x_1 + a_2 x_2 + \cdots + a_n x_n)^2$ in the form $X^T A X$, where A is a symmetric matrix. What is the rank of A?

8. If $m_x = 1/n \sum_1^n x_j$, rewrite the quadratic form

$$s_x^2 = \frac{1}{n-1} \sum_{j=1}^n (x_j - m_x)^2$$

in the form $X^T A X$, where A is symmetric. What is the rank of A? For what sets of values of the x's will this form have the value zero? (This form is the **variance**, much used in statistics.)

9. Show that the matrix of a sum of quadratic forms is the sum of their matrices, and that the rank of their sum is equal to or less than the sum of their ranks.

10. Show that a quadratic form $X^T A X$ over a field $\mathscr{F}$ may be factored into a product $(X^T V_1)(X^T V_2)$, where V_1 and V_2 are n-vectors over $\mathscr{F}$ if and only if its rank is 1 or 0.

11. Find the matrix and the rank of the quadratic form

$$(A_1^T X)^2 + (A_2^T X)^2 + \cdots + (A_k^T X)^2,$$

where $A_1, A_2, \ldots, A_k$ are linearly independent n-vectors.

12. Examine for definiteness:

(a) $\begin{bmatrix} 1 & -1 & -1 \\ -1 & 2 & 4 \\ -1 & 4 & 6 \end{bmatrix}$, (b) $\begin{bmatrix} 4 & 2 & -2 \\ 2 & 4 & 2 \\ -2 & 2 & 4 \end{bmatrix}$, (c) $\begin{bmatrix} 1 & 2 & 0 \\ 2 & 1 & 3 \\ 0 & 3 & -3 \end{bmatrix}$,

(d) $\begin{bmatrix} 2 & 1 & 1 \\ 1 & 2 & 1 \\ 1 & 1 & \frac{2}{3} \end{bmatrix}$, (e) $\begin{bmatrix} x & 1 & 0 \\ 1 & x & 1 \\ 0 & 1 & x \end{bmatrix}$.

***13.** Show that for every real matrix A, $A^T A$ is positive semidefinite or positive definite. When is it positive definite?

14. What is the condition that the quadratic form $Ax^2 + 2Bxy + Cy^2$ be positive definite? Test $x^2 + xy + y^2$, $x^2 + 2xy + y^2$, $x^2 + 4xy + y^2$, and $x^2 + 2kxy + my^2$ for definiteness.

15. Show that the quadratic form in $x_1, x_2, \ldots, x_n$ defined by

$$q = x_1^2 + (x_1 + x_2)^2 + \cdots + (x_1 + x_2 + \cdots + x_n)^2$$

is positive definite.

16. Show that the quadratic form in $x_1, x_2, \ldots, x_n$ defined by

$$q = (x_1 - x_2)^2 + (x_2 - x_3)^2 + \cdots + (x_{n-1} - x_n)^2 + (x_n - x_1)^2$$

is positive semidefinite, but not positive definite. Find its rank.

17. Let $f_1(X), f_2(X), \ldots, f_k(X)$ be real linear functions of $x_1, x_2, \ldots, x_n$. Under what conditions is the quadratic form defined by

$$q = (f_1(X))^2 + (f_2(X))^2 + \cdots + (f_k(X))^2$$

(a) positive definite, (b) positive semidefinite?

18. Under what conditions on x and a is the real matrix

$$\begin{bmatrix} x & a & a & \cdots & a \\ a & x & a & \cdots & a \\ \vdots & & & & \\ a & a & a & \cdots & x \end{bmatrix}_n$$

positive definite?

19. Show that if $\alpha > 1$, the matrix A of order n for which $a_{i,i \pm k} = \alpha^{(n-1)-k}$ is positive definite.

*20. Prove that if A and B are real, positive definite, symmetric matrices of order n, then so is $A + B$.

*21. Show that a real symmetric matrix A is negative definite if and only if all its characteristic roots are negative.

22. A real, symmetric matrix A is negative definite if and only if its principal minor determinants alternate in sign thus:

$$p_0 = 1, \ p_1 < 0, \ p_2 > 0, \ p_3 < 0, \dots.$$

(a) Show that these matrices are negative definite:

$$A = [-2]_{1 \times 1}, \quad B = \begin{bmatrix} -2 & 1 \\ 1 & -2 \end{bmatrix}, \quad C = \begin{bmatrix} -2 & 1 & -1 \\ 1 & -2 & 1 \\ -1 & 1 & -2 \end{bmatrix}.$$

(b) Given that A is a real, symmetric, negative definite matrix, show how to write $X^T A X$ as a linear combination of squares of linearly independent linear functions of x_1, x_2, and x_3, the coefficients of combination all being negative.

23. Prove that, if A is real, positive definite, and symmetric, and if C has rank r, then $C^T A C$ has rank r also.

24. Identify each of the following conics and sketch a figure:

(a) $x_1^2 + 6x_1 x_2 + x_2^2 = 4$,

(b) $x_1^2 + 2x_1 x_2 + 4x_2^2 = 6$,

(c) $4x_1^2 - x_1 x_2 + 4x_2^2 = 12$.

25. Identify each quadric and sketch a figure:

(a) $\quad X^T \begin{bmatrix} 2 & 1 & 0 \\ 1 & 2 & 0 \\ 0 & 0 & -2 \end{bmatrix} X = 0$, \qquad (b) $\quad X^T \begin{bmatrix} 2 & 1 & 0 \\ 1 & 2 & 0 \\ 0 & 0 & 2 \end{bmatrix} X = 24$.

26. The potential energy of a certain mechanical system is given by

$$V = \tfrac{1}{2}(k_1 x_1^2 + k_2 x_2^2 + k_3 x_3^2 + l_1(x_1 - x_2)^2 + l_2(x_1 - x_3)^2 \\ + l_3(x_2 - x_3)^2 + m(x_1 + x_2 + x_3)^2).$$

Write this quadratic form in matrix form. Given that $k_1, k_2, k_3, l_1, l_2, l_3, m$ are all positive, prove that $V(X) > 0$ unless $X = 0$.

27. By rearranging the quadratic form in four variables,

$$q = x_1^2 + 2x_1 x_2 - 2x_2 x_3 + 2x_3 x_4,$$

as a linear combination of squares, determine a nonsingular transformation which will diagonalize it.

28. Prove that, if a symmetric matrix

$$\begin{bmatrix} P & Q \\ Q^\mathsf{T} & S \end{bmatrix},$$

where S is nonsingular, is a real, positive semidefinite matrix, then $P - QS^{-1}Q^\mathsf{T}$ is also positive semidefinite.

29. Let A, B be of order n and symmetric, B positive definite, $A - B$ positive definite or semidefinite. Prove that $\det A \geq \det B$ and that A is positive definite.

30. A student asserted that, if $R = [r_{ij}]_{n \times n}$ is symmetric and real with

$$r_{ii} > 0,$$

$$r_{ii} > \sum_{j \neq i} |r_{ij}|, \qquad i = 1, 2, \ldots, n,$$

then R is positive definite. Prove or disprove his assertion.

31. Prove that the tridiagonal matrix

$$M_n = \begin{bmatrix} 2 & -1 & 0 & \cdots & 0 & 0 \\ -1 & 2 & -1 & \cdots & 0 & 0 \\ 0 & -1 & 2 & \cdots & 0 & 0 \\ \vdots & & & & & \\ 0 & 0 & 0 & \cdots & 2 & -1 \\ 0 & 0 & 0 & \cdots & -1 & 2 \end{bmatrix}_{n \times n}$$

is positive definite.

32. Under what conditions on α will the following tridiagonal matrix be positive definite?

$$\begin{bmatrix} \alpha & -1 & 0 & \cdots & 0 & 0 \\ -1 & \alpha & -1 & \cdots & 0 & 0 \\ 0 & -1 & \alpha & \cdots & 0 & 0 \\ \vdots & & & & & \\ 0 & 0 & 0 & \cdots & \alpha & -1 \\ 0 & 0 & 0 & \cdots & -1 & \alpha \end{bmatrix}_{n \times n}$$

33. Let A and B be real n-square matrices. Define $A > B$ if and only if $A - B$ is a positive definite symmetric matrix. Prove that if $A > B > 0$, then $B^{-1} > A^{-1}$.

34. Prove that if A is a real, symmetric matrix, there exist non-negative definite matrices $P_1, P_2, \ldots, P_n$ such that $P_i P_j = \delta_{ij} P_j^2$ and such that $A = \sum_{i=1}^{n} \lambda_i P_i^2$. (This is the **spectral decomposition** of A.)

35. Given that the terminal point of the vector X_0 is on the quadric $X^\mathsf{T} A X = b$, under what conditions will the line determined by X_0 and $X_1 \neq X_0$ be tangent to the quadric? (A line is tangent to a quadric if the two points in which it intersects the quadric coincide.)

36. Given that $X_0^T A X_0 = 1$, show that $A X_0$ is parallel to the normal to the quadric surface $X^T A X = 1$ at the terminal point of the vector X_0.

10.6 Lagrange's Reduction

There are methods of diagonalizing a quadratic form which, in contrast to orthogonal reduction, require only linear operations on the coefficients and apply to quadratic forms over any field $\mathcal{F}$. We consider first **Lagrange's reduction**, which involves basically just a repeated completing of the square. Before stating the general theorem, we illustrate the process with two examples.

Consider first the quadratic form of rank 3,

$$q = 2x_1^2 + x_1 x_2 - 3x_1 x_3 + 2x_2 x_3 - x_3^2.$$

We first group all the terms containing x_1 and factor out the coefficient of x_1^2:

$$q = 2(x_1^2 + (\tfrac{1}{2}x_2 - \tfrac{3}{2}x_3)x_1) + 2x_2 x_3 - x_3^2.$$

Now we complete the square on x_1 and pay for the inserted terms:

$$q = 2(x_1^2 + (\tfrac{1}{2}x_2 - \tfrac{3}{2}x_3)x_1 + (\tfrac{1}{4}x_2 - \tfrac{3}{4}x_3)^2) - 2(\tfrac{1}{4}x_2 - \tfrac{3}{4}x_3)^2 + 2x_2 x_3 - x_3^2,$$

so that

$$q = 2(x_1 + \tfrac{1}{4}x_2 - \tfrac{3}{4}x_3)^2 - \tfrac{1}{8}(x_2^2 - 22x_2 x_3) - \tfrac{17}{8}x_3^2.$$

Next we complete the square on x_2, thus obtaining

$$q = 2(x_1 + \tfrac{1}{4}x_2 - \tfrac{3}{4}x_3)^2 - \tfrac{1}{8}(x_2 - 11x_3)^2 + 13x_3^2.$$

The substitution

$$
\begin{aligned}
y_1 &= x_1 + \tfrac{1}{4}x_2 - \tfrac{3}{4}x_3 \\
y_2 &= \phantom{x_1 + \tfrac{1}{4}} x_2 - 11x_3 \\
y_3 &= \phantom{x_1 + \tfrac{1}{4}x_2 - 11} x_3
\end{aligned}
$$

has determinant 1 and is, therefore, nonsingular. It gives

$$q = 2y_1^2 - \tfrac{1}{8}y_2^2 + 13y_3^2,$$

which also has rank 3.

In matrix form, we have

$$Y = \begin{bmatrix} 1 & \frac{1}{4} & -\frac{3}{4} \\ 0 & 1 & -11 \\ 0 & 0 & 1 \end{bmatrix} X \quad \text{or} \quad X = \begin{bmatrix} 1 & -\frac{1}{4} & -2 \\ 0 & 1 & 11 \\ 0 & 0 & 1 \end{bmatrix} Y,$$

so that

$$X^T \begin{bmatrix} 2 & \frac{1}{2} & -\frac{3}{2} \\ \frac{1}{2} & 0 & 1 \\ -\frac{3}{2} & 1 & -1 \end{bmatrix} X = Y^T \begin{bmatrix} 1 & 0 & 0 \\ -\frac{1}{4} & 1 & 0 \\ -2 & 11 & 1 \end{bmatrix}$$

$$\times \begin{bmatrix} 2 & \frac{1}{2} & -\frac{3}{2} \\ \frac{1}{2} & 0 & 1 \\ -\frac{3}{2} & 1 & -1 \end{bmatrix} \begin{bmatrix} 1 & -\frac{1}{4} & -2 \\ 0 & 1 & 11 \\ 0 & 0 & 1 \end{bmatrix} Y$$

$$= Y^T \begin{bmatrix} 2 & 0 & 0 \\ 0 & -\frac{1}{8} & 0 \\ 0 & 0 & 13 \end{bmatrix} Y.$$

If there had been no x_1^2 term in q, our completing the square on x_1 would have been impossible. Similarly, at the second stage, if no x_2^2 term had appeared, we could not have completed the square on x_2. The same difficulty may, of course, arise at any stage. To show how to deal with this situation, we consider the form in three variables,

$$q = 2x_1x_2 + 2x_2x_3 + x_3^2.$$

Let us first put

(10.6.1)
$$\begin{aligned} x_1 &= \tilde{x}_1 \\ x_2 &= \tilde{x}_1 + \tilde{x}_2 \\ x_3 &= \quad\quad \tilde{x}_3. \end{aligned}$$

This transformation has determinant 1 and is therefore nonsingular. It gives

$$q = 2\tilde{x}_1^2 + 2\tilde{x}_1\tilde{x}_2 + 2\tilde{x}_1\tilde{x}_3 + 2\tilde{x}_2\tilde{x}_3 + \tilde{x}_3^2,$$

from which, proceeding as before, we obtain

$$q = 2(\tilde{x}_1 + \tfrac{1}{2}\tilde{x}_2 + \tfrac{1}{2}\tilde{x}_3)^2 - \tfrac{1}{2}(\tilde{x}_2 - \tilde{x}_3)^2 + \tilde{x}_3^2.$$

The transformation with determinant 1 defined by

(10.6.2)
$$\begin{aligned} y_1 &= \tilde{x}_1 + \tfrac{1}{2}\tilde{x}_2 + \tfrac{1}{2}\tilde{x}_3 \\ y_2 &= \qquad \tilde{x}_2 - \tilde{x}_3 \\ y_3 &= \qquad\qquad \tilde{x}_3, \end{aligned}$$

then gives

$$q = 2y_1^2 - \tfrac{1}{2}y_2^2 + y_3^2,$$

so that the reduction is now complete. (We could, of course, have begun by completing the square on x_3 in order to avoid using a transformation of type (10.6.1).)

Combining (10.6.1) and (10.6.2), we obtain the transformation of determinant unity,

$$\begin{aligned} y_1 &= \tfrac{1}{2}x_1 + \tfrac{1}{2}x_2 + \tfrac{1}{2}x_3 & x_1 &= y_1 - \tfrac{1}{2}y_2 - y_3 \\ y_2 &= -x_1 + x_2 - x_3 \quad\text{or}\quad & x_2 &= y_1 + \tfrac{1}{2}y_2 \\ y_3 &= \qquad\qquad x_3, & x_3 &= \qquad\qquad y_3, \end{aligned}$$

which effects the reduction in one step. The reader should check this assertion by the matrix method used in the previous example.

The device used in this problem may be introduced at any stage of the reduction process. A systematic scheme permitting this is contained in the proof of

Theorem 10.6.1: *Every not identically zero quadratic form over a field $\mathscr{F}$ can be reduced by a nonsingular transformation with coefficients in $\mathscr{F}$ to the form*

$$c_1 y_1^2 + c_2 y_2^2 + \cdots + c_r y_r^2,$$

where the c's are not zero but are in $\mathscr{F}$ and where r is necessarily the rank of the given quadratic form.

We prove the theorem by induction.

When $n = 1$, $r = 1$ also. The quadratic form is simply

$$a_{11}x_1^2, \qquad a_{11} \neq 0,$$

and the identity transformation

$$y_1 = x_1$$

is the transformation mentioned in the theorem.

Suppose now that the theorem is true for quadratic forms in $n - 1$ or fewer variables and consider a quadratic form

$$q = \sum_{i,\,j=1}^{n} a_{ij} x_i x_j$$

of rank r in n variables. In the argument to follow it is necessary to have an x_1^2 term present. If $a_{11} \neq 0$, that is good enough. If, however, $a_{11} = 0$, but $a_{kk} \neq 0$ for some $k > 1$, then we may put

(10.6.3)
$$\begin{aligned} x_1 &= \tilde{x}_k, \\ x_k &= \tilde{x}_1, \\ x_j &= \tilde{x}_j, \qquad j \neq 1, k. \end{aligned}$$

This is a transformation with determinant -1 which takes the term $a_{kk} x_k^2$ into the term $a_{kk} \tilde{x}_1^2$. Since no other $\tilde{x}_1^2$ can appear, $\tilde{x}_1^2$ has a nonzero coefficient in the new form. The new form has the same rank as the old, and its coefficients are also in the same field.

If every coefficient a_{kk} is zero, then some coefficient a_{ij}, $i < j$, is not zero, and we put

(10.6.4)
$$\begin{aligned} x_j &= \tilde{x}_i + \tilde{x}_j, \\ x_k &= \tilde{x}_k, \qquad k \neq j. \end{aligned}$$

As a result of this nonsingular transformation (determinant 1) the term $a_{ij} x_i x_j$ is replaced by the terms $a_{ij} \tilde{x}_i^2 + a_{ij} \tilde{x}_i \tilde{x}_j$, and similarly for the symmetric term $a_{ji} x_j x_i$. Hence, since $a_{ii} = a_{jj} = 0$ by hypothesis, the total coefficient of $\tilde{x}_i^2$ in the new form will be $2a_{ij}$, which is not zero. Then an application of the first transformation discussed will give us a quadratic form with an $\tilde{x}_1^2$-type term actually present. Again, the new form will have the same rank as the original form, and its coefficients will be in the same field.

We may thus assume that we are in fact dealing with a quadratic form q_1 equivalent to q in which $a_{11} \neq 0$. Using the fact that $a_{ij} = a_{ji}$, we rearrange q_1 and complete the square on x_1 as follows:

$$\begin{aligned} q_1 &= \sum_{i,\,j=1}^{n} a_{ij} x_i x_j = a_{11} x_1^2 + 2 \sum_{j=2}^{n} a_{1j} x_1 x_j + \sum_{i,\,j=2}^{n} a_{ij} x_i x_j \\ &= a_{11} \left[x_1^2 + 2 \left\{ \sum_{j=2}^{n} \frac{a_{1j}}{a_{11}} x_j \right\} x_1 \right] + \sum_{i,\,j=2}^{n} a_{ij} x_i x_j \\ &= a_{11} \left[x_1^2 + 2 \left\{ \sum_{j=2}^{n} \frac{a_{1j}}{a_{11}} x_j \right\} x_1 + \left\{ \sum_{j=2}^{n} \frac{a_{1j}}{a_{11}} x_j \right\}^2 \right] \\ &\quad + \left[\sum_{i,\,j=2}^{n} a_{ij} x_i x_j - a_{11} \left\{ \sum_{j=2}^{n} \frac{a_{1j}}{a_{11}} x_j \right\}^2 \right] \\ &= a_{11} \left(x_1 + \frac{a_{12}}{a_{11}} x_2 + \cdots + \frac{a_{1n}}{a_{11}} x_n \right)^2 + q_2. \end{aligned}$$

Here q_2 is a quadratic form in $x_2, \ldots, x_n$ only, with coefficients in $\mathscr{F}$. We now make the transformation

(10.6.5)
$$\tilde{x}_1 = x_1 + \frac{a_{12}}{a_{11}} x_2 + \cdots + \frac{a_{1n}}{a_{11}} x_n,$$
$$\tilde{x}_i = x_i, \qquad i = 2, 3, \ldots, n,$$

which is over $\mathscr{F}$ and also has determinant 1, thus obtaining

(10.6.6)
$$q_1 = a_{11}\tilde{x}_1^2 + q_2(\tilde{x}_2, \ldots, \tilde{x}_n).$$

Now by the induction hypothesis, there exists over $\mathscr{F}$ a nonsingular transformation taking $\tilde{x}_2, \tilde{x}_3, \ldots, \tilde{x}_n$ into $y_2, y_3, \ldots, y_n$ and such that

$$q_2 = c_2 y_2^2 + c_3 y_3^2 + \cdots + c_k y_k^2,$$

where $k - 1$ is the rank of q_2. Let the matrix of order $n - 1$ of this transformation be B. Then the transformation

(10.6.7)
$$Y = \left[\begin{array}{c|c} 1 & 0 \\ \hline 0 & B \end{array} \right] \tilde{X}$$

is also nonsingular and over $\mathscr{F}$. Applying this to (10.6.6) and putting $a_{11} = c_1$ for the sake of uniformity, we obtain

(10.6.8)
$$q_1 = c_1 y_1^2 + c_2 y_2^2 + \cdots + c_k y_k^2.$$

Similarly, starting with an arbitrary quadratic form q, we may combine any transformations of the types (10.6.3), (10.6.4), (10.6.5), and (10.6.7) we may have used and thus obtain a single transformation $X = MY$ which reduces q at once to a form of the type (10.6.8). Furthermore, M is nonsingular and has elements in $\mathscr{F}$.

Hence

(10.6.9)
$$q = X^{\mathsf{T}} A X = Y^{\mathsf{T}} (M^{\mathsf{T}} A M) Y = \sum_{j=1}^{k} c_j y_j^2,$$

where

$$M^{\mathsf{T}} A M = D[c_1, c_2, \ldots, c_k, 0, 0, \ldots, 0].$$

But since M is nonsingular, the diagonal matrix on the right must have the same rank as A. Hence $k = r$ and the theorem is proved.

For certain applications, the following observations are important. In the preceding reduction, only transformations with determinant 1 or -1 were used. Since the determinant of a product of matrices is the product of their determinants, an additional transformation exchanging two variables and having determinant -1 may be employed if necessary to guarantee that the matrix M of (10.6.9) has determinant 1. Hence we have

Corollary 10.6.2: *The nonsingular transformation of Theorem 10.6.1 may be chosen so that its determinant is 1.*

10.7 Kronecker's Reduction

By each of the preceding methods of reduction, a quadratic form of rank $r < n$ is reduced to a form in which only r variables appear. Another way of accomplishing this is by means of **Kronecker's reduction**, by which we prove

Theorem 10.7.1: *Let the quadratic form $q = \sum_{i,j=1}^{n} a_{ij}x_i x_j$ over a given field $\mathscr{F}$ have rank $r < n$. Then if the leading principal minor of order r is not zero, there exists a nonsingular linear transformation over $\mathscr{F}$ with determinant equal to 1 which reduces q to the form $q = \sum_{i,j=1}^{r} a_{ij} y_i y_j$.*

The unique aspect of Kronecker's reduction is that the coefficients a_{ij}, $1 \le i, j \le r$, of the reduced form are precisely the same as the corresponding coefficients in the original form.

We write the matrix of the given form partitioned into columns thus: $A = [A_1, A_2, \ldots, A_n]$. By hypothesis, A has rank r with $A_1, A_2, \ldots, A_r$ linearly independent. Hence there exist relations

$$(10.7.1) \qquad A_k = \sum_{j=1}^{r} c_{kj} A_j, \qquad k = r + 1, \ldots, n,$$

expressing the dependence of all later columns on the first r independent ones. Then, if we postmultiply A by the matrix

$$C = \left[\begin{array}{cccc:cccc}
1 & 0 & \cdots & 0 & -c_{r+1,1} & -c_{r+2,1} & \cdots & -c_{n1} \\
0 & 1 & \cdots & 0 & -c_{r+1,2} & -c_{r+2,2} & \cdots & -c_{n2} \\
\vdots & & & & \vdots & & & \\
0 & 0 & \cdots & 1 & -c_{r+1,r} & -c_{r+2,r} & \cdots & -c_{nr} \\
\hdashline
 & & & & 1 & 0 & \cdots & 0 \\
 & \mathbf{0} & & & 0 & 1 & \cdots & 0 \\
 & & & & \vdots & & & \\
 & & & & 0 & 0 & \cdots & 1
\end{array}\right]_n,$$

we will obtain as the product the matrix $[A_1, \ldots, A_r, 0, \ldots, 0]$. Next, because of the symmetry of A, if we premultiply this product by C^T, the corresponding result will be effected with respect to the rows. We therefore have

$$
C^\mathsf{T}AC = \begin{bmatrix} a_{11} & \cdots & a_{1r} & \vdots & \\ \vdots & & & \vdots & 0 \\ a_{r1} & \cdots & a_{rr} & \vdots & \\ \hline & 0 & & \vdots & 0 \end{bmatrix}.
$$

From this it follows that the transformation

$$
X = CY
$$

effects the desired reduction of the given quadratic form. Moreover, $\det C = 1$. Also, since finding the coefficients c_{kj} in (10.7.1) involves solving linear equations with coefficients in $\mathscr{F}$, it follows that the c_{kj}'s may be chosen in $\mathscr{F}$, so that C is over $\mathscr{F}$.

If now $X^\mathsf{T}AX$ is any quadratic form of rank r, with $r < n$, we know by Theorem 7.12.4 that the symmetric matrix A has at least one nonvanishing principal minor of order r. Then a suitable symmetric rearrangement of the rows and columns of A, which amounts to renaming the variables of the quadratic form, may be used to bring the rows and columns of this minor into the leading position. A little thought will reveal that a suitable renaming of the variables can always be effected with the aid of a permutation matrix—in fact, a matrix with determinant $+1$ if desired. Hence any quadratic form of rank r may be reduced in the manner of the theorem to a form of rank r in r variables only. Moreover, the net transformation that does the job may be chosen so as to have determinant $+1$ if desired.

We illustrate by reducing the form of rank 2,

$$
X^\mathsf{T}AX = X^\mathsf{T} \begin{bmatrix} 1 & -1 & 0 \\ -1 & 1 & 0 \\ 0 & 0 & 2 \end{bmatrix} X.
$$

Since the leading principal minor of order 2 is zero, but the lower right one is not, we rename the variables thus:

$$
\begin{cases} x_1 = z_2 \\ x_2 = z_3 \\ x_3 = z_1 \end{cases} \quad \text{or} \quad X = \begin{bmatrix} 0 & 1 & 0 \\ 0 & 0 & 1 \\ 1 & 0 & 0 \end{bmatrix} Z.
$$

This gives

$$X^{\mathsf{T}}AX = Z^{\mathsf{T}} \begin{bmatrix} 2 & 0 & 0 \\ 0 & 1 & -1 \\ 0 & -1 & 1 \end{bmatrix} Z,$$

which meets the conditions of the theorem. Using the notation of the proof,

$$A_3 = 0 \cdot A_1 + (-1)A_2.$$

Hence

$$C = \begin{bmatrix} 1 & 0 & 0 \\ 0 & 1 & 1 \\ 0 & 0 & 1 \end{bmatrix}.$$

Putting $Z = CY$, we obtain the form

$$Y^{\mathsf{T}} \begin{bmatrix} 1 & 0 & 0 \\ 0 & 1 & 0 \\ 0 & 1 & 1 \end{bmatrix} \cdot \begin{bmatrix} 2 & 0 & 0 \\ 0 & 1 & -1 \\ 0 & -1 & 1 \end{bmatrix} \cdot \begin{bmatrix} 1 & 0 & 0 \\ 0 & 1 & 1 \\ 0 & 0 & 1 \end{bmatrix} Y$$

$$= Y^{\mathsf{T}} \begin{bmatrix} 2 & 0 & 0 \\ 0 & 1 & 0 \\ 0 & 0 & 0 \end{bmatrix} Y = 2y_1^2 + y_2^2.$$

Combining the two transformations, we have the single transformation

$$X = \begin{bmatrix} 0 & 1 & 1 \\ 0 & 0 & 1 \\ 1 & 0 & 0 \end{bmatrix} Y,$$

whose determinant is $+1$ and which effects the reduction in one step.

10.8   Sylvester's Law of Inertia for Real Quadratic Forms

Suppose now that a quadratic form q of rank r is over the real field, so that, in the equivalent form

$$q = c_1 y_1^2 + c_2 y_2^2 + \cdots + c_r y_r^2$$

of Theorem 10.6.1, or of (10.2.2), in which case the c's are the λ's, each coefficient c_i is also real. It is not hard to see that, in effecting the reduction of q to this form, we can assign the names of the variables in such a way that all the positive terms appear first, followed by all the negative terms. We may therefore assume that q is given by

$$(10.8.1) \qquad q = h_1 y_1^2 + \cdots + h_p y_p^2 - h_{p+1} y_{p+1}^2 - \cdots - h_r y_r^2,$$

where $h_i > 0$ in every case. Then p is called the **index** of the form and the difference between the numbers of positive and negative terms, $p - (r - p)$, is called the **signature** of the form. These definitions are given significance by

Theorem 10.8.1: *Every quadratic form q over the field of real numbers may be reduced by a real nonsingular transformation to the form*

$$(10.8.2) \qquad z_1^2 + \cdots + z_p^2 - z_{p+1}^2 - \cdots - z_r^2,$$

where p is the index of the form and r is its rank, p and r being uniquely determined integers for a given form q.

The reduction of (10.8.1) to (10.8.2) is effected by the real, nonsingular transformation

$$(10.8.3) \qquad \begin{aligned} z_i &= h_i^{1/2} y_i, & i &= 1, 2, \ldots, r, \\ z_i &= y_i, & i &= r + 1, \ldots, n. \end{aligned}$$

Since r is unique by Theorem 10.6.1, it remains only to establish the uniqueness of p. Suppose that a different chain of real transformations had reduced the form q to the form

$$(10.8.4) \qquad w_1^2 + \cdots + w_k^2 - w_{k+1}^2 - \cdots - w_r^2.$$

We examine first the possibility that $p > k$. From preceding results we know there exist nonsingular transformations

$$Z = CX \qquad \text{and} \qquad W = DX$$

which reduce q to (10.8.2) and (10.8.4) respectively. These transformations express each z and each w as a linear function of the x's, so that, when the z's and w's are replaced by their expressions in terms of the x's, we have the *identities* in $x_1, x_2, \ldots, x_n$,

$$z_1^2 + \cdots + z_p^2 - z_{p+1}^2 - \cdots - z_r^2 \equiv q \equiv w_1^2 + \cdots + w_k^2 - w_{k+1}^2 - \cdots - w_r^2.$$

That is, the values taken on by these three quadratic forms will be equal for each given set of values of $x_1, x_2, \ldots, x_n$.

Now consider the possibility of picking a set of values of $x_1, x_2, \ldots, x_n$ to satisfy the set of $k + (n - p)$ simultaneous equations,

$$
\begin{array}{ll}
w_1 = 0 & z_{p+1} = 0 \\
w_2 = 0 & z_{p+2} = 0 \\
\vdots & \vdots \\
w_k = 0, & z_n = 0.
\end{array}
$$

Suppose first that $p > k$. Then we have here $k + (n - p) = n - (p - k) < n$ equations, which necessarily have real, nontrivial, simultaneous solutions. For each such solution we have

$$
z_1^2 + \cdots + z_p^2 = -w_{k+1}^2 - \cdots - w_r^2.
$$

Since we are dealing with *squares* of real numbers, this equation then implies the vanishing of each term on both left and right, so that we have also

$$
\begin{array}{lll}
z_1 = 0 & & w_{k+1} = 0 \\
z_2 = 0 & \text{and} & w_{k+2} = 0 \\
\vdots & & \vdots \\
z_p = 0, & & w_r \quad = 0.
\end{array}
$$

Thus the n equations in the x's

$$
z_1 = z_2 = \cdots = z_n = 0,
$$

or, more briefly, since $Z = CX$,

$$
CX = 0,
$$

have a nontrivial solution. This, in turn, implies that $\det C = 0$, which is not true since the transformation $Z = CX$ is nonsingular. Hence $p \not> k$. Similarly, $k \not> p$, so that $p = k$. Thus p is a uniquely defined integer, and the proof of the theorem is complete.

Putting the conclusion another way, since every real reduction of q to the form (10.8.2) must lead to the same p and the same r, *the rank and the index are invariants of the real quadratic form q with respect to real, nonsingular transformations.* This fact is known as **Sylvester's law of inertia.**

It is now simple to prove

Theorem 10.8.2: *Two real quadratic forms are equivalent under nonsingular real transformations if and only if they have the same rank r and the same index p.*

Suppose first that $X^\mathsf{T} A_1 X$ and $\tilde{X}^\mathsf{T} A_2 \tilde{X}$ have the same rank r and the same index p. Then there exist, by Theorem 10.8.1, real, nonsingular transformations

$$X = B_1 Y \qquad \text{and} \qquad \tilde{X} = B_2 Y$$

such that

$$Y^\mathsf{T}(B_1^\mathsf{T} A_1 B_1) Y \qquad \text{and} \qquad Y^\mathsf{T}(B_2^\mathsf{T} A_2 B_2) Y$$

are the same form:

$$y_1^2 + y_2^2 + \cdots + y_p^2 - y_{p+1}^2 - \cdots - y_r^2.$$

Hence we must have

$$B_1^\mathsf{T} A_1 B_1 = B_2^\mathsf{T} A_2 B_2,$$

so that

$$A_1 = (B_2 B_1^{-1})^\mathsf{T} A_2 (B_2 B_1^{-1}),$$

and the forms are equivalent.

On the other hand, if $X^\mathsf{T} A_1 X$ and $\tilde{X}^\mathsf{T} A_2 \tilde{X}$ are equivalent under a real, nonsingular linear transformation $X = B\tilde{X}$, we will have

$$(10.8.5) \qquad\qquad A_2 = B^\mathsf{T} A_1 B.$$

Suppose now that $\tilde{X} = CY$ reduces $\tilde{X}^\mathsf{T} A_2 \tilde{X}$ to a form

$$y_1^2 + y_2^2 + \cdots + y_p^2 - y_{p+1}^2 - \cdots - y_r^2.$$

Then, from (10.8.5),

$$\tilde{X}^\mathsf{T} A_2 \tilde{X} = Y^\mathsf{T}(C^\mathsf{T} A_2 C) Y = Y^\mathsf{T}(BC)^\mathsf{T} A_1 (BC) Y.$$

We conclude that the transformation

$$X = (BC)Y$$

will reduce $X^\mathsf{T} A_1 X$ to the same result. Thus p and r are the same for the two forms and the theorem is proved.

10.9   A Necessary and Sufficient Condition for Positive Definiteness

In a matrix A of order n, the **leading principal minor determinants** were defined thus:

$$p_0 = 1, \qquad p_1 = a_{11},$$

$$p_2 = \begin{vmatrix} a_{11} & a_{12} \\ a_{21} & a_{22} \end{vmatrix}, \qquad p_3 = \begin{vmatrix} a_{11} & a_{12} & a_{13} \\ a_{21} & a_{22} & a_{23} \\ a_{31} & a_{32} & a_{33} \end{vmatrix}, \dots, p_n = \det A.$$

These minors play an important role in Lagrange's method of reduction, as we shall now show.

Let us suppose that at each stage of the process the necessary square term is present, so that we can first complete the square on x_1, then on x_2, and so on until the task is complete. Then we can write q as a linear combination of squares of linear forms

$$(10.9.1) \quad q = c_1(x_1 + \alpha_{12} x_2 + \cdots + \alpha_{1n} x_n)^2 + c_2(x_2 + \alpha_{23} x_3 + \cdots$$
$$+ \alpha_{2n} x_n)^2 + \cdots + c_r(x_r + a_{r,\,r+1} x_{r+1} + \cdots + \alpha_{r,\,n} x_n)^2,$$

where $r \leq n$ and where the coefficient of x_j in the jth linear form is 1. The nonsingular transformation

$$(10.9.2) \quad \begin{aligned} y_1 &= x_1 + \alpha_{12} x_2 + \quad \cdots \quad + \alpha_{1n} x_n \\ y_2 &= \qquad\quad x_2 + \quad \cdots \quad + \alpha_{2n} x_n \\ &\vdots \\ y_r &= \qquad\qquad\qquad\quad x_r + \cdots + \alpha_{rn} x_n \\ y_{r+1} &= \qquad\qquad\qquad\qquad\quad x_{r+1} \\ &\vdots \\ y_n &= \qquad\qquad\qquad\qquad\qquad\qquad x_n \end{aligned}$$

then gives

$$q = c_1 y_1^2 + c_2 y_2^2 + \cdots + c_r y_r^2.$$

We have, of course,

$$c_1 = a_{11} = \frac{p_1}{p_0} \neq 0.$$

The coefficient of x_2^2 in q is a_{22}, but from (10.9.1) it is also $c_1\alpha_{12}^2 + c_2$. Now from Section 10.6,

$$\alpha_{12} = \frac{a_{12}}{a_{11}},$$

so that after a little manipulation we have

$$c_2 = a_{22} - c_1\alpha_{12}^2 = \frac{a_{11}a_{22} - a_{12}^2}{a_{11}} = \frac{p_2}{p_1}.$$

The pattern suggested here is perfectly general:

Theorem 10.9.1: *Let A be symmetric and of rank r. Then there exists an identity*

$$(10.9.3) \qquad X^{\mathsf{T}}AX \equiv \frac{p_1}{p_0}y_1^2 + \frac{p_2}{p_1}y_2^2 + \cdots + \frac{p_r}{p_{r-1}}y_r^2,$$

where $p_0 = 1$ and where y_j has the form $y_j = x_j + \alpha_{j,j+1}x_{j+1} + \cdots + \alpha_{jn}x_n$, if and only if $p_j \neq 0$, $j = 1, 2, \ldots, r$.

By inspection of (10.9.3), we see that it is necessary that $p_1, p_2, \ldots, p_{r-1}$ be different from zero in order for an identity of the form (10.9.3) to exist, and that p_r be different from zero for the form on the right to have the rank r.

The sufficiency of the condition is proved by induction. Suppose that $p_1, p_2, \ldots, p_r$ are all different from zero. Then $a_{11} \neq 0$, and we can write

$$q = X^{\mathsf{T}}AX = a_{11}\left(x_1 + \frac{a_{12}}{a_{11}}x_2 + \cdots + \frac{a_{1n}}{a_{11}}x_n\right)^2 + q_2(x_2, \ldots, x_n),$$

as in Section 10.6, or

$$q = \frac{p_1}{p_0}y_1^2 + q_2(x_2, \ldots, x_n),$$

where

$$y_1 = x_1 + \frac{a_{12}}{a_{11}}x_2 + \cdots + \frac{a_{1n}}{a_{11}}x_n.$$

That is, since $p_1 \neq 0$, the first step of the reduction can be carried out. Let us suppose then that we have been able to write

$$q = \frac{p_1}{p_0}y_1^2 + \frac{p_2}{p_1}y_2^2 + \cdots + \frac{p_{k-1}}{p_{k-2}}y_{k-1}^2 + q_k(x_k, \ldots, x_n),$$

where

(10.9.4) $\quad y_j = x_j + \alpha_{j,j+1} x_{j+1} + \cdots + \alpha_{jn} x_n, \qquad j = 1, 2, \ldots, k-1.$

Let us put

$$q_k = \sum_{i,j=k}^{n} c_{ij} x_i x_j, \qquad c_{ij} = c_{ji}.$$

Then we have $k-1$ equations of the form

(10.9.5a) $\quad \dfrac{1}{2} \dfrac{\partial q}{\partial x_s} \equiv \sum_{j=1}^{n} a_{sj} x_j \equiv \sum_{j=1}^{k-1} \dfrac{p_j}{p_{j-1}} y_j \dfrac{\partial y_j}{\partial x_s}, \qquad s = 1, 2, \ldots, k-1,$

and one equation of the form

(10.9.5b) $\quad \dfrac{1}{2} \dfrac{\partial q}{\partial x_k} \equiv \sum_{j=1}^{n} a_{kj} x_j \equiv \sum_{j=1}^{k-1} \dfrac{p_j}{p_{j-1}} y_j \dfrac{\partial y_j}{\partial x_k} + \sum_{j=k}^{n} c_{kj} x_j.$

In these identities and in (10.9.4) let us put $x_{k+1} = \cdots = x_n = 0$, and then let us choose $x_1, x_2, \ldots, x_k$ so that, in (10.9.4),

$$y_1 = y_2 = \cdots = y_{k-1} = 0.$$

We have here $k-1$ equations in the k remaining unknowns, $x_1, x_2, \ldots, x_k$, so that a nontrivial solution certainly exists. For such a solution, together with $x_{k+1} = \cdots = x_n = 0$, we deduce from (10.9.5a, b) that

$$\sum_{j=1}^{k} a_{sj} x_j = 0, \qquad s = 1, 2, \ldots, k-1,$$

$$\sum_{j=1}^{k} a_{kj} x_j = c_{kk} x_k.$$

Here we have k homogeneous equations in k unknowns which have by hypothesis a nontrivial solution. Hence the determinant of the system must vanish:

$$\begin{vmatrix} a_{11} & a_{12} & \cdots & a_{1k} \\ a_{21} & a_{22} & \cdots & a_{2k} \\ \vdots & & & \\ a_{k1} & a_{k2} & \cdots & (a_{kk} - c_{kk}) \end{vmatrix} = 0.$$

Expanding, we have

$$p_k - c_{kk} p_{k-1} = 0,$$

or, since $p_{k-1} \neq 0$ and $p_k \neq 0$,

$$c_{kk} = \frac{p_k}{p_{k-1}} \neq 0.$$

We can therefore proceed with the next step of the reduction, obtaining $(p_k/p_{k-1})y_k^2$ as the next term. Finally, since A has rank r, the process must stop with the term in y_r^2, so the proof of the theorem is complete.

We are now able to prove the following result:

Theorem 10.9.2: *A real quadratic form q of rank r is positive semidefinite, or definite if $r = n$, and actually contains the variables $x_1, x_2, \ldots, x_r$ if $p_1, p_2, \ldots, p_r$ are all positive.*

Since $p_1, p_2, \ldots, p_r$ are all $\neq 0$ and q has rank r, we have, by the preceding theorem,

$$q \equiv \frac{p_1}{p_0} y_1^2 + \frac{p_2}{p_1} y_2^2 + \cdots + \frac{p_r}{p_{r-1}} y_r^2.$$

Hence, all the p's being positive, q is positive semidefinite if $r < n$ and positive definite if $r = n$. Furthermore, q must contain $x_1, x_2, \ldots, x_r$, for the absence of x_j, say, with $j \leq r$, would imply the vanishing of the jth row and the jth column in the matrix of q, so that $p_j, p_{j+1}, \ldots, p_r$ would all vanish, contrary to hypothesis.

Combining results from Theorems 10.4.3 and 10.9.2, we have

Theorem 10.9.3: *A real quadratic form is positive definite if and only if $p_1 > 0, p_2 > 0, \ldots, p_n > 0$, that is, if and only if the leading principal minors of the matrix of the form are all positive.*

10.10   An Important Example

Consider the real quadratic form $q = X^{\mathsf{T}}(A^{\mathsf{T}}A)X$. This form is seen at once to be positive definite or semidefinite because $q = (AX)^{\mathsf{T}}AX$, which is a non-negative scalar since AX is a real vector.

For an illustration of Theorem 10.9.2, let A be $m \times n$ and let its rank be r. In addition, let the first r columns of A be linearly independent. Then the leading principal minors of the n-square matrix $A^{\mathsf{T}}A$ are given by

(10.10.1)

$$p_i = \det \begin{bmatrix} A_1^\mathsf{T} \\ \vdots \\ A_i^\mathsf{T} \end{bmatrix} [A_1, \ldots, A_i] = \det \begin{bmatrix} A_1^\mathsf{T} A_1 & \cdots & A_1^\mathsf{T} A_i \\ \vdots & & \\ A_i^\mathsf{T} A_1 & \cdots & A_i^\mathsf{T} A_i \end{bmatrix}, \quad i = 1, 2, \ldots, n.$$

For $i = 1, 2, \ldots, r$, this determinant is positive because it is a sum of squares of major determinants of $[A_1, \ldots, A_i]$, and not all these are zero because the columns are independent. However, if $i > r$, $p_i = 0$ since every such major determinant is zero because now the columns are dependent. Since the rank of $A^\mathsf{T}A$ cannot exceed r and since $p_r \neq 0$, the rank of $A^\mathsf{T}A$ is exactly r. Then, by Theorem 10.9.2, q is positive semidefinite, or positive definite if $r = n$, and $x_1, x_2, \ldots, x_r$ all actually appear in q. Indeed, since $A_j^\mathsf{T}A_j > 0$ for $j = 1, 2, \ldots, r$, it follows that $x_1^2, x_2^2, \ldots, x_r^2$ all appear with nonzero coefficients.

When A is of order n and nonsingular, the nonsingular transformation $Y = AX$ gives $X^\mathsf{T}A^\mathsf{T}AX = Y^\mathsf{T}Y = \sum_{i=1}^{n} y_i^2$, which again shows that the form is positive definite. This same reduction to a sum of n squares works even if A is singular, but in this case $A^\mathsf{T}A$ is only semidefinite. This illustrates the fact that a form of rank r may be written as a sum of more than r squares of linear forms. When this is done, however, the linear forms— the y's in this example—are *not linearly independent*.

Let $X = [x_{ij}]_{(n, p)}$ be a matrix of p sets of n numerical observations each, each column of X comprising a set of observations. Let $Y = [(x_{ij} - m_j)]_{(n, p)}$, where m_j is the mean of the observations in the jth column. Then

$$\frac{1}{n-1} Y^\mathsf{T}Y = \left[\frac{1}{n-1} \sum_{k=1}^{n} (x_{ki} - m_i)(x_{kj} - m_j) \right]_p$$

is known as the **covariance matrix** of the given observations. From the previous example, we see that this matrix is positive definite if the rank of Y is p, but that otherwise it is positive semidefinite.

The preceding discussion of the real matrix $A^\mathsf{T}A$ generalizes at once to the complex case. Given k vectors $A_1, A_2, \ldots, A_n$ of $\mathcal{U}^n$, the **Gramian matrix** of these vectors is defined to be the matrix of inner products

$$(10.10.2) \qquad G(A_1, A_2, \ldots, A_k) = \begin{bmatrix} A_1^*A_1 & \cdots & A_1^*A_k \\ \vdots & & \\ A_k^*A_1 & \cdots & A_k^*A_k \end{bmatrix} = A^*A,$$

where

$$A = [A_1, A_2, \ldots, A_k].$$

The **Gramian determinant** of these vectors is defined to be $\det G$. Since G is Hermitian, $\det G$ is real.

If $k > n$, $\det G = 0$ since G is a k-square matrix but its rank cannot exceed n. If $k \leq n$, $\det G$ is the sum of the products of corresponding majors of A^* and A. Since in this case, pairs of corresponding majors are conjugate complex numbers, their products are non-negative and we have

Theorem 10.10.1: *The Gramian determinant of k vectors of $\mathcal{U}^n$ is a non-negative real number and hence every Gramian matrix is positive semi-definite or positive definite.*

If now $A_1, A_2, \ldots, A_k$ are independent, each of A^* and A has rank k. For some pair of corresponding majors, $\det M^*$ and $\det M$, we therefore have $\det M^* \det M > 0$, so $\det G > 0$. Conversely, suppose that $\det G > 0$. Then for some pair of corresponding majors $\det M^*$ and $\det M$, we must have $\det M^* \det M > 0$. Since therefore $\det M \neq 0$, it follows that A has rank k and $A_1, A_2, \ldots, A_k$ are independent. Hence we have

Theorem 10.10.2: *A necessary and sufficient condition that k vectors A_1, $A_2, \ldots, A_k$ of $\mathcal{U}^n$ be independent is that their Gramian determinant be positive, that is, that their Gramian matrix be positive definite.*

10.11   Exercises

*1. Prove that, over the field of complex numbers, every quadratic form of rank r may be reduced to the form

$$z_1^2 + z_2^2 + \cdots + z_r^2$$

by a nonsingular transformation. (Begin by assuming the result of Theorem 10.6.1.)

*2. Prove that, over the complex field, two quadratic forms in n variables are equivalent if and only if they have the same rank.

3. Reduce by Kronecker's method:

(a) $X^\mathsf{T} \begin{bmatrix} 1 & 2 & 4 \\ 2 & 4 & 8 \\ 4 & 8 & 16 \end{bmatrix} X$,   (b) $X^\mathsf{T} \begin{bmatrix} 1 & -1 & 0 & 2 \\ -1 & 2 & 1 & -3 \\ 0 & 1 & 1 & -1 \\ 2 & -3 & -1 & 5 \end{bmatrix} X$.

4. Reduce by Lagrange's method:

(a) $X^\mathsf{T} \begin{bmatrix} 4 & 2 & 1 \\ 2 & 4 & 2 \\ 1 & 2 & 4 \end{bmatrix} X$,   (b) $X^\mathsf{T} \begin{bmatrix} 0 & 1 & 0 & 0 \\ 1 & 0 & 2 & 0 \\ 0 & 2 & 0 & 3 \\ 0 & 0 & 3 & 0 \end{bmatrix} X$.

*5. Show that Lagrange's reduction can always be effected by a single transformation whose determinant is 1. (*Hint:* Use an extra interchange of two variables and (or) a change of sign of one variable, if necessary.)

6. Show how the result of Exercise 5 is useful in the evaluation over a region R in $\mathscr{E}^n$ of the multiple integral

$$\iint_R \cdots \int \left(\sum_{i,j} \alpha_{ij} x_i x_j \right) dx_1 dx_2 \cdots dx_n .$$

7. Using Exercise 5, show how Lagrange's method of reduction may be used to evaluate the determinant of a symmetric matrix.

8. Diagonalize the forms

$$q_1(X) = 2x_1 x_2 + 2x_2 x_3 \qquad (n = 3),$$

$$q_2(X) = 2x_1 x_2 + 2x_2 x_3 + 2x_3 x_1 \qquad (n = 3),$$

$$q_3(X) = 2(x_1 x_2 + x_2 x_3 + \cdots + x_{n-1} x_n),$$

$$q_4(X) = 2(x_1 x_2 + x_2 x_3 + \cdots + x_{n-1} x_n + x_n x_1).$$

9. Prove that the number of positive characteristic roots of a real symmetric matrix A of rank r is equal to the index p of A, and that the number of negative characteristic roots is equal to the difference $r - p$.

10. Use the Gramian matrix to prove that if k nonzero vectors of $\mathscr{U}^n$ are mutually orthogonal, they are linearly independent.

11. Use the Gramian matrix to prove that if a nonzero vector X is orthogonal to each of k linearly independent vectors $A_1, A_2, \ldots, A_k$, then $X, A_1, A_2, \ldots, A_k$ are linearly independent.

10.12 Pairs of Quadratic Forms

In certain applications it is necessary to effect the simultaneous reduction of two real symmetric matrices or two real quadratic forms, at least one of which is, ordinarily, positive definite. We begin with the problem of finding the scalars λ and the vectors X which satisfy the equation

$$(10.12.1) \qquad\qquad AX = \lambda BX.$$

This is a generalization of the characteristic value problem of Chapter 9, to which this reduces if $B = I$. The solutions λ and X of this problem are respectively called the **characteristic roots** and the **characteristic vectors** of the pair of matrices A and B *in that order*.

As before, there will exist vectors X satisfying this equation if and only if λ is a root of the **characteristic equation**

$$(10.12.2) \qquad\qquad \det [A - \lambda B] = 0.$$

For our present purposes, we *assume that A is symmetric and that B is positive definite and symmetric.* Then there exists an orthogonal matrix U such that

$$U^{\mathsf{T}}BU = D[\mu_1, \mu_2, \ldots, \mu_n],$$

where $\mu_1, \mu_2, \ldots, \mu_n$ are the characteristic roots of B and are all positive, since B is positive definite. If we now put

$$R = D[\mu_1^{-1/2}, \mu_2^{-1/2}, \ldots, \mu_n^{-1/2}],$$

we have

$$R^{\mathsf{T}}(U^{\mathsf{T}}BU)R = R^{\mathsf{T}}D[\mu_1, \mu_2, \ldots, \mu_n]R = I,$$

which, by putting $S = UR$, we may write as

$$S^{\mathsf{T}}BS = I.$$

Now S is nonsingular and therefore we may replace (10.12.2) by the equivalent equation

$$\det S^{\mathsf{T}}[A - \lambda B]S = 0,$$

that is, by

(10.12.3) $\det[S^{\mathsf{T}}AS - \lambda I] = 0.$

Thus we have reduced our characteristic value problem to one of the basic type treated in Chapter 9.

In (10.12.3), $S^{\mathsf{T}}AS$ is a real symmetric matrix. Hence all its characteristic roots are real. We have, therefore,

Theorem 10.12.1: *The roots $\lambda_1, \lambda_2, \ldots, \lambda_n$ of the equation $\det[A - \lambda B] = 0$, where A and B are symmetric and B is positive definite, are all real.*

Next, since $S^{\mathsf{T}}AS$ is symmetric, there exists an orthogonal matrix Q which will diagonalize it:

$$Q^{\mathsf{T}}(S^{\mathsf{T}}AS)Q = D[\lambda_1, \lambda_2, \ldots, \lambda_n],$$

that is,

$$(SQ)^{\mathsf{T}}A(SQ) = D[\lambda_1, \lambda_2, \ldots, \lambda_n].$$

If we now transform B by the matrix SQ, we obtain

$$(SQ)^\mathsf{T} B(SQ) = Q^\mathsf{T}(S^\mathsf{T} BS)Q = Q^\mathsf{T} IQ = I,$$

since Q is orthogonal. If we write $V = SQ$, we may summarize thus:

Theorem 10.12.2: *If A and B are real matrices of order n and if A is symmetric and B is positive definite and symmetric, then there exists a real nonsingular matrix V such that $V^\mathsf{T} AV$ is diagonal and $V^\mathsf{T} BV$ is the identity matrix.*

Now consider the pair of quadratic forms $X^\mathsf{T} AX$ and $X^\mathsf{T} BX$ associated with the real symmetric matrices A and B. If we put $X = VY$, we have

$$X^\mathsf{T} AX = Y^\mathsf{T}(V^\mathsf{T} AV)Y = \sum_{i=1}^{n} \lambda_i y_i^2,$$

and

$$X^\mathsf{T} BX = Y^\mathsf{T}(V^\mathsf{T} BV)Y = \sum_{i=1}^{n} y_i^2.$$

Thus a restatement of the preceding theorem is

Theorem 10.12.3: *If $X^\mathsf{T} AX$ is an arbitrary quadratic form and if $X^\mathsf{T} BX$ is any positive definite quadratic form in the same number of variables, then there exists a nonsingular transformation $X = VY$ which reduces $X^\mathsf{T} AX$ to the form $\sum \lambda_i y_i^2$, where the λ's are the roots of the equation $\det[A - \lambda B] = 0$, and which reduces $X^\mathsf{T} BX$ to the unit form $\sum y_i^2$.*

In the theory of small vibrations, the quadratic form $X^\mathsf{T} AX$ is a positive definite form representing the potential energy, $X^\mathsf{T} BX$ is a positive definite form representing the kinetic energy, and the λ's are used to compute the "normal modes of vibration."

10.13 Values of Quadratic Forms

Consider an arbitrary real quadratic form $q(X) = X^\mathsf{T} AX$. It is often important to know certain properties of the set of values of $q(X)$ when X is a unit vector.

Let $U_1, U_2, \ldots, U_n$ be an orthonormal set of characteristic vectors of A, associated with the characteristic roots $\lambda_1, \lambda_2, \ldots, \lambda_n$ respectively, and let X be an arbitrary unit vector. Let θ_j denote the angle between U_j and X so that

(10.13.1) $$\cos \theta_j = U_j^\mathsf{T} X.$$

Also, let $U = [U_1, U_2, \ldots, U_n]$, and define Y by the equation $X = UY$ or $Y = U^{\mathsf{T}}X$. Then,

$$Y = \begin{bmatrix} U_1^{\mathsf{T}} \\ U_2^{\mathsf{T}} \\ \vdots \\ U_n^{\mathsf{T}} \end{bmatrix} X = \begin{bmatrix} \cos\theta_1 \\ \cos\theta_2 \\ \vdots \\ \cos\theta_n \end{bmatrix}.$$

Hence,

$$q(X) = X^{\mathsf{T}}AX = Y^{\mathsf{T}}U^{\mathsf{T}}AUY = Y^{\mathsf{T}}D[\lambda_1, \lambda_2, \ldots, \lambda_n]Y$$
$$= \sum_{j=1}^{n} \lambda_j \cos^2\theta_j.$$

We have, therefore, what is known as **Euler's theorem**:

Theorem 10.13.1: *The value of a quadratic form* $q(X) = X^{\mathsf{T}}AX$ *at a unit vector* X *is* $\sum \lambda_j \cos^2\theta_j$, *where the* λ*'s are the characteristic roots of* A, *and the angles* θ_j *are the angles between* X *and the vectors* U_j *of an orthonormal set of characteristic vectors of* A *corresponding to the* λ_j*'s respectively.*

Again, let $B_1, B_2, \ldots, B_n$ denote an arbitrary orthonormal reference system in $\mathscr{E}^n$, and let θ_{ij} denote the angle between B_i and U_j, where U_j is as defined above. Then, by the preceding theorem, $q(B_i) = \sum_{j=1}^{n} \lambda_j \cos^2\theta_{ij}$. Also, since the B's are an orthonormal reference system, $\sum_{i=1}^{n} \cos^2\theta_{ij} = 1$. Hence,

$$\sum_{i=1}^{n} q(B_i) = \sum_{i=1}^{n} \left(\sum_{j=1}^{n} \lambda_j \cos^2\theta_{ij} \right)$$
$$= \sum_{j=1}^{n} \lambda_j \left(\sum_{i=1}^{n} \cos^2\theta_{ij} \right) = \sum_{j=1}^{n} \lambda_j.$$

We thus have

Theorem 10.13.2: *The sum of the values of a real quadratic form at the* n *vectors of an arbitrary orthogonal reference system is equal to the sum of the characteristic roots of* A; *that is, it is equal to the trace of* A.

Now let the characteristic roots of A be indexed in descending order:

$$\lambda_1 \geq \lambda_2 \geq \cdots \geq \lambda_n,$$

and let a corresponding set of orthonormal characteristic vectors be U_1, $U_2, \ldots, U_n$. Let X be an arbitrary unit vector making the angle θ_i with U_i, so that

$$X^{\mathsf{T}} U_i = \cos \theta_i,$$

and, by Euler's theorem,

$$q(X) = \sum_{i=1}^{n} \lambda_i \cos^2 \theta_i.$$

Suppose we choose X so that $X^{\mathsf{T}} U_i = 0$ for $i = 1, 2, \ldots, m - 1$. Then $\cos \theta_i = 0$, $i = 1, 2, \ldots, m - 1$, and hence, for each such X,

$$q(X) = \lambda_m \cos^2 \theta_m + \cdots + \lambda_n \cos^2 \theta_n.$$

Because of the ordering of the λ's and because $\sum_{j=1}^{n} \cos^2 \theta_j = 1$, we now have

$$q(X) \leq \lambda_m(\cos^2 \theta_m + \cdots + \cos^2 \theta_n) = \lambda_m.$$

On the other hand, if we choose $X = U_m$, we have $\cos \theta_m = 1$, $\cos \theta_i = 0$, $i \neq m$. Hence, $q(U_m) = \lambda_m$. Thus the value λ_m is actually attainable. In summary, we have

Theorem 10.13.3: *Given the real quadratic form $q(X) = X^{\mathsf{T}} A X$, let the characteristic roots of A be $\lambda_1 \geq \lambda_2 \geq \cdots \geq \lambda_n$, and let an associated set of orthonormal characteristic vectors be $U_1, U_2, \ldots, U_n$. Then λ_m is the maximum value of q over the set of unit vectors which are orthogonal to $U_1, U_2, \ldots, U_{m-1}$. In particular, λ_1 is the maximum value of q over the set of all unit vectors.*

Thus appropriate knowledge of the values of a quadratic form gives us important knowledge about the characteristic roots of the corresponding symmetric matrix.

Another important aspect of the values of a quadratic function is simply the sign of $q(X)$. Let $q(X) = X^{\mathsf{T}} A X$ be a real quadratic form in n variables. Let the rank be r and the index be p. Let U be an orthogonal matrix such that

$$U^{\mathsf{T}} A U = D[\alpha_1, \alpha_2, \ldots, \alpha_p, \beta_1, \beta_2, \ldots, \beta_{r-p}, 0, 0, \ldots, 0],$$

where each α_i is positive and each β_j is negative. Now let

$$(10.13.2) \qquad X = U\tilde{X}, \qquad \tilde{X} = \sum_{i=1}^{p} b_i E_i.$$

If not all the b's are 0, that is, if $X \neq 0$, we have

$$X^\mathsf{T} A X = \tilde{X}^\mathsf{T} U^\mathsf{T} A U \tilde{X} = \left(\sum_{i=1}^{p} b_i E_i^\mathsf{T} \right) D \left(\sum_{i=1}^{p} b_i E_i \right)$$

$$= \sum_{i=1}^{p} \alpha_i b_i^2 > 0.$$

Thus the given quadratic form is positive for every nonzero vector X of the p-dimensional subspace $\mathcal{V}$ defined by (10.13.2).

Similarly, if we let

(10.13.3) $Y = U \tilde{Y}$,   where $\tilde{Y} = \sum_{j=1}^{r-p} c_j E_{j+p}$,

we obtain for any $Y \neq 0$,

$$Y^\mathsf{T} A Y = \tilde{Y}^\mathsf{T} U^\mathsf{T} A U \tilde{Y} = \left(\sum_{j=1}^{r-p} c_j E_{j+p}^\mathsf{T} \right) D \left(\sum_{j=1}^{r-p} c_j E_{j+p} \right)$$

$$= \sum_{j=1}^{r-p} \beta_j c_j^2 < 0.$$

The given quadratic form is therefore negative for every nonzero vector Y of the $(r - p)$-dimensional subspace $\mathcal{W}$ defined by (10.13.3).

Note next that, for arbitrary vectors X and Y of the two spaces just defined,

$$X^\mathsf{T} Y = \tilde{X}^\mathsf{T} U^\mathsf{T} U \tilde{Y} = \tilde{X}^\mathsf{T} \tilde{Y} = \left(\sum_{i=1}^{p} b_i E_i^\mathsf{T} \right) \left(\sum_{j=1}^{r-p} c_j E_{j+p} \right) = 0;$$

that is, the subspaces $\mathcal{V}$ and $\mathcal{W}$ are orthogonal. Hence, in the case when $r = n$, we have two orthogonal spaces whose dimensions total n so that $\mathcal{E}^n$ is the sum of these spaces. (It should not be overlooked that these spaces vary with the orthogonal transformation U). In summary, we have

Theorem 10.13.4: *Given a real quadratic form q of rank r and index p, there exist real spaces of dimension p on which q is positive if $p > 0$, and real spaces of dimension $r - p$ on which q is negative if $r - p > 0$. One can choose (but not uniquely) pairs of spaces, one of each kind, which are orthogonal. Any two such orthogonal spaces span $\mathcal{E}^n$ in the case when $r = n$.*

As a simple example in $\mathcal{E}^2$, consider the form $q = x_1^2 - x_2^2$, for which $p = 1$ and $r = n = 2$. This form is positive on the subspace $\mathcal{V}$ with basis E_1, and negative on the subspace $\mathscr{W}$ with basis E_2. E_1 is orthogonal to E_2 and, together, E_1 and E_2 span $\mathscr{E}^2$. The form is also positive on any subspace with basis $[a, b]^\mathsf{T}$, where $|a| > |b|$, and negative on any subspace with basis $[a, b]^\mathsf{T}$, where $|a| < |b|$. Suppose that $|a| > |b|$. Then $[a, b]^\mathsf{T}$ and $[-b, a]^\mathsf{T}$ span orthogonal spaces on which q is positive and negative respectively.

A simple and important application of values of quadratic forms occurs in the theory of maxima and minima of differentiable functions. Let $f(x_1, x_2, \ldots, x_n) \equiv f(X)$ be differentiable with continuous third partial derivatives. Then the point P_0 (endpoint of the vector X_0) is called a **critical point** of f if

$$\left(\frac{\partial f}{\partial x_i}\right)_{X_0} = 0, \qquad i = 1, 2, \ldots, n.$$

The symmetric matrix,

$$H(X_0) = \left[\frac{\partial^2 f}{\partial x_i \, \partial x_j}\right]_{X_0},$$

is called the **Hessian matrix of** f **at** X_0.

The point P_0 is called a **nondegenerate critical point of index** $n - p$ if the quadratic form $Y^\mathsf{T} H(X_0) Y$ has rank n and index p.

At any point P_0, the finite Taylor expansion up to the second order can be written in the form

$$f(X) = f(X_0) + \left[\frac{\partial f}{\partial x_1}, \frac{\partial f}{\partial x_2}, \ldots, \frac{\partial f}{\partial x_n}\right]_{X_0} \Delta X + \frac{1}{2}(\Delta X)^\mathsf{T} H(X_0)\Delta X + \sigma(|\Delta X|^3),$$

where $\Delta X = X - X_0$ and $\sigma(|\Delta X|^3)$ is a quantity which approaches 0 at least as fast as $|\Delta X|^3$. It follows that the approximate behavior of $f(X) - f(X_0)$ in a neighborhood of a nondegenerate critical point P_0 is the same as the behavior of $(\Delta X)^\mathsf{T} H(X_0)\Delta X$ in a neighborhood of $\Delta X = 0$, for here

$$f(X) - f(X_0) = \tfrac{1}{2}(\Delta X)^\mathsf{T} H(X_0)\Delta X + \sigma(|\Delta X|^3).$$

In particular:

(1) If $p = 0$, P_0 is a relative maximum point, for $H(X_0)$ is negative definite;
(2) if $p = n$, P_0 is a relative minimum point, for $H(X_0)$ is positive definite;
(3) if $0 < p < n$, P_0 is neither a relative maximum point nor a relative minimum point, for every neighborhood of P_0 contains points P, P' for which $f(X) > f(X_0)$ and $f(X') < f(X_0)$, as follows from Theorem 10.13.4.

10.14   Exercises

1. Prove that if A and B are both positive definite, symmetric matrices of order n, then all the roots of the equation $\det[A - \lambda B] = 0$ are positive. The positive square roots of $\lambda_1, \lambda_2, \ldots, \lambda_n$ are the normal modes (natural frequencies) referred to above.

2. Find the characteristic roots and vectors of the pair of positive definite matrices

$$A = \begin{bmatrix} 2 & 1 \\ 1 & 2 \end{bmatrix}, \qquad B = \begin{bmatrix} 2 & 1 \\ 1 & 1 \end{bmatrix}.$$

3. Find a linear transformation that will simultaneously diagonalize the quadratic forms

$$X^{\mathsf{T}} \begin{bmatrix} 1 & 2 \\ 2 & 1 \end{bmatrix} X \quad \text{and} \quad X^{\mathsf{T}} \begin{bmatrix} 2 & 1 \\ 1 & 2 \end{bmatrix} X.$$

4. Show that there is no linear transformation which will simultaneously diagonalize the quadratic forms

$$X^{\mathsf{T}} \begin{bmatrix} 1 & 0 \\ 0 & -1 \end{bmatrix} X \quad \text{and} \quad X^{\mathsf{T}} \begin{bmatrix} 0 & 1 \\ 1 & 0 \end{bmatrix} X.$$

5. Let A and B be symmetric n-square matrices, let B be positive definite, and let $A - B$ be positive definite or semidefinite. Prove that $\det A \geq \det B$ and that A is positive definite.

6. Define nonzero vectors X_1 and X_2 to be A-**orthogonal** if and only if $X_1^{\mathsf{T}} A X_2 = 0$, where A is a real, positive definite, symmetric matrix. Prove that if B is a real, symmetric matrix and if $BX_1 = \lambda_1 A X_1$, $BX_2 = \lambda_2 A X_2$, $\lambda_1 \neq \lambda_2$, then X_1 and X_2 are A-orthogonal.

7. Prove that if $Z = X + Y$ and if X and Y are A-orthogonal, then $Z^{\mathsf{T}} A Z = X^{\mathsf{T}} A X + Y^{\mathsf{T}} A Y$. Note that this generalizes the Pythagorean theorem in $\mathscr{E}^n$, to which this reduces if $A = I_n$.

8. Prove that mutually A-orthogonal vectors are necessarily linearly independent.

9. Let A be a real, positive definite, symmetric matrix. Define X to be an A-**unit vector** if and only if $X^{\mathsf{T}} A X = 1$. Given that $X_1, X_2, \ldots, X_k$ are a basis for a subspace $\mathscr{V}$ of $\mathscr{E}^n$, show how to construct an A-**orthonormal basis** (mutually A-orthogonal, A-unit vectors) for $\mathscr{V}$. (The ideas and results of Exercises 6, 7, 8, and 9 are useful in the theory of vibrations.)

10. Given

$$A = \begin{bmatrix} 2 & 1 & 0 \\ 1 & 2 & 1 \\ 0 & 1 & 2 \end{bmatrix}, \qquad X_1 = \begin{bmatrix} 1 \\ 2 \\ -1 \end{bmatrix}, \qquad X_2 = \begin{bmatrix} 2 \\ -1 \\ 0 \end{bmatrix},$$

   (a) determine all vectors Y which are A-orthogonal to X_1,
   (b) determine all vectors Y which are A-orthogonal to both X_1 and X_2.

11. Let $X_{m \times 1}$, $A_{m \times n}$, and $Y_{n \times 1}$ be over the complex field. Define X and Y to be **conjugate with respect to the operator** A if and only if $X^*AY = 0$. Show that a given vector Y of $\mathcal{U}^n$ is conjugate to every vector X of $\mathcal{U}^m$ if and only if $AY = 0$. (Note that A-conjugacy is a natural generalization of A-orthogonality.)

12. Show that the range of values of a real quadratic form over the set of all unit vectors remains the same in all orthonormal reference systems.

13. Show that in the notation of Theorem 10.13.3, λ_m is the *minimum* value assumed by q over the set of all unit vectors X such that

$$X^T U_{m+1} = 0, \ X^T U_{m+2} = 0, \dots, X^T U_n = 0.$$

In particular, λ_n is the minimum value of q over the set of all unit vectors.

14. Given the quadratic form in $\mathcal{E}^2$, $x_1^2 + 2x_1 x_2$, find a subspace $\mathcal{V}$ over which the form is positive, and an orthogonal subspace $\mathcal{W}$ over which the form is negative. Show that together these spaces span $\mathcal{E}^2$.

15. Proceed as in Exercise 14 for the quadratic form in $\mathcal{E}^3$, $x_1^2 + x_2^2 - 2x_2 x_3$.

16. Determine the nature of the critical points of the functions:
    (a) $4x_1 x_2 - x_1^2 - x_2^2 + 12x_1$,
    (b) $x_1^2 + x_2^2 + x_3^2 + x_1 x_2 + x_2 x_3 + 4x_1 - 8x_3$.

10.15 Bilinear Forms

In Chapter 8 we saw that every linear mapping from $\mathcal{F}^n$ to $\mathcal{F}^m$ could be represented in the form $Y = AX$ where the columns of A are the images of the elementary vectors of $\mathcal{F}^n$. We now consider a function that maps pairs of vectors onto scalars.

A **bilinear form** b over a field $\mathcal{F}$ is defined to be a scalar-valued function of two vector variables, X with domain $\mathcal{F}^m$ and Y with domain $\mathcal{F}^n$, with the following properties:

(a) To each pair of vectors X and Y it relates a unique scalar $b(X, Y)$ of $\mathcal{F}$.

(b) For each scalar α of $\mathcal{F}$, $b(\alpha X, Y) = b(X, \alpha Y) = \alpha b(X, Y)$.

(c) For vectors X, X_1, X_2 of $\mathcal{F}^m$ and Y, Y_1, Y_2 of $\mathcal{F}^n$,

$$b(X_1 + X_2, Y) = b(X_1, Y) + b(X_2, Y),$$

and

$$b(X, Y_1 + Y_2) = b(X, Y_1) + b(X, Y_2).$$

That is, b possesses the characteristic properties of linearity with respect to both of the variables X and Y.

Let $E_1, E_2, \dots, E_m$ denote the elementary vectors of $\mathcal{F}^m$ and E'_1, $E'_2, \dots, E'_n$ denote those of $\mathcal{F}^n$. Then, by (a), there exist unique scalars

a_{ij} such that $b(E_i, E'_j) = a_{ij}$, $i = 1, 2, \ldots, m$; $j = 1, 2, \ldots, n$. Then, since $X = \sum_{i=1}^{m} x_i E_i$ and $Y = \sum_{j=1}^{n} y_j E'_j$, repeated application of (b) and (c) yields

$$(10.15.1) \qquad b(X, Y) = \sum_{i=1}^{m} \sum_{j=1}^{n} x_i y_j a_{ij} = X^{\mathsf{T}} A Y,$$

where $A = [a_{ij}]_{m \times n}$. For example,

$$[x_1, x_2, x_3] \cdot \begin{bmatrix} 1 & 0 \\ 0 & 2 \\ -1 & -2 \end{bmatrix} \cdot \begin{bmatrix} y_1 \\ y_2 \end{bmatrix} = x_1 y_1 + 2x_2 y_2 - x_3 y_1 - 2x_3 y_2$$

is a bilinear form over the real field, or simply a **real bilinear form.**

The $m \times n$ matrix A which contains all the coefficients of a bilinear form is called the **matrix of the form** and if the x's and y's are independent variables, the rank of A is called the **rank of the form**. In the example just given, the rank of the form is 2. When the matrix of the form is symmetric, the form itself is called a **symmetric bilinear form**. The most important example of a symmetric bilinear form is the inner product $X^{\mathsf{T}} Y = X^{\mathsf{T}} I Y$, where $\mathscr{F}$ is the field of real numbers or a subfield thereof.

Since every $m \times n$ matrix may be used as the matrix of a bilinear form in $m + n$ variables, many definitions and theorems about matrices have simple counterparts in the theory of bilinear forms, as will appear in following paragraphs.

It is important to note that in (10.15.1), a_{ij} is the coefficient of the product $x_i y_j$, for by this observation we are enabled to go from the form to its matrix representation and vice versa *by inspection*. A useful example is provided by a form much used in statistics, the **covariance** of X and Y,

$$(10.15.2) \qquad \mathrm{cov}\,(X, Y) = \frac{1}{n-1} \sum_{j=1}^{n} (x_j - m_x)(y_j - m_y),$$

where

$$m_x = \frac{1}{n} \sum_{j=1}^{n} x_j \qquad \text{and} \qquad m_y = \frac{1}{n} \sum_{j=1}^{n} y_j$$

are the means of the x's and the y's respectively. This form may be regarded most simply as a bilinear form in the $2n$ "deviations" $x_j - m_x$ and $y_j - m_y$, its matrix being the identity matrix divided by $n - 1$. However, since $\sum (x_j - m_x) = 0$ and $\sum (y_j - m_y) = 0$, the deviations are not independent variables. To determine the rank of the form, we therefore expand the product in (10.15.2), sum the individual terms, and substitute for m_x and m_y, thus obtaining the expansion

(10.15.3) $\text{cov}(X, Y) = \dfrac{1}{n-1}\left(\dfrac{n-1}{n}\sum\limits_{j=1}^{n} x_j\, y_j - \dfrac{1}{n}\sum\limits_{j \neq k} x_j\, y_k\right).$

Hence the matrix of the covariance is

(10.15.4) $\dfrac{1}{n(n-1)}\begin{bmatrix} (n-1) & -1 & \cdots & -1 \\ -1 & (n-1) & \cdots & -1 \\ \vdots & & & \\ -1 & -1 & \cdots & (n-1) \end{bmatrix}_n ,$

which may be shown to have rank $n - 1$. Assuming the x's and y's to be independent, $n - 1$ is then the rank of the form.

10.16   The Equivalence of Bilinear Forms

Frequently it is necessary or desirable to introduce new variables into a bilinear form in place of X and Y, that is, to effect linear transformations of coordinates in the spaces $\mathscr{F}^m$ and $\mathscr{F}^n$.

We next investigate the effect of this operation and some of the results which can be accomplished in this way.

Let $X^{\mathsf{T}}AY$ be a bilinear form over a field $\mathscr{F}$. Let $X = B\tilde{X}$ and $Y = C\tilde{Y}$ be nonsingular linear transformations relating X and Y to new variables $\tilde{X}$ and $\tilde{Y}$, the matrices B and C also being over $\mathscr{F}$. Then we have

$$X^{\mathsf{T}}AY = (B\tilde{X})^{\mathsf{T}} A(C\tilde{Y}) = \tilde{X}^{\mathsf{T}}(B^{\mathsf{T}}AC)\tilde{Y},$$

which gives the representation of the form in the new reference systems. The matrix $B^{\mathsf{T}}AC$ has the same rank as A since B^{T} and C are nonsingular.

Alternatively, we can regard the transformations $X = B\tilde{X}$, $Y = C\tilde{Y}$ as nonsingular operators on their vector spaces, the effect of which is to take the bilinear form $X^{\mathsf{T}}AY$ into the new form $\tilde{X}^{\mathsf{T}}(B^{\mathsf{T}}AC)\tilde{Y}$, which always has the same value as the given one at the corresponding pair of vectors.

We have then, in summary,

Theorem 10.16.1: The rank of a bilinear form is the same in all reference systems. Nonsingular linear operators on $\mathscr{F}^m$ and $\mathscr{F}^n$ take a bilinear form over $\mathscr{F}$ into a bilinear form of the same rank and also over $\mathscr{F}$. The values of the two forms are always the same at corresponding pairs of vectors (X, Y) and $(\tilde{X}, \tilde{Y})$.

For example, let us apply the real transformations of coordinates

$$
X = \begin{bmatrix} 1 & 1 & 0 \\ 0 & 1 & 0 \\ 0 & 0 & -1 \end{bmatrix} \tilde{X} \quad \text{and} \quad Y = \begin{bmatrix} 1 & 1 & 0 \\ 0 & 1 & 0 \\ 0 & 0 & -1 \end{bmatrix} \tilde{Y}
$$

to the symmetric bilinear form

$$
X^T A Y = X^T \begin{bmatrix} 1 & -1 & 0 \\ -1 & 2 & 0 \\ 0 & 0 & 1 \end{bmatrix} Y = x_1 y_1 - x_1 y_2 - x_2 y_1 + 2 x_2 y_2 + x_3 y_3 .
$$

We obtain

$$
X^T A Y = \tilde{X}^T \begin{bmatrix} 1 & 0 & 0 \\ 1 & 1 & 0 \\ 0 & 0 & -1 \end{bmatrix} \cdot \begin{bmatrix} 1 & -1 & 0 \\ -1 & 2 & 0 \\ 0 & 0 & 1 \end{bmatrix} \cdot \begin{bmatrix} 1 & 1 & 0 \\ 0 & 1 & 0 \\ 0 & 0 & -1 \end{bmatrix} \tilde{Y}
$$

$$
= \tilde{X}^T \begin{bmatrix} 1 & 0 & 0 \\ 0 & 1 & 0 \\ 0 & 0 & 1 \end{bmatrix} \tilde{Y} = \tilde{x}_1 \tilde{y}_1 + \tilde{x}_2 \tilde{y}_2 + \tilde{x}_3 \tilde{y}_3 ,
$$

so that, in the new reference systems, the bilinear form is diagonal; that is, it has a diagonal matrix.

This example also illustrates the concept of equivalent bilinear forms. We shall say that two bilinear forms in $m + n$ variables, whose matrices A_1 and A_2 have elements in a field $\mathscr{F}$, are **equivalent** over $\mathscr{F}$ if and only if there exist nonsingular matrices B and C over $\mathscr{F}$, of orders m and n respectively, such that $B^T A_1 C = A_2$, that is, if and only if the matrices of the two forms are equivalent. If we write the forms as $X^T A_1 Y$ and $\tilde{X}^T A_2 \tilde{Y}$, the definition amounts to saying that the two forms are equivalent over $\mathscr{F}$ if and only if there exist nonsingular transformations $X = B\tilde{X}$ and $Y = C\tilde{Y}$ over $\mathscr{F}$ which transform the first form into the second. The reader should show that the inverse transformations will then carry the second form into the first, so that equivalence is actually symmetric in character, even though it is not symmetrically defined. Because the equivalence of forms as thus defined is identical to the equivalence of the corresponding matrices, it is in fact an equivalence relation as defined in Chapter 1.

If we choose to regard the equivalent forms as being related by linear transformations of coordinates, it follows that, in this case, *distinct but equivalent forms are representations of the same bilinear function but in different reference systems.*

In the event that the transformations are regarded as operators, two forms are equivalent if and only if there exist nonsingular operators on

$\mathscr{F}^m$ and $\mathscr{F}^n$ such that the forms always have equal values at corresponding pairs of vectors. Since two matrices over a field $\mathscr{F}$ are equivalent if and only if they have the same order and the same rank, we may conclude

Theorem 10.16.2: *Two bilinear forms over a field $\mathscr{F}$, each with an $m \times n$ matrix, are equivalent over $\mathscr{F}$ if and only if they have the same rank.*

In particular, every bilinear form $X^{\mathsf{T}}AY$ in $m + n$ variables and of rank r is equivalent to the **canonical form**

$$(10.16.1) \qquad \tilde{X}^{\mathsf{T}}\left[\begin{array}{c|c} I_r & 0 \\ \hline 0 & 0 \end{array}\right]\tilde{Y} = \tilde{x}_1\tilde{y}_1 + \tilde{x}_2\tilde{y}_2 + \cdots + \tilde{x}_r\tilde{y}_r.$$

In fact, if B and C are matrices such that $B^{\mathsf{T}}AC$ is the normal form of A, then the transformations $X = B\tilde{X}$ and $Y = C\tilde{Y}$ effect the reduction of $X^{\mathsf{T}}AY$ to the canonical form.

By determining first what transformations reduce each of two equivalent bilinear forms to the canonical form, we can determine by what transformations either may be transformed into the other.

10.17 Cogredient and Contragredient Transformations

Suppose that a bilinear form has a matrix A of order n and that X and Y are both n-vectors which belong to $\mathscr{F}^n$, the problem now being to determine the effect of a linear transformation of coordinates in $\mathscr{F}^n$ or of an operator on $\mathscr{F}^n$. That is, we now wish to subject both X and Y to the *same* transformation: $X = B\tilde{X}$ and $Y = B\tilde{Y}$. In this case, we say that X and Y are **transformed cogrediently**. The effect of a cogredient transformation of X and Y is to take the form $X^{\mathsf{T}}AY$ into the form $\tilde{X}^{\mathsf{T}}(B^{\mathsf{T}}AB)\tilde{Y}$. If A and B are over $\mathscr{F}$ and B in addition is nonsingular, the matrices A and $B^{\mathsf{T}}AB$ are equivalent over $\mathscr{F}$, but in a special way. We recognize this by introducing a special term. In general, if A_1, A_2 of order n are over $\mathscr{F}$ and if there exists an $n \times n$ nonsingular matrix B over $\mathscr{F}$ such that $B^{\mathsf{T}}A_1B = A_2$, then we call A_1 and A_2 **congruent** over $\mathscr{F}$. The reader may verify that congruence is an equivalence relation. Using this terminology, we summarize in

Theorem 10.17.1: *Two bilinear forms over a field $\mathscr{F}$ are equivalent under cogredient transformation of the variables if and only if their matrices are congruent over $\mathscr{F}$.*

When A over $\mathscr{F}$ is square and X and Y both belong to $\mathscr{F}^n$, then $X^{\mathsf{T}}AY$ is called a **bilinear form on** $\mathscr{F}^n$. In this case, X and Y are, of course, transformed cogrediently by any transformation of coordinates.

However, when A is square it is not necessary, nor is it always useful, to require that the vectors X and Y of a bilinear form be transformed cogrediently. Indeed, at times they are regarded as belonging to two distinct spaces $\mathscr{F}^n$ (for example, two planes), and in such a case they may be subjected to distinct linear transformations.

An instance of this which is of special importance is the following. We assume that X and Y need *not* be transformed cogrediently. Under this condition, we wish to know what transformations leave identically invariant the matrix of the canonical bilinear form

$$X^{\mathsf{T}} I_n Y = x_1 y_1 + x_2 y_2 + \cdots + x_n y_n.$$

If we put $X = B\tilde{X}$ and $Y = C\tilde{Y}$, we have

$$X^{\mathsf{T}} I_n Y = \tilde{X}^{\mathsf{T}} B^{\mathsf{T}} C \tilde{Y},$$

in which we require that $B^{\mathsf{T}} C = I_n$. From this it follows that B and C must be nonsingular and that $C = (B^{\mathsf{T}})^{-1}$. The transformations $X = B\tilde{X}$ and $Y = (B^{\mathsf{T}})^{-1}\tilde{Y}$ are called **contragredient transformations**.

In particular, if B is a real orthogonal matrix, we have $B^{\mathsf{T}} B = I_n$, or $B = (B^{\mathsf{T}})^{-1}$. Hence an orthogonal transformation is contragredient to itself. Consequently, when one subjects $\mathscr{E}^n$ to an orthogonal transformation of coordinates or to an orthogonal operator, the variables X and Y of a bilinear form $X^{\mathsf{T}} A Y$ are transformed both cogrediently and contragrediently, a fact which is of particular geometrical importance.

For bilinear forms with real, symmetric matrices we can prove the following analogues of Theorems 10.8.1 and 10.8.2:

Theorem 10.17.2: *Every real, symmetric bilinear form may be reduced by cogredient, nonsingular transformations to the form*

$$x_1 y_1 + \cdots + x_p y_p - x_{p+1} y_{p+1} - \cdots - x_r y_r$$

where p and r are uniquely determined.

Here p is the **index** and r is the rank of the form.

Theorem 10.17.3: *Two real bilinear forms with symmetric matrices of the same order are equivalent under real, nonsingular, cogredient transformations if and only if they may both be reduced to the same form:*

$$x_1 y_1 + x_2 y_2 + \cdots + x_p y_p - x_{p+1} y_{p+1} - \cdots - x_r y_r.$$

Thus the equivalence over the real field of the two bilinear forms also rests on the equality of the two numbers r (the rank) and p (the index) which are uniquely defined for each form.

The forms

$$y_1^2 + y_2^2 + \cdots + y_p^2 - y_{p+1}^2 - \cdots - y_r^2$$

and

$$(10.17.1) \quad x_1 y_1 + x_2 y_2 + \cdots + x_p y_p - x_{p+1} y_{p+1} - \cdots - x_r y_r$$

are called canonical forms for real quadratic and bilinear forms with symmetric matrices. Correspondingly,

$$(10.17.2) \quad \begin{bmatrix} I_p & 0 & 0 \\ \hline 0 & -I_{r-p} & 0 \\ \hline 0 & 0 & 0 \end{bmatrix}$$

is a canonical form for a real, symmetric matrix A. That is, given A, there always exists a real, nonsingular matrix B such that $B^{\mathsf{T}} A B$ is in the stated canonical form.

The normal form,

$$\begin{bmatrix} I_r & 0 \\ \hline 0 & 0 \end{bmatrix},$$

developed in Chapter 6 for an arbitrary matrix of rank r over a field $\mathscr{F}$ is another canonical form. There are other canonical forms for matrices and for quadratic and bilinear forms, each designed to exhibit certain important properties thereof.

10.18   Exercises

*1. Given that the products $B_1 A_1 C_1$ and $B_2 A_2 C_2$ are in the same normal form and that the B's and the C's are nonsingular, write the transformations that will take the bilinear form $X^{\mathsf{T}} A_1 Y$ into the bilinear form $\tilde{X}^{\mathsf{T}} A_2 \tilde{Y}$.

2. Use the methods and theorems developed in Chapter 6 to reduce the bilinear form

$$(a) \quad X^{\mathsf{T}} \begin{bmatrix} 1 & 1 & 0 & 3 \\ 2 & 0 & 2 & 2 \\ 3 & -2 & 5 & -1 \end{bmatrix} Y$$

to the canonical form defined in Section 10.16.

3. Supply the details of the reduction of (10.15.2) to (10.15.3).

4. Under what conditions are the real transformations $X = B\tilde{X}$ and $Y = C\tilde{Y}$ simultaneously cogredient and contragredient?

5. Show that the matrix (10.15.4) has rank $n - 1$.

6. Show that, if A is symmetric, then $X^T A Y = Y^T A X$.

7. Show that, if $X^T A Y$ is symmetric, and if X and Y are transformed cogrediently, then the new form is also symmetric.

8. Determine the transformation contragredient to

$$X = \begin{bmatrix} 1 & 0 & 1 \\ 0 & 1 & 1 \\ 1 & 1 & 1 \end{bmatrix} \tilde{X}.$$

9. Reduce the bilinear form

$$X^T \begin{bmatrix} 1 & 1 & 1 & 2 \\ 2 & 2 & 2 & 3 \\ 3 & 3 & 3 & 4 \end{bmatrix} Y$$

to the canonical form defined in (10.17.1).

10. Show that a real $n \times n$ matrix A is symmetric if and only if $X^T A Y = Y^T A X$ for all real n-vectors X and Y.

11. Let $X^T A Y$ be a *real* bilinear form with A of order n and nonsingular. Then $X^T A^{-1} Y$ is called the **reciprocal bilinear form** of the given one. Show that

$$X^T A^{-1} Y = -\det \begin{bmatrix} 0 & | & X^T \\ \text{---} & | & \text{---} \\ Y & | & A \end{bmatrix} \det A^{-1}.$$

Show also that if reciprocal bilinear forms are transformed cogrediently by the same orthogonal transformation, reciprocal bilinear forms result.

12. Show that an identity in the y's, $Y^T A X \equiv Y^T B$, where X, Y, and B are vectors, can hold if and only if X is a solution of the system of equations $AX = B$.

13. Prove that the matrix A of a bilinear form on $\mathscr{F}^n$ is symmetric if and only if $X^T A Y = Y^T A X$ for all X and Y of $\mathscr{F}^n$.

14. If B is a square matrix over $\mathscr{F}$ such that $B^T A B = A$, then the operator with matrix B is said to **preserve** the bilinear form $X^T A Y$. Show that the set of all nonsingular operators over $\mathscr{F}$ which preserve a fixed bilinear form constitutes a group. What group is it when $A = I$?

15. Let A, B be $m \times n$ matrices over a field $\mathscr{F}$. Show that tr $(A^T B)$ is a symmetric bilinear form in the elements of A and B.

16. Prove that the set of all bilinear forms over a field $\mathscr{F}$, which have $m \times n$ matrices, constitutes a vector space. What is its dimension? What set of bilinear forms constitutes a particularly simple basis for this space?

17. Let $L_1(X)$ and $L_2(Y)$ denote linear forms over $\mathscr{F}$. Show that $L_1(X)L_2(Y)$ is a bilinear form on $\mathscr{F}^n$, and that every bilinear form on $\mathscr{F}^n$ is representable as a sum of at most n such products. What is the significance of the *least* number required for such a representation?

18. Given the bilinear form $b = X^{\mathsf{T}}AY$, in which A is square and X and Y belong to $\mathscr{F}^n$, we say that b is **nonsingular** if and only if A is nonsingular. Prove that b is nonsingular if and only if for each $X \neq 0$ there exists a Y such that $X^{\mathsf{T}}AY \neq 0$ and for each $Y \neq 0$ there exists an X such that $X^{\mathsf{T}}AY \neq 0$.

*19. Let f be a function of n vector variables whose common domain is $\mathscr{F}^n$. Assume f has these properties for all X of $\mathscr{F}^n$:

(a) $f(X_1, X_2, \ldots, X_n)$ is a scalar in $\mathscr{F}$.

(b) $f(X_1, \ldots, X_{j-1}, \alpha X_j, X_{j+1}, \ldots, X_n) = \alpha f(X_1, \ldots, X_j, \ldots, X_n)$ for $j = 1, 2, \ldots,$ n and α any scalar in $\mathscr{F}$.

(c) $f(X_1, \ldots, X_{j-1}, X_j + X_j', X_{j+1}, \ldots, X_n) = f(X_1, \ldots, X_{j-1}, X_j, X_{j+1}, \ldots, X_n) + f(X_1, \ldots, X_{j-1}, X_j', X_{j+1}, \ldots, X_n), j = 1, 2, \ldots, n.$

(d) $f(E_1, E_2, \ldots, E_n) = 1.$

(e) If $X_1, X_2, \ldots, X_n$ are linearly dependent, $f(X_1, X_2, \ldots, X_n) = 0.$

Prove that:

(α) $f(E_1, \ldots, E_{j-1}, X, E_{j+1}, \ldots, E_n) = x_j.$

(β) $f(E_{j_1}, E_{j_2}, \ldots, E_{j_n}) = \varepsilon_{j_1 j_2 \ldots j_n}.$

(γ) $f(X_1, X_2, \ldots, X_n) = \det[X_1, X_2, \ldots, X_n].$

Properties (a), (b), (c) say that f is a **multilinear form** over $\mathscr{F}$. Properties (d), (e) restrict the form enough to make it, in fact, the determinant. Prove that (e) can be replaced by (e'): If any two of $X_1, X_2, \ldots, X_n$ are the same, $f(X_1, X_2, \ldots, X_n) = 0$.

*20. Let $q(X)$ denote a non-negative definite quadratic form $X^{\mathsf{T}}AX$, and let $b(X, Y)$ denote the associated bilinear form $X^{\mathsf{T}}AY$. Prove that, for all real t, X, and Y,

$$q(X + tY) = t^2 q(Y) + 2t\, b(X, Y) + q(X) \geq 0,$$

and, hence, that the **Cauchy-Schwarz inequality**

$$b(X, Y)^2 \leq q(X)q(Y)$$

holds. To what does this reduce when $A = I$?

10.19 Hermitian Forms

There is an analogue of the real quadratic form which is useful in applications in which the complex number field is basic. It is the **Hermitian form**, a function from $\mathscr{C}^n$ to $\mathscr{C}$, defined by

(10.19.1) $$h = X^*HX = \sum_{i,j=1}^{n} h_{ij} \bar{x}_i x_j,$$

where H is a Hermitian matrix and the components of X are in the complex field. If X and H are real, the Hermitian form (10.19.1) reduces to a real quadratic form as a special case. This explains why many results here parallel those of earlier sections. The rank of H is called the **rank of the form**. The form is called **singular** or **nonsingular** according as its rank is less than n or equal to n respectively.

Since h is a scalar, we have $h = h^*$. Hence, since H is Hermitian,

$$\bar{h} = h^* = (X^*HX)^* = X^*HX = h.$$

Thus h, being equal to its own conjugate, is real for every choice of X.

Conversely, suppose h is real for every choice of X. Then $h = \bar{h} = h^*$, so that we have the identity

$$X^*HX \equiv (X^*HX)^* \equiv X^*H^*X.$$

Hence $H^* = H$ and H is Hermitian (see Exercise 6, Section 10.21). Summing up, we have

Theorem 10.19.1: *A matrix H is Hermitian if and only if the form X^*HX is real for every choice of the vector X.*

In Chapter 9 we saw that for every Hermitian matrix H, there exists a unitary matrix U such that

$$U^*HU = D[\lambda_1, \lambda_2, \ldots, \lambda_n],$$

where the λ's are the characteristic roots of H and are all real. For a Hermitian form (10.19.1) this implies

Theorem 10.19.2: *By a suitable unitary transformation $X = UY$, X^*HX can be reduced to the form*

$$Y^*D[\lambda_1, \lambda_2, \ldots, \lambda_n]Y = \sum_{j=1}^{n} \lambda_j \bar{y}_j y_j,$$

where the λ's are the characteristic roots of the Hermitian matrix H.

It is possible to use other nonsingular transformations to reduce a given Hermitian form to the diagonal form

$$Y^*D[g_1, g_2, \ldots, g_n]Y = \sum_{j=1}^{n} g_j \bar{y}_j y_j,$$

where the g's are not necessarily the characteristic roots of H. Thus, for example, the nonunitary transformation

$$X = \begin{bmatrix} 1 & -i \\ 0 & 1 \end{bmatrix} Y$$

applied to the Hermitian form

$$X^* \begin{bmatrix} 1 & i \\ -i & 0 \end{bmatrix} X$$

with characteristic roots $(1 \pm \sqrt{5})/2$ yields the Hermitian form

$$Y^* \begin{bmatrix} 1 & 0 \\ 0 & -1 \end{bmatrix} Y = \bar{y}_1 y_1 - \bar{y}_2 y_2,$$

as the reader may verify.

Concerning such transformations, we have

Theorem 10.19.3: *If X^*HX is Hermitian and if the substitution $X = AY$ yields $X^*HX = Y^*(A^*HA)Y = \sum_{j=1}^{n} g_j \bar{y}_j y_j$, then the g's are all real.*

Indeed, $Y^*(A^*HA)Y$ is also Hermitian and is, therefore, real for all Y. If $y_j = 1$ and all other y's $= 0$, then $Y^*(A^*HA)Y = g_j$. Thus all the g_j's must be real.

If A in the preceding theorem is *nonsingular*, we define the number p of positive terms in $\sum g_j \bar{y}_j y_j$ to be the **index** of the form. The number of nonzero g's is evidently the rank r of the form. The **signature** is then defined to be $2p - r$. As in the case of quadratic forms, **Sylvester's law of inertia** applies:

Theorem 10.19.4: *No matter by what nonsingular transformation a given Hermitian form is reduced to a form*

$$g_1 \bar{z}_1 z_1 + \cdots + g_p \bar{z}_p z_p - g_{p+1} \bar{z}_{p+1} z_{p+1} - \cdots - g_r \bar{z}_r z_r,$$

where the g's are all positive, the integers p and r will be the same.

The proof of this theorem is like that of Theorem 10.8.1 and is left to the reader to supply in detail.

We define two Hermitian forms X^*H_1X and Y^*H_2Y to be **equivalent** if and only if there exists a nonsingular transformation $X = AY$ such that $X^*H_1X \equiv Y^*(A^*H_1A)Y \equiv Y^*H_2 Y$, that is, if and only if for some nonsingular A, $H_2 = A^*H_1A$ (see Exercise 6, Section 10.21). Then, analogously to Theorem 10.8.2, we have

Theorem 10.19.5: *Two Hermitian forms are equivalent if and only if they have the same rank and the same index.*

10.20   Definite Hermitian Forms

A *nonsingular* Hermitian form is called **positive definite** if and only if its rank and its index are equal; that is, $p = r = n$. A *singular* Hermitian form is called **positive semidefinite** if and only if its rank and its index are equal; that is, $p = r < n$.

Concerning definite Hermitian forms, we have the following theorems, all proved in much the same way as the corresponding theorems for quadratic forms:

Theorem 10.20.1: *If X^*HX is positive definite (semidefinite), then its value is > 0 (≥ 0) for all nonzero vectors X.*

Theorem 10.20.2: *If X^*HX is positive definite, then* det H, *and every principal minor determinant of H, is positive.*

Theorem 10.20.3: *If X^*HX is positive semidefinite, every principal minor determinant of H is ≥ 0.*

Theorem 10.20.4: *If X^*HX is positive semidefinite and x_i actually appears in X^*HX, then $h_{ii} > 0$.*

Theorem 10.20.5: *A Hermitian matrix H is positive definite if and only if any one of the following conditions is satisfied:*

(a) *A^*HA is positive definite for arbitrary, nonsingular A;*
(b) *H^p is positive definite for every integer p;*
(c) *there exists a nonsingular matrix A such that $H = A^*A$;*
(d) *the principal minor determinants $p_0, p_1, \ldots, p_n$ are all positive.*

10.21   Exercises

1. Prove Theorems 10.19.4 and 10.19.5 in detail.
2. Write proofs for Theorems 10.20.1 through 10.20.5.
3. Show that every positive definite Hermitian form may be reduced to the form $\sum_{j=1}^{n} |z_j|^2$ by a nonsingular transformation.
4. Determine the unitary transformation which will diagonalize the Hermitian form

$$X^* \begin{bmatrix} 1 & \sqrt{42}i \\ -\sqrt{42}i & 2 \end{bmatrix} X.$$

5. Determine whether or not the matrix

$$\begin{bmatrix} 4 & i & 0 \\ -i & 8 & -i \\ 0 & i & 4 \end{bmatrix}$$

is positive definite

6. Prove in detail that $X^*H_1X = X^*H_2X$ for all vectors X if and only if $H_1 = H_2$.

7. Prove that if H is positive definite and Hermitian, then $H^{1/p}$ is defined for each positive integer p.

8. Prove that for arbitrary A, the matrix A^*A has a Hermitian square root.

*9. Let H be a non-negative definite Hermitian matrix such that

$$H^2 = D[\lambda_1, \lambda_2, \ldots, \lambda_n].$$

Prove that each λ_j is ≥ 0 and that

$$H = D[\lambda_1^{1/2}, \lambda_2^{1/2}, \ldots, \lambda_n^{1/2}].$$

Thus a non-negative definite diagonal matrix has a unique non-negative definite square root and this root is also diagonal.

10. Prove that a complex matrix H of order n is a positive definite Hermitian matrix if and only if C^*HC is a positive definite Hermitian matrix for every nonsingular complex matrix C of order n.

11. Restate the theorems of Section 10.12 for Hermitian matrices and prove the results.

*12. Show that, if $A = B^2 = C^2$, where B and C are non-negative definite and Hermitian, then $B = C$; that is, if a non-negative definite Hermitian square root exists, it is unique. Exercise 9 will help.

13. Investigate for definiteness where $\omega^3 = 1$ but $\omega \neq 1$:

$$\begin{bmatrix} 1 & \omega & \omega^2 \\ \omega^2 & 1 & \omega \\ \omega & \omega^2 & 1 \end{bmatrix}.$$

*14. Represent the Hermitian form X^*A^*AX as a sum of squares of absolute values of linear forms in the x's. Here A is an arbitrary $m \times n$ matrix. (*Hint:* Let $Y = AX$, etc.)

15. Let X and Y be vectors of $\mathscr{C}^n$. Let the complex matrix H be Hermitian. Define X and Y to be **conjugate** with respect to H if and only if $X^*HY = 0$. Prove that a fixed Y is conjugate to *all* X if and only if $HY = 0$. To what does conjugacy reduce if $H = I$?

*16. Given that A is nonsingular, prove that there exist unique positive definite Hermitian matrices R and S and a unique unitary matrix U such that $A = RU = US$. This is called the **polar representation** of A. (First prove that if $A = RU = US$ with R, S, U, as described, then $AA^* = R^2$ and $A^*A = S^2$, so that R and S, and hence U, are unique (Exercise 12). To prove the converse, define R by $A^*A = R^2$ (Exercise 8), U by $A = RU$. Show that U is unitary and then define $S = U^*RU$. Show that S is positive definite and Hermitian and that $A = US$.)

APPENDIX
I

The Notations Σ and Π

Through his previous work in mathematics, the reader may already have become somewhat familiar with the notations $\sum$ and $\prod$ for sums and products respectively, but there are operations with these symbols which we use rather frequently in this book and which may well be new to him. For his convenience, we therefore provide a discussion of these symbols.

The $\sum$ Notation

I.1 Definitions

The $\sum$ notation is simply a shorthand method for designating sums. Thus, for example, we write

$$x_1 + x_2 + x_3 + x_4 + x_5 = \sum_{j=1}^{5} x_j.$$

Here j is a variable ranging over the integers 1, 2, 3, 4, and 5. The symbols $j = 1$ below the $\sum$ sign indicate that 1 is the initial value taken on by j, and the 5 written above the $\sum$ sign indicates that 5 is the terminal value of j. We call j the **index of summation**. The **summand**, x_j, is a function of j which takes on the values x_1, x_2, x_3, x_4, and x_5 respectively, as j takes on successively the values 1, 2, 3, 4, and 5. Finally, the $\sum$ sign denotes the fact that the values x_1, x_2, x_3, x_4, and x_5 taken on by x_j are to be *added*. The entire symbol $\sum_{j=1}^{5} x_j$ is read, "the summation of x_j as j ranges from 1 to 5."

In the same way, we have

$$x_6 + x_7 + x_8 = \sum_{j=6}^{8} x_j,$$

where now the initial value of j is 6 and the terminal value is 8. Combining these two examples we have

$$x_1 + x_2 + x_3 + x_4 + x_5 + x_6 + x_7 + x_8 = \sum_{j=1}^{5} x_j + \sum_{j=6}^{8} x_j,$$

so

$$\sum_{j=1}^{8} x_j = \sum_{j=1}^{5} x_j + \sum_{j=6}^{8} x_j.$$

Our first example above is an illustration of the basic definition:

(I.1.1) $$\sum_{j=1}^{n} x_j = x_1 + x_2 + \cdots + x_n.$$

Our third example above is an illustration of the theorem:

(I.1.2) $$(x_1 + x_2 + \cdots + x_p) + (x_{p+1} + \cdots + x_n) = \sum_{j=1}^{n} x_j = \sum_{j=1}^{p} x_j + \sum_{j=p+1}^{n} x_j.$$

A familiar function of n quantities x_1, x_2, ..., x_n is their "average" or arithmetic mean m_x, their sum divided by n. Using the above notation, we can write

$$m_x = \frac{x_1 + x_2 + \cdots + x_n}{n} = \frac{1}{n} \sum_{j=1}^{n} x_j.$$

The compactness of the $\sum$ notation, as here demonstrated, is one indication of its value.

In the above examples, the values actually represented by $x_1, x_2, \ldots, x_n$, of course, have to be given before the sums can be evaluated. Sometimes, however, the notation is such as to designate the values of the various terms. An example of such a sum is

$$1^2 + 2^2 + 3^2 + 4^2 + 5^2 = \sum_{k=1}^{5} k^2,$$

or, more generally,

$$1^2 + 2^2 + \cdots + n^2 = \sum_{k=1}^{n} k^2.$$

Here the index of summation k ranges over the values 1, 2, 3, 4, and 5 in the first case, while the summand k^2 ranges over the values 1^2, 2^2, 3^2, 4^2, and 5^2. In the second case the range of the index k is from 1 to n; that of the summand k^2 is from 1^2 to n^2, inclusive.

It should also be pointed out that sometimes the initial value of the summation index is zero or a negative integer. For example,

$$\sum_{j=0}^{k} \frac{1}{2^j} = \frac{1}{2^0} + \frac{1}{2^1} + \cdots + \frac{1}{2^k}$$

and

$$\sum_{j=-n}^{n} a_j x^j = a_{-n} x^{-n} + a_{-n+1} x^{-n+1} + \cdots + a_{-1} x^{-1}$$
$$+ a_0 x^0 + a_1 x^1 + a_2 x^2 + \cdots + a_n x^n \qquad (x \neq 0).$$

As a final illustration, we recall that infinite series are also commonly written with a $\sum$ sign. Thus, for example, we might have

$$\frac{1}{1^p} + \frac{1}{2^p} + \frac{1}{3^p} + \cdots + \frac{1}{n^p} + \cdots = \sum_{n=1}^{\infty} \frac{1}{n^p}$$

or

$$\frac{x}{1+2} + \frac{2x^2}{1+2^2} + \frac{3x^3}{1+2^3} + \cdots + \frac{nx^n}{1+2^n} + \cdots = \sum_{n=1}^{\infty} \frac{nx^n}{1+2^n}.$$

In each case we can obtain the first three terms on the left by substituting $n = 1, 2, 3$ respectively into the **general term** of the series, that is, the term containing the index n which appears on both the left and the right in the appropriate equation. As many more terms as may be desired may, of course, be found in the same way. Here the $\sum$ sign denotes a *purely formal sum* which may or may not represent a number depending on whether the series does or does not converge.

I.2   Exercises

1. Given that $x_1 = -2$, $x_2 = 1$, $x_3 = -1$, $x_4 = 3$, $x_5 = 7$, $x_6 = -8$, find

$$\sum_{j=1}^{6} x_j, \qquad \sum_{j=1}^{6} x_j^2, \qquad \sum_{j=1}^{6} (2x_j + 3), \qquad \text{and} \qquad \sum_{j=1}^{6} (x_j + 2)(x_j - 2).$$

2. Rewrite in the $\sum$ notation:

(a) $2t + 4t^2 + 8t^3 + 16t^4 + 32t^5 + 64t^6$.
(b) $1 + 3 + 5 + \cdots + (2n - 1)$.
(c) $1 \cdot 2 + 2 \cdot 3 + \cdots + n(n + 1)$.
(d) $(x_1 - m_x)(y_1 - m_y) + (x_2 - m_x)(y_2 - m_y) + \cdots + (x_n - m_x)(y_n - m_y)$.

3. Rewrite in the ordinary notation:

(a) $\displaystyle\sum_{k=0}^{5} k(k - 1)$; $\displaystyle\sum_{k=2}^{5} k(k - 1)$; $\displaystyle\sum_{k=-5}^{5} k(k - 1)$.

(b) $\displaystyle\sum_{j=1}^{n} a_j x_j$; $\displaystyle\sum_{j=1}^{n} a_j x^j$.

(c) $\displaystyle\sum_{n=0}^{\infty} \frac{x^n}{n!}$ (0! is defined to be 1, in case you have forgotten, and $x^0 = 1$ here).

(d) $\displaystyle\sum_{n=0}^{\infty} \left(\frac{x^n}{n!} + n(n - 1)(n - 2) \right)$. [Compare with (c).]

(e) $\displaystyle\sum_{n=0}^{\infty} \left(\frac{x^n}{n!} \right)(1 + \sin n(n - 1)(n - 2)x)$. [Compare with (d) and (c).]

4. Show that

(a) $\left(\displaystyle\sum_{j=1}^{n} x_j \right) + x_{n+1} = \displaystyle\sum_{j=1}^{n+1} x_j$,

(b) $\displaystyle\sum_{j=p+1}^{n} x_j = \displaystyle\sum_{j=1}^{n} x_j - \displaystyle\sum_{j=1}^{p} x_j$, $n \geq p + 1$,

(c) $\displaystyle\sum_{j=1}^{k} x_j + \displaystyle\sum_{j=1}^{n-k} x_{k+j} = \displaystyle\sum_{j=1}^{n} x_j$.

I.3   Basic Rules of Operation

In each of the examples given in Section I.1, the symbol used for the index of summation is entirely arbitrary, so it is called a **dummy index**. Thus we have, for example,

$$x_1 + x_2 + \cdots + x_n = \sum_{i=1}^{n} x_i = \sum_{j=1}^{n} x_j = \sum_{p=1}^{n} x_p = \cdots,$$

$$1^2 + 2^2 + \cdots + n^2 = \sum_{j=1}^{n} j^2 = \sum_{k=1}^{n} k^2 = \sum_{v=1}^{n} v^2 = \cdots.$$

There is another kind of arbitrariness in the summation index; this is indicated by the following examples, which the student should examine carefully.

$$\sum_{j=1}^{n} x_j = \sum_{j=0}^{n-1} x_{j+1} = \sum_{j=2}^{n+1} x_{j-1} = \cdots$$

and

$$\sum_{n=0}^{\infty} \frac{x^n}{n!} = \sum_{n=1}^{\infty} \frac{x^{n-1}}{(n-1)!} = \sum_{n=-1}^{\infty} \frac{x^{n+1}}{(n+1)!} = \cdots.$$

Here we have altered the initial value of the index of summation, but we have altered the function being summed in a compensating way, so the net sum remains unaltered. Can you write in words a rule for how this is to be done? This sort of shift in the range of summation is often useful.

Let us consider again the sum

(I.3.1)
$$\sum_{j=1}^{n} x_j = x_1 + x_2 + \cdots + x_n.$$

If each of the x's here is equal to the same fixed quantity c, we have

$$\sum_{j=1}^{n} x_j = c + c + \cdots + c = nc,$$

or, as we write it in this case,

(I.3.2)
$$\sum_{j=1}^{n} c = nc.$$

For example,

$$\sum_{j=1}^{5} 10 = 50.$$

Equation (I.3.2) is, in fact, a *definition* of the symbol $\sum_{j=1}^{n} c$, which is a priori meaningless since the constant c does not depend on the index of summation j.

Next let us suppose that in (I.3.1) we have $x_j = ky_j$, where k is a constant. Then

$$\sum_{j=1}^{n} x_j = \sum_{j=1}^{n} (ky_j) = ky_1 + ky_2 + \cdots + ky_n = k(y_1 + y_2 + \cdots + y_n)$$

$$= k \sum_{j=1}^{n} y_j.$$

Thus we have our second basic rule,

(I.3.3) $$\sum_{j=1}^{n} (ky_j) = k \sum_{j=1}^{n} y_j.$$

Finally, let us suppose that $x_j = y_j + z_j$ in (I.3.1). Then we have

$$\sum_{j=1}^{n} x_j = \sum_{j=1}^{n} (y_j + z_j) = (y_1 + z_1) + (y_2 + z_2) + \cdots + (y_n + z_n)$$

$$= (y_1 + y_2 + \cdots + y_n) + (z_1 + z_2 + \cdots + z_n) = \sum_{j=1}^{n} y_j + \sum_{j=1}^{n} z_j,$$

which gives our third basic rule,

(I.3.4) $$\sum_{j=1}^{n} (y_j + z_j) = \sum_{j=1}^{n} y_j + \sum_{j=1}^{n} z_j.$$

We make one more observation in this section. When there is no possible misinterpretation, the index of summation is often omitted. Thus we write simply $\sum x$ in place of $\sum_{j=1}^{n} x_j$ if the range of summation is clearly indicated by the context. Similarly, we could write $\sum x^2 - (\sum x)^2$ in place of $\sum_{j=1}^{n} x_j^2 - (\sum_{j=1}^{n} x_j)^2$, and so on.

I.4 Exercises

1. Use (I.3.3) and (I.3.4) to show that

$$\sum_{j=1}^{n}(ax_j + by_j) = a\sum_{j=1}^{n}x_j + b\sum_{j=1}^{n}y_j.$$

2. Show that

$$\sum_{j=1}^{n}x_j(x_j - 1) = \sum_{j=1}^{n}x_j^2 - \sum_{j=1}^{n}x_j,$$

and that

$$\sum_{j=1}^{n}(x_j - 1)(x_j + 1) = \left(\sum_{j=1}^{n}x_j^2\right) - n.$$

3. Using the fact that $m_x = (\sum x)/n$, (I.3.2), and Exercise 1, show that $\sum(x_j - m_x) = 0$. (Fill in the missing indices of summation first. The differences $x_j - m_x$ are called the *deviations of the x's from their mean*. You are thus to prove that the sum of the deviations of a set of quantities from their mean is zero.)

4. Show that

(a) $\sum_{j=1}^{k}(x_j + 1)^2 f_j = \sum_{j=1}^{k}x_j^2 f_j + 2\sum_{j=1}^{k}x_j f_j + \sum_{j=1}^{k}f_j,$

(b) $\sum_{j=1}^{n}(x_j - m_x)^2 = \left(\sum_{j=1}^{n}x_j^2\right) - nm_x^2.$

These two results are used in deriving various formulas in statistics.

I.5 Finite Double Sums

We shall now consider the matter of **double sums**. Let us suppose that we have a set of nm quantities U_{ij}, where $i = 1, 2, \ldots, n$ and $j = 1, 2, \ldots, m$. We arrange these in a rectangular pattern, thus:

$$\begin{matrix} U_{11} & U_{12} & \cdots & U_{1m} \\ U_{21} & U_{22} & \cdots & U_{2m} \\ \vdots & & & \\ U_{n1} & U_{n2} & \cdots & U_{nm}. \end{matrix}$$

If we wish to add all the U's, we may add first the various rows and then add the row totals to get the desired result:

$$\sum_{j=1}^{m} U_{1j} + \sum_{j=1}^{m} U_{2j} + \cdots + \sum_{j=1}^{m} U_{nj},$$

which may be written more compactly by using a second summation sign thus:

$$\sum_{i=1}^{n} \left(\sum_{j=1}^{m} U_{ij} \right).$$

If we had found the column totals first instead of the row totals, we would have obtained in the same way the result

$$\sum_{j=1}^{m} \left(\sum_{i=1}^{n} U_{ij} \right).$$

Since the sum will be the same in either case, we have

(I.5.1) $$\sum_{i=1}^{n} \left(\sum_{j=1}^{m} U_{ij} \right) = \sum_{j=1}^{m} \left(\sum_{i=1}^{n} U_{ij} \right),$$

which says that *in a finite double sum, the order of summation is immaterial.* This result does not necessarily hold for sums of infinitely many terms.

Such double sums are usually written without parentheses:

$$\sum_{i=1}^{n} \sum_{j=1}^{m} U_{ij} = \sum_{j=1}^{m} \sum_{i=1}^{n} U_{ij}.$$

The indices of summation here are, of course, dummy indices, just as in the case of simple sums.

An important kind of double sum is obtained when we put

$$U_{ij} = a_{ij} x_i y_j \begin{cases} i = 1, 2, \ldots, n, \\ j = 1, 2, \ldots, m, \end{cases}$$

and obtain

$$\sum_{i=1}^{n} \sum_{j=1}^{m} a_{ij} x_i y_j.$$

The expanded form of this sum is a polynomial in the $m + n$ variables x_1,

$x_2, \ldots, x_n, y_1, y_2, \ldots, y_m$. Since each term of this polynomial is of the first degree in the x variables as well as in the y variables, we call it a *bilinear form* in these variables.

If we had, for example, $n = 2$, $m = 3$ and $a_{11} = a_{12} = a_{13} = 1$, $a_{21} = a_{22} = a_{23} = -1$, the bilinear form would be

$$\sum_{i=1}^{2} \sum_{j=1}^{3} a_{ij} x_i y_j = a_{11} x_1 y_1 + a_{12} x_1 y_2 + a_{13} x_1 y_3$$

$$+ a_{21} x_2 y_1 + a_{22} x_2 y_2 + a_{23} x_2 y_3$$

$$= x_1 y_1 + x_1 y_2 + x_1 y_3 - x_2 y_1 - x_2 y_2 - x_2 y_3.$$

Another special situation of prime importance is obtained when $m = n$ and $U_{ij} = a_{ij} x_i x_j; i, j = 1, 2, \ldots, n$. We have then

$$\sum_{i=1}^{n} \sum_{j=1}^{n} a_{ij} x_i x_j,$$

or, as it is more commonly written,

$$\sum_{i, j=1}^{n} a_{ij} x_i x_j.$$

(When several indices of summation have the same range, as here, it is convenient to write them on one summation sign. This is permissible because the order of summation is irrelevant in a finite sum.) If, for example, $n = 2$, the expanded form of the sum is

$$a_{11} x_1 x_1 + a_{12} x_1 x_2 + a_{21} x_2 x_1 + a_{22} x_2 x_2$$

$$= a_{11} x_1^2 + (a_{12} + a_{21}) x_1 x_2 + a_{22} x_2^2.$$

A polynomial of this kind is called a *quadratic form* in $x_1, x_2, \ldots, x_n$, since every term in it is of the second degree in those variables. In most applications, the requirement $a_{ij} = a_{ji}; i, j = 1, 2, \ldots, n$, is useful. In this case we call the quadratic form *symmetric*.

I.6   Exercises

1. Write out in full the *trilinear* form

$$\sum_{i=1}^{2} \sum_{j=1}^{2} \sum_{k=1}^{3} a_{ijk} x_i y_j z_k.$$

2. Show in Exercise 1 that the same result is obtained independently of the order in which the various summations are carried out.
3. Write out the quadratic form for which $a_{ij} = 0$, $i \neq j$, and $a_{ii} = 1$; $i, j = 1, 2, \ldots, n$.
4. In how many different orders may the summation in

$$\sum_{i_1=1}^{n_1} \sum_{i_2=1}^{n_2} \cdots \sum_{i_k=1}^{n_k} U_{i_1 i_2 \cdots i_k}$$

be carried out? Are the results all equal? By what method of proof would you establish your answer to this last question?

5. Show that

$$\left(\sum_{j=1}^{n} x_j\right)^2 - \sum_{j=1}^{n} x_j^2 = \sum_{\substack{i, j=1 \\ i \neq j}}^{n} x_i x_j = 2 \sum_{\substack{i, j=1 \\ i < j}}^{n} x_i x_j.$$

(When, as here, a restriction is imposed on a summation process, the intention is that the summation should proceed as usual except that only those terms satisfying the restriction are to be written.)

6. Show that, if $\sum_{j=1}^{n} x_j = 0$, then

$$\sum_{j=1}^{n} x_j^2 = - \sum_{\substack{i, j=1 \\ i \neq j}}^{n} x_i x_j.$$

7. Given that $n_i m_{x_i} = \sum_{j=1}^{n_i} x_{ij}$, $i = 1, 2, \ldots, k$ and that

$$\left(\sum_{i=1}^{k} n_i\right) m_x = \sum_{i=1}^{k} \sum_{j=1}^{n_i} x_{ij},$$

show that

$$\sum_{i=1}^{k} n_i(m_{x_i} - m_x) = 0$$

and

$$\sum_{i=1}^{k} \sum_{j=1}^{n_i} (x_{ij} - m_x)^2 = \sum_{i=1}^{k} \sum_{j=1}^{n_i} (x_{ij} - m_{x_i})^2 + \sum_{i=1}^{k} n_i(m_{x_i} - m_x)^2.$$

These are more formulas useful in statistics.

8. Write out in full the "triangular" sums

(a) $\displaystyle\sum_{\substack{i, j=1 \\ i < j}}^{6} U_{ij}$ or $\displaystyle\sum_{1 \le i < j \le 6} U_{ij}$,

(b) $\displaystyle\sum_{\substack{i, j=1 \\ i \le j}}^{6} U_{ij}$.

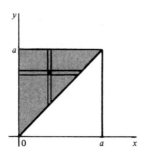

FIGURE I-1.

9. Show that the sum of the elements in the triangular array

$$a_{11}$$
$$a_{12} \quad a_{22}$$
$$a_{13} \quad a_{23} \quad a_{33}$$
$$\vdots$$
$$a_{1n} \quad a_{2n} \quad a_{3n} \quad \cdots \quad a_{nn}$$

may be represented as either

$$\sum_{i=1}^{n} \left(\sum_{j=i}^{n} a_{ij} \right) \qquad \text{or} \qquad \sum_{j=1}^{n} \left(\sum_{i=1}^{j} a_{ij} \right).$$

It is instructive to compare this with the change of order in a double integration (Figure I-1):

$$\int_{0}^{a} \int_{x}^{a} f(x, y) \, dy \, dx = \int_{0}^{a} \int_{0}^{y} f(x, y) \, dx \, dy.$$

10. Show that, if

$$\frac{1}{N} \left(\sum_{i=1}^{N} \alpha_{ik} \right) = \mu, \qquad k = 1, 2, \dots, M,$$

and

$$\frac{1}{M} \left(\sum_{k=1}^{M} \beta_{kj} \right) = \nu, \qquad j = 1, 2, \dots, R,$$

then

$$\frac{1}{NRM} \sum_{i=1}^{N} \sum_{j=1}^{R} \sum_{k=1}^{M} \alpha_{ik} \beta_{kj} = \mu\nu.$$

The $\prod$ Notation

I.7 Definitions and Basic Properties

We have for **products** a notation, analogous to the $\sum$ notation for sums, the definition of which is contained in the equation

$$(I.7.1) \qquad \prod_{j=1}^{n} x_j = x_1 x_2 \cdots x_n.$$

Here again, j is an index whose range is indicated by the notations on the $\prod$ symbol, and x_j is a function of j, just as before. The values taken on by x_j are, however, multiplied in this case, as the symbol $\prod$ (for "product") is intended to imply.

We have the further definition

$$(I.7.2) \qquad \prod_{j=1}^{n} c = c^n,$$

and the properties

$$(I.7.3) \qquad \prod_{j=1}^{n} (kx_j) = k^n \left(\prod_{j=1}^{n} x_j \right),$$

$$(I.7.4) \qquad \prod_{j=1}^{n} x_j y_j = \left(\prod_{j=1}^{n} x_j \right) \left(\prod_{j=1}^{n} y_j \right),$$

$$(I.7.5) \qquad \prod_{i=1}^{n} \left(\prod_{j=1}^{m} U_{ij} \right) = \prod_{j=1}^{m} \left(\prod_{i=1}^{n} U_{ij} \right).$$

We also have **triangular products**, just as we have triangular sums. (See Exercises 8 and 9, Section I.6.) An important example of this type of product is

$$\prod_{1 \le i < j \le n} (x_i - x_j),$$

or, as it is also written,

$$\prod_{\substack{i=1, \ldots, n-1 \\ j=i+1, \ldots, n}} (x_i - x_j).$$

The notations under the $\prod$'s here mean that we are to use all factors of the form $(x_i - x_j)$ as i and j range over the values 1, 2, ..., n, subject to the restriction that i be always less than j. We have, therefore,

$$\prod_{1 \leq i < j \leq n} (x_i - x_j) = (x_1 - x_2)(x_1 - x_3) \cdots (x_1 - x_n)$$
$$(x_2 - x_3) \cdots (x_2 - x_n)$$
$$\vdots$$
$$(x_{n-1} - x_n).$$

How many factors are there in this product?

This function of $x_1, x_2, \ldots, x_n$ is known as the **alternating function**. It is zero unless the x's are all distinct, and when it is not zero, it changes its sign if any two x's are interchanged. (Verify this last statement for x_1 and x_2.)

For other ways of writing triangular products, see Exercise 5 in the next section.

I.8   Exercises

1. Prove rules (I.7.3), (I.7.4), and (I.7.5).

2. Show that

(a) $\left(\prod_{j=1}^{n} x_j\right) x_{n+1} = \prod_{j=1}^{n+1} x_j,$

(b) $\prod_{j=1}^{k} x_j \cdot \prod_{j=k+1}^{n} x_j = \prod_{j=1}^{n} x_j, n > k,$

(c) $\prod_{i=1}^{k} x_i \cdot \prod_{j=1}^{n-k} x_{k+j} = \prod_{p=1}^{n} x_p, n > k.$

3. Show by examples that the index in the $\prod$ notation is a dummy index and that the range of the index may be shifted if desired.

4. Simplify

$$\left(\prod_{k=1}^{n} (a^{1/2^k} + b^{1/2^k})\right) (a^{1/2^n} - b^{1/2^n}), \qquad a, b > 0.$$

5. Show that

$$\prod_{j=1}^{n} \left(\prod_{i=1}^{j} a_{ij}\right) = \prod_{i=1}^{n} \left(\prod_{j=i}^{n} a_{ij}\right).$$

6. Write in expanded notation:

(a) $\displaystyle\sum_{i=1}^{n}\left(\prod_{j=1}^{n}x_{ij}\right).$

(b) $\displaystyle\prod_{j=1}^{n}\left(\sum_{i=1}^{n}x_{ij}\right).$

(c) $\displaystyle\sum_{i=1}^{n}\left\{\frac{y_i}{x_0-x_i}\prod_{\substack{1\le j\le n \\ j\ne i}}\left(\frac{x-x_j}{x_0-x_j}\right)\right\}.$

7. Show that

$$\prod_{\substack{i,\,j=1,\,2,\,\ldots,\,n \\ i\ne j}}(x_i-x_j)=(-1)^{n(n-1)/2}\left[\prod_{\substack{i,\,j=1,\,2,\,\ldots,\,n \\ i<j}}(x_i-x_j)\right]^2.$$

8. Show that

$$\prod_{i=1}^{n}\left(\sum_{j=1}^{m}x_{ij}\right)=\sum_{j_1,\,\ldots,\,j_n=1}^{m}\left(\prod_{i=1}^{n}x_{ij_i}\right).$$

APPENDIX

II

The Algebra of Complex Numbers

Since complex numbers are used rather extensively in this book, and since some readers may have only a passing acquaintance with them, we give in this appendix a brief review of their more important algebraic properties.

II.1 Definitions and Fundamental Operations

If a and b are real numbers, symbols of the form $a + bi$ (subject to the rules of operation listed below) are called **complex numbers**. The real number a is called the **real part** of $a + bi$, and the real number b is called its **imaginary part**. It is often convenient to denote complex numbers by single letters from the end of the alphabet: $a + bi = z$, etc.

Two complex numbers $a + bi$ and $c + di$ are defined to be equal if and only if the real and imaginary parts of one are respectively equal to the real and imaginary parts of the other, that is, if and only if $a = c$ and $b = d$.

The complex numbers which are the **sum** and the **product** of two complex numbers are defined as follows:

$$(\text{II.1.1}) \qquad (a + bi) + (c + di) = (a + c) + (b + d)i,$$

$$(\text{II.1.2}) \qquad (a + bi)(c + di) = (ac - bd) + (ad + bc)i.$$

From the preceding equations, we note that

$$(a + 0i) + (c + 0i) = (a + c) + 0i,$$
$$(a + 0i)(c + 0i) = (ac) + 0i.$$

Thus complex numbers of the form $a + 0i$ behave just like the corresponding real numbers a with respect to addition and multiplication. This fact leads us to redefine the symbol $a + 0i$ as the real number a:

(II.1.3) $a + 0i = a$

for every real number a. In particular, we have

$$0 + 0i = 0 \quad \text{and} \quad 1 + 0i = 1.$$

The first of these special cases leads to:

Theorem II.1.1: *A complex number is zero if and only if its real and imaginary parts are both zero.*

The preceding definitions have the consequence that the set of real numbers is contained in the set of complex numbers: Every real number a is a complex number of the special form $a + 0i$. Every statement true for complex numbers in general is thus true in particular for real numbers, but not conversely, of course.

Since

$$(a + bi) + 0 = (a + bi) + (0 + 0i) = (a + 0) + (b + 0)i = a + bi,$$

we see that 0 is an **identity element for addition**.

We define the **negative** of $a + bi$ to be the complex number

$$-(a + bi) = (-a) + (-b)i,$$

so

$$(a + bi) + (-(a + bi)) = (a + bi) + ((-a) + (-b)i) = 0.$$

Subtraction is next defined thus:

$$(a + bi) - (c + di) = (a + bi) + (-(c + di)),$$

so

(II.1.4) $(a + bi) - (c + di) = (a - c) + (b - d)i.$

This makes subtraction the operation inverse to addition; that is, we will have $(z + w) - w = z$ for all complex numbers z and w.

It is consistent with these definitions of addition and subtraction to define also

(II.1.5)
$$a + (-b)i = a - bi.$$

We could establish easily, but not in little space, that the associative, commutative, and distributive laws, the laws of signs and parentheses, and the laws of positive integral exponents apply in the above operations with complex numbers just as in the case of real numbers. We shall simply assume these results and proceed.

The equation

$$(a + bi)1 = (a + bi)(1 + 0i) = a + bi,$$

which follows from (II.1.2), shows that 1 is an **identity element for multiplication.**

For economy of representation, we define

(II.1.6)
$$0 + bi = bi.$$

A complex number of this form is called a **pure imaginary number.** The pure imaginary number $1i = i$ is called the **imaginary unit.** From (II.1.2), by putting $a = c = 0$, $b = d = 1$, we obtain the equation

$$i^2 = -1.$$

Thus, when multiplying complex numbers, we proceed as though we were multiplying polynomials of the form $a + bx$, except that now we replace i^2 by -1 whenever it appears. It is helpful in this connection to observe that $i^3 = -i$, $i^4 = 1$, etc. For some purposes it is helpful to rewrite the relation $i^2 = -1$ in the form $i = \sqrt{-1}$, in which case $-\sqrt{-1} = -i$.

Using (II.1.2) we may now verify that

$$(c + di)\left(\left(\frac{c}{c^2 + d^2}\right) + \left(\frac{-d}{c^2 + d^2}\right)i\right) = 1,$$

provided that $c^2 + d^2 \neq 0$. Hence we define the **inverse** or **reciprocal** of $c + di$ as follows:

(II.1.7)
$$(c + di)^{-1} = \left(\frac{c}{c^2 + d^2}\right) + \left(\frac{-d}{c^2 + d^2}\right)i = \frac{c - di}{c^2 + d^2}.$$

Since c^2 and d^2 are both nonnegative, $c^2 + d^2 = 0$ if and only if $c = d = 0$. Thus *the only complex number $c + di$ which has no reciprocal is zero.*

This definition of the reciprocal leads us to define **division** of complex numbers thus:

(II.1.8) $$\frac{a + bi}{c + di} = (a + bi)(c + di)^{-1}, \qquad c^2 + d^2 \neq 0.$$

Substituting and expanding, we obtain

$$\frac{a + bi}{c + di} = \frac{(a + bi)(c - di)}{c^2 + d^2} = \left(\frac{ac + bd}{c^2 + d^2}\right) + \left(\frac{bc - ad}{c^2 + d^2}\right)i,$$

which is again a complex number. With this definition, we have $(zw) \div w = z$ for all z and w except $w = 0$, so division as defined here is indeed the inverse of multiplication.

Any collection $\mathscr{F}$ of complex numbers which has the property that the sum, difference, product, and quotient (division by zero excepted) of any two numbers of $\mathscr{F}$ also belongs to $\mathscr{F}$ is called a **number field**. In particular, the set of all complex numbers is a field since, as has been shown above, the four operations applied to any two complex numbers each yield a complex number. The set of all complex numbers of the form $a + 0i$, namely the set of all real numbers, is likewise a number field since

$$(a + 0i) \pm (b + 0i) = (a \pm b) + 0i,$$

$$(a + 0i)(b + 0i) = (ab) + 0i,$$

$$(a + 0i) \div (b + 0i) = \frac{a}{b} + 0i, \qquad b \neq 0;$$

that is, the four operations applied to two real numbers again result in real numbers. The real field is a **subfield** of the complex number field. A third example of a number field is the set of all rational real numbers, that is, the set of all real numbers of the form a/b, where a and b are integers ($b \neq 0$). There are many other examples of number fields.

II.2   Exercises

1. Simplify $(1 - i)^3 - (1 + i)^3$.

2. If x and y are real, what can we conclude about them from the equation

$$(x - y + 2) + (2x + y)i = 3 + 5i?$$

3. Let n be any integer, so that $n = 4q + r$, where q is an integer and $r = 0, 1, 2,$ or 3. Give a rule for evaluating i^n.

4. Show that the complex numbers $1, -1, i, -i$ form a group with respect to the operation of multiplication.

5. Use (II.1.1) and (II.1.2) to show that, for arbitrary complex numbers $z_j = x_j + iy_j, j = 1, 2, 3$, the commutative law

$$z_1 z_2 = z_2 z_1$$

and the distributive law

$$z_1(z_2 + z_3) = z_1 z_2 + z_1 z_3$$

hold.

6. Show that, for all complex numbers z, $z \cdot 0 = 0$. [Use (II.1.2) and the fact that $0 = 0 + 0i$.]

7. Prove that, if $z_1 + w = z_2 + w$, then $z_1 = z_2$, and that, if $z_1 w = z_2 w$ and $w \neq 0$, then $z_1 = z_2$. (**Cancellation laws** for addition and multiplication.)

8. Given that

$$\omega = \frac{-1 + \sqrt{3}i}{2},$$

show that

$$\omega^2 = \frac{-1 - \sqrt{3}i}{2}$$

and that

$$\omega^3 = 1.$$

Show also that

$$\omega^2 + \omega + 1 = 0.$$

9. Show that, for every complex number $x_1 + x_2 i$, there exist unique real numbers y_1 and y_2 such that

$$x_1 + x_2 i = y_1 \frac{1 + i}{\sqrt{2}} + y_2 \frac{1 - i}{\sqrt{2}}.$$

II.3 Conjugate Complex Numbers

The complex numbers $a + bi$ and $a - bi$ are called **conjugate complex numbers**, each being the conjugate of the other. A real number is its own conjugate. The conjugate of a complex number z is denoted by $\bar{z}$.

If $z = a + bi$, then $\bar{z} = a - bi$, and

$$z + \bar{z} = 2a, \qquad z - \bar{z} = 2bi, \qquad z\bar{z} = a^2 + b^2.$$

We have, therefore,

Theorem II.3.1: *The sum, difference, and product of conjugate complex numbers are respectively a real number, a pure imaginary number, and a non-negative real number.*

Again, by the definition of equality, if $z = \bar{z}$, we have $b = -b$, so $b = 0$ and $z = a$. If $z = -\bar{z}$, then $a = -a$ so $a = 0$ and $z = bi$. This yields

Theorem II.3.2: *If a complex number equals its conjugate, it is a real number, but if it equals the negative of its conjugate, it is a pure imaginary number.*

An important use of the conjugate of a complex number is in the evaluation of a quotient according to the process

$$\frac{a + bi}{c + di} = \frac{(a + bi)(c - di)}{(c + di)(c - di)} = \frac{(ac + bd) + (bc - ad)i}{c^2 + d^2}.$$

The rule "multiply numerator and denominator by the conjugate of the denominator" is easier to use and remember than is the formula for the quotient.

By direct computation, we can show readily that for any two complex numbers w and z, we have

(II.3.1)
$$\overline{(\bar{z})} = z,$$
$$\overline{w + z} = \bar{w} + \bar{z},$$
$$\overline{wz} = \bar{w}\bar{z},$$
$$\overline{\left(\frac{w}{z}\right)} = \frac{\bar{w}}{\bar{z}}, \qquad z \neq 0.$$

For example, if

$$w = a + bi, \qquad z = c + di,$$

then

$$\overline{wz} = \overline{(ac - bd) + (ad + bc)i} = (ac - bd) - (ad + bc)i$$

and

$$\bar{w}\bar{z} = (a - bi)(c - di) = (ac - bd) + (-ad - bc)i,$$

so

$$\overline{wz} = \overline{w}\overline{z}.$$

Similar procedures apply in the other cases.

The **absolute value** of the complex number $z = a + bi$, denoted by $|z|$, is by definition the nonnegative real number $\sqrt{z\overline{z}} = \sqrt{a^2 + b^2}$. The square of the absolute value, $z\overline{z}$ or $a^2 + b^2$, appears in the process of division:

$$z^{-1} = \frac{1}{z} = \frac{\overline{z}}{z\overline{z}}, \qquad z \neq 0,$$

and similarly

$$\frac{w}{z} = \frac{w\overline{z}}{z\overline{z}}, \qquad z \neq 0.$$

In particular, the reciprocal of a complex number is its conjugate divided by the square of its absolute value.

Concerning the absolute value, we may show, again by direct computation, that

(II.3.2)
$$|\overline{z}| = |z|,$$
$$|wz| = |w| \cdot |z|,$$
$$\left|\frac{w}{z}\right| = \frac{|w|}{|z|}.$$

For example, if $w = a + bi$, $z = c + di$, we have

$$|wz| = |(ac - bd) + (ad + bc)i|$$
$$= \sqrt{(ac - bd)^2 + (ad + bc)^2} = \sqrt{(a^2 + b^2)(c^2 + d^2)}$$
$$= |w| \cdot |z|.$$

Alternatively,

$$(wz) \cdot (\overline{wz}) = wz\overline{w}\overline{z} = (w\overline{w})(z\overline{z}).$$

Hence, taking positive square roots on both sides,

$$\sqrt{(wz)(\overline{wz})} = \sqrt{w\overline{w}} \cdot \sqrt{z\overline{z}},$$

or

$$|wz| = |w| \cdot |z|.$$

Similar proofs may be developed for the other rules.

The absolute value of a complex number also appears in some important inequalities:

$$\text{If } z = a + bi, \text{ then } |a| \leq |z| \text{ and } |b| \leq |z|.$$

(II.3.3)
$$|w + z| \leq |w| + |z|.$$

The first two of these are left to the reader to prove. To prove the last one, we note that

$$(w + z)\overline{(w + z)} = (w + z)(\overline{w} + \overline{z}) = w\overline{w} + z\overline{z} + \overline{w}z + w\overline{z},$$

or

$$|w + z|^2 = |w|^2 + |z|^2 + (\overline{w}z + \overline{w}\overline{z}).$$

The quantity in parentheses is twice the real part of $\overline{w}z$ and hence, by the first inequality of (II.3.3), we have

$$\overline{w}z + \overline{\overline{w}z} \leq |\overline{w}z + \overline{\overline{w}z}| = 2|\overline{w}z| = 2|w||z|.$$

Therefore,

$$|w + z|^2 \leq |w|^2 + 2|w||z| + |z|^2 = (|w| + |z|)^2.$$

Taking the positive square root on both sides, we then have the desired result:

$$|w + z| \leq |w| + |z|.$$

II.4 Exercises

1. Complete the proof of (II.3.1) and (II.3.2).
*2. Prove that, if f is a polynomial with *real ceofficients*, $f(\overline{z}) = \overline{f(z)}$.
3. Give an example of a function g such that $g(\overline{z}) \neq \overline{g(z)}$.
*4. Prove that, if f is a polynomial with real coefficients and if $f(z) = 0$, then $f(\overline{z}) = 0$ also. In words, this says that in the case of polynomial equations with real coefficients, complex roots occur in conjugate pairs.
5. Prove that $|w + z| \geq ||w| - |z||$.
6. For what complex numbers do we have $|z| = z$? $|z| = -z$?
7. Simplify $(1 + i)^3/(1 - i)^3$.
*8. Show that $\sum_{j=1}^{n} |w_j|^2 \geq 0$ for arbitrary complex numbers w_j, and that, if either

$$\sum_{j=1}^{n} |w_j|^2 = 0 \quad \text{or} \quad \sum_{j=1}^{n} |w_j| = 0,$$

then

$$w_j = 0, \qquad j = 1, 2, \ldots, n.$$

9. Show that, if $wz = 1$ and $|w| = 1$, where w and z are complex numbers, then $|z| = 1$ and $w^{-1} = \bar{w}$.

A Bibliography
of Linear Algebra and
Its Applications

This list of books, though extensive, is by no means exhaustive. It provides references to many areas of application as well as to different, more mature, more extensive, or classical treatments of linear algebra.

Journal

Linear Algebra and its Applications. New York: American Elsevier, 1968 ——.

Linear algebra, matrices, and determinants

F. V. Atkinson. "Multiparameter Eigenvalue Problems." New York: Academic Press, Vol. I: *Matrices and Compact Operators*. 1972.

R. Bellman. *Introduction to Matrix Analysis*. 2nd ed. New York, McGraw-Hill, 1970.

J. N. Franklin. *Matrix Theory*. Englewood Cliffs, N.J.: Prentice-Hall, 1968.

F. R. Gantmacher. *The Theory of Matrices*, Vols. I and II, New York: Chelsea, 1959.

———. *Matrizenrechnung*, 3rd ed., Vol. 1, 1970; Vol. 2, 1971. Berlin: VEB Deutscher Verlag der Wissenschaften.

W. H. Greub. *Linear Algebra*, 3rd ed. New York: Springer-Verlag, 1967.

P. R. Halmos. *Finite Dimensional Vector Spaces*, 2nd ed. Princeton: Van Nostrand, 1958.

K. Hoffman and R. Kunze. *Linear Algebra*, 2nd ed. Englewood Cliffs, N.J.: Prentice-Hall, 1971.

P. Lancaster. *Theory of Matrices*. New York: Academic Press, 1969.

C. C. McDuffee. *The Theory of Matrices*, 2nd ed. New York: Chelsea, 1946.

T. Muir and W. H. Metzler. *A Treatise on the Theory of Determinants*. New York: Longmans, 1933.

M. Newman. *Integral Matrices*. New York: Academic Press, 1972.

M. C. Pease, III. *Methods of Matrix Algebra*. New York: Academic Press, 1965.

G. E. Shilov. *Linear Algebra*. Englewood Cliffs, N.J.: Prentice-Hall, 1971.

J. H. M. Wedderburn. *Lectures on Matrices*. Providence, R.I.: American Mathematical Society, 1934.

J. H. Wilkinson. *The Algebraic Eigenvalue Problem*. New York: Oxford University Press, 1965.

Linear and matrix inequalities

M. Marcus and H. Minc. *A Survey of Matrix Theory and Matrix Inequalities*. Boston: Allyn and Bacon, 1964.

S. N. Tschernikow. *Lineare Ungleichungen*. Berlin: VEB Deutscher Verlag der Wissenschaften, 1971.

Numerical linear algebra

B. W. Arden and K. N. Astill. *Numerical Algorithms: Origins and Applications*. Reading, Mass.: Addison-Wesley, 1970.

S. Conte and C. de Boor. *Elementary Numerical Analysis: An Algorithmic Approach*, 2nd ed. New York: Academic Press, 1972.

D. K. Faddeev and V. N. Faddeeva. *Computational Methods of Linear Algebra*. San Francisco: Freeman, 1963.

G. Forsythe and C. B. Moler. *Computer Solution of Linear Algebraic Systems*. Englewood Cliffs, N.J.: Prentice-Hall, 1967.

L. Fox. *An Introduction to Numerical Linear Algebra*. New York: Oxford University Press, 1964.

A. S. Householder. *The Theory of Matrices in Numerical Analysis*. New York: Blaisdell, 1964.

E. Isaacson and H. B. Kettler. *Analysis of Numerical Methods*. New York: Wiley, 1966.

J. M. Ortega. *Numerical Analysis: A Second Course*. New York: Academic Press, 1972.

J. M. Ortega and W. C. Rheinboldt. *Iterative Solution of Nonlinear Equations in Several Variables*. New York: Academic Press, 1970.

A. M. Ostrowski. *Solution of Equations and Systems of Equations*, 2nd ed. New York: Academic Press, 1966.

R. S. Varga. *Matrix Iterative Analysis*. Englewood Cliffs, N.J.: Prentice-Hall, 1962.

D. M. Young. *Iterative Solution of Large Linear Systems*. New York: Academic Press, 1971.

Applied linear algebra

R. A. Frazer, W. J. Duncan, and A. R. Collar. *Elementary Matrices and Some Applications to Dynamics and Differential Equations.* Cambridge: Cambridge University Press, 1950.

B. Higman. *Applied Group-Theoretic and Matrix Methods.* Oxford: Clarendon Press, 1955.

F. B. Hildebrand. *Methods of Applied Mathematics*, 2nd ed. New York: Prentice Hall, 1965.

C. A. Hollingsworth. *Vectors, Matrices, and Group Theory for Scientists and Engineers.* New York: McGraw-Hill, 1967.

J. Indritz. *Methods in Analysis.* New York: Macmillan, 1963.

H. Jeffreys and B. S. Jeffreys. *Methods of Mathematical Physics*, 3rd ed. Cambridge: Cambridge University Press, 1966.

B. Noble. *Applied Linear Algebra.* Englewood Cliffs, N.J.: Prentice-Hall, 1969.

L. Pipes. *Matrix Methods for Engineering.* Englewood Cliffs, N.J.: Prentice-Hall, 1962.

Generalized inverse matrices

T. L. Boullion and P. L. Odell. *Generalized Inverse Matrices.* New York: Wiley-Interscience, 1971.

R. M. Pringle and A. A. Rayner. *Generalized Inverse Matrices and Their Statistical Applications.* New York: Hafner, 1971.

C. R. Rao and S. K. Mitra. *Generalized Inverse of Matrices and its Applications.* New York: Wiley-Interscience, 1971.

Abstract algebra

G. Birkhoff and S. MacLane. *A Survey of Modern Algebra*, 3rd ed. New York: Macmillan, 1965.

M. Bôcher. *Introduction to Higher Algebra.* New York: Dover, 1964.

D. K. Faddeev and I. S. Sominskii. *Problems in Higher Algebra.* San Francisco: Freeman, 1965.

N. Jacobson. "Lectures in Abstract Algebra," New York: Van Nostrand, Vol. I, *Basic Concepts*, 1951; Vol. II, *Linear Algebra*, 1953; Vol. III, *Theory of Fields and Galois Theory*, 1964.

S. MacLane and G. Birkhoff. *Algebra.* New York: Macmillan, 1967.

H. Paley and P. M. Weichsel. Elements of Abstract and Linear Algebra. New York: Holt, Rinehart and Winston, 1972.

B. L. van der Waerden. *Algebra*, Vol. I, 7th ed., 1966; Vol. II, 5th ed., 1967. New York: Springer-Verlag. (In German)

Analysis

G. F. Feeman and N. R. Grabois. *Linear Algebra and Multivariable Calculus.* New York: McGraw-Hill, 1970.

S. Karlin. *Total Positivity.* Stanford, Cal.: Stanford University Press, Vol. 1, 1968.

H. Rossi. *Advanced Calculus, with Problems and Applications to Science and Engineering.* New York: Benjamin, 1970.

R. Sikorski. *Advanced Calculus: Functions of Several Variables.* Warsaw: Polish Scientific Publishers, 1969.

M. Spivak, *Calculus on Manifolds.* New York: Benjamin, 1965.

R. E. Williamson, R. H. Crowell, and H. F. Trotter. *Calculus of Vector Functions,* 3rd ed. Englewood Cliffs, N. J.: Prentice-Hall, 1972.

Combinatorial mathematics

E. F. Beckenbach, Editor. *Applied Combinatorial Mathematics.* New York: Wiley, 1964.

R. C. Bose and T. A. Dowling. *Combinatorial Mathematics and its Applications.* Chapel Hill: University of North Carolina Press, 1969.

M. Hall, Jr. *Combinatorial Theory.* Waltham, Mass.: Blaisdell, 1967.

H. J. Ryser. *Combinatorial Mathematics.* Carus Mathematical Monographs, No. 14. New York: Wiley, 1963.

Differential, difference, and integral equations

S. Barnett and C. Storey. *Matrix Methods in Stability Theory.* New York: Barnes and Noble, 1970.

F. Brauer and J. A. Nobel. *Qualitative Theory of Ordinary Differential Equations.* New York: Benjamin, 1969.

C. D. Green. *Integral Equation Methods.* New York: Barnes and Noble, 1969.

J. L. Goldberg and A. L. Schwartz. *Systems of Ordinary Differential Equations.* New York: Harper and Row, 1972.

S. Goldberg. *Introduction to Difference Equations.* New York: Wiley, 1958.

F. B. Hildebrand. *Finite Difference Equations and Simulations.* Englewood Cliffs, N. J.: Prentice-Hall, 1968.

H. Hochstadt. *Differential Equations, A Modern Approach.* New York: Holt, Rinehart and Winston, 1964.

J. LaSalle and S. Lefschetz. *Stability by Liapunov's Direct Method, with Applications.* New York: Academic Press, 1961.

W. W. Lovitt. *Linear Integral Equations.* New York: Dover, 1950.

R. C. MacCamy and V. J. Mizel. *Linear Analysis and Differential Equations.* New York: Macmillan, 1969.

W. T. Reid. *Riccati Differential Equations.* New York: Academic Press, 1972.

W. Schmeidler: *Integralgleichungen mit Anwendungen in Physik und Technik.* Leipzig: Akademische Verlagsgesellschaft, 1950.

Geometry

A. R. Amir-Moez. *Extreme Properties of Linear Transformations and Geometry in Unitary Spaces.* Lubbock, Texas: Texas Tech. University, 1970.

K. Borsuk. *Multidimensional Analytic Geometry.* Warsaw: Polish Scientific Publishers, 1969.

J. Dieudonné. *Linear Algebra and Geometry.* Boston: Houghton-Mifflin 1969.

K. W. Gruenberg and A. J. Weir. *Linear Geometry.* Princeton: Van Nostrand, 1967.

J. H. C. Gerretsen. *Tensor Calculus and Differential Geometry.* Groningen, Netherlands: Noordhoff, 1962.

I. Kaplansky. *Linear Algebra and Geometry: A Second Course.* Boston: Allyn and Bacon, 1969.

H. Levy. *Projective and Related Geometries.* New York: Macmillan, 1964.

E. Snapper and R. J. Troyer. *Metric Affine Geometry.* New York: Academic Press, 1971.

D. M. Y. Sommerville. *Analytic Geometry of Three Dimensions.* Cambridge: Cambridge University Press, 1934.

Graph theory

R. G. Busacker and T. L. Saaty. *Finite Graphs and Networks: An Introduction with Applications.* New York: McGraw-Hill, 1965.

F. Harary. *Graph Theory.* Reading, Mass.: Addison-Wesley, 1969.

F. Harary, R. Z. Norman, and D. Cartwright. *Structural Models: An Introduction to the Theory of Directed Graphs.* New York: Wiley, 1965.

D. E. Johnson and J. R. Johnson. *Graph Theory with Engineering Applications.* New York: Ronald, 1972.

C. W. Marshall. *Applied Graph Theory.* New York: Wiley-Interscience, 1971.

W. Mayeda. *Graph Theory.* New York: Wiley-Interscience, 1972.

S. Seshu and M. B. Reed. *Linear Graphs and Electrical Networks.* Reading, Mass.: Addison-Wesley, 1961.

Group theory

C. W. Curtis and I. Reiner. *Representation Theory of Finite Groups.* New York: Interscience, 1963.

L. Dornhoff. *Group Representation Theory.* Part A, 1971; Part B, 1972. New York: Marcel Dekker.

M. Hall. *The Theory of Groups.* New York: Macmillan, 1959.

D. E. Littlewood. *The Theory of Group Characters and Matrix Representations of Groups.* Oxford: Clarendon Press, 1950.

J. S. Lomont. *Applications of Finite Groups.* New York: Academic Press, 1959.

R. McWeeny. *Symmetry: An Introduction to Group Theory and its Applications.* New York: Pergamon, 1963.

Hilbert space and infinite matrices

S. Berberian. *Introduction to Hilbert Space.* New York: Oxford University Press, 1961.

R. G. Cooke. *Infinite Matrices and Sequence Spaces.* London: Macmillan, 1950.

P. R. Halmos. *Introduction to Hilbert Space.* New York: Chelsea, 1951.

———. *A Hilbert Space Problem Book.* Princeton: Van Nostrand, 1967.

G. Helmberg. *Introduction to Spectral Theory in Hilbert Space.* New York: Wiley-Interscience, 1969.

M. Schechter. *Principles of Functional Analysis.* New York: Academic Press, 1971.

Number theory

V. E. Hoggatt, Jr. *Fibonacci and Lucas Numbers.* Mathematics Enrichment Series. Boston: Houghton-Mifflin, 1969.

I. Niven and H. S. Zuckerman. *Introduction to the Theory of Numbers,* 3rd ed. New York: Wiley, 1972.

B. M. Stewart. *Theory of Numbers,* 2nd ed. New York: Macmillan, 1964.

Probability and statistics

T. W. Anderson. *An Introduction to Multivariate Statistical Analysis.* New York: Wiley, 1965.

———. *The Statistical Analysis of Time Series.* New York: Wiley, 1971.

A. P. Dempster. *Elements of Continuous Multivariate Analysis.* Reading, Mass.: Addison-Wesley, 1969.

N. R. Draper and H. Smith. *Applied Regression Analysis.* New York: Wiley, 1966.

V. V. Fedorov. *Theory of Optimal Experiments.* New York: Academic Press, 1972.

W. Feller. *An Introduction to Probability Theory and its Applications.* Vol. I, 3rd ed., 1968; Vol. II, 2nd ed., 1971. New York: Wiley.

T. S. Ferguson. *Mathematical Statistics, A Decision-Theoretic Approach.* New York: Academic Press, 1967.

F. A. Graybill. *An Introduction to Linear Statistical Models,* Vol. I, New York: McGraw-Hill, 1961.

S. Karlin. *A First Course in Stochastic Processes.* New York: Academic Press, 1966.

J. G. Kemeny and J. L. Snell. *Finite Markov Chains.* Princeton: Van Nostrand, 1960.

———, ———, and A. W. Knapp. *Denumerable Markov Chains.* Princeton: Van Nostrand, 1966.

B. W. Lindgren. *Statistical Theory,* 2nd ed. New York: Macmillan, 1968.

J. Overall and J. Klett. *Applied Multivariate Analysis.* New York: McGraw-Hill, 1972.

E. Parzen. *Stochastic Processes.* San Francisco: Holden-Day, 1962.

C. R. Rao. *Linear Statistical Inference and Its Applications.* New York: Wiley, 1965.

H. Scheffé. *The Analysis of Variance.* New York: Wiley, 1959.

Automata, coding, and switching theory

C. R. Berlekamp. *Algebraic Coding Theory*, New York: McGraw-Hill, 1968.

S. H. Caldwell. *Switching Circuits and Logical Design.* New York: Wiley, 1958.

A. Gill. *Introduction to the Theory of Finite State Machines.* New York: McGraw-Hill, 1962.

————. *Linear Sequential Circuits.* New York: McGraw-Hill, 1966.

M. A. Harrison. *Lectures on Linear Sequential Machines.* New York: Academic Press, 1969.

Z. Kohavi. *Switching and Finite Automata Theory.* New York: McGraw-Hill, 1970.

S. Muroga. *Threshold Logic and Its Applications.* New York: Wiley, 1971.

R. M. M. Oberman. *Disciplines in Combinational and Sequential Circuit Design.* New York: McGraw-Hill, 1970.

A. Paz. *Introduction to Probabilistic Automata.* New York: Academic Press, 1971.

W. W. Peterson and E. J. Weldon, Jr. *Error-Correcting Codes*, 2nd ed. Boston: M.I.T. Press, 1972.

Celestial mechanics

D. Brouwer and G. M. Clemence. *Methods of Celestial Mechanics.* New York: Academic Press, 1961.

P. M. Fitzpatrick. *Principles of Celestial Mechanics.* New York: Academic Press, 1970.

A. Wintner. *The Analytical Foundations of Celestial Mechanics.* Princeton: Princeton University Press, 1947.

Chemistry and crystallography

N. R. Amundson. *Mathematical Methods in Chemical Engineering; Matrices and Their Application.* Englewood Cliffs, N. J.: Prentice-Hall, 1966.

F. A. Cotton. *Chemical Applications of Group Theory*, 2nd ed. New York: Wiley, 1971.

M. A. Jaswon. *Introduction to Mathematical Crystallography.* New York: American Elsevier, 1965.

J. F. Nye. *Physical Properties of Crystals: Their Representation by Tensors and Matrices*, 2nd ed. Oxford: Clarendon Press, 1964.

F. L. Pilar. *Elementary Quantum Chemistry.* New York: McGraw-Hill, 1968.

D. S. Shonland. *Molecular Symmetry.* London: Van Nostrand, 1965.

Color

G. Wyszecki and W. S. Stiles. *Color Science; Concepts and Methods, Quantitative Data and Formulas.* New York: Wiley, 1967.

Control and systems theory

M. Athens and P. L. Falb. *Optimal Control. An Introduction to the Theory and Its Applications.* New York: McGraw-Hill, 1966.

R. W. Brockett. *Finite Dimensional Control Systems.* New York: Wiley, 1970.

P. M. Derusso, R. J. Roy, and C. M. Close. *State Variables for Engineers.* New York: Wiley, 1965.

H. Frank and I. T. Frisch. *Communication, Transmission, and Transportation Networks.* Reading, Mass.: Addison-Wesley, 1971.

B. C. Kuo. *Discrete-Data Control Systems.* Englewood Cliffs, N. J.: Prentice-Hall, 1970.

E. B. Lee and L. Markus. *Foundations of Optimal Control Theory.* New York: Wiley, 1967.

H. S. White and S. Tauber. *Systems Analysis.* Philadelphia: Saunders, 1969.

D. M. Wiberg. *State Space and Linear Systems.* New York: McGraw-Hill, 1971.

W-J. Yang and M. Masubuchi. *Dynamics for Systems and Process Control.* New York: Gordon and Breach, 1970.

L. A. Zadeh and C. A. Desoer. *Linear System Theory; The State Space Approach.* New York: McGraw-Hill, 1963.

Economics and operations research

C. Almon, Jr. *Matrix Methods in Economics.* Reading, Mass.: Addison-Wesley, 1967.

P. J. Dhrymes. *Econometrics; Statistical Foundations and Applications.* New York: Harper and Row, 1970.

D. Gale. *The Theory of Linear Economic Models.* New York: McGraw-Hill, 1960.

R. L. Gue and M. E. Thomas. *Mathematical Methods in Operations Research.* New York: Macmillan, 1968.

T. Harder. *Introduction to Mathematical Models in Market and Opinion Research.* New York: Gordon and Breach, 1969.

J. M. Henderson and R. E. Quandt. *Microeconomic Theory; A Mathematical Approach,* 2nd ed. New York: McGraw-Hill, 1971.

F. S. Hillier and G. J. Lieberman. *Introduction to Operations Research.* San Francisco: Holden-Day, 1967.

M. D. Intrilligator. *Mathematical Optimization and Economic Theory.* Englewood Cliffs, N. J.: Prentice-Hall, 1971.

J. Kmenta. *Elements of Econometrics.* New York: Macmillan, 1971.

K. Lancaster. *Mathematical Economics.* New York: Macmillan, 1968.

H. A. Taha. *Operations Research: An Introduction.* New York: Macmillan, 1971.

H. Theil. *Principles of Econometrics.* New York: John Wiley, 1971.

Elasticity

R. L. Bisplinghoff, H. Ashley, and R. L. Halfman. *Aeroelasticity.* Reading, Mass.: Addison-Wesley, 1955.

A. E. Green and J. E. Adkins. *Large Elastic Deformations.* Oxford: Clarendon Press, 1970.

F. D. Murnaghan. *Finite Deformation of an Elastic Solid*. New York: Wiley, 1951.

E. C. Pestel and F. A. Leckie. *Matrix Methods in Elastomechanics*. New York: McGraw-Hill, 1963.

I. S. Sokolnikoff. *Mathematical Theory of Elasticity*, 2nd ed. New York: McGraw-Hill, 1956.

Electrical engineering

N. Balabanian and T. A. Bickart. *Electrical Network Theory*. New York: Wiley, 1969.

D. A. Callahan. *Computer-Aided Network Design*. Revised Ed. New York: McGraw-Hill, 1972.

C. A. Desoer and E. S. Kuh. *Basic Circuit Theory*. New York, McGraw-Hill, 1969.

W. H. Kim and H. E. Meadows, Jr. *Modern Network Analysis*. New York: Wiley, 1971.

B. C. Kuo. *Linear Systems and Networks*. New York: McGraw-Hill, 1967.

R. W. Newcomb. *Linear Multiport Synthesis*. New York: McGraw-Hill, 1966.

M. B. Reed. *Alternating Current Circuit Theory*, 2nd ed. New York: Harper, 1956.

P. Slepian. *Mathematical Foundations of Network Analysis*. New York: Springer-Verlag, 1968.

O. Wing. *Circuit Theory with Computer Methods*. New York: Holt, Rinehart and Winston, 1968.

Game theory

S. Karlin. *Mathematical Methods and Theory in Games, Programming, and Economics*, Vols. I and II. Reading, Mass.: Addison-Wesley, 1959.

J. C. C. McKinsey. *Introduction to the Theory of Games*. New York: McGraw-Hill, 1952.

G. Owen. *Game Theory*. Philadelphia: Saunders, 1968.

J. von Neumann and O. Morgenstern. *Theory of Games and Economic Behavior*. Princeton: Princeton University Press, 1944.

Geography, geology, and meteorology

J. W. Harbaugh and G. Bonham-Carter. *Computer Simulation in Geology*. New York: Wiley, 1970.

———— and D. F. Merriam. *Computer Applications in Stratigraphic Analysis*. New York, Wiley, 1968.

L. H. King. *Statistical Analysis in Geography*. Englewood Cliffs, N. J.: Prentice-Hall, 1969.

W. C. Krumbein and F. A. Graybill. *An Introduction to Statistical Models in Geology*. New York: McGraw-Hill, 1965.

R. G. Miller. *Statistical Prediction by Discriminant Analysis.* Meteorological Monographs, Vol. IV, No. 25. Boston: American Meteorological Society, 1962.
A. B. Vistelius. *Studies in Mathematical Geology.* New York: Consultants Bureau, 1967.

Life sciences

N. T. Bailey. *Introduction to the Mathematical Theory of Genetic Linkage.* Oxford: Clarendon Press, 1961.
E. Batschelet. *Introduction to Mathematics for Life Scientists.* New York: Springer-Verlag, 1971.
N. Jardine and R. Gibson. *Mathematical Taxonomy.* New York: Wiley, 1970.
N. Keyfitz. *Introduction to the Mathematics of Population.* Reading, Mass.: Addison-Wesley, 1968.
H. M. Nahikian. *A Modern Algebra for Biologists.* Chicago: University of Chicago Press, 1964.
E. C. Pielou. *An Introduction to Mathematical Ecology.* New York: Wiley-Interscience, 1969.
A. Rogers. *Matrix Analysis of Interregional Population Growth and Distribution.* Berkeley: University of California Press, 1968.
S. Searle. *Matrix Algebra for the Biological Sciences.* New York: Wiley, 1967.

Mechanics

T. C. Bradbury. *Theoretical Mechanics.* New York: Wiley, 1968.
H. C. Corben and P. Stehle. *Classical Mechanics,* 2nd ed. New York: Wiley, 1960.
W. Hauser. *Introduction to the Principles of Mechanics.* Reading, Mass.: Addison-Wesley, 1965.
W. Jaunzemis. *Continuum Mechanics.* New York: Macmillan, 1967.
H. L. Langhaar. *Dimensional Analysis and Theory of Models.* New York: Wiley, 1951.
J. B. Marion. *Classical Dynamics of Particles and Systems,* 2nd ed. New York: Academic Press, 1970.

Optics

W. Brouwer. *Matrix Methods in Optical Instrument Design.* New York: Benjamin, 1964.

Optimization theory; linear and nonlinear programming

M. Aoki. *Introduction to Optimization Techniques; Fundamentals and Applications of Nonlinear Programming.* New York: Macmillan, 1971.
L. Cooper and D. Sternberg. *Introduction to Methods of Optimization.* Philadelphia: Saunders, 1970.

R. L. Fox. *Optimization Methods for Engineering Design.* Reading, Mass.: Addison-Wesley, 1971.

S. I. Gass. *Linear Programming,* 3rd ed. New York: McGraw-Hill, 1969.

G. Hadley. *Linear Programming.* Reading, Mass.: Addison-Wesley, 1962.

———. *Nonlinear and Dynamic Programming.* New York: Wiley, 1967.

D. M. Himmelbau. *Applied Nonlinear Programming.* New York: McGraw-Hill, 1972.

M. Iri. *Network Flow, Transportation, and Scheduling: Theory and Algorithms.* New York: Academic Press, 1969.

D. G. Luenberger. *Optimization by Vector Space Methods.* New York: Wiley, 1968.

O. L. Mangasarian. *Nonlinear Programming.* New York: McGraw-Hill, 1969.

G. L. Nemhauser. *Introduction to Dynamic Programming.* New York: Wiley, 1967.

D. Russell. *Optimization Theory.* New York: Benjamin, 1970.

D. M. Simmons. *Linear Programming for Operations Research.* San Francisco: Holden-Day, 1972.

G. Uebe. *Optimale Fahrpläne.* New York: Springer-Verlag, 1970.

S. Vadja. *Introduction to Linear Programming and the Theory of Games.* New York: Wiley, 1960.

S. Vadja. *Mathematical Programming.* Reading, Mass.: Addison-Wesley, 1961.

D. J. Wilde and C. S. Beightler. *Foundations of Optimization.* Englewood Cliffs, N. J.: Prentice-Hall, 1967.

Psychology and sociology

R. R. Bush and F. Mosteller. *Stochastic Models for Learning.* New York: Wiley, 1955.

W. W. Cooley and P. R. Lohnes. *Multivariate Data Analysis.* New York: Wiley, 1971.

H. H. Harman. *Modern Factor Analysis,* 2nd ed. Chicago: University of Chicago Press, 1967.

P. F. Lazarsfeld and N. W. Henry. *Latent Structure Analysis.* Boston: Houghton-Mifflin, 1968.

M. F. Norman. *Markov Processes and Learning Models.* New York: Academic Press, 1972.

M. M. Tatsuoka. *Multivariate Analysis; Techniques for Educational and Psychological Research.* New York: Wiley, 1971.

Quantum mechanics

E. E. Anderson. *Modern Physics and Quantum Mechanics.* Philadelphia: Saunders, 1971.

T. Kahan. *Theory of Groups in Classical and Quantum Physics.* New York: American Elsevier, 1966.

M. L. Mehta. *Random Matrices and the Statistical Theory of Energy Levels.* New York: Academic Press, 1967.

E. Merzbacher. *Quantum Mechanics,* 2nd ed. New York: Wiley, 1970.

A. Messiah. *Quantum Mechanics.* Amsterdam: North Holland, Vol. I, 1965; Vol. II, 1966.

M. Tinkham. *Group Theory and Quantum Mechanics.* New York: McGraw-Hill, 1964.

R. L. White. *Basic Quantum Mechanics.* New York: McGraw-Hill, 1966.

Structures

W. C. Hurty and M. F. Rubinstein. *Dynamics of Structures.* Englewood Cliffs, N. J.: Prentice-Hall, 1964.

R. K. Livesley. *Matrix Methods of Structural Analysis.* Oxford: Pergamon, 1964.

J. S. Przemieniecki. *Theory of Matrix Structural Analysis.* New York: McGraw-Hill, 1968.

M. R. Rubinstein. *Matrix Computer Analysis of Structures.* Englewood Cliffs, N. J.: Prentice-Hall, 1966.

P-C. Wang. *Numerical and Matrix Methods in Structural Mechanics.* New York: Wiley, 1966.

O. C. Zienkiewicz. *The Finite Element Method in Structural and Continuum Mechanics.* New York: McGraw-Hill, 1967.

Vibrations

R. A. Anderson. *Fundamentals of Vibrations.* New York: Macmillan, 1967.

P. Lancaster. *Lambda-Matrices and Vibrating Strings.* Oxford: Pergamon, 1966.

L. Meirovich. *Analytical Methods in Vibrations.* New York: Macmillan, 1967.

R. H. Scanlon and R. Rosenbaum. *Aircraft Vibration and Flutter.* New York: Dover, 1968.

R. F. Steidel, Jr. *An Introduction to Mechanical Vibrations.* New York: Wiley, 1971.

W. T. Thomson. *Vibration Theory and Applications.* Englewood Cliffs, N. J.: Prentice-Hall, 1965.

INDEX

A

a_{ij}, 1
$[a_{ij}]$, $[a_{ij}]_{m \times n}$, 2
$A^\mathbf{T}$, 20
$\overline{A}$, A^*, 23
A_{ij}, 247
$A_{(k)(p)}$, 294
$A^{(H)}$, $A^{(SH)}$, 35 (Ex. 27)
$A^{(S)}$, $A^{(SS)}$, 33 (Ex. 10)
$\mathscr{A}$, 271, 300
$\mathscr{A}_{pk}$, 302
A-orthogonal vectors, 462 (Ex. 6)
A-orthonormal basis, 462 (Ex. 9)
A-unit vector, 462 (Ex. 9)
Abelian group, 342
Absolute value of a complex number, 497
Addition of
  geometric vectors, 91
  matrices, 3
Additive inverse, 5
Adjacent transposition, 296
Adjoint matrix and determinant, 271, 300
Algebraic complement, 294
Alternating function, 263 (Ex. 53), 489
Angle between two vectors, 97, 132
Angle between a vector and a subspace, 373
Anticommutative matrices, 17 (Ex. 22)
Area of a triangle, 240, 286

Arithmetic n-space, 121
Associative laws for
  fields, 86
  groups, 342
  matrix operations, 3, 6, 11, 14 (Ex. 7)
  operators, 340
  rings, 86, 87
  vector spaces, 149, 150
Augmented matrix, 2, 218

B

Basic solution, 234
  degenerate and nondegenerate, 235
Basis for
  $\mathscr{E}^3$, 117; $\mathscr{E}^n$, 134
  a vector space, 168, 323, 325
Bilinear form, 164 (Ex. 38), 463, 467
Boolean field, 86
Bordered matrices, 305 (Ex. 4), 306 (Ex. 5)
Boundary of a half-space, 142

C

$\mathscr{C}$ (complex field), 150
$\mathscr{C}^n$, 150
cov (X, Y), 464
Cancellation laws, 4, 88 (Ex. 6, 7)

513

Canonical bilinear form, 467
Cauchy-Schwarz inequality, 99, 136,
  280 (Ex. 3), 471 (Ex. 20)
Cayley-Hamilton theorem, 411
Center of similitude, 334 (Ex. 3)
Change of basis, 325
Characteristic
  equation of a matrix, 376
  equation of a pair of matrices, 455
  polynomial of a matrix, 376, 387
  roots of a
    Hermitian matrix, 395
    matrix, 375, 377
    pair of matrices (forms), 455
    polynomial function of a matrix,
      409
    projection, 385 (Ex. 22)
    real symmetric matrix, 396
  subspace, 377
  value problem, 375
  value problem, generalized, 420
    (Ex. 8)
  vector of a matrix, 375
  vector of a pair of matrices, 455
Checking by row sums, 66
Closed
  half-line, 135
  half-space, 142
Closure property
  for addition of matrices, 3
  for addition of vectors, 104, 126, 149
  for multiplication by scalars, 150
Codomain, 311
Coefficient matrix, 2
Coefficients of the characteristic
    polynomial, 388
Cofactor
  of an element of a matrix, 247
  expressed as a determinant, 251
  of a minor, 294
Cofactors of elements of parallel lines,
    290
Cogredient transformations, 467
Collinear points, 123
Collinear vectors, 92, 125
Column
  echelon form, 173
  expansion of a determinant, 243

matrix, 2
  of a matrix, 1
  rank, 191
  space of a matrix, 174
Commutative
  group, 342
  law of addition, 3, 149
  matrices, 16 (Ex. 20), 89 (Ex. 15),
    394 (Ex. 11, 12)
  ring, 87
Commutator of two matrices, 17
    (Ex. 23, 24)
Companion matrix, 384 (Ex. 8)
Complement, algebraic, 294
Complementary
  minors, 294
  orthogonal vector spaces, 363
  sets of indices, 294
  subspaces, 185
Complete solution, 60
  of a homogeneous system, 212
  of a nonhomogeneous system, 219,
    231
Complex
  number, 491
  number field, 83, 494
Components of a vector, 19
Computation
  of determinants, 250
  of inverse of a matrix, 75, 78, 271
  of minimum polynomial, 416
  of powers of a matrix, 418
Cone, 427
Conditions for definiteness, 433, 436
    (Ex. 22), 452
Conformability, 3, 9
Conformably partitioned matrices, 45
Congruent matrices, 423, 467
Conjugate
  complex numbers, 495
  of a matrix, 23
  vectors, 463 (Ex. 11), 475 (Ex. 15)
Consistent system of equations, 52,
    224 (Ex. 21, 22, 23)
Contraction, 334 (Ex. 1)
Contragredient transformations, 468
Coordinates
  of a point, 121

of a vector, with respect to a basis, 323

Coplanar vectors, 92, 111, 128 (Ex. 10)

Corresponding major determinants, 268

Counterimage, 310

Covariance, 464
matrix, 453, 465

Cramer's rule, 276

Critical point, 461

Cross product, 323 (Ex. 25), 339 (Ex. 37)

Cubic curve, 206 (Ex. 16)

Cyclic permutation matrix, 347 (Ex. 22)

D

$d(A,B)$, 130

δ_{ik}, 249

$D[\lambda_1, \lambda_2, \ldots, \lambda_n]$, 397

det A, 239

Decomposable matrices, 47 (Ex. 6), 48 (Ex. 13), 49 (Ex. 17)

Decomposition of a matrix, 33 (Ex. 10), 35 (Ex. 27)

Decomposition of a vector, 185, 364, 370

Deficiency of a matrix, 215

Definiteness, conditions for, 433, 436 (Ex. 22), 452

Degenerate basic solution, 235

Derivative of a determinant, 264 (Ex. 57)

Detached coefficients, 66

Determinant, 239, 242
of a product of matrices, 264, 299

Determinantal equation of a
circle, 286 (Ex. 38)
hyperplane, 278, 286 (Ex. 37)
line, 278
parabola, 287 (Ex. 44)
plane, 279

Determinantal solution of systems of equations, 274, 276

Determinative property, 3

Deviations from the mean, 483 (Ex. 3)

Diagonal
expansion of a determinant, 303
form of a Hermitian matrix, 397
matrix, 18 (Ex. 30)
of a matrix, 2

Diagonalizable matrix, 420 (Ex. 5)

Diagonalization
of a Hermitian matrix, 400
of a quadratic form, 424
impossibility of, in general, 407 (Ex. 13)

Difference of two vectors, 97

Dimension of a vector space, 170, 184

Dimensional identity, 184

Direct sum of vector spaces, 187 (Ex. 10)

Directed distance
from a hyperplane to a hyperplane, 148 (Ex. 10)
from a hyperplane to a point, 148 (Ex. 10)
on a line, 135
from the origin to a hyperplane, 148 (Ex. 9)

Directed line, 135
segment in $\mathscr{R}^n$, 121

Direction
angles of a directed segment, 95, 134
angles of a vector, 133
cosines, 95, 104, 133
of a line, 104
numbers of a line, 105

Discriminant of a quadratic form, 423

Distance between two points, 94, 130

Distributive laws of multiplication, 12, 86, 87, 150

Division ring, 87

Divisors of zero, 10, 39 (Ex. 54)

Domain of a mapping, 311

Double sums, 483

Dummy index, 481

E

$\mathscr{E}^2$, $\mathscr{E}^3$, 90; $\mathscr{E}^n$, 130

E_j, 34 (Ex. 16, 19)

E_{ij}, 35 (Ex. 21)

$\varepsilon_{j_1 j_2 \ldots j_n}$, 263 (Ex. 54, 55, 56)

$\varepsilon^{k_1 k_2 \ldots k_m}_{j_1 j_2 \ldots j_m}$, 269

Echelon form, 65

Eigenvalue, 375
  problem, 375

Eigenvector, 375

Element of a
  group, 342
  matrix, 1

Elementary
  matrices, 200
  n-vectors, 34 (Ex. 16, 19)
  operations, 172
  transformations in matrix form, 200,
    208 (Ex. 36), 209 (Ex. 37)
  unit vectors, 130

Ellipse, 426

Ellipsoid, 427, 428

Elongation, 334 (Ex. 1)

Entry of a matrix, 1

Epsilon symbols, 263 (Ex. 54, 55, 56),
  269

Equal
  directed segments, 97, 122
  matrices, 2, 35 (Ex. 23)
  partitioned matrices, 42

Equation of a
  line, 102, 122
  hyperplane, 124, 128 (Ex. 13), 278,
    286 (Ex. 37)
  k-flat, 123
  plane, 105, 108, 123

Equivalence relation, 3

Equivalent
  bilinear forms, 466
  Hermitian forms, 473
  matrices, 203
  quadratic forms, 423
  systems of equations, 60
  systems of inequalities, 148

Euclidean n-space, 130

Euler's theorem, 458

Eulerian angles, 354

Even permutation, 241, 263 (Ex. 52)

Expansion of a determinant, 242, 243,
  247, 302, 303

Exponent laws for matrices, 15
  (Ex. 12), 38 (Ex. 47)

F

$\mathscr{F}$, 85, $\mathscr{F}^n$, 150

Factorization using
  elementary matrices, 202, 209
    (Ex. 41)
  plane rotations, 357, 359
  unitary and definite matrices, 475
    (Ex. 16)

Field, 85

Finite dimensional vector space, 167

Finite group, 342

First minor, 293

Fixed space
  of an operator, 335 (Ex. 12)
  of a projection, 369

Flat spaces, 123

Four rational operations, 83

Free subscripts, 21

Full linear group, 343

Function, 310

G

Gauss-Jordan elimination, 65

Generalized characteristic value
  problem, 420 (Ex. 8)

Generators of a
  group, 346 (Ex. 9)
  vector space, 167

Geometric vector, 90, 121

Gram-Schmidt process, 177

Gramian matrix and determinant, 453,
  454

Group, 342
  of operators, 342
  of orthogonal transformations, 353
  of permutations, 343, 344
  of unitary transformations, 352

H

Half-line, 134

Hermitian
  form, 159 (Ex. 5), 471
  matrix, 23, 347
  unit form, 352

Hessian matrix, 461
Higher traces of a matrix, 390
Homogeneous system of equations, 70, 210, 212
Hyperbola, 426
Hyperboloid, 427
Hyperplane, 124, 278, 286 (Ex. 37)
Hyperquadric, 427

I

i (imaginary unit), 493
I_n, 26
i-axis of $\mathcal{R}^n$, 121
Idempotent matrix, 18 (Ex. 32)
Identically partitioned matrices, 43
Identity element
  for addition, 4
  of a group, 342
  for multiplication, 26
Identity matrix, 26
Identity operator, 340
Image, 187, 310, 317
Imaginary part of a complex number, 491
Imaginary unit, 493
Improper orthogonal transformation (determinant-1), 360 (Ex. 3)
Inclusion relation for projections, 374 (Ex. 11)
Inconsistent system of equations, 52
Index of
  a bilinear form, 468
  a Hermitian form, 473
  a quadratic form, 446
  summation, 21, 478
Initial point of a segment, 121
Inner product, 30, 339 (Ex. 40)
Intercept form, 110 (Ex. 8), 128 (Ex. 13)
Intersection of two vector spaces, 183
Into mapping, 312
Invariance of
  determinant and trace, 393
  Hermitian unit form, 352
  inner product, 354
  length, 328, 351
  orthogonality, 353, 354

Invariant, 328
  subspace, 340
Invariants of an operator, 393
Inverse
  of a complex number, 493
  computation of, 75, 78
  determinantal formula for, 271
  of an elementary transformation, 200
  of a field element, 86
  of a mapping, 312
  of a matrix, 27, 75, 78, 271
  of a non-symmetric matrix, 78
  by partitioning, 78
  of a permutation matrix, 346 (Ex. 17), 347 (Ex. 19)
  of a ring element, 87
  of a scalar matrix, 27
  of a symmetric matrix, 78
Inversion, 242
Invertible
  function or mapping, 313
  matrix, 28
Isomorphic
  groups, 346 (Ex. 8)
  systems, 94
  vector spaces, 188
Isomorphism, 25, 94

K

k-flat, 123
Kernel of an operator, 319
Kronecker delta, 249
Kronecker's reduction, 443

L

λ (characteristic root), 375
Lagrange's reduction, 438
Laplace's expansion, 293, 298
Latent roots and vectors, 375
Laws of exponents for matrices, 15 (Ex. 12), 38 (Ex. 47)
Leading principal minors, 293, 432, 433, 449

Left
  identity, 36 (Ex. 36), 89 (Ex. 18), 226 (Ex. 41)
  inverse, 39 (Ex. 53), 226 (Ex. 40)
Length of a vector, 94, 129, 136
Line
  in $\mathscr{E}^3$, 102
  of a matrix, 1
  in $\mathscr{R}^n$, 122
  segment in $\mathscr{R}^n$, 121
Linear
  equations, 18, 19, 51, 85, 210, 217, 229
  form, 163 (Ex. 34)
  homogeneous transformation, 7, 315
  mapping, 313
  operator, 315, 319
  transformation, 315, 351
  transformation of coordinates, 325, 329
Linear combination, 111, 125, 153, 157
Linearly dependent and independent
  characteristic vectors, 381
  equations, 232
  vectors, 114, 125, 154, 155
Lower matrix, 40 (Ex. 69)
Lower triangular matrix, 89 (Ex. 13)

M

$m(\lambda)$, 414
$m \times n$, (m,n), 2
m_x (mean), 483 (Ex. 3)
M_{ij}, 293
$M_{(i)(j)}$, 294
$\mathscr{M}_{(p)(k)}$, 301
Main diagonal, 2
Major determinants, 267, 288
Many-to-one mapping, 312
Mapping, 310, 312
Matrix, 1
  of a bilinear form, 464
  notation for linear equations, 19
  of a quadratic form, 423
Maximum value of a quadratic form, 459
Method of detached coefficients, 66

Minimum polynomial of a matrix, 414
Minimum value of a quadratic form, 463 (Ex. 13)
Minor determinant, 272, 293
  of order r, 272
  of order zero, 293
Multilinear form, 471 (Ex. 19)
Multiplication
  by a diagonal matrix, 18 (Ex. 30, 31), 48 (Ex. 13)
  of matrices, 7, 8
  of partitioned matrices, 43
  by a scalar, 6, 14 (Ex. 7), 92
  by a scalar matrix, 24
Mutually orthogonal vector spaces, 361

N

$N(A)$, 215
n-quadric, 427
n-sphere, 141
n-square matrix, 2
n-vector, 19
Natural basis, 323
Negative
  of a field element, 86
  half of a line, 135
  half space, 142
  of a matrix, 4
  of a ring element, 86
  sense on a line, 135
  of a vector, 150
Negative definite or semidefinite
  matrix, 436 (Ex. 22)
  quadratic form, 429
Nilpotent matrix, 17 (Ex. 25), 39 (Ex. 54), 41 (Ex. 75–78)
Noncommutativity of multiplication, 10, 340
Nondegenerate
  basic solution, 235
  critical point, 461
Nondiagonalizable matrix, 407 (Ex. 13)
Nonhomogeneous equations, 217, 219, 229, 231
Non-negative definite form, 429

Non-negative orthant, 147
Nonsingular
  bilinear form, 471 (Ex. 18)
  Hermitian form, 472
  matrix, 28
  operator, 319
Nontrivial solution, 70
Norm of a vector, 129
Normal
  form of a matrix, 194
  to a hyperplane, 139 (Ex. 9)
  matrices, 404
  to a plane, 108
  to a quadric, 438 (Ex. 36)
Normalization, 133
Normalized vector, 101 (Ex. 8), 133
Null space
  of a matrix, 211
  of an operator, 319, 335 (Ex. 13)
  of a projection, 367
Nullity, 215, 319
Number field, 83, 494

O

o (operation), 17 (Ex. 24), 342
$\overrightarrow{OP}$, 90, 121
Odd permutation, 241, 263 (Ex. 52)
One-to-one mapping, 312
Onto mapping, 312
Open half-line, half-space, 135, 142
Operator. *See* Linear operator
Opposite
  parity, 242
  senses, 133
  sides of a hyperplane, 141
Order
  of a finite group, 342
  of a matrix, 2
  of summation, 484
Ordered differences, 97, 121
Origin of $\mathscr{R}^n$, 121
Orthocomplements, 363
Orthogonal
  diagonalization of a quadratic form,
    424
  matrix, 348

  projection, 371
  subspaces, 361
  transformation, 353
  vectors, 99, 134, 136
Orthogonality condition, 99, 134, 136
Orthonormal basis (reference system),
    118, 134, 177, 348

P

p (index of a form), 446, 468, 473
$\phi(\lambda)$, 387
π (product notation), 488
Pairs of matrices (quadratic forms),
    455
Parabola, 287 (Ex. 44)
Parallel
  hyperplanes, 148 (Ex. 10)
  line and hyperplane, 124
  lines, 135
Parallelogram law of addition, 92
Parametric
  equation for a hyperplane, 123
  equation(s) for a line, 102, 103
  equation for a plane, 105
  solution of a system of equations,
    56
Parity index, 241
Particular solution, 60
Pauli spin matrices, 17 (Ex. 22), 163
    (Ex. 31, 32), 351 (Ex. 27)
Permutation, 241
  group, 344
  matrix, 344, 346 (Ex. 17), 347
    (Ex. 19, 22)
Perpendicular bisector, 141 (Ex. 28)
Plane
  in $\mathscr{E}^3$, 105
  in $\mathscr{R}^n$, 123
  rotation, 359
Point of $\mathscr{R}^n$, 121
Polar representation of a matrix, 475
    (Ex. 16)
Polynomial function of a matrix, 30,
    409
Polynomial matrix equation, 31, 411,
    414

Positive
   half of a line, 135
   half space, 142
   orthant, 147
   sense on a line, 135
Positive definite or semidefinite
   Hermitian form, 474
   matrix, 431
   quadratic form, 429, 433, 452
Postmultiplication, 9
Powers of matrices, 418
Premultiplication, 9
Preservation of a bilinear form, 470
      (Ex. 4)
Preservation of an operation, 94
Principal
   axes, 401, 426, 427, 428
   axis transformation, 401, 426
   diagonal, 2
   minors, 291, 293
Principle of superposition, 225
Product
   of complex numbers, 491
   of matrices, 8
   of operators, 340
   of partitioned matrices, 43
Projection, 368
   matrix, 370, 406 (Ex. 8)
   of a vector on a subspace, 364
   of a vector on a unit vector, 182
      (Ex. 5)
Proper orthogonal matrix (determinant
      1), 359, 360 (Ex. 2, 6)
Proportionality of
   cofactors, 290
   major determinants, 288
Pseudodiagonal matrix, 47 (Ex. 6)
Pure imaginary number, 493

Q

Quadratic form, 164 (Ex. 37), 422
Quadric, 427
Quadric surface, 427
Quasidiagonal matrix, 47 (Ex. 6)
Quaternion rings, 387 (Ex. 32, 33)

R

$r(A)$, 194
$\mathcal{R}$ (real number field), 150
$\mathcal{R}^n$, 121
Range of a mapping, 311
Range of a projection, 368
Rank
   of a bilinear form, 464
   of a Hermitian form, 472
   of a matrix, 194
   normal form, 194
   of a product, 196, 197, 217
   of a quadratic form, 423
   of a sum, 198, 217
   of a system of equations, 218
Rational
   function field, 86
   function of a matrix, 89
   number field, 83
   operations, 83
Real
   bilinear form, 464
   matrix, 23
   number field, 83
   part of a complex number, 491
Reciprocal
   bilinear form, 470 (Ex. 11)
   of a complex number, 493
   of a field element, 86
   of a ring element, 87
Reduced echelon form, 65
Reduction to normal form, 194, 201
Reference system, 348
Reflection, 344 (Ex. 4), 359
Reflexive property, 3
Replacement theorems, 170, 171
Resultant, 261 (Ex. 41)
Right identity, 36 (Ex. 36), 89 (Ex. 18),
      226 (Ex. 41)
Right inverse, 39 (Ex. 53), 226 (Ex. 40)
Ring, 86, 387 (Ex. 33)
   with unit, 87
Row
   expansion of a determinant, 242
   matrix, 19
   of a matrix, 1

rank, 192
space, 191
Row-and-column expansion, 302

S

Σ (summation notation), 477
s_x^2 (variance), 434 (Ex. 8)
Same parity, 242
Scalar, 3
    matrix, 24, 35 (Ex. 24)
    multiple of a matrix, 6, 14 (Ex. 7)
    product of vectors, 29, 98
    solution of a matrix equation, 31
    triple product of vectors, 323
        (Ex. 25)
Scattering matrices, 349 (Ex. 9)
Secondary diagonal, 2
Secular equation, 376
Shear transformation, 335 (Ex. 8)
Signature
    of a Hermitian form, 473
    of a real quadratic form, 446
Similar matrices, 332, 392
Similarity transformation, 332, 334
        (Ex. 3), 361 (Ex. 17)
Singular
    Hermitian form, 472
    mapping, 319
    matrix, 28
Skew-Hermitian matrix, 35 (Ex. 27),
        406 (Ex. 1)
Skew-symmetric matrix, 22, 226
        (Ex. 42)
Solution of a system of equations, 20,
        60, 212, 219, 231
Span, 167
Sparse matrix, 274
Spectral decomposition, 409 (Ex. 33),
        437 (Ex. 34)
Stable equation, 262 (Ex. 42)
Standard basis, 117, 323
Steinitz replacement theorem, 171
Strictly-into mapping, 312
Subfield, 83, 494
Subgroup, 342

Submatrix, 41, 293
Subspace of a vector space, 153
Subtraction of
    matrices, 4
    vectors, 97
Sum
    column for checking, 66
    of matrices, 3
    of vector spaces, 183
Summand, 478
Summation
    index, 21, 478
    notation, 477
Superdiagonals, 41 (Ex. 74)
Superposition, 225 (Ex. 33)
Sweepout process, 54
    for computing determinants, 252
    for computing rank, 194
    for inverting matrices, 75
    for solving equations, 66
Sylvester's law of inertia, 447, 473
Sylvester's law of nullity, 217
Symmetric
    bilinear form, 464
    group, 343
    matrix, 22, 33 (Ex. 7, 12)
    property, 3
Symmetrically partitioned matrices, 46
        (Ex. 3)
Synthetic elimination, 66
System of
    homogeneous equations, 70, 210,
        212
    nonhomogeneous equations, 65, 217,
        229

T

tr A, 17 (Ex. 28)
Tangent to a quadric, 437 (Ex. 35)
Terminal point of a segment, 121
Tetrahedron (volume of), 287 (Ex. 43)
Total number of inversions, 242
Trace, 17 (Ex. 28), 18 (Ex. 29)
Tranjugate, 23, 347

Transformation
  of coordinates, 323
  of a linear operator, 331
Transitive property, 3, 392
Translation, 221, 334 (Ex. 5)
Transpose of a matrix, 20
Transposed conjugate, 23
Transposed homogeneous system, 366
Transposition, 296
Triangle inequality, 130, 131
Triangular product, 488
Triangular sum, 486 (Ex. 9)
Triangularization
  of a complex matrix, 403
  of a real matrix, 356
Tridiagonal matrix, 259 (Ex. 34)
Trilinear form, 485 (Ex. 1)
Trivial solution, 70

U

$\mathcal{U}^n$, 136
Uniqueness of
  identity, 6, 26
  inverse, 6, 27
  solution, 221 (Ex. 2)
  unit element, 88 (Ex. 5)
  zero element, 88 (Ex. 5)
Unimodular matrix, 348
Unit
  element of a field, 86
  element of a ring, 87
  matrix, 26
  n-sphere, 133, 141 (Ex. 29)
  point, 135
  sphere, 95
  vector, 94, 129
Unitary
  matrix, 347
  n-space, 136
  operator, 351
  transformation, 351

Upper matrix, 40 (Ex. 68), 41 (Ex. 74)
Upper triangular matrix, 89 (Ex. 13),
  162 (Ex. 21)

V

Values of quadratic forms, 457
Vandermonde matrix, 257 (Ex. 23),
  258
Variance, 434 (Ex. 8)
Variables one can solve for, 233
Vector, 2, 19, 90
  equation of a line, 102
  equation of a hyperplane, 123, 124
  equation of a plane, 105
  orthogonal to a subspace, 361
  space, 104, 107, 126, 149
Volume of a
  parallelepiped, 339 (Ex. 39)
  tetrahedron, 287 (Ex. 43)

W

Weighted permutation matrix, 347

X

$|X|$, 94, 129, 136
$\langle X, Y \rangle$ (inner product), 30

Z

$|z|$, 497
$\mathcal{Z}$ (zero space), 150, 167
Zero
  element, 86
  matrix, 4
  space, 150, 167
  vector, 90, 150

A CATALOG OF SELECTED
DOVER BOOKS
IN SCIENCE AND MATHEMATICS

A CATALOG OF SELECTED

DOVER BOOKS

IN SCIENCE AND MATHEMATICS

Astronomy

BURNHAM'S CELESTIAL HANDBOOK, Robert Burnham, Jr. Thorough guide to the stars beyond our solar system. Exhaustive treatment. Alphabetical by constellation: Andromeda to Cetus in Vol. 1; Chamaeleon to Orion in Vol. 2; and Pavo to Vulpecula in Vol. 3. Hundreds of illustrations. Index in Vol. 3. 2,000pp. 6⅛ x 9¼.
23567-X, 23568-8, 23673-0 Three-vol. set

THE EXTRATERRESTRIAL LIFE DEBATE, 1750–1900, Michael J. Crowe. First detailed, scholarly study in English of the many ideas that developed from 1750 to 1900 regarding the existence of intelligent extraterrestrial life. Examines ideas of Kant, Herschel, Voltaire, Percival Lowell, many other scientists and thinkers. 16 illustrations. 704pp. 5⅜ x 8½. 40675-X

A HISTORY OF ASTRONOMY, A. Pannekoek. Well-balanced, carefully reasoned study covers such topics as Ptolemaic theory, work of Copernicus, Kepler, Newton, Eddington's work on stars, much more. Illustrated. References. 521pp. 5⅜ x 8½.
65994-1

AMATEUR ASTRONOMER'S HANDBOOK, J. B. Sidgwick. Timeless, comprehensive coverage of telescopes, mirrors, lenses, mountings, telescope drives, micrometers, spectroscopes, more. 189 illustrations. 576pp. 5⅝ x 8¼. (Available in U.S. only.)
24034-7

STARS AND RELATIVITY, Ya. B. Zel'dovich and I. D. Novikov. Vol. 1 of *Relativistic Astrophysics* by famed Russian scientists. General relativity, properties of matter under astrophysical conditions, stars, and stellar systems. Deep physical insights, clear presentation. 1971 edition. References. 544pp. 5⅝ x 8¼. 69424-0

Chemistry

CHEMICAL MAGIC, Leonard A. Ford. Second Edition, Revised by E. Winston Grundmeier. Over 100 unusual stunts demonstrating cold fire, dust explosions, much more. Text explains scientific principles and stresses safety precautions. 128pp. 5⅜ x 8½. 67628-5

THE DEVELOPMENT OF MODERN CHEMISTRY, Aaron J. Ihde. Authoritative history of chemistry from ancient Greek theory to 20th-century innovation. Covers major chemists and their discoveries. 209 illustrations. 14 tables. Bibliographies. Indices. Appendices. 851pp. 5⅜ x 8½. 64235-6

CATALYSIS IN CHEMISTRY AND ENZYMOLOGY, William P. Jencks. Exceptionally clear coverage of mechanisms for catalysis, forces in aqueous solution, carbonyl- and acyl-group reactions, practical kinetics, more. 864pp. 5⅜ x 8½.
65460-5

THE HISTORICAL BACKGROUND OF CHEMISTRY, Henry M. Leicester. Evolution of ideas, not individual biography. Concentrates on formulation of a coherent set of chemical laws. 260pp. 5⅜ x 8½.   61053-5

A SHORT HISTORY OF CHEMISTRY, J. R. Partington. Classic exposition explores origins of chemistry, alchemy, early medical chemistry, nature of atmosphere, theory of valency, laws and structure of atomic theory, much more. 428pp. 5⅜ x 8½. (Available in U.S. only.)   65977-1

GENERAL CHEMISTRY, Linus Pauling. Revised 3rd edition of classic first-year text by Nobel laureate. Atomic and molecular structure, quantum mechanics, statistical mechanics, thermodynamics correlated with descriptive chemistry. Problems. 992pp. 5⅜ x 8½.   65622-5

Engineering

DE RE METALLICA, Georgius Agricola. The famous Hoover translation of greatest treatise on technological chemistry, engineering, geology, mining of early modern times (1556). All 289 original woodcuts. 638pp. 6¾ x 11.   60006-8

FUNDAMENTALS OF ASTRODYNAMICS, Roger Bate et al. Modern approach developed by U.S. Air Force Academy. Designed as a first course. Problems, exercises. Numerous illustrations. 455pp. 5⅜ x 8½.   60061-0

DYNAMICS OF FLUIDS IN POROUS MEDIA, Jacob Bear. For advanced students of ground water hydrology, soil mechanics and physics, drainage and irrigation engineering and more. 335 illustrations. Exercises, with answers. 784pp. 6⅛ x 9¼.
65675-6

ANALYTICAL MECHANICS OF GEARS, Earle Buckingham. Indispensable reference for modern gear manufacture covers conjugate gear-tooth action, gear-tooth profiles of various gears, many other topics. 263 figures. 102 tables. 546pp. 5⅜ x 8½.
65712-4

MECHANICS, J. P. Den Hartog. A classic introductory text or refresher. Hundreds of applications and design problems illuminate fundamentals of trusses, loaded beams and cables, etc. 334 answered problems. 462pp. 5⅜ x 8½.   60754-2

MECHANICAL VIBRATIONS, J. P. Den Hartog. Classic textbook offers lucid explanations and illustrative models, applying theories of vibrations to a variety of practical industrial engineering problems. Numerous figures. 233 problems, solutions. Appendix. Index. Preface. 436pp. 5⅜ x 8½.   64785-4

STRENGTH OF MATERIALS, J. P. Den Hartog. Full, clear treatment of basic material (tension, torsion, bending, etc.) plus advanced material on engineering methods, applications. 350 answered problems. 323pp. 5⅜ x 8½.   60755-0

A HISTORY OF MECHANICS, René Dugas. Monumental study of mechanical principles from antiquity to quantum mechanics. Contributions of ancient Greeks, Galileo, Leonardo, Kepler, Lagrange, many others. 671pp. 5⅜ x 8½.   65632-2

METAL FATIGUE, N. E. Frost, K. J. Marsh, and L. P. Pook. Definitive, clearly written, and well-illustrated volume addresses all aspects of the subject, from the historical development of understanding metal fatigue to vital concepts of the cyclic stress that causes a crack to grow. Includes 7 appendixes. 544pp. 5⅜ x 8½. 40927-9

STATISTICAL MECHANICS: Principles and Applications, Terrell L. Hill. Standard text covers fundamentals of statistical mechanics, applications to fluctuation theory, imperfect gases, distribution functions, more. 448pp. 5⅜ x 8½. 65390-0

THE VARIATIONAL PRINCIPLES OF MECHANICS, Cornelius Lanczos. Graduate level coverage of calculus of variations, equations of motion, relativistic mechanics, more. First inexpensive paperbound edition of classic treatise. Index. Bibliography. 418pp. 5⅜ x 8½. 65067-7

THE VARIOUS AND INGENIOUS MACHINES OF AGOSTINO RAMELLI: A Classic Sixteenth-Century Illustrated Treatise on Technology, Agostino Ramelli. One of the most widely known and copied works on machinery in the 16th century. 194 detailed plates of water pumps, grain mills, cranes, more. 608pp. 9 x 12. 28180-9

ORDINARY DIFFERENTIAL EQUATIONS AND STABILITY THEORY: An Introduction, David A. Sánchez. Brief, modern treatment. Linear equation, stability theory for autonomous and nonautonomous systems, etc. 164pp. 5⅜ x 8¼. 63828-6

ROTARY WING AERODYNAMICS, W. Z. Stepniewski. Clear, concise text covers aerodynamic phenomena of the rotor and offers guidelines for helicopter performance evaluation. Originally prepared for NASA. 537 figures. 640pp. 6⅛ x 9¼. 64647-5

INTRODUCTION TO SPACE DYNAMICS, William Tyrrell Thomson. Comprehensive, classic introduction to space-flight engineering for advanced undergraduate and graduate students. Includes vector algebra, kinematics, transformation of coordinates. Bibliography. Index. 352pp. 5⅜ x 8½. 65113-4

HISTORY OF STRENGTH OF MATERIALS, Stephen P. Timoshenko. Excellent historical survey of the strength of materials with many references to the theories of elasticity and structure. 245 figures. 452pp. 5⅜ x 8½. 61187-6

ANALYTICAL FRACTURE MECHANICS, David J. Unger. Self-contained text supplements standard fracture mechanics texts by focusing on analytical methods for determining crack-tip stress and strain fields. 336pp. 6⅛ x 9¼. 41737-9

Mathematics

HANDBOOK OF MATHEMATICAL FUNCTIONS WITH FORMULAS, GRAPHS, AND MATHEMATICAL TABLES, edited by Milton Abramowitz and Irene A. Stegun. Vast compendium: 29 sets of tables, some to as high as 20 places. 1,046pp. 8 x 10½. 61272-4

FUNCTIONAL ANALYSIS (Second Corrected Edition), George Bachman and Lawrence Narici. Excellent treatment of subject geared toward students with background in linear algebra, advanced calculus, physics and engineering. Text covers introduction to inner-product spaces, normed, metric spaces, and topological spaces; complete orthonormal sets, the Hahn-Banach Theorem and its consequences, and many other related subjects. 1966 ed. 544pp. 6⅛ x 9¼. 40251-7

ASYMPTOTIC EXPANSIONS OF INTEGRALS, Norman Bleistein & Richard A. Handelsman. Best introduction to important field with applications in a variety of scientific disciplines. New preface. Problems. Diagrams. Tables. Bibliography. Index. 448pp. 5⅜ x 8½. 65082-0

FAMOUS PROBLEMS OF GEOMETRY AND HOW TO SOLVE THEM, Benjamin Bold. Squaring the circle, trisecting the angle, duplicating the cube: learn their history, why they are impossible to solve, then solve them yourself. 128pp. 5⅜ x 8½. 24297-8

VECTOR AND TENSOR ANALYSIS WITH APPLICATIONS, A. I. Borisenko and I. E. Tarapov. Concise introduction. Worked-out problems, solutions, exercises. 257pp. 5⅝ x 8¼. 63833-2

THE ABSOLUTE DIFFERENTIAL CALCULUS (CALCULUS OF TENSORS), Tullio Levi-Civita. Great 20th-century mathematician's classic work on material necessary for mathematical grasp of theory of relativity. 452pp. 5⅝ x 8¼. 63401-9

AN INTRODUCTION TO ORDINARY DIFFERENTIAL EQUATIONS, Earl A. Coddington. A thorough and systematic first course in elementary differential equations for undergraduates in mathematics and science, with many exercises and problems (with answers). Index. 304pp. 5⅜ x 8½. 65942-9

FOURIER SERIES AND ORTHOGONAL FUNCTIONS, Harry F. Davis. An incisive text combining theory and practical example to introduce Fourier series, orthogonal functions and applications of the Fourier method to boundary-value problems. 570 exercises. Answers and notes. 416pp. 5⅜ x 8½. 65973-9

COMPUTABILITY AND UNSOLVABILITY, Martin Davis. Classic graduate-level introduction to theory of computability, usually referred to as theory of recurrent functions. New preface and appendix. 288pp. 5⅜ x 8½. 61471-9

ASYMPTOTIC METHODS IN ANALYSIS, N. G. de Bruijn. An inexpensive, comprehensive guide to asymptotic methods–the pioneering work that teaches by explaining worked examples in detail. Index. 224pp. 5⅜ x 8½ 64221-6

ESSAYS ON THE THEORY OF NUMBERS, Richard Dedekind. Two classic essays by great German mathematician: on the theory of irrational numbers; and on transfinite numbers and properties of natural numbers. 115pp. 5⅜ x 8½. 21010-3

CATALOG OF DOVER BOOKS

APPLIED COMPLEX VARIABLES, John W. Dettman. Step-by-step coverage of fundamentals of analytic function theory–plus lucid exposition of five important applications: Potential Theory; Ordinary Differential Equations; Fourier Transforms; Laplace Transforms; Asymptotic Expansions. 66 figures. Exercises at chapter ends. 512pp. 5⅜ x 8½. 64670-X

INTRODUCTION TO LINEAR ALGEBRA AND DIFFERENTIAL EQUATIONS, John W. Dettman. Excellent text covers complex numbers, determinants, orthonormal bases, Laplace transforms, much more. Exercises with solutions. Undergraduate level. 416pp. 5⅜ x 8½. 65191-6

MATHEMATICAL METHODS IN PHYSICS AND ENGINEERING, John W. Dettman. Algebraically based approach to vectors, mapping, diffraction, other topics in applied math. Also generalized functions, analytic function theory, more. Exercises. 448pp. 5⅜ x 8½. 65649-7

CALCULUS OF VARIATIONS WITH APPLICATIONS, George M. Ewing. Applications-oriented introduction to variational theory develops insight and promotes understanding of specialized books, research papers. Suitable for advanced undergraduate/graduate students as primary, supplementary text. 352pp. 5⅜ x 8½. 64856-7

COMPLEX VARIABLES, Francis J. Flanigan. Unusual approach, delaying complex algebra till harmonic functions have been analyzed from real variable viewpoint. Includes problems with answers. 364pp. 5⅜ x 8½. 61388-7

AN INTRODUCTION TO THE CALCULUS OF VARIATIONS, Charles Fox. Graduate-level text covers variations of an integral, isoperimetrical problems, least action, special relativity, approximations, more. References. 279pp. 5⅜ x 8½. 65499-0

CATASTROPHE THEORY FOR SCIENTISTS AND ENGINEERS, Robert Gilmore. Advanced-level treatment describes mathematics of theory grounded in the work of Poincaré, R. Thom, other mathematicians. Also important applications to problems in mathematics, physics, chemistry and engineering. 1981 edition. References. 28 tables. 397 black-and-white illustrations. xvii + 666pp. 6⅛ x 9¼. 67539-4

INTRODUCTION TO DIFFERENCE EQUATIONS, Samuel Goldberg. Exceptionally clear exposition of important discipline with applications to sociology, psychology, economics. Many illustrative examples; over 250 problems. 260pp. 5⅜ x 8½. 65084-7

NUMERICAL METHODS FOR SCIENTISTS AND ENGINEERS, Richard Hamming. Classic text stresses frequency approach in coverage of algorithms, polynomial approximation, Fourier approximation, exponential approximation, other topics. Revised and enlarged 2nd edition. 721pp. 5⅜ x 8½. 65241-6

INTRODUCTION TO NUMERICAL ANALYSIS (2nd Edition), F. B. Hildebrand. Classic, fundamental treatment covers computation, approximation, interpolation, numerical differentiation and integration, other topics. 150 new problems. 669pp. 5⅜ x 8½. 65363-3

THE FUNCTIONS OF MATHEMATICAL PHYSICS, Harry Hochstadt. Comprehensive treatment of orthogonal polynomials, hypergeometric functions, Hill's equation, much more. Bibliography. Index. 322pp. 5⅜ x 8½.   65214-9

THREE PEARLS OF NUMBER THEORY, A. Y. Khinchin. Three compelling puzzles require proof of a basic law governing the world of numbers. Challenges concern van der Waerden's theorem, the Landau-Schnirelmann hypothesis and Mann's theorem, and a solution to Waring's problem. Solutions included. 64pp. 5⅜ x 8½.
40026-3

CALCULUS REFRESHER FOR TECHNICAL PEOPLE, A. Albert Klaf. Covers important aspects of integral and differential calculus via 756 questions. 566 problems, most answered. 431pp. 5⅜ x 8½.   20370-0

THE PHILOSOPHY OF MATHEMATICS: An Introductory Essay, Stephan Körner. Surveys the views of Plato, Aristotle, Leibniz & Kant concerning propositions and theories of applied and pure mathematics. Introduction. Two appendices. Index. 198pp. 5⅜ x 8½.   25048-2

INTRODUCTORY REAL ANALYSIS, A.N. Kolmogorov, S. V. Fomin. Translated by Richard A. Silverman. Self-contained, evenly paced introduction to real and functional analysis. Some 350 problems. 403pp. 5⅜ x 8½.   61226-0

APPLIED ANALYSIS, Cornelius Lanczos. Classic work on analysis and design of finite processes for approximating solution of analytical problems. Algebraic equations, matrices, harmonic analysis, quadrature methods, much more. 559pp. 5⅜ x 8½.
65656-X

AN INTRODUCTION TO ALGEBRAIC STRUCTURES, Joseph Landin. Superb self-contained text covers "abstract algebra": sets and numbers, theory of groups, theory of rings, much more. Numerous well-chosen examples, exercises. 247pp. 5⅜ x 8½.
65940-2

SPECIAL FUNCTIONS, N. N. Lebedev. Translated by Richard Silverman. Famous Russian work treating more important special functions, with applications to specific problems of physics and engineering. 38 figures. 308pp. 5⅜ x 8½.   60624-4

QUALITATIVE THEORY OF DIFFERENTIAL EQUATIONS, V. V. Nemytskii and V.V. Stepanov. Classic graduate-level text by two prominent Soviet mathematicians covers classical differential equations as well as topological dynamics and ergodic theory. Bibliographies. 523pp. 5⅜ x 8½.   65954-2

NUMBER THEORY AND ITS HISTORY, Oystein Ore. Unusually clear, accessible introduction covers counting, properties of numbers, prime numbers, much more. Bibliography. 380pp. 5⅜ x 8½.   65620-9

THEORY OF MATRICES, Sam Perlis. Outstanding text covering rank, nonsingularity and inverses in connection with the development of canonical matrices under the relation of equivalence, and without the intervention of determinants. Includes exercises. 237pp. 5⅜ x 8½.   66810-X

INTRODUCTION TO ANALYSIS, Maxwell Rosenlicht. Unusually clear, accessible coverage of set theory, real number system, metric spaces, continuous functions, Riemann integration, multiple integrals, more. Wide range of problems. Undergraduate level. Bibliography. 254pp. 5⅜ x 8½. 65038-3

MODERN NONLINEAR EQUATIONS, Thomas L. Saaty. Emphasizes practical solution of problems; covers seven types of equations. ". . . a welcome contribution to the existing literature...."–*Math Reviews.* 490pp. 5⅜ x 8½. 64232-1

MATRICES AND LINEAR ALGEBRA, Hans Schneider and George Phillip Barker. Basic textbook covers theory of matrices and its applications to systems of linear equations and related topics such as determinants, eigenvalues and differential equations. Numerous exercises. 432pp. 5⅜ x 8½. 66014-1

MATHEMATICS APPLIED TO CONTINUUM MECHANICS, Lee A. Segel. Analyzes models of fluid flow and solid deformation. For upper-level math, science and engineering students. 608pp. 5⅜ x 8½. 65369-2

ELEMENTS OF REAL ANALYSIS, David A. Sprecher. Classic text covers fundamental concepts, real number system, point sets, functions of a real variable, Fourier series, much more. Over 500 exercises. 352pp. 5⅜ x 8½. 65385-4

AN INTRODUCTION TO MATRICES, SETS AND GROUPS FOR SCIENCE STUDENTS, G. Stephenson. Concise, readable text introduces sets, groups, and most importantly, matrices to undergraduate students of physics, chemistry, and engineering. Problems. 164pp. 5⅜ x 8½. 65077-4

SET THEORY AND LOGIC, Robert R. Stoll. Lucid introduction to unified theory of mathematical concepts. Set theory and logic seen as tools for conceptual understanding of real number system. 496pp. 5⅜ x 8¼. 63829-4

TENSOR CALCULUS, J.L. Synge and A. Schild. Widely used introductory text covers spaces and tensors, basic operations in Riemannian space, non-Riemannian spaces, etc. 324pp. 5⅜ x 8¼. 63612-7

ORDINARY DIFFERENTIAL EQUATIONS, Morris Tenenbaum and Harry Pollard. Exhaustive survey of ordinary differential equations for undergraduates in mathematics, engineering, science. Thorough analysis of theorems. Diagrams. Bibliography. Index. 818pp. 5⅜ x 8½. 64940-7

INTEGRAL EQUATIONS, F. G. Tricomi. Authoritative, well-written treatment of extremely useful mathematical tool with wide applications. Volterra Equations, Fredholm Equations, much more. Advanced undergraduate to graduate level. Exercises. Bibliography. 238pp. 5⅜ x 8½. 64828-1

FOURIER SERIES, Georgi P. Tolstov. Translated by Richard A. Silverman. A valuable addition to the literature on the subject, moving clearly from subject to subject and theorem to theorem. 107 problems, answers. 336pp. 5⅜ x 8½. 63317-9

POPULAR LECTURES ON MATHEMATICAL LOGIC, Hao Wang. Noted logician's lucid treatment of historical developments, set theory, model theory, recursion theory and constructivism, proof theory, more. 3 appendixes. Bibliography. 1981 edition. ix + 283pp. 5⅜ x 8½.   67632-3

CALCULUS OF VARIATIONS, Robert Weinstock. Basic introduction covering isoperimetric problems, theory of elasticity, quantum mechanics, electrostatics, etc. Exercises throughout. 326pp. 5⅜ x 8½.   63069-2

THE CONTINUUM: A Critical Examination of the Foundation of Analysis, Hermann Weyl. Classic of 20th-century foundational research deals with the conceptual problem posed by the continuum. 156pp. 5⅜ x 8½.   67982-9

CHALLENGING MATHEMATICAL PROBLEMS WITH ELEMENTARY SOLUTIONS, A. M. Yaglom and I. M. Yaglom. Over 170 challenging problems on probability theory, combinatorial analysis, points and lines, topology, convex polygons, many other topics. Solutions. Total of 445pp. 5⅜ x 8½. Two-vol. set.
Vol. I: 65536-9   Vol. II: 65537-7

A SURVEY OF NUMERICAL MATHEMATICS, David M. Young and Robert Todd Gregory. Broad self-contained coverage of computer-oriented numerical algorithms for solving various types of mathematical problems in linear algebra, ordinary and partial, differential equations, much more. Exercises. Total of 1,248pp. 5⅜ x 8½. Two volumes.   Vol. I: 65691-8   Vol. II: 65692-6

INTRODUCTION TO PARTIAL DIFFERENTIAL EQUATIONS WITH APPLICATIONS, E. C. Zachmanoglou and Dale W. Thoe. Essentials of partial differential equations applied to common problems in engineering and the physical sciences. Problems and answers. 416pp. 5⅜ x 8½.   65251-3

THE THEORY OF GROUPS, Hans J. Zassenhaus. Well-written graduate-level text acquaints reader with group-theoretic methods and demonstrates their usefulness in mathematics. Axioms, the calculus of complexes, homomorphic mapping, p-group theory, more. Many proofs shorter and more transparent than older ones. 276pp. 5⅜ x 8½.   40922-8

DISTRIBUTION THEORY AND TRANSFORM ANALYSIS: An Introduction to Generalized Functions, with Applications, A. H. Zemanian. Provides basics of distribution theory, describes generalized Fourier and Laplace transformations. Numerous problems. 384pp. 5⅜ x 8½.   65479-6

Math–Decision Theory, Statistics, Probability

ELEMENTARY DECISION THEORY, Herman Chernoff and Lincoln E. Moses. Clear introduction to statistics and statistical theory covers data processing, probability and random variables, testing hypotheses, much more. Exercises. 364pp. 5⅜ x 8½.   65218-1

STATISTICS MANUAL, Edwin L. Crow et al. Comprehensive, practical collection of classical and modern methods prepared by U.S. Naval Ordnance Test Station. Stress on use. Basics of statistics assumed. 288pp. 5⅜ x 8½. 60599-X

SOME THEORY OF SAMPLING, William Edwards Deming. Analysis of the problems, theory and design of sampling techniques for social scientists, industrial managers and others who find statistics important at work. 61 tables. 90 figures. xvii +602pp. 5⅜ x 8½. 64684-X

STATISTICAL ADJUSTMENT OF DATA, W. Edwards Deming. Introduction to basic concepts of statistics, curve fitting, least squares solution, conditions without parameter, conditions containing parameters. 26 exercises worked out. 271pp. 5⅜ x 8½. 64685-8

LINEAR PROGRAMMING AND ECONOMIC ANALYSIS, Robert Dorfman, Paul A. Samuelson and Robert M. Solow. First comprehensive treatment of linear programming in standard economic analysis. Game theory, modern welfare economics, Leontief input-output, more. 525pp. 5⅜ x 8½. 65491-5

DICTIONARY/OUTLINE OF BASIC STATISTICS, John E. Freund and Frank J. Williams. A clear concise dictionary of over 1,000 statistical terms and an outline of statistical formulas covering probability, nonparametric tests, much more. 208pp. 5⅜ x 8½. 66796-0

PROBABILITY: An Introduction, Samuel Goldberg. Excellent basic text covers set theory, probability theory for finite sample spaces, binomial theorem, much more. 360 problems. Bibliographies. 322pp. 5⅜ x 8½. 65252-1

GAMES AND DECISIONS: Introduction and Critical Survey, R. Duncan Luce and Howard Raiffa. Superb nontechnical introduction to game theory, primarily applied to social sciences. Utility theory, zero-sum games, n-person games, decision-making, much more. Bibliography. 509pp. 5⅜ x 8½. 65943-7

FIFTY CHALLENGING PROBLEMS IN PROBABILITY WITH SOLUTIONS, Frederick Mosteller. Remarkable puzzlers, graded in difficulty, illustrate elementary and advanced aspects of probability. Detailed solutions. 88pp. 5⅜ x 8½. 65355-2

PROBABILITY THEORY: A Concise Course, Y. A. Rozanov. Highly readable, self-contained introduction covers combination of events, dependent events, Bernoulli trials, etc. 148pp. 5⅜ x 8¼. 63544-9

STATISTICAL METHOD FROM THE VIEWPOINT OF QUALITY CONTROL, Walter A. Shewhart. Important text explains regulation of variables, uses of statistical control to achieve quality control in industry, agriculture, other areas. 192pp. 5⅜ x 8½. 65232-7

THE COMPLEAT STRATEGYST: Being a Primer on the Theory of Games of Strategy, J. D. Williams. Highly entertaining classic describes, with many illustrated examples, how to select best strategies in conflict situations. Prefaces. Appendices. 268pp. 5⅜ x 8½. 25101-2

Math–Geometry and Topology

ELEMENTARY CONCEPTS OF TOPOLOGY, Paul Alexandroff. Elegant, intuitive approach to topology from set-theoretic topology to Betti groups; how concepts of topology are useful in math and physics. 25 figures. 57pp. 5⅜ x 8½. 60747-X

COMBINATORIAL TOPOLOGY, P. S. Alexandrov. Clearly written, well-organized, three-part text begins by dealing with certain classic problems without using the formal techniques of homology theory and advances to the central concept, the Betti groups. Numerous detailed examples. 654pp. 5⅜ x 8½. 40179-0

EXPERIMENTS IN TOPOLOGY, Stephen Barr. Classic, lively explanation of one of the byways of mathematics. Klein bottles, Moebius strips, projective planes, map coloring, problem of the Koenigsberg bridges, much more, described with clarity and wit. 43 figures. 210pp. 5⅜ x 8½. 25933-1

CONFORMAL MAPPING ON RIEMANN SURFACES, Harvey Cohn. Lucid, insightful book presents ideal coverage of subject. 334 exercises make book perfect for self-study. 55 figures. 352pp. 5⅝ x 8¼. 64025-6

THE GEOMETRY OF RENÉ DESCARTES, René Descartes. The great work founded analytical geometry. Original French text, Descartes's own diagrams, together with definitive Smith-Latham translation. 244pp. 5⅜ x 8½. 60068-8

THE THIRTEEN BOOKS OF EUCLID'S ELEMENTS, translated with introduction and commentary by Sir Thomas L. Heath. Definitive edition. Textual and linguistic notes, mathematical analysis. 2,500 years of critical commentary. Unabridged. 1,414pp. 5⅜ x 8½.. Three-vol. set.
Vol. I: 60088-2 Vol. II: 60089-0 Vol. III: 60090-4

GEOMETRY OF COMPLEX NUMBERS, Hans Schwerdtfeger. Illuminating, widely praised book on analytic geometry of circles, the Moebius transformation, and two-dimensional non-Euclidean geometries. 200pp. 5⅝ x 8¼. 63830-8

DIFFERENTIAL GEOMETRY, Heinrich W. Guggenheimer. Local differential geometry as an application of advanced calculus and linear algebra. Curvature, transformation groups, surfaces, more. Exercises. 62 figures. 378pp. 5⅜ x 8½. 63433-7

CURVATURE AND HOMOLOGY: Enlarged Edition, Samuel I. Goldberg. Revised edition examines topology of differentiable manifolds; curvature, homology of Riemannian manifolds; compact Lie groups; complex manifolds; curvature, homology of Kaehler manifolds. New Preface. Four new appendixes. 416pp. 5⅜ x 8½.
40207-X

TOPOLOGY, John G. Hocking and Gail S. Young. Superb one-year course in classical topology. Topological spaces and functions, point-set topology, much more. Examples and problems. Bibliography. Index. 384pp. 5⅝ x 8¼. 65676-4

LECTURES ON CLASSICAL DIFFERENTIAL GEOMETRY, Second Edition, Dirk J. Struik. Excellent brief introduction covers curves, theory of surfaces, fundamental equations, geometry on a surface, conformal mapping, other topics. Problems. 240pp. 5⅜ x 8½. 65609-8

Math–History of

A SHORT ACCOUNT OF THE HISTORY OF MATHEMATICS, W. W. Rouse Ball. One of clearest, most authoritative surveys from the Egyptians and Phoenicians through 19th-century figures such as Grassman, Galois, Riemann. Fourth edition. 522pp. 5⅜ x 8½. 20630-0

THE HISTORY OF THE CALCULUS AND ITS CONCEPTUAL DEVELOPMENT, Carl B. Boyer. Origins in antiquity, medieval contributions, work of Newton, Leibniz, rigorous formulation. Treatment is verbal. 346pp. 5⅜ x 8½. 60509-4

THE HISTORICAL ROOTS OF ELEMENTARY MATHEMATICS, Lucas N. H. Bunt, Phillip S. Jones, and Jack D. Bedient. Fundamental underpinnings of modern arithmetic, algebra, geometry and number systems derived from ancient civilizations. 320pp. 5⅜ x 8½. 25563-8

A HISTORY OF MATHEMATICAL NOTATIONS, Florian Cajori. This classic study notes the first appearance of a mathematical symbol and its origin, the competition it encountered, its spread among writers in different countries, its rise to popularity, its eventual decline or ultimate survival. Original 1929 two-volume edition presented here in one volume. xxviii+820pp. 5⅜ x 8½. 67766-4

GAMES, GODS & GAMBLING: A History of Probability and Statistical Ideas, F. N. David. Episodes from the lives of Galileo, Fermat, Pascal, and others illustrate this fascinating account of the roots of mathematics. Features thought-provoking references to classics, archaeology, biography, poetry. 1962 edition. 304pp. 5⅜ x 8½. (Available in U.S. only.) 40023-9

OF MEN AND NUMBERS: The Story of the Great Mathematicians, Jane Muir. Fascinating accounts of the lives and accomplishments of history's greatest mathematical minds–Pythagoras, Descartes, Euler, Pascal, Cantor, many more. Anecdotal, illuminating. 30 diagrams. Bibliography. 256pp. 5⅜ x 8½. 28973-7

HISTORY OF MATHEMATICS, David E. Smith. Nontechnical survey from ancient Greece and Orient to late 19th century; evolution of arithmetic, geometry, trigonometry, calculating devices, algebra, the calculus. 362 illustrations. 1,355pp. 5⅜ x 8½. Two-vol. set. Vol. I: 20429-4 Vol. II: 20430-8

A CONCISE HISTORY OF MATHEMATICS, Dirk J. Struik. The best brief history of mathematics. Stresses origins and covers every major figure from ancient Near East to 19th century. 41 illustrations. 195pp. 5⅜ x 8½. 60255-9

Physics

OPTICAL RESONANCE AND TWO-LEVEL ATOMS, L. Allen and J. H. Eberly. Clear, comprehensive introduction to basic principles behind all quantum optical resonance phenomena. 53 illustrations. Preface. Index. 256pp. 5⅜ x 8½.   65533-4

ULTRASONIC ABSORPTION: An Introduction to the Theory of Sound Absorption and Dispersion in Gases, Liquids and Solids, A. B. Bhatia. Standard reference in the field provides a clear, systematically organized introductory review of fundamental concepts for advanced graduate students, research workers. Numerous diagrams. Bibliography. 440pp. 5⅜ x 8½.   64917-2

QUANTUM THEORY, David Bohm. This advanced undergraduate-level text presents the quantum theory in terms of qualitative and imaginative concepts, followed by specific applications worked out in mathematical detail. Preface. Index. 655pp. 5⅜ x 8½.   65969-0

ATOMIC PHYSICS (8th edition), Max Born. Nobel laureate's lucid treatment of kinetic theory of gases, elementary particles, nuclear atom, wave-corpuscles, atomic structure and spectral lines, much more. Over 40 appendices, bibliography. 495pp. 5⅜ x 8½.   65984-4

AN INTRODUCTION TO HAMILTONIAN OPTICS, H. A. Buchdahl. Detailed account of the Hamiltonian treatment of aberration theory in geometrical optics. Many classes of optical systems defined in terms of the symmetries they possess. Problems with detailed solutions. 1970 edition. xv + 360pp. 5⅜ x 8½.   67597-1

THIRTY YEARS THAT SHOOK PHYSICS: The Story of Quantum Theory, George Gamow. Lucid, accessible introduction to influential theory of energy and matter. Careful explanations of Dirac's anti-particles, Bohr's model of the atom, much more. 12 plates. Numerous drawings. 240pp. 5⅜ x 8½.   24895-X

ELECTRONIC STRUCTURE AND THE PROPERTIES OF SOLIDS: The Physics of the Chemical Bond, Walter A. Harrison. Innovative text offers basic understanding of the electronic structure of covalent and ionic solids, simple metals, transition metals and their compounds. Problems. 1980 edition. 582pp. 6⅛ x 9¼.   66021-4

HYDRODYNAMIC AND HYDROMAGNETIC STABILITY, S. Chandrasekhar. Lucid examination of the Rayleigh-Benard problem; clear coverage of the theory of instabilities causing convection. 704pp. 5⅜ x 8¼.   64071-X

INVESTIGATIONS ON THE THEORY OF THE BROWNIAN MOVEMENT, Albert Einstein. Five papers (1905–8) investigating dynamics of Brownian motion and evolving elementary theory. Notes by R. Fürth. 122pp. 5⅜ x 8½.   60304-0

THE PHYSICS OF WAVES, William C. Elmore and Mark A. Heald. Unique overview of classical wave theory. Acoustics, optics, electromagnetic radiation, more. Ideal as classroom text or for self-study. Problems. 477pp. 5⅜ x 8½.   64926-1

CATALOG OF DOVER BOOKS

PHYSICAL PRINCIPLES OF THE QUANTUM THEORY, Werner Heisenberg. Nobel Laureate discusses quantum theory, uncertainty, wave mechanics, work of Dirac, Schroedinger, Compton, Wilson, Einstein, etc. 184pp. 5⅜ x 8½. 60113-7

ATOMIC SPECTRA AND ATOMIC STRUCTURE, Gerhard Herzberg. One of best introductions; especially for specialist in other fields. Treatment is physical rather than mathematical. 80 illustrations. 257pp. 5⅜ x 8½. 60115-3

AN INTRODUCTION TO STATISTICAL THERMODYNAMICS, Terrell L. Hill. Excellent basic text offers wide-ranging coverage of quantum statistical mechanics, systems of interacting molecules, quantum statistics, more. 523pp. 5⅜ x 8½. 65242-4

THEORETICAL PHYSICS, Georg Joos, with Ira M. Freeman. Classic overview covers essential math, mechanics, electromagnetic theory, thermodynamics, quantum mechanics, nuclear physics, other topics. First paperback edition. xxiii + 885pp. 5⅜ x 8½. 65227-0

PROBLEMS AND SOLUTIONS IN QUANTUM CHEMISTRY AND PHYSICS, Charles S. Johnson, Jr. and Lee G. Pedersen. Unusually varied problems, detailed solutions in coverage of quantum mechanics, wave mechanics, angular momentum, molecular spectroscopy, more. 280 problems plus 139 supplementary exercises. 430pp. 6½ x 9¼. 65236-X

THEORETICAL SOLID STATE PHYSICS, Vol. 1: Perfect Lattices in Equilibrium; Vol. II: Non-Equilibrium and Disorder, William Jones and Norman H. March. Monumental reference work covers fundamental theory of equilibrium properties of perfect crystalline solids, non-equilibrium properties, defects and disordered systems. Appendices. Problems. Preface. Diagrams. Index. Bibliography. Total of 1,301pp. 5⅜ x 8½. Two volumes. Vol. I: 65015-4 Vol. II: 65016-2

A TREATISE ON ELECTRICITY AND MAGNETISM, James Clerk Maxwell. Important foundation work of modern physics. Brings to final form Maxwell's theory of electromagnetism and rigorously derives his general equations of field theory. 1,084pp. 5⅜ x 8½. Two-vol. set. Vol. I: 60636-8 Vol. II: 60637-6

OPTICKS, Sir Isaac Newton. Newton's own experiments with spectroscopy, colors, lenses, reflection, refraction, etc., in language the layman can follow. Foreword by Albert Einstein. 532pp. 5⅜ x 8½. 60205-2

THEORY OF ELECTROMAGNETIC WAVE PROPAGATION, Charles Herach Papas. Graduate-level study discusses the Maxwell field equations, radiation from wire antennas, the Doppler effect and more. xiii + 244pp. 5⅜ x 8½. 65678-5

INTRODUCTION TO QUANTUM MECHANICS With Applications to Chemistry, Linus Pauling & E. Bright Wilson, Jr. Classic undergraduate text by Nobel Prize winner applies quantum mechanics to chemical and physical problems. Numerous tables and figures enhance the text. Chapter bibliographies. Appendices. Index. 468pp. 5⅜ x 8½. 64871-0

CATALOG OF DOVER BOOKS

METHODS OF THERMODYNAMICS, Howard Reiss. Outstanding text focuses on physical technique of thermodynamics, typical problem areas of understanding, and significance and use of thermodynamic potential. 1965 edition. 238pp. 5⅜ x 8½. 69445-3

TENSOR ANALYSIS FOR PHYSICISTS, J. A. Schouten. Concise exposition of the mathematical basis of tensor analysis, integrated with well-chosen physical examples of the theory. Exercises. Index. Bibliography. 289pp. 5⅜ x 8½. 65582-2

RELATIVITY IN ILLUSTRATIONS, Jacob T. Schwartz. Clear nontechnical treatment makes relativity more accessible than ever before. Over 60 drawings illustrate concepts more clearly than text alone. Only high school geometry needed. Bibliography. 128pp. 6⅛ x 9¼. 25965-X

THE ELECTROMAGNETIC FIELD, Albert Shadowitz. Comprehensive undergraduate text covers basics of electric and magnetic fields, builds up to electromagnetic theory. Also related topics, including relativity. Over 900 problems. 768pp. 5⅜ x 8¼. 65660-8

GREAT EXPERIMENTS IN PHYSICS: Firsthand Accounts from Galileo to Einstein, edited by Morris H. Shamos. 25 crucial discoveries: Newton's laws of motion, Chadwick's study of the neutron, Hertz on electromagnetic waves, more. Original accounts clearly annotated. 370pp. 5⅜ x 8½. 25346-5

RELATIVITY, THERMODYNAMICS AND COSMOLOGY, Richard C. Tolman. Landmark study extends thermodynamics to special, general relativity; also applications of relativistic mechanics, thermodynamics to cosmological models. 501pp. 5⅜ x 8½. 65383-8

LIGHT SCATTERING BY SMALL PARTICLES, H. C. van de Hulst. Comprehensive treatment including full range of useful approximation methods for researchers in chemistry, meteorology and astronomy. 44 illustrations. 470pp. 5⅜ x 8½. 64228-3

STATISTICAL PHYSICS, Gregory H. Wannier. Classic text combines thermodynamics, statistical mechanics and kinetic theory in one unified presentation of thermal physics. Problems with solutions. Bibliography. 532pp. 5⅜ x 8½. 65401-X